Ford Capri II Owners Workshop Manual

A K Legg T Eng MIMI

Models covered
Capri 2.8 Injection, Injection Special and 280; 2792 cc
Capri 3.0 GT, S and Ghia; 2994 cc

Does not cover 2.8 Turbo conversion

(1309–3P1)

Haynes Publishing Group
Sparkford Nr Yeovil
Somerset BA22 7JJ England

Haynes Publications, Inc
861 Lawrence Drive
Newbury Park
California 91320 USA

Acknowledgements

Thanks are due to the Champion Sparking Plug Company Limited who supplied the illustrations showing the spark plug conditions, and to Duckhams Oils who provided lubrication data. Certain other illustrations are the copyright of Ford Motor Company Limited and are used with their permission. Thanks are also due to Sykes-Pickavant who supplied some of the workshop tools, and to all those people at Sparkford who helped in the production of this manual.

© Haynes Publishing Group 1987, 1988

A book in the **Haynes Owners Workshop Manual Series**

Printed by J. H. Haynes & Co. Ltd, Sparkford, Nr Yeovil, Somerset BA22 7JJ, England

All rights reserved. No part of this book may be reproduced or transmitted in any form or by any means, electronic or mechanical, including photocopying, recording or by any information storage or retrieval system, without permission in writing from the copyright holder.

ISBN 1 85010 309 7

British Library Cataloguing in Publication Data
Legg, A. K.
    Ford Capri II (& III) 2.8 and 3.0 owners
workshop manual. – (Owners Workshop Manuals).
1. Capri automobile
I. Title   II. Series
629.28'722   TL215.C34
ISBN 1-85010-309-7

Whilst every care is taken to ensure that the information in this manual is correct, no liability can be accepted by the authors or publishers for loss, damage or injury caused by any errors in, or omissions from, the information given.

Restoring and Preserving our Motoring Heritage

Few people can have had the luck to realise their dreams to quite the same extent and in such a remarkable fashion as John Haynes, Founder and Chairman of the Haynes Publishing Group.

Since 1965 his unique approach to workshop manual publishing has proved so successful that millions of Haynes Manuals are now sold every year throughout the world, covering literally thousands of different makes and models of cars, vans and motorcycles.

A continuing passion for cars and motoring led to the founding in 1985 of a Charitable Trust dedicated to the restoration and preservation of our motoring heritage. To inaugurate the new Museum, John Haynes donated virtually his entire private collection of 52 cars.

Now with an unrivalled international collection of over 210 veteran, vintage and classic cars and motorcycles, the Haynes Motor Museum in Somerset is well on the way to becoming one of the most interesting Motor Museums in the world.

A 70 seat video cinema, a cafe and an extensive motoring bookshop, together with a specially constructed one kilometre motor circuit, make a visit to the Haynes Motor Museum a truly unforgettable experience.

Every vehicle in the museum is preserved in as near as possible mint condition and each car is run every six months on the motor circuit.

Enjoy the picnic area set amongst the rolling Somerset hills. Peer through the William Morris workshop windows at cars being restored, and browse through the extensive displays of fascinating motoring memorabilia.

From the 1903 Oldsmobile through such classics as an MG Midget to the mighty 'E' Type Jaguar, Lamborghini, Ferrari Berlinetta Boxer, and Graham Hill's Lola Cosworth, there is something for everyone, young and old alike, at this Somerset Museum.

Haynes Motor Museum

Situated mid-way between London and Penzance, the Haynes Motor Museum is located just off the A303 at Sparkford, Somerset (home of the Haynes Manual) and is open to the public 7 days a week all year round, except Christmas Day and Boxing Day.

Contents

Introductory pages
   About this manual   5
   Introduction to the Capri V6   5
   General dimensions, weights and capacities   6
   Jacking and towing   7
   Buying spare parts and vehicle identification numbers   8
   General repair procedures   9
   Tools and working facilities   10
   Safety first!   12
   Routine maintenance   13
   Recommended lubricants and fluids   17
   Conversion factors   18
   Fault diagnosis   19

Chapter 1
Engine   23

Chapter 2
Cooling system   57

Chapter 3
Fuel and exhaust system   66

Chapter 4
Ignition system   87

Chapter 5
Clutch   96

Chapter 6
Manual gearbox and automatic transmission   101

Chapter 7
Propeller shaft   127

Chapter 8
Rear axle   131

Chapter 9
Braking system   141

Chapter 10
Suspension and steering   156

Chapter 11
Bodywork and fittings   172

Chapter 12
Electrical system   195

Index   247

Spark plug condition and bodywork repair colour section between pages 32 and 33

Ford Capri 2.8 Injection

Ford Capri 3.0 Ghia

About this manual

Its aim

The aim of this manual is to help you get the best value from your vehicle. It can do so in several ways. It can help you decide what work must be done (even should you choose to get it done by a garage), provide information on routine maintenance and servicing, and give a logical course of action and diagnosis when random faults occur. However, it is hoped that you will use the manual by tackling the work yourself. On simpler jobs it may even be quicker than booking the car into a garage and going there twice, to leave and collect it. Perhaps most important, a lot of money can be saved by avoiding the costs a garage must charge to cover its labour and overheads.

The manual has drawings and descriptions to show the function of the various components so that their layout can be understood. Then the tasks are described and photographed in a step-by-step sequence so that even a novice can do the work.

Its arrangement

The manual is divided into twelve Chapters, each covering a logical sub-division of the vehicle. The Chapters are each divided into Sections, numbered with single figures, eg 5; and the Sections into paragraphs (or sub-sections), with decimal numbers following on from the Section they are in, eg 5.1, 5.2, 5.3 etc.

It is freely illustrated, especially in those parts where there is a detailed sequence of operations to be carried out. There are two forms of illustration: figures and photographs. The figures are numbered in sequence with decimal numbers, according to their position in the Chapter – eg Fig. 6.4 is the fourth drawing/illustration in Chapter 6. Photographs carry the same number (either individually or in related groups) as the Section or sub-section to which they relate.

There is an alphabetical index at the back of the manual as well as a contents list at the front. Each Chapter is also preceded by its own individual contents list.

References to the 'left' or 'right' of the vehicle are in the sense of a person in the driver's seat facing forwards.

Unless otherwise stated, nuts and bolts are removed by turning anti-clockwise, and tightened by turning clockwise.

Vehicle manufacturers continually make changes to specifications and recommendations, and these, when notified, are incorporated into our manuals at the earliest opportunity.

Whilst every care is taken to ensure that the information in this manual is correct, no liability can be accepted by the authors or publishers for loss, damage or injury caused by any errors in, or omissions from, the information given.

Project vehicle

The vehicle used in the preparation of this manual, and appearing in many of the photographic sequences, was a Ford Capri 2.8 Injection.

Introduction to the Capri V6

The Capri II models were first introduced in the UK in February 1974 using a wide range of engines previously used on other Ford vehicles. The models covered in this manual use the V6 3.0 litre Essex and V6 2.8 litre Cologne engines.

The restyled Series III models were introduced in March 1978 with a front spoiler and four headlamps, and following this the 2.8 Injection version was launched in July 1981. September 1984 saw the introduction of the high performance 2.8 Injection Special which featured a limited slip differential, lockable spoked alloy wheels and a higher level of interior trim. In 1987 the last Capris were made – the limited edition 280.

The car is conventional in mechanical layout, drive from the engine being transmitted to the rear axle through a 4 or 5-speed manual gearbox or alternatively on Series II models, a 3-speed automatic transmission.

General dimensions, weights and capacities

	Series II	Series III
Dimensions (typical)		
Overall length	4340 mm (170.9 in)	4439 mm (174.8 in)
Overall width	1698 mm (66.9 in)	1698 mm (66.9 in)
Overall height	1357 mm (53.4 in)	1323 mm (52.1 in)
Wheelbase	2559 mm (100.7 in)	2563 mm (100.9 in)
Track:		
Front:		
3.0 litre	1353 mm (53.3 in)	1353 mm (53.5 in)
2.8 litre	–	1400 mm (55.1 in)
Rear:		
3.0 litre	1384 mm (54.5 in)	1384 mm (54.5 in)
2.8 litre	–	1431 mm (56.3 in)
Turning circle (between kerbs):		
3.0 litre	10.0 m (32.8 ft)	9.8 m (32.2 ft)
2.8 litre	–	9.9 m (32.5 ft)

Weights (typical)

3.0 S:	
Manual transmission	1155 kg (2546 lb)
Automatic transmission	1145 kg (2524 lb)
3.0 Ghia:	
Manual transmission	1185 kg (2612 lb)
Automatic transmission	1174 kg (2588 lb)
2.8 Injection	1190 kg (2623 lb)
Roof rack load	75 kg (165 lb) max

Capacities (typical)

Engine:	
3.0 litre:	
With filter	5.0 litre (8.8 pint)
Without filter	4.25 litre (7.5 pint)
2.8 litre:	
With filter	4.25 litre (7.5 pint)
Without filter	4.0 litre (7.0 pint)
Cooling system:	
3.0 litre	9.3 litre (16.4 pint)
2.8 litre	8.7 litre (15.3 pint)
Fuel system	58 litre (12.7 gal)
Transmission:	
Manual:	
4-speed	2.0 litre (3.5 pint)
5-speed	1.9 litre (3.3 pint)
Automatic (approx)	7.5 litre (13.2 pint)
Rear axle	1.1 litre (1.9 pint)
Steering gear:	
Manual	0.15 litre (0.26 pint)
Power-assisted:	
Gear oil	0.20 litre (0.35 pint)
Fluid:	
Separate reservoir	2.2 litre (3.87 pint)
Integral reservoir	0.50 litre (0.88 pint)

Jacking and towing

Jacking points

To change a wheel in an emergency, use the jack supplied with the vehicle. Ensure that the roadwheel nuts are released before jacking up the car and make sure that the arm of the jack is fully engaged with the body bracket and that the base of the jack is standing on a firm surface.

The jack supplied with the vehicle is not suitable for use when raising the vehicle for maintenance or repair operations. For this work, use a trolley, hydraulic or screw-type jack located under the front crossmember, bodyframe side-members or rear axle casing, as illustrated. Always supplement the jack with axle-stands or blocks before crawling beneath the car.

Towing points

If your vehicle is being towed, make sure that the tow rope is attached to the front crossmember or, where fitted, the towing eye attached to the front crossmember. If the vehicle is equipped with automatic transmission, the distance towed must not exceed 15 miles (24 km), nor the speed 30 mph (48 km/h), otherwise serious damage to the transmission may result. If these limits are likely to be exceeded disconnect and remove the propeller shaft.

If you are towing another vehicle, attach a tow rope to the lower shock absorber mounting bracket at the axle tube or, if fitted, the rear towing eye.

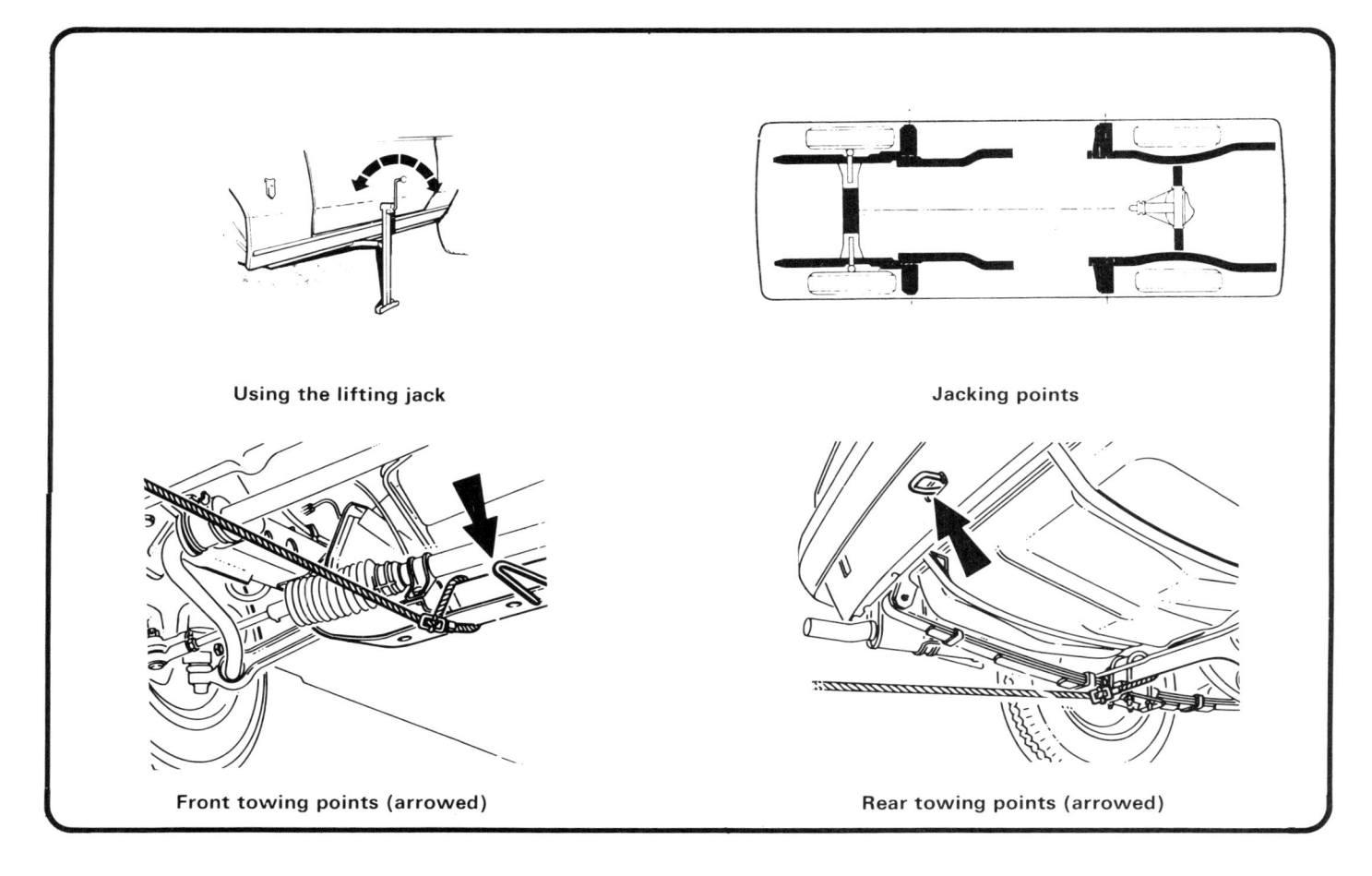

Using the lifting jack

Jacking points

Front towing points (arrowed)

Rear towing points (arrowed)

Buying spare parts and vehicle identification numbers

Buying spare parts

Spare parts are available from many sources, for example: Ford garages, other garages and accessory shops, and motor factors. Our advice regarding spare part sources is as follows:

Officially appointed Ford garages – This is the best source of parts which are peculiar to your car and are otherwise not generally available (eg complete cylinder heads, internal gearbox components, badges, interior trim etc). It is also the only place at which you should buy parts if your car is still under warranty – non-Ford components may invalidate the warranty. To be sure of obtaining the correct parts it will always be necessary to give the storeman your car's vehicle identification number, and if possible, to take the 'old' part along for positive identification. Remember that many parts are available on a factory exchange scheme – any parts returned should always be clean! It obviously makes good sense to go straight to the specialists on your car for this type of part as they are best equipped to supply you.

Other garages and accessory shops – These are often very good places to buy materials and components needed for the maintenance of your car (eg oil filters, spark plugs, bulbs, drive belts, oils and greases, touch-up paint, filler paste etc). They also sell general accessories, usually have convenient opening hours, charge lower prices and can often be found not far from home.

Motor factors – Good factors will stock all the more important components which wear out relatively quickly (eg clutch components, pistons, valves, exhaust systems, brake cylinders/pipes/hoses/seals/shoes and pads etc). Motor factors will often provide new or reconditioned components on a part exchange basis – this can save a considerable amount of money!

Vehicle identification numbers

Although many individual parts, and in some cases sub-assemblies, fit a number of different models it is dangerous to assume that just because they look the same, they are the same. Differences are not always easy to detect except by serial numbers. Make sure therefore that the appropriate identify number for the model or sub-assembly is known and quoted when a spare part is obtained.

The vehicle identification plate is mounted on the right-hand front wing apron on Series II models or on the right-hand side of the front crossmember on Series III models.

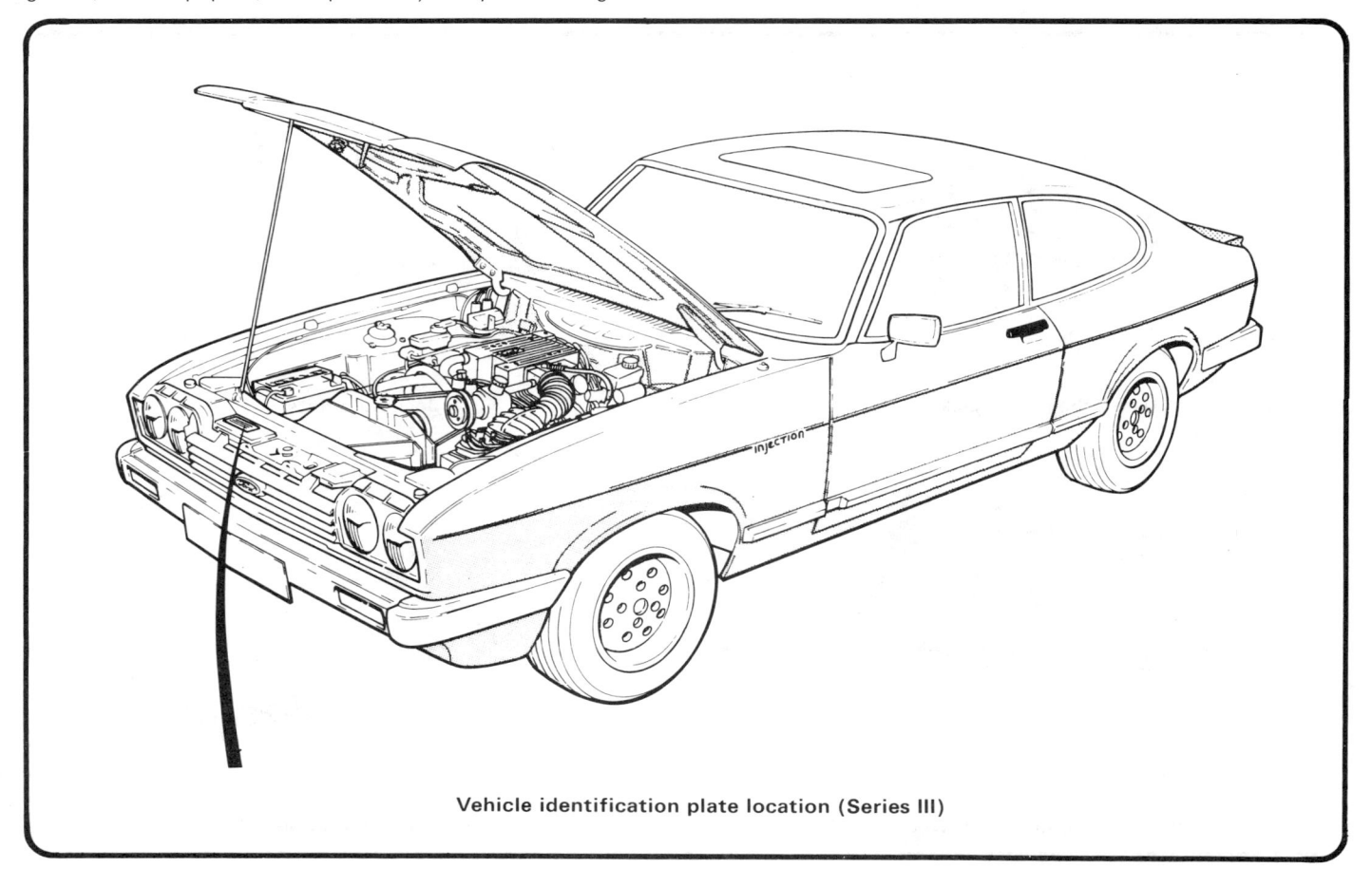

Vehicle identification plate location (Series III)

General repair procedures

Whenever servicing, repair or overhaul work is carried out on the car or its components, it is necessary to observe the following procedures and instructions. This will assist in carrying out the operation efficiently and to a professional standard of workmanship.

Joint mating faces and gaskets

Where a gasket is used between the mating faces of two components, ensure that it is renewed on reassembly, and fit it dry unless otherwise stated in the repair procedure. Make sure that the mating faces are clean and dry with all traces of old gasket removed. When cleaning a joint face, use a tool which is not likely to score or damage the face, and remove any burrs or nicks with an oilstone or fine file.

Make sure that tapped holes are cleaned with a pipe cleaner, and keep them free of jointing compound if this is being used unless specifically instructed otherwise.

Ensure that all orifices, channels or pipes are clear and blow through them, preferably using compressed air.

Oil seals

Whenever an oil seal is removed from its working location, either individually or as part of an assembly, it should be renewed.

The very fine sealing lip of the seal is easily damaged and will not seal if the surface it contacts is not completely clean and free from scratches, nicks or grooves. If the original sealing surface of the component cannot be restored, the component should be renewed.

Protect the lips of the seal from any surface which may damage them in the course of fitting. Use tape or a conical sleeve where possible. Lubricate the seal lips with oil before fitting and, on dual lipped seals, fill the space between the lips with grease.

Unless otherwise stated, oil seals must be fitted with their sealing lips toward the lubricant to be sealed.

Use a tubular drift or block of wood of the appropriate size to install the seal and, if the seal housing is shouldered, drive the seal down to the shoulder. If the seal housing is unshouldered, the seal should be fitted with its face flush with the housing top face.

Screw threads and fastenings

Always ensure that a blind tapped hole is completely free from oil, grease, water or other fluid before installing the bolt or stud. Failure to do this could cause the housing to crack due to the hydraulic action of the bolt or stud as it is screwed in.

When tightening a castellated nut to accept a split pin, tighten the nut to the specified torque, where applicable, and then tighten further to the next split pin hole. Never slacken the nut to align a split pin hole unless stated in the repair procedure.

When checking or retightening a nut or bolt to a specified torque setting, slacken the nut or bolt by a quarter of a turn, and then retighten to the specified setting.

Locknuts, locktabs and washers

Any fastening which will rotate against a component or housing in the course of tightening should always have a washer between it and the relevant component or housing.

Spring or split washers should always be renewed when they are used to lock a critical component such as a big-end bearing retaining nut or bolt.

Locktabs which are folded over to retain a nut or bolt should always be renewed.

Self-locking nuts can be reused in non-critical areas, providing resistance can be felt when the locking portion passes over the bolt or stud thread.

Split pins must always be replaced with new ones of the correct size for the hole.

Special tools

Some repair procedures in this manual entail the use of special tools such as a press, two or three-legged pullers, spring compressors etc. Wherever possible, suitable readily available alternatives to the manufacturer's special tools are described, and are shown in use. In some instances, where no alternative is possible, it has been necessary to resort to the use of a manufacturer's tool and this has been done for reasons of safety as well as the efficient completion of the repair operation. Unless you are highly skilled and have a thorough understanding of the procedure described, never attempt to bypass the use of any special tool when the procedure described specifies its use. Not only is there a very great risk of personal injury, but expensive damage could be caused to the components involved.

Tools and working facilities

Introduction

A selection of good tools is a fundamental requirement for anyone contemplating the maintenance and repair of a motor vehicle. For the owner who does not possess any, their purchase will prove a considerable expense, offsetting some of the savings made by doing-it-yourself. However, provided that the tools purchased meet the relevant national safety standards and are of good quality, they will last for many years and prove an extremely worthwhile investment.

To help the average owner to decide which tools are needed to carry out the various tasks detailed in this manual, we have compiled three lists of tools under the following headings: *Maintenance and minor repair*, *Repair and overhaul*, and *Special*. The newcomer to practical mechanics should start off with the *Maintenance and minor repair* tool kit and confine himself to the simpler jobs around the vehicle. Then, as his confidence and experience grow, he can undertake more difficult tasks, buying extra tools as, and when, they are needed. In this way, a *Maintenance and minor repair* tool kit can be built-up into a *Repair and overhaul* tool kit over a considerable period of time without any major cash outlays. The experienced do-it-yourselfer will have a tool kit good enough for most repair and overhaul procedures and will add tools from the *Special* category when he feels the expense is justified by the amount of use to which these tools will be put.

It is obviously not possible to cover the subject of tools fully here. For those who wish to learn more about tools and their use there is a book entitled *How to Choose and Use Car Tools* available from the publishers of this manual.

Maintenance and minor repair tool kit

The tools given in this list should be considered as a minimum requirement if routine maintenance, servicing and minor repair operations are to be undertaken. We recommend the purchase of combination spanners (ring one end, open-ended the other); although more expensive than open-ended ones, they do give the advantages of both types of spanner.

Combination spanners - 10, 11, 12, 13, 14 & 17 mm
Adjustable spanner - 9 inch
Gearbox/rear axle drain/filler plug key
Spark plug spanner (with rubber insert)
Spark plug gap adjustment tool
Set of feeler gauges
Brake bleed nipple spanner
Screwdriver - 4 in long x 1/4 in dia (flat blade)
Screwdriver - 4 in long x 1/4 in dia (cross blade)
Combination pliers - 6 inch
Hacksaw (junior)
Tyre pump
Tyre pressure gauge
Grease gun
Oil can
Fine emery cloth (1 sheet)
Wire brush (small)
Funnel (medium size)

Repair and overhaul tool kit

These tools are virtually essential for anyone undertaking any major repairs to a motor vehicle, and are additional to those given in the *Maintenance and minor repair* list. Included in this list is a comprehensive set of sockets. Although these are expensive they will be found invaluable as they are so versatile - particularly if various drives are included in the set. We recommend the ½ in square-drive type, as this can be used with most proprietary torque wrenches. If you cannot afford a socket set, even bought piecemeal, then inexpensive tubular box spanners are a useful alternative.

The tools in this list will occasionally need to be supplemented by tools from the *Special* list.

Sockets (or box spanners) to cover range in previous list
Reversible ratchet drive (for use with sockets)
Extension piece, 10 inch (for use with sockets)
Universal joint (for use with sockets)
Torque wrench (for use with sockets)
'Mole' wrench - 8 inch
Ball pein hammer
Soft-faced hammer, plastic or rubber
Screwdriver - 6 in long x 5/16 in dia (flat blade)
Screwdriver - 2 in long x 5/16 in square (flat blade)
Screwdriver - 1 1/2 in long x 1/4 in dia (cross blade)
Screwdriver - 3 in long x 1/8 in dia (electricians)
Pliers - electricians side cutters
Pliers - needle nosed
Pliers - circlip (internal and external)
Cold chisel - 1/2 inch
Scriber
Scraper
Centre punch
Pin punch
Hacksaw
Valve grinding tool
Steel rule/straight-edge
Allen keys
Selection of files
Wire brush (large)
Axle-stands
Jack (strong trolley or hydraulic type)

Special tools

The tools in this list are those which are not used regularly, are expensive to buy, or which need to be used in accordance with their manufacturers' instructions. Unless relatively difficult mechanical jobs are undertaken frequently, it will not be economic to buy many of these tools. Where this is the case, you could consider clubbing together with friends (or joining a motorists' club) to make a joint purchase, or borrowing the tools against a deposit from a local garage or tool hire specialist.

The following list contains only those tools and instruments freely available to the public, and not those special tools produced by the

Tools and working facilities

vehicle manufacturer specifically for its dealer network. You will find occasional references to these manufacturers' special tools in the text of this manual. Generally, an alternative method of doing the job without the vehicle manufacturers' special tool is given. However, sometimes, there is no alternative to using them. Where this is the case and the relevant tool cannot be bought or borrowed, you will have to entrust the work to a franchised garage.

Valve spring compressor (where applicable)
Piston ring compressor
Balljoint separator
Universal hub/bearing puller
Impact screwdriver
Micrometer and/or vernier gauge
Dial gauge
Stroboscopic timing light
Dwell angle meter/tachometer
Universal electrical multi-meter
Cylinder compression gauge
Lifting tackle
Trolley jack
Light with extension lead

Buying tools

For practically all tools, a tool factor is the best source since he will have a very comprehensive range compared with the average garage or accessory shop. Having said that, accessory shops often offer excellent quality tools at discount prices, so it pays to shop around.

There are plenty of good tools around at reasonable prices, but always aim to purchase items which meet the relevant national safety standards. If in doubt, ask the proprietor or manager of the shop for advice before making a purchase.

Care and maintenance of tools

Having purchased a reasonable tool kit, it is necessary to keep the tools in a clean serviceable condition. After use, always wipe off any dirt, grease and metal particles using a clean, dry cloth, before putting the tools away. Never leave them lying around after they have been used. A simple tool rack on the garage or workshop wall, for items such as screwdrivers and pliers is a good idea. Store all normal wrenches and sockets in a metal box. Any measuring instruments, gauges, meters, etc, must be carefully stored where they cannot be damaged or become rusty.

Take a little care when tools are used. Hammer heads inevitably become marked and screwdrivers lose the keen edge on their blades from time to time. A little timely attention with emery cloth or a file will soon restore items like this to a good serviceable finish.

Working facilities

Not to be forgotten when discussing tools, is the workshop itself. If anything more than routine maintenance is to be carried out, some form of suitable working area becomes essential.

It is appreciated that many an owner mechanic is forced by circumstances to remove an engine or similar item, without the benefit of a garage or workshop. Having done this, any repairs should always be done under the cover of a roof.

Wherever possible, any dismantling should be done on a clean, flat workbench or table at a suitable working height.

Any workbench needs a vice: one with a jaw opening of 4 in (100 mm) is suitable for most jobs. As mentioned previously, some clean dry storage space is also required for tools, as well as for lubricants, cleaning fluids, touch-up paints and so on, which become necessary.

Another item which may be required, and which has a much more general usage, is an electric drill with a chuck capacity of at least 5/16 in (8 mm). This, together with a good range of twist drills, is virtually essential for fitting accessories such as mirrors and reversing lights.

Last, but not least, always keep a supply of old newspapers and clean, lint-free rags available, and try to keep any working area as clean as possible.

Spanner jaw gap comparison table

Jaw gap (in)	Spanner size
0.250	1/4 in AF
0.276	7 mm
0.313	5/16 in AF
0.315	8 mm
0.344	11/32 in AF; 1/8 in Whitworth
0.354	9 mm
0.375	3/8 in AF
0.394	10 mm
0.433	11 mm
0.438	7/16 in AF
0.445	3/16 in Whitworth; 1/4 in BSF
0.472	12 mm
0.500	1/2 in AF
0.512	13 mm
0.525	1/4 in Whitworth; 5/16 in BSF
0.551	14 mm
0.563	9/16 in AF
0.591	15 mm
0.600	5/16 in Whitworth; 3/8 in BSF
0.625	5/8 in AF
0.630	16 mm
0.669	17 mm
0.686	11/16 in AF
0.709	18 mm
0.710	3/8 in Whitworth; 7/16 in BSF
0.748	19 mm
0.750	3/4 in AF
0.813	13/16 in AF
0.820	7/16 in Whitworth; 1/2 in BSF
0.866	22 mm
0.875	7/8 in AF
0.920	1/2 in Whitworth; 9/16 in BSF
0.938	15/16 in AF
0.945	24 mm
1.000	1 in AF
1.010	9/16 in Whitworth; 5/8 in BSF
1.024	26 mm
1.063	1 1/16 in AF; 27 mm
1.100	5/8 in Whitworth; 11/16 in BSF
1.125	1 1/8 in AF
1.181	30 mm
1.200	11/16 in Whitworth; 3/4 in BSF
1.250	1 1/4 in AF
1.260	32 mm
1.300	3/4 in Whitworth; 7/8 in BSF
1.313	1 5/16 in AF
1.390	13/16 in Whitworth; 15/16 in BSF
1.417	36 mm
1.438	1 7/16 in AF
1.480	7/8 in Whitworth; 1 in BSF
1.500	1 1/2 in AF
1.575	40 mm; 15/16 in Whitworth
1.614	41 mm
1.625	1 5/8 in AF
1.670	1 in Whitworth; 1 1/8 in BSF
1.688	1 11/16 in AF
1.811	46 mm
1.813	1 13/16 in AF
1.860	1 1/8 in Whitworth; 1 1/4 in BSF
1.875	1 7/8 in AF
1.969	50 mm
2.000	2 in AF
2.050	1 1/4 in Whitworth; 1 3/8 in BSF
2.165	55 mm
2.362	60 mm

Safety first!

Professional motor mechanics are trained in safe working procedures. However enthusiastic you may be about getting on with the job in hand, do take the time to ensure that your safety is not put at risk. A moment's lack of attention can result in an accident, as can failure to observe certain elementary precautions.

There will always be new ways of having accidents, and the following points do not pretend to be a comprehensive list of all dangers; they are intended rather to make you aware of the risks and to encourage a safety-conscious approach to all work you carry out on your vehicle.

Essential DOs and DON'Ts

DON'T rely on a single jack when working underneath the vehicle. Always use reliable additional means of support, such as axle stands, securely placed under a part of the vehicle that you know will not give way.

DON'T attempt to loosen or tighten high-torque nuts (e.g. wheel hub nuts) while the vehicle is on a jack; it may be pulled off.

DON'T start the engine without first ascertaining that the transmission is in neutral (or 'Park' where applicable) and the parking brake applied.

DON'T suddenly remove the filler cap from a hot cooling system – cover it with a cloth and release the pressure gradually first, or you may get scalded by escaping coolant.

DON'T attempt to drain oil until you are sure it has cooled sufficiently to avoid scalding you.

DON'T grasp any part of the engine, exhaust or catalytic converter without first ascertaining that it is sufficiently cool to avoid burning you.

DON'T allow brake fluid or antifreeze to contact vehicle paintwork.

DON'T syphon toxic liquids such as fuel, brake fluid or antifreeze by mouth, or allow them to remain on your skin.

DON'T inhale dust – it may be injurious to health (see *Asbestos* below).

DON'T allow any spilt oil or grease to remain on the floor – wipe it up straight away, before someone slips on it.

DON'T use ill-fitting spanners or other tools which may slip and cause injury.

DON'T attempt to lift a heavy component which may be beyond your capability – get assistance.

DON'T rush to finish a job, or take unverified short cuts.

DON'T allow children or animals in or around an unattended vehicle.

DO wear eye protection when using power tools such as drill, sander, bench grinder etc, and when working under the vehicle.

DO use a barrier cream on your hands prior to undertaking dirty jobs – it will protect your skin from infection as well as making the dirt easier to remove afterwards; but make sure your hands aren't left slippery. Note that long-term contact with used engine oil can be a health hazard.

DO keep loose clothing (cuffs, tie etc) and long hair well out of the way of moving mechanical parts.

DO remove rings, wristwatch etc, before working on the vehicle – especially the electrical system.

DO ensure that any lifting tackle used has a safe working load rating adequate for the job.

DO keep your work area tidy – it is only too easy to fall over articles left lying around.

DO get someone to check periodically that all is well, when working alone on the vehicle.

DO carry out work in a logical sequence and check that everything is correctly assembled and tightened afterwards.

DO remember that your vehicle's safety affects that of yourself and others. If in doubt on any point, get specialist advice.

IF, in spite of following these precautions, you are unfortunate enough to injure yourself, seek medical attention as soon as possible.

Asbestos

Certain friction, insulating, sealing, and other products – such as brake linings, brake bands, clutch linings, torque converters, gaskets, etc – contain asbestos. *Extreme care must be taken to avoid inhalation of dust from such products since it is hazardous to health.* If in doubt, assume that they *do* contain asbestos.

Fire

Remember at all times that petrol (gasoline) is highly flammable. Never smoke, or have any kind of naked flame around, when working on the vehicle. But the risk does not end there – a spark caused by an electrical short-circuit, by two metal surfaces contacting each other, by careless use of tools, or even by static electricity built up in your body under certain conditions, can ignite petrol vapour, which in a confined space is highly explosive.

Always disconnect the battery earth (ground) terminal before working on any part of the fuel or electrical system, and never risk spilling fuel on to a hot engine or exhaust.

It is recommended that a fire extinguisher of a type suitable for fuel and electrical fires is kept handy in the garage or workplace at all times. Never try to extinguish a fuel or electrical fire with water.

Note: *Any reference to a 'torch' appearing in this manual should always be taken to mean a hand-held battery-operated electric lamp or flashlight. It does NOT mean a welding/gas torch or blowlamp.*

Fumes

Certain fumes are highly toxic and can quickly cause unconsciousness and even death if inhaled to any extent. Petrol (gasoline) vapour comes into this category, as do the vapours from certain solvents such as trichloroethylene. Any draining or pouring of such volatile fluids should be done in a well ventilated area.

When using cleaning fluids and solvents, read the instructions carefully. Never use materials from unmarked containers – they may give off poisonous vapours.

Never run the engine of a motor vehicle in an enclosed space such as a garage. Exhaust fumes contain carbon monoxide which is extremely poisonous; if you need to run the engine, always do so in the open air or at least have the rear of the vehicle outside the workplace.

If you are fortunate enough to have the use of an inspection pit, never drain or pour petrol, and never run the engine, while the vehicle is standing over it; the fumes, being heavier than air, will concentrate in the pit with possibly lethal results.

The battery

Never cause a spark, or allow a naked light, near the vehicle's battery. It will normally be giving off a certain amount of hydrogen gas, which is highly explosive.

Always disconnect the battery earth (ground) terminal before working on the fuel or electrical systems.

If possible, loosen the filler plugs or cover when charging the battery from an external source. Do not charge at an excessive rate or the battery may burst.

Take care when topping up and when carrying the battery. The acid electrolyte, even when diluted, is very corrosive and should not be allowed to contact the eyes or skin.

If you ever need to prepare electrolyte yourself, always add the acid slowly to the water, and never the other way round. Protect against splashes by wearing rubber gloves and goggles.

When jump starting a car using a booster battery, for negative earth (ground) vehicles, connect the jump leads in the following sequence: First connect one jump lead between the positive (+) terminals of the two batteries. Then connect the other jump lead first to the negative (−) terminal of the booster battery, and then to a good earthing (ground) point on the vehicle to be started, at least 18 in (45 cm) from the battery if possible. Ensure that hands and jump leads are clear of any moving parts, and that the two vehicles do not touch. Disconnect the leads in the reverse order.

Mains electricity and electrical equipment

When using an electric power tool, inspection light etc, always ensure that the appliance is correctly connected to its plug and that, where necessary, it is properly earthed (grounded). Do not use such appliances in damp conditions and, again, beware of creating a spark or applying excessive heat in the vicinity of fuel or fuel vapour. Also ensure that the appliances meet the relevant national safety standards.

Ignition HT voltage

A severe electric shock can result from touching certain parts of the ignition system, such as the HT leads, when the engine is running or being cranked, particularly if components are damp or the insulation is defective. Where an electronic ignition system is fitted, the HT voltage is much higher and could prove fatal.

Routine maintenance

Maintenance is essential for ensuring safety, and desirable for the purpose of getting the best in terms of performance and economy from your car. Over the years the need for periodic lubrication – oiling, greasing and so on – has been drastically reduced, if not totally eliminated. This has unfortunately tended to lead some owners to think that because no such action is required, components either no longer exist, or will last forever. This is a serious delusion. If follows therefore that the largest initial element of maintenance is visual examination and a general sense of awareness. This may lead to repairs or renewals, but should help to avoid roadside breakdowns. Other neglect results in unreliability, increased running costs, more rapid wear and depreciation of the vehicle in general.

The following service schedules are a list of the maintenance requirements and the intervals at which they should be carried out. Where applicable these procedures are covered in greater detail throughout this manual near the beginning of each Chapter or Chapter division.

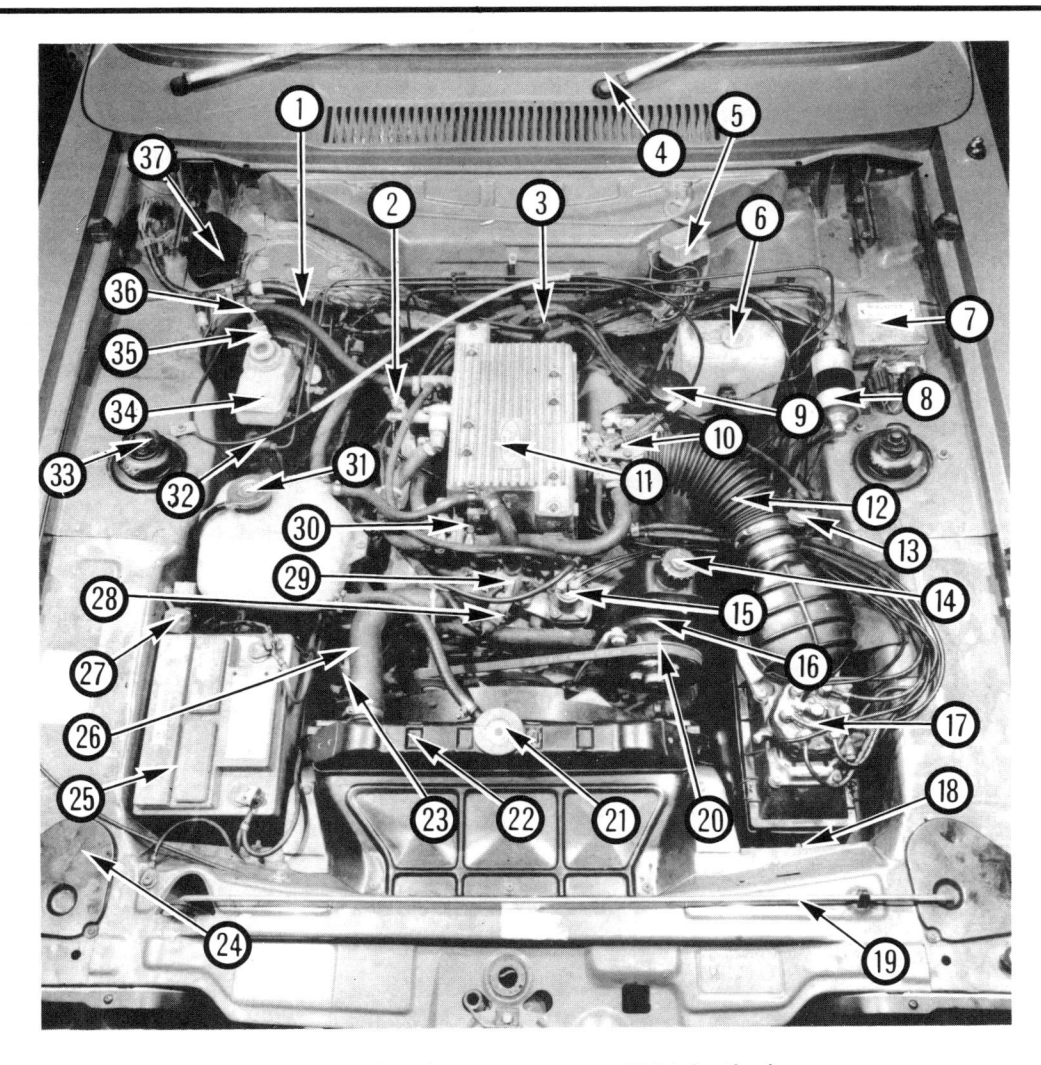

View of engine compartment (2.8 Injection)

1	Brake servo unit
2	Start valve
3	Distributor cap
4	Windscreen wiper arm
5	Headlamp relay
6	Windscreen washer reservoir
7	Ignition amplifier module
8	Fuel filter
9	Engine oil filler cap
10	Throttle housing
11	Air chamber
12	Air inlet duct
13	Ignition coil
14	Power steering fluid reservoir
15	Warm-up regulator
16	Power steering pump
17	Distributor/mixture control unit
18	Air cleaner body
19	Bonnet stay
20	Power steering pump drivebelt
21	Radiator filler cap
22	Radiator
23	Alternator
24	Headlight access cover
25	Battery
26	Top hose
27	Driving light relay/fuse unit
28	Thermotime switch
29	Coolant temperature sender unit
30	Auxiliary air device
31	Cooling system pressure cap and expansion tank
32	Brake master cylinder
33	Front suspension top mounting
34	Brake fluid reservoir
35	Brake fluid low level checking membrane
36	Vacuum hose
37	Fusebox

View of front underbody (2.8 injection)

1. Exhaust system front pipe
2. Clutch cable and rubber gaiter
3. Oil filter cartridge
4. Brake caliper
5. Brake flexible hydraulic hose
6. Front suspension track control arm
7. Track rod end balljoint
8. Rubber bellows
9. Power steering gear
10. Front stabiliser bar
11. Drivebelts
12. Fan
13. Sump oil drain plug
14. Bottom hose
15. Starter motor solenoid
16. Starter motor
17. Fuel feed and return pipes
18. Manual gearbox
19. Propeller shaft
20. Mounting (engine rear)
21. Speedometer cable

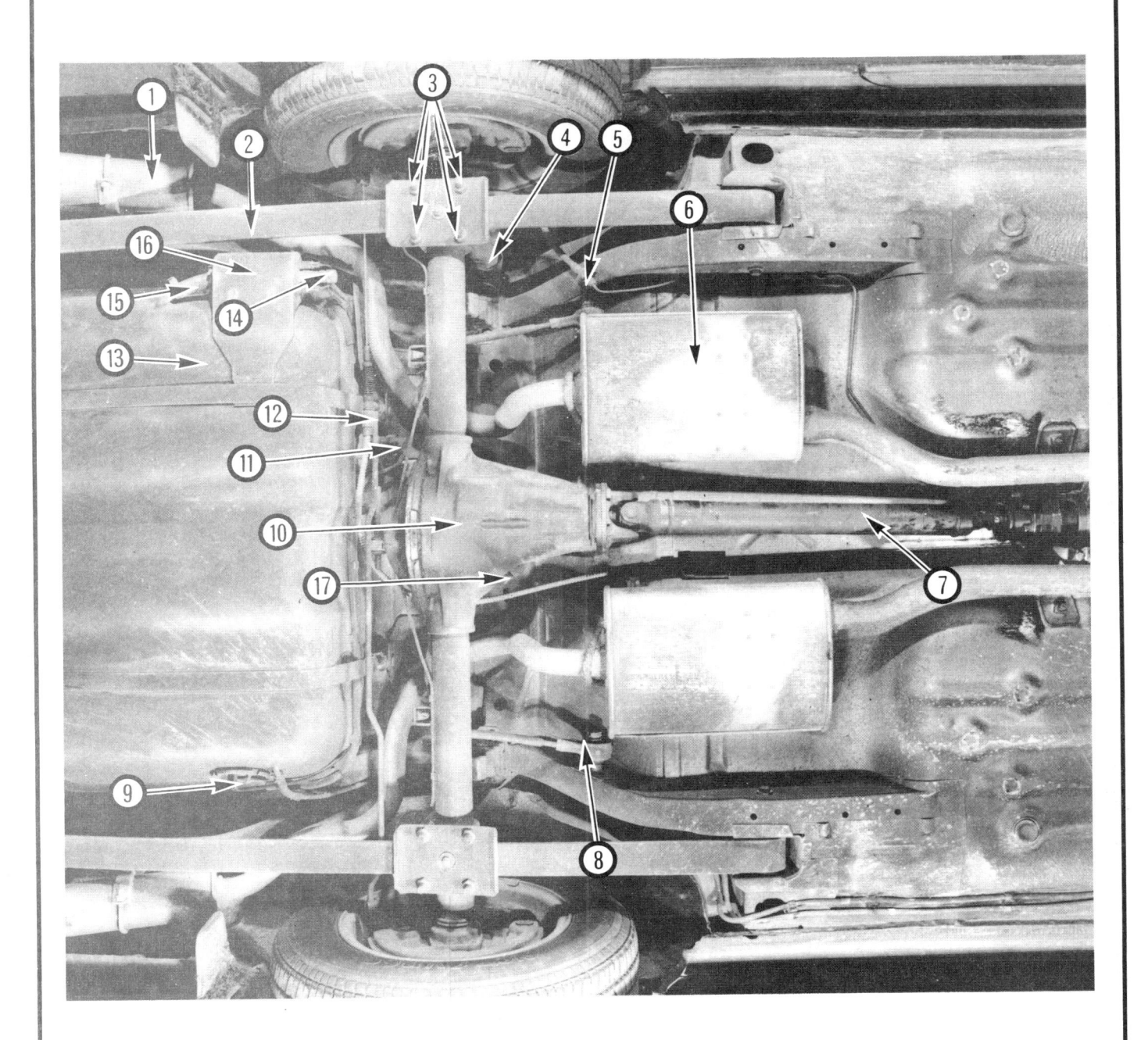

View of rear underbody (2.8 Injection)

1 Rear silencer and tail pipe
2 Rear spring
3 Rear spring U-bolts
4 Rear shock absorber
5 Rear brake hydraulic line connection
6 Intermediate silencer
7 Propeller shaft
8 Rear stabiliser bar
9 Fuel gauge sender unit
10 Rear axle casing and differential unit
11 Fuel accumulator
12 Handbrake linkage
13 Fuel tank
14 Outlet union
15 Inlet hose
16 Fuel pump and rubber mounting
17 Rear axle filler/level plug

Every 250 miles (400 km) or weekly – whichever comes first

Check the engine oil level and top up if necessary (Chapter 1)
Check and adjust the tyre pressures (Chapter 10)
Clean the windscreen and windows
Clean the headlamps

Every 6000 miles (10 000 km) or six months – whichever comes first

Engine (Chapter 1)
Change engine oil and filter
Clean oil filler cap
Check for any fluid leakage in engine compartment

Cooling system (Chapter 2)
Check coolant level

Fuel system (Chapter 3)
Check idling mixture (carburettor models only) at first service after overhaul
Check idling speed

Ignition system (Chapter 4)
Clean and regap spark plugs (contact breaker models only)
Clean and check distributor cap, rotor, HT leads and coil (contact breaker models only)
Check dwell angle (contact breaker models only)
Check ignition timing (contact breaker models only)

Clutch (Chapter 5)
Check adjustment

Braking system (Chapter 9)
Check servo
Check hydraulic fluid level
Check pads and shoes for wear
Check brake lines and hoses

Suspension and steering (Chapter 10)
Check tyres for wear and damage
Check steering components for wear and damage

Bodywork and fittings (Chapter 11)
Check seat belts for wear, damage and security

Electrical system (Chapter 12)
Check operation of lights
Check washer fluid levels
Check battery electrolyte level

Every 12 000 miles (20 000 km) or 12 months – whichever comes first

In addition to the 6000 mile schedule

Engine (Chapter 1)
Check and adjust valve clearances

Cooling system (Chapter 2)
Check condition and tension of alternator drivebelt

Fuel system (Chapter 3)
Tighten inlet manifold bolts/nuts
Check vacuum hoses
Check exhaust system
Check air cleaner temperature control (if applicable)
Renew fuel filter (fuel injection models only)

Ignition system (Chapter 4)
Lubricate the distributor (contact breaker models only)
Renew the contact breaker points (if applicable)
Renew the spark plugs

Transmission (Chapter 6)
Check oil/fluid level
Lubricate automatic transmission linkage

Rear axle (Chapter 8)
Check oil level

Braking system (Chapter 9)
Check vacuum hose
Lubricate the handbrake linkage

Suspension and steering (Chapter 10)
Check condition and tension of power steering drivebelt (if applicable)
Check suspension for wear and damage

Bodywork and fittings (Chapter 11)
Check operation of all locks
Check security of door check straps
Lubricate locks, door check straps and fuel filler cap flap
Check underbody protective coating

Electrical system (Chapter 12)
Check operation of all electrical items

Every 24 000 miles (40 000 km) or 2 years – whichever comes first

In addition to the 12 000 mile schedule

Engine (Chapter 1)
Renew crankcase emission valve

Cooling system (Chapter 2)
Renew coolant

Fuel system (Chapter 3)
Renew air filter element

Transmission (Chapter 6)
Adjust automatic transmission brake band

Every 36 000 miles (60 000 km) or 3 years – whichever comes first

In addition to the previous schedules (where applicable)

Braking system (Chapter 9)
Renew hydraulic fluid

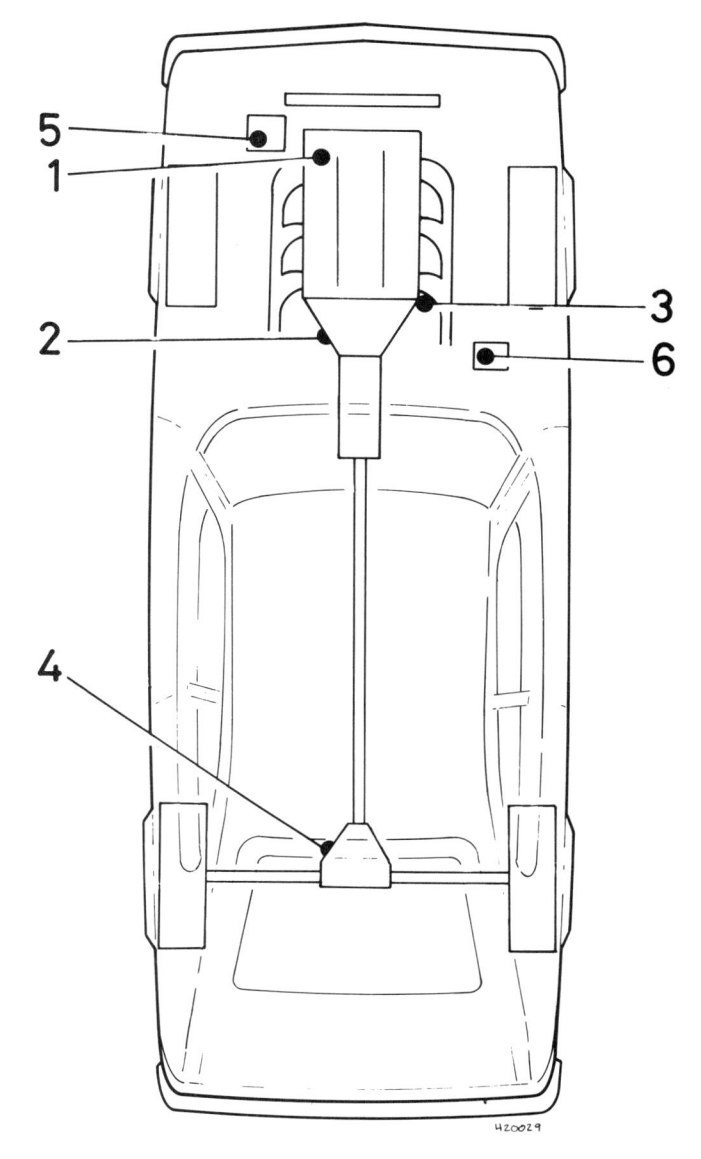

Recommended lubricants and fluids

Component or system	Lubricant type/specification	Duckhams recommendation
1 Engine	Multigrade engine oil, viscosity SAE 20W/40 or 20W/50 to Ford spec SSM-2C 9001-AA	Duckhams QXR, Hypergrade, or 10W/40 Motor Oil
2 Manual Gearbox		
All models except 5-speed 2.8 Injection	Hypoid gear oil, viscosity SAE 80EP to Ford spec SQM-2C 9008-A	Duckhams Hypoid 80
5-speed 2.8 Injection models	Semi-synthetic gear oil to Ford spec ESD-M2C-175-A	Duckhams Hypoid 75W/90S
3 Automatic transmission		
Black dipstick	ATF to Ford spec SQM-2C 9007-AA	Duckhams Q-Matic
Red dipstick	ATF to Ford spec SQM-2C 9010-A	Duckhams D-Matic
4 Rear axle	Gear oil, viscosity SAE 90 to Ford spec SQM-2C 9002-AA	Duckhams Hypoid 90S (Hypoid 90DL for limited slip differential)
5 Power steering	ATF to Ford spec SQM-2C 9007-AA	Duckhams Q-Matic
6 Braking system	Hydraulic fluid Ford spec SAM-6C 9103-A	Duckhams Universal Brake and Clutch Fluid

Conversion factors

Length (distance)
Inches (in)	X	25.4	= Millimetres (mm)	X	0.0394	= Inches (in)
Feet (ft)	X	0.305	= Metres (m)	X	3.281	= Feet (ft)
Miles	X	1.609	= Kilometres (km)	X	0.621	= Miles

Volume (capacity)
Cubic inches (cu in; in^3)	X	16.387	= Cubic centimetres (cc; cm^3)	X	0.061	= Cubic inches (cu in; in^3)
Imperial pints (Imp pt)	X	0.568	= Litres (l)	X	1.76	= Imperial pints (Imp pt)
Imperial quarts (Imp qt)	X	1.137	= Litres (l)	X	0.88	= Imperial quarts (Imp qt)
Imperial quarts (Imp qt)	X	1.201	= US quarts (US qt)	X	0.833	= Imperial quarts (Imp qt)
US quarts (US qt)	X	0.946	= Litres (l)	X	1.057	= US quarts (US qt)
Imperial gallons (Imp gal)	X	4.546	= Litres (l)	X	0.22	= Imperial gallons (Imp gal)
Imperial gallons (Imp gal)	X	1.201	= US gallons (US gal)	X	0.833	= Imperial gallons (Imp gal)
US gallons (US gal)	X	3.785	= Litres (l)	X	0.264	= US gallons (US gal)

Mass (weight)
Ounces (oz)	X	28.35	= Grams (g)	X	0.035	= Ounces (oz)
Pounds (lb)	X	0.454	= Kilograms (kg)	X	2.205	= Pounds (lb)

Force
Ounces-force (ozf; oz)	X	0.278	= Newtons (N)	X	3.6	= Ounces-force (ozf; oz)
Pounds-force (lbf; lb)	X	4.448	= Newtons (N)	X	0.225	= Pounds-force (lbf; lb)
Newtons (N)	X	0.1	= Kilograms-force (kgf; kg)	X	9.81	= Newtons (N)

Pressure
Pounds-force per square inch (psi; lbf/in^2; lb/in^2)	X	0.070	= Kilograms-force per square centimetre (kgf/cm^2; kg/cm^2)	X	14.223	= Pounds-force per square inch (psi; lbf/in^2; lb/in^2)
Pounds-force per square inch (psi; lbf/in^2; lb/in^2)	X	0.068	= Atmospheres (atm)	X	14.696	= Pounds-force per square inch (psi; lbf/in^2; lb/in^2)
Pounds-force per square inch (psi; lbf/in^2; lb/in^2)	X	0.069	= Bars	X	14.5	= Pounds-force per square inch (psi; lbf/in^2; lb/in^2)
Pounds-force per square inch (psi; lbf/in^2; lb/in^2)	X	6.895	= Kilopascals (kPa)	X	0.145	= Pounds-force per square inch (psi; lbf/in^2; lb/in^2)
Kilopascals (kPa)	X	0.01	= Kilograms-force per square centimetre (kgf/cm^2; kg/cm^2)	X	98.1	= Kilopascals (kPa)
Millibar (mbar)	X	100	= Pascals (Pa)	X	0.01	= Millibar (mbar)
Millibar (mbar)	X	0.0145	= Pounds-force per square inch (psi; lbf/in^2, lb/in^2)	X	68.947	= Millibar (mbar)
Millibar (mbar)	X	0.75	= Millimetres of mercury (mmHg)	X	1.333	= Millibar (mbar)
Millibar (mbar)	X	1.40	= Inches of water (inH$_2$O)	X	0.714	= Millibar (mbar)
Millimetres of mercury (mmHg)	X	1.868	= Inches of water (inH$_2$O)	X	0.535	= Millimetres of mercury (mmHg)
Inches of water (inH$_2$O)	X	27.68	= Pounds-force per square inch (psi; lbf/in^2, lb/in^2)	X	0.036	= Inches of water (inH$_2$O)

Torque (moment of force)
Pounds-force inches (lbf in; lb in)	X	1.152	= Kilograms-force centimetre (kgf cm; kg cm)	X	0.868	= Pounds-force inches (lbf in; lb in)
Pounds-force inches (lbf in; lb in)	X	0.113	= Newton metres (Nm)	X	8.85	= Pounds-force inches (lbf in; lb in)
Pounds-force inches (lbf in; lb in)	X	0.083	= Pounds-force feet (lbf ft; lb ft)	X	12	= Pounds-force inches (lbf in; lb in)
Pounds-force feet (lbf ft; lb ft)	X	0.138	= Kilograms-force metres (kgf m; kg m)	X	7.233	= Pounds-force feet (lbf ft; lb ft)
Pounds-force feet (lbf ft; lb ft)	X	1.356	= Newton metres (Nm)	X	0.738	= Pounds-force feet (lbf ft; lb ft)
Newton metres (Nm)	X	0.102	= Kilograms-force metres (kgf m; kg m)	X	9.804	= Newton metres (Nm)

Power
Horsepower (hp)	X	745.7	= Watts (W)	X	0.0013	= Horsepower (hp)

Velocity (speed)
Miles per hour (miles/hr; mph)	X	1.609	= Kilometres per hour (km/hr; kph)	X	0.621	= Miles per hour (miles/hr; mph)

Fuel consumption*
Miles per gallon, Imperial (mpg)	X	0.354	= Kilometres per litre (km/l)	X	2.825	= Miles per gallon, Imperial (mpg)
Miles per gallon, US (mpg)	X	0.425	= Kilometres per litre (km/l)	X	2.352	= Miles per gallon, US (mpg)

Temperature

Degrees Fahrenheit = (°C x 1.8) + 32   Degrees Celsius (Degrees Centigrade; °C) = (°F - 32) x 0.56

*It is common practice to convert from miles per gallon (mpg) to litres/100 kilometres (l/100km), where mpg (Imperial) x l/100 km = 282 and mpg (US) x l/100 km = 235

Fault diagnosis

Introduction

The vehicle owner who does his or her own maintenance according to the recommended schedules should not have to use this section of the manual very often. Modern component reliability is such that, provided those items subject to wear or deterioration are inspected or renewed at the specified intervals, sudden failure is comparatively rare. Faults do not usually just happen as a result of sudden failure, but develop over a period of time. Major mechanical failures in particular are usually preceded by characteristic symptoms over hundreds or even thousands of miles. Those components which do occasionally fail without warning are often small and easily carried in the vehicle.

With any fault finding, the first step is to decide where to begin investigations. Sometimes this is obvious, but on other occasions a little detective work will be necessary. The owner who makes half a dozen haphazard adjustments or replacements may be successful in curing a fault (or its symptoms), but he will be none the wiser if the fault recurs and he may well have spent more time and money than was necessary. A calm and logical approach will be found to be more satisfactory in the long run. Always take into account any warning signs or abnormalities that may have been noticed in the period preceding the fault – power loss, high or low gauge readings, unusual noises or smells, etc – and remember that failure of components such as fuses or spark plugs may only be pointers to some underlying fault.

The pages which follow here are intended to help in cases of failure to start or breakdown on the road. There is also a Fault Diagnosis Section at the end of each Chapter which should be consulted if the preliminary checks prove unfruitful. Whatever the fault, certain basic principles apply. These are as follows:

Verify the fault. This is simply a matter of being sure that you know what the symptoms are before starting work. This is particularly important if you are investigating a fault for someone else who may not have described it very accurately.

Don't overlook the obvious. For example, if the vehicle won't start, is there petrol in the tank? (Don't take anyone else's word on this particular point, and don't trust the fuel gauge either!) If an electrical fault is indicated, look for loose or broken wires before digging out the test gear.

Cure the disease, not the symptom. Substituting a flat battery with a fully charged one will get you off the hard shoulder, but if the underlying cause is not attended to, the new battery will go the same way. Similarly, changing oil-fouled spark plugs for a new set will get you moving again, but remember that the reason for the fouling (if it wasn't simply an incorrect grade of plug) will have to be established and corrected.

Don't take anything for granted. Particularly, don't forget that a 'new' component may itself be defective (esrecially if it's been rattling round in the boot for months), and don't leave components out of a fault diagnosis sequence just because they are new or recently fitted. When you do finally diagnose a difficult fault, you'll probably realise that all the evidence was there from the start.

Electrical faults

Electrical faults can be more puzzling than straightforward mechanical failures, but they are no less susceptible to logical analysis if the basic principles of operation are understood. Vehicle electrical wiring exists in extremely unfavourable conditions – heat, vibration and chemical attack – and the first things to look for are loose or corroded connections and broken or chafed wires, especially where the wires pass through holes in the bodywork or are subject to vibration.

All metal-bodied vehicles in current production have one pole of the battery 'earthed', ie connected to the vehicle bodywork, and in nearly all modern vehicles it is the negative (–) terminal. The various electrical components – motors, bulb holders etc – are also connected to earth, either by means of a lead or directly by their mountings. Electric current flows through the component and then back to the battery via the bodywork. If the component mounting is loose or corroded, or if a good path back to the battery is not available, the circuit will be incomplete and malfunction will result. The engine and/or gearbox are also earthed by means of flexible metal straps to the body or subframe; if these straps are loose or missing, starter motor, generator and ignition trouble may result.

Assuming the earth return to be satisfactory, electrical faults will be due either to component malfunction or to defects in the current supply. Individual components are dealt with in Chapter 12. If supply wires are broken or cracked internally this results in an open-circuit, and the easiest way to check for this is to bypass the suspect wire temporarily with a length of wire having a crocodile clip or suitable connector at each end. Alternatively, a 12V test lamp can be used to verify the presence of supply voltage at various points along the wire and the break can be thus isolated.

If a bare portion of a live wire touches the bodywork or other earthed metal part, the electricity will take the low-resistance path thus formed back to the battery: this is known as a short-circuit. Hopefully a short-circuit will blow a fuse, but otherwise it may cause burning of the insulation (and possibly further short-circuits) or even a fire. This is why it is inadvisable to bypass persistently blowing fuses with silver foil or wire.

Spares and tool kit

Most vehicles are supplied only with sufficient tools for wheel changing; the *Maintenance and minor repair* tool kit detailed in *Tools and working facilities*, with the addition of a hammer, is probably sufficient for those repairs that most motorists would consider attempting at the roadside. In addition a few items which can be fitted

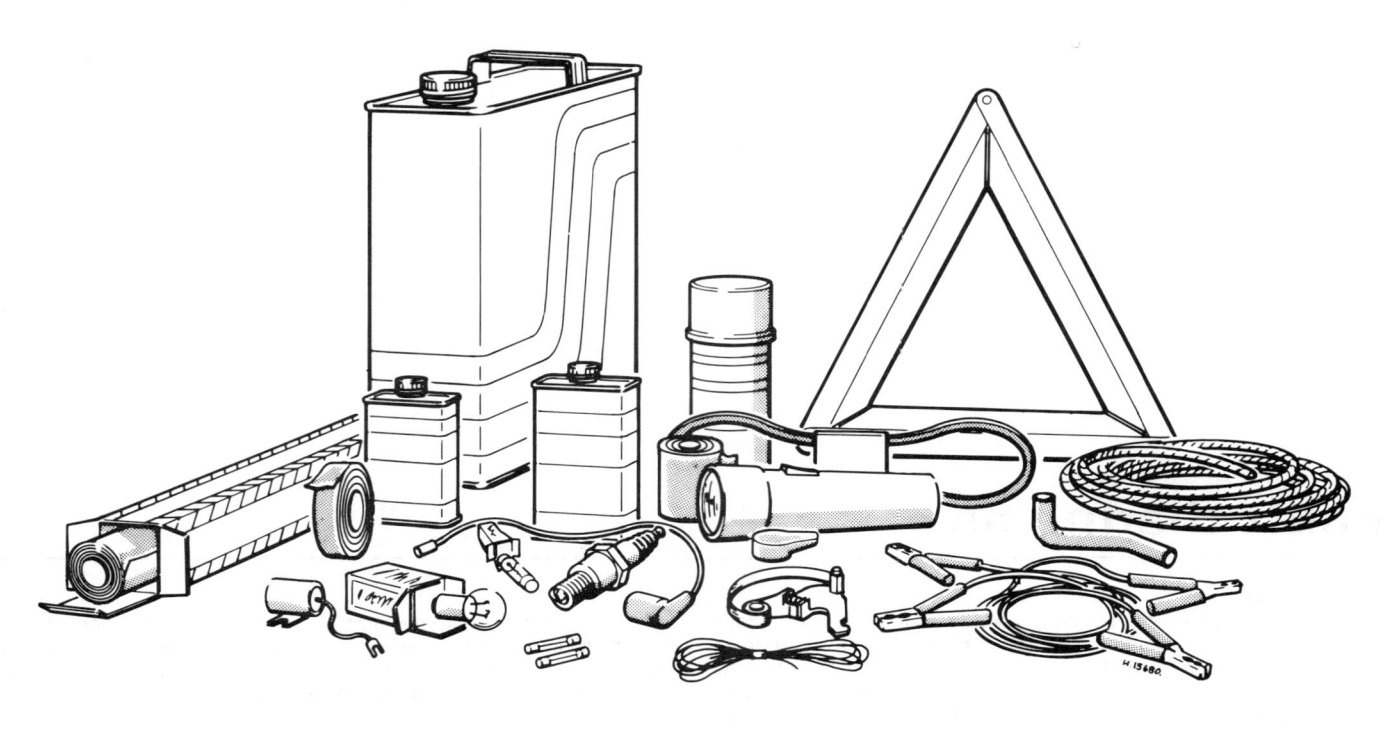

Carrying a few spares can save you a long walk!

without too much trouble in the event of a breakdown should be carried. Experience and available space will modify the list below, but the following may save having to call on professional assistance:

Spark plugs, clean and correctly gapped
HT lead and plug cap – long enough to reach the plug furthest from the distributor
Distributor rotor, condenser and contact breaker points (as applicable)
Drivebelt(s) – emergency type may suffice
Spare fuses
Set of principal light bulbs
Tin of radiator sealer and hose bandage
Exhaust bandage
Roll of insulating tape
Length of soft iron wire
Length of electrical flex
Torch or inspection lamp (can double as test lamp)
Battery jump leads
Tow-rope
Ignition waterproofing aerosol
Litre of engine oil
Sealed can of hydraulic fluid
Emergency windscreen
Worm drive clips
Tube of filler paste

If spare fuel is carried, a can designed for the purpose should be used to minimise risks of leakage and collision damage. A first aid kit and a warning triangle, whilst not at present compulsory in the UK, are obviously sensible items to carry in addition to the above.

When touring abroad it may be advisable to carry additional spares which, even if you cannot fit them yourself, could save having to wait while parts are obtained. The items below may be worth considering:

Clutch and throttle cables
Cylinder head gasket
Alternator brushes
Fuel pump repair kit
Tyre valve core

One of the motoring organisations will be able to advise on availability of fuel etc in foreign countries.

Engine will not start

Engine fails to turn when starter operated
    Flat battery (recharge, use jump leads, or push start)
    Battery terminals loose or corroded
    Battery earth to body defective
    Engine earth strap loose or broken
    Starter motor (or solenoid) wiring loose or broken
    Automatic transmission selector in wrong position, or inhibitor switch faulty
    Ignition/starter switch faulty
    Major mechanical failure (seizure)
    Starter or solenoid internal fault (see Chapter 12)

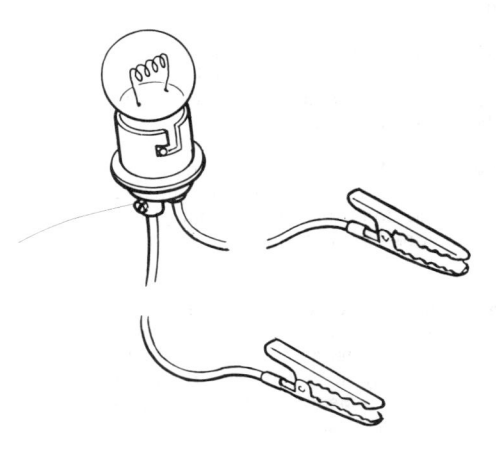

A simple test lamp is useful for checking electrical faults

Fault diagnosis

Starter motor turns engine slowly
- Partially discharged battery (recharge, use jump leads, or push start)
- Battery terminals loose or corroded
- Battery earth to body defective
- Engine earth strap loose
- Starter motor (or solenoid) wiring loose
- Starter motor internal fault (see Chapter 12)

Engine turns normally but fails to start
- Damp or dirty HT leads and distributor cap (crank engine and check for spark)
- Dirty or incorrectly gapped distributor points (if applicable)
- No fuel in tank (check for delivery at carburettor)
- Excessive choke (hot engine) or insufficient choke (cold engine)
- Fouled or incorrectly gapped spark plugs (remove, clean and regap)
- Other ignition system fault (see Chapter 4)
- Other fuel system fault (see Chapter 3)
- Poor compression
- Major mechanical failure (eg camshaft gear teeth stripped)

Engine fires but will not run
- Insufficient choke (cold engine)
- Air leaks at carburettor or inlet manifold
- Fuel starvation (see Chapter 3)
- Ballast resistor defective, or other ignition fault (see Chapter 4)

Engine cuts out and will not restart

Engine cuts out suddenly – ignition fault
- Loose or disconnected LT wires
- Wet HT leads or distributor cap (after traversing water splash)
- Coil or condenser failure (check for spark)
- Other ignition fault (see Chapter 4)

Engine misfires before cutting out – fuel fault
- Fuel tank empty
- Fuel pump defective or filter blocked (check for delivery)
- Fuel tank filler vent blocked (suction will be evident on releasing cap)
- Carburettor needle valve sticking (where applicable)
- Carburettor jets blocked (fuel contaminated)
- Other fuel system fault (see Chapter 3)

Engine cuts out – other causes
- Serious overheating
- Major mechanical failure (eg camshaft drive)

Engine overheats

Ignition (no-charge) warning light illuminated
- Slack or broken drivebelt – retension or renew (Chapter 2)

Ignition warning light not illuminated
- Coolant loss due to internal or external leakage (see Chapter 2)
- Thermostat defective
- Low oil level
- Brakes binding
- Radiator clogged externally or internally
- Engine waterways clogged
- Ignition timing incorrect or automatic advance malfunctioning
- Mixture too weak

Note: *Do not add cold water to an overheated engine or damage may result*

Low engine oil pressure

Gauge reads low or warning light illuminated with engine running
- Oil level low or incorrect grade
- Defective gauge or sender unit
- Wire to sender unit earthed
- Engine overheating
- Oil filter clogged or bypass valve defective
- Oil pressure relief valve defective
- Oil pick-up strainer clogged
- Oil pump worn or mountings loose
- Worn main or big-end bearings

Note: *Low oil pressure in a high-mileage engine at tickover is not necessarily a cause for concern. Sudden pressure loss at speed is far more significant. In any event, check the gauge or warning light sender before condemning the engine.*

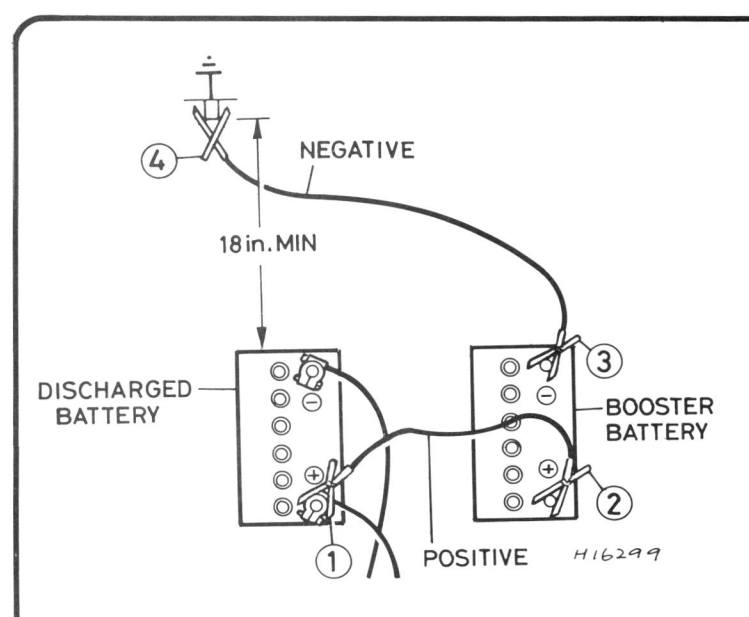

Jump start lead connections for negative earth vehicles – connect leads in order shown

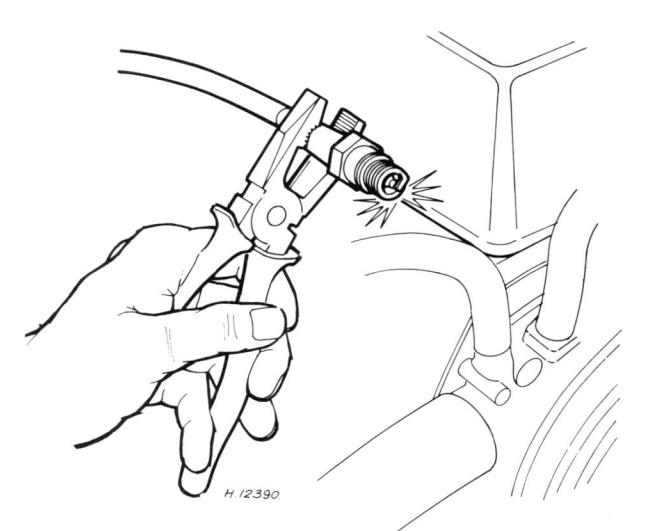

Crank engine and check for spark. Note use of insulated tool to hold plug lead

Engine noises

Pre-ignition (pinking) on acceleration
    Incorrect grade of fuel
    Ignition timing incorrect
    Distributor faulty or worn
    Worn or maladjusted carburettor or idling settings
    Excessive carbon build-up in engine

Whistling or wheezing noises
    Leaking vacuum hose
    Leaking carburettor or manifold gasket
    Blowing head gasket

Tapping or rattling
    Incorrect valve clearances
    Worn valve gear
    Broken piston ring (ticking noise)

Knocking or thumping
    Unintentional mechanical contact (eg fan blades)
    Worn drivebelt
    Peripheral component fault (generator, water pump etc)
    Worn big-end bearings (regular heavy knocking, perhaps less under load)
    Worn main bearings (rumbling and knocking, perhaps worsening under load)
    Piston slap (most noticeable when cold)

Chapter 1 Engine

Contents

Part A: 2.8 litre engine
Camshaft and camshaft bearings – examination and renovation ...	14
Camshaft and front intermediate plate – refitting	28
Connecting rods and gudgeon pins – examination and renovation	18
Crankcase ventilation system – description	25
Crankshaft – examination and renovation	19
Crankshaft – refitting	27
Cylinder bores – examination and renovation	16
Cylinder heads – dismantling, renovation and reassembly	15
Cylinder heads, rocker gear and inlet manifold – refitting	34
Engine – dismantling	9
Engine – initial start-up after major overhaul or repair	38
Engine – refitting	37
Engine – removal	6
Engine ancillaries – refitting	36
Enigne ancillaries – removal	8
Engine components – examination for wear	11
Engine dismantling – general	7
Engine reassembly – general	26
Flywheel – examination and renovation	22
Flywheel and clutch – refitting	31
General description	1
Lubrication system – description	24
Main and big-end bearings – examination and renovation	20
Major operations possible with engine in car	3
Major operations requiring engine removal	4
Methods of engine removal	5
Oil pump – dismantling, examination and reassembly	23
Oil pump – refitting	30
Pistons and connecting rods – refitting	29
Pistons and piston rings – examination and renovation	17
Rocker assembly – dismantling, examination and reassembly	12
Routine maintenance	2
Sump – removal and refitting with engine in car	10
Sump, water pump and crankshaft pulley – refitting	33
Tappets and pushrods – examination and renovation	13
Timing gears – examination and renovation	21
Timing gears and timing cover – refitting	32
Valve clearances – checking and adjustment	35

Part B: 3.0 litre engine
Camshaft refitting	79
Camshaft – removal	57
Camshaft and camshaft bearings – examination and renovation ...	65
Connecting rods and gudgeon pins – examination and renovation	68
Crankcase ventilation system – description	77
Crankshaft – examination and renovation	69
Crankshaft – refitting	80

Crankshaft and main bearings – removal	61
Crankshaft pulley – removal	52
Crankshaft rear oil seal – removal	60
Cylinder bores – examination and renovation	66
Cylinder heads – reassembly	86
Cylinder heads – refitting	87
Cylinder heads – removal with engine in car	47
Cylinder heads – removal with engine out of car	48
Decarbonising – general	74
Engine – initial start-up after major overhaul or repair	92
Engine – refitting	91
Engine – removal	44
Engine ancillaries – refitting	90
Enigne ancillaries – removal	46
Engine components – examination for wear	62
Engine dismantling – general	45
Engine reassembly – general	78
Flywheel – removal	53
Flywheel/driveplate – examination and renovation	72
Front cover – removal	55
General description	39
Lubrication system – description	76
Main and big-end bearings – examination and renovation	70
Major operations possible with engine in car	41
Major operations requiring engine removal	42
Methods of engine removal	43
Oil pump – dismantling, examination and reassembly	75
Oil pump – refitting	82
Oil pump – removal	58
Pistons and connecting rods – refitting	81
Pistons and piston rings – examination and renovation	67
Pistons, connecting rods and bearings – removal	59
Rear oil seal and flywheel – refitting	84
Rocker arms – examination and renovation	63
Rocker arms – removal	50
Rocker arms and covers – refitting	88
Routine maintenance	40
Sump – refitting	85
Sump – removal	54
Tappets – removal	51
Tappets and pushrods – examination and renovation	64
Timing gears – examination and renovation	71
Timing gears – removal	56
Timing gears and front plate – refitting	83
Valve clearances – checking and adjustment	89
Valves – removal	49
Valves and valve seats – examination and renovation	73

Part C: Fault diagnosis
Fault diagnosis – engine	93

Specifications

Part A: 2.8 litre engine

General
Type	6-cylinder ohv in 60° V formation
Bore	93.03 mm (3.663 in)
Stroke	68.50 mm (2.697 in)
Cubic capacity	2792 cc
Compression ratio	9.2 : 1

Cylinder block
Bore diameter mm (in):
  Standard grades:
    1 .... 93.010 to 93.020 (3.6618 to 3.6622)
    2, 3 and 4 .... Increments of + 0.010 (0.0004)
  Oversize grades:
    A .... 93.520 to 93.530 (3.6819 to 3.6823)
    B and C .... Increments of 0.010 (0.0004)

Crankshaft
Endfloat mm (in) .... 0.080 to 0.280 (0.0031 to 0.0110)

Camshaft
Thrust plate thickness mm (in):
  Red .... 3.960 to 3.985 (0.1559 to 0.1569)
  Blue .... 3.986 to 4.011 (0.1569 to 0.1579)
Endfloat arm (in) .... 0.02 to 0.10 (0.0008 to 0.0039)

Pistons
Clearance in bore mm (in) .... 0.028 to 0.048 (0.0011 to 0.0019)
Ring gap mm (in):
  Top .... 0.38 to 0.58 (0.0150 to 0.0228)
  Centre .... 0.38 to 0.58 (0.0150 to 0.0228)
  Bottom .... 0.38 to 1.40 (0.0150 to 0.0551)

Connecting rods
Big-end bearing running clearance mm (in) .... 0.006 to 0.064 (0.0002 to 0.0025)

Valves
Seat angle .... 46°
Valve clearance (cold) mm (in)
  Inlet 0.35 (0.014)
  Exhaust 0.40 (0.016)
Valve timing:
  Inlet opens .... 24° BTDC
  Inlet closes .... 72° ABDC
  Exhaust opens .... 73° BBDC
  Exhaust closes .... 25° ATDC
Tappet side clearance mm (in) .... 0.023 to 0.060 (0.0009 to 0.0024)
Valve spring free length mm (in) 52.5 (2.0669)
Stem-to-guide clearance mm (in)
  Inlet .... 0.020 to 0.063 (0.0008 to 0.0025)
  Exhaust .... 0.046 to 0.089 (0.0018 to 0.0035)

Lubrication system
Oil type/specification .... Multigrade engine oil, viscosity SAE 20W/40 or 20W/50 to Ford spec SSM-2C 9001-AA (Duckhams QXR, Hypergrade, or 10W/40 Motor Oil)
Capacity:
  With filter .... 4.25 litre (7.5 pint)
  Without filter .... 4.00 litre (7.0 pint)
Oil pressure (minimum at 80°C)
  750 rpm .... 1.0 kgf/cm² (14.2 lbf/in²)
  2000 rpm .... 2.8 kgf/cm² (39.8 lbf/in²)

Torque wrench settings

	Nm	lbf ft
Main bearing caps	90 to 104	66 to 77
Big-end caps	29 to 34	21 to 25
Crankshaft pulley bolt	42 to 50	31 to 37
Camshaft gear	42 to 50	31 to 37
Camshaft thrust plate	17 to 21	13 to 15
Flywheel	65 to 71	48 to 52
Timing cover:		
To block	17 to 21	13 to 15
To adaptor plate	13 to 17	10 to 13
Oil pump	14 to 17	10 to 13
Oil pump cover	9 to 13	7 to 10
Rocker shaft	59 to 67	44 to 49
Sump:		
Stage 1	4 to 7	3 to 5
Stage 2	7 to 10	5 to 7
Oil drain plug	21 to 28	15 to 21
Cylinder head bolts:		
Stage 1	39 to 54	29 to 40
Stage 2	54 to 69	40 to 51
Stage 3 (after 10 to 20 minutes wait)	95 to 115	70 to 85
Stage 4 (after running engine for 15 minutes at 1000 rpm)	Back off 1/2 turn and retorque one bolt at a time to Stage 3 value	
Rocker cover	4 to 7	3 to 5

Chapter 1 Engine

Part B: 3.0 litre engine

General
Type	6-cylinder ohv in 60° V formation
Bore	93.67 mm (3.688 in)
Stroke	72.42 mm (2.851 in)
Cubic capacity	2993 cc
Compression ratio	8.9 : 1

Cylinder block
Bore diameter mm (in):
- Standard grades:
  - A .... 93.647 to 93.658 (3.6869 to 3.6873)
  - B, C and D .... Increments of + 0.010 (0.0004)
- Oversize grade A .... 94.036 to 94.091 (3.7022 to 3.7044)

Crankshaft
Endfloat mm (in) .... 0.080 to 0.280 (0.0031 to 0.0110)

Camshaft
Thrust plate thickness mm (in)	5.334 to 5.384 (0.210 to 0.212)
Endfloat mm (in)	0.153 to 0.254 (0.006 to 0.010)

Pistons
Clearances in bore mm (in) .... 0.036 to 0.051 (0.0014 to 0.0020)

Ring gap mm (in)
- Top .... 0.254 to 0.508 (0.010 to 0.020)
- Centre .... 0.254 to 0.508 (0.010 to 0.020)
- Bottom .... 0.254 to 0.381 (0.010 to 0.015)

Connecting rods
Big-end bearing running clearance mm (in) .... 0.006 to 0.064 (0.0002 to 0.0025)

Valves
Seat angle	46°

Valve clearance (cold) mm (in):
- Inlet 0.32 (0.013)
- Exhaust 0.55 (0.022)

Valve timing:
- Inlet opens .... 29° BTDC
- Inlet closes .... 67° ABDC
- Exhaust opens .... 70° BBDC
- Exhaust closes .... 14° ATDC

Tappet side clearance mm (in) .... 0.013 to 0.051 (0.0005 to 0.0020)
Valve spring free length mm (in) 47.955 (1.8880)

Stem-to-guide clearance mm (in):
- Inlet .... 0.020 to 0.063 (0.0008 to 0.0025)
- Exhaust .... 0.048 to 0.050 (0.0019 to 0.0020)

Lubrication system
Oil type/specification .... Multigrade engine oil, viscosity SAE 20W/40 or 20W/50 to Ford spec SSM-2C 9001-AA (Duckhams QXR, Hypergrade, or 10W/40 Motor Oil)

Capacity:
- With filter .... 5.00 litre (8.8 pint)
- Without filter .... 4.25 litre (7.5 pint)

Oil pressure (minimum at 80°C):
- 750 rpm .... 1.0 kgf/cm² (14.2 lbf/in²)
- 2000 rpm .... 2.8 kgf/cm² (39.8 lbf/in²)

Torque wrench settings

	Nm	lbf ft
Main bearing caps	68 to 76	50 to 56
Big-end caps	53 to 59	39 to 44
Crankshaft pulley	55 to 62	41 to 46
Camshaft gear	55 to 61	41 to 45
Flywheel/driveplate	69 to 76	51 to 56
Front cover	15 to 18	11 to 13
Oil pump and pick-up	17 to 21	13 to 15
Oil pump cover	8 to 12	6 to 9
Sump	8 to 11	6 to 8
Cylinder head bolts:		
Stage 1	10	7
Stage 2	30 to 50	22 to 37
Stage 3 (after 10 to 20 minutes wait)	82 to 92	61 to 68
Stage 4	110 to 117	81 to 86
Stage 5 (after running engine for 15 minutes at 1000 rpm)	110 to 117	81 to 86
Rocker cover	3 to 5	2 to 4
Oil drain plug	27 to 34	20 to 25
Rear oil seal carrier	16 to 20	12 to 15

Chapter 1 Engine

PART A: 2.8 LITRE ENGINE

1 General description

The 2.8 litre engine is of German V6 overhead valve type. The camshaft is located centrally in the cylinder block and is driven by helical gears from the crankshaft. The valves are operated by tappets, pushrods and rocker arms.

The cylinder heads are of the crossflow design with the inlet manifold located on top of the cylinder block between the two cylinder heads and the exhaust manifolds on the outside of the heads. The combined crankcase and cylinder block is made of cast iron. Attached to the bottom of the crankcase is a pressed steel sump which acts as a reservoir for the engine oil.

Aluminium alloy pistons are connected to the crankshaft by H-section forged steel connecting rods and gudgeon pins. Two compression rings and one oil control ring, all located above the gudgeon pin, are fitted.

The forged crankshaft runs in four main bearings, and endfloat is controlled by thrust washers on No 3 main bearing. The drive gear for the distributor and oil pump is located in front of the rear camshaft bearing.

Lubrication is by means of a bi-rotor oil pump.

2 Routine maintenance

At the intervals specified in the Routine Maintenance section in the front of the manual carry out the following procedures.

Change engine oil, renew oil filter and clean oil filler cap

1 Apply the handbrake then jack up the front of the car and support on axle stands. Alternatively position the car over an inspection pit.
2 Place a suitable container beneath the sump then unscrew the drain plug (photos) and drain the oil.
3 While the oil is draining unscrew the oil filter cartridge using a strap wrench.
4 Clean the drain plug and sump then refit the plug and tighten it.
5 Clean the filter contact area of the extension bracket.
6 Smear a film of engine oil on the new oil filter rubber gasket, then screw it on until it just contacts the bracket face (photo). From this

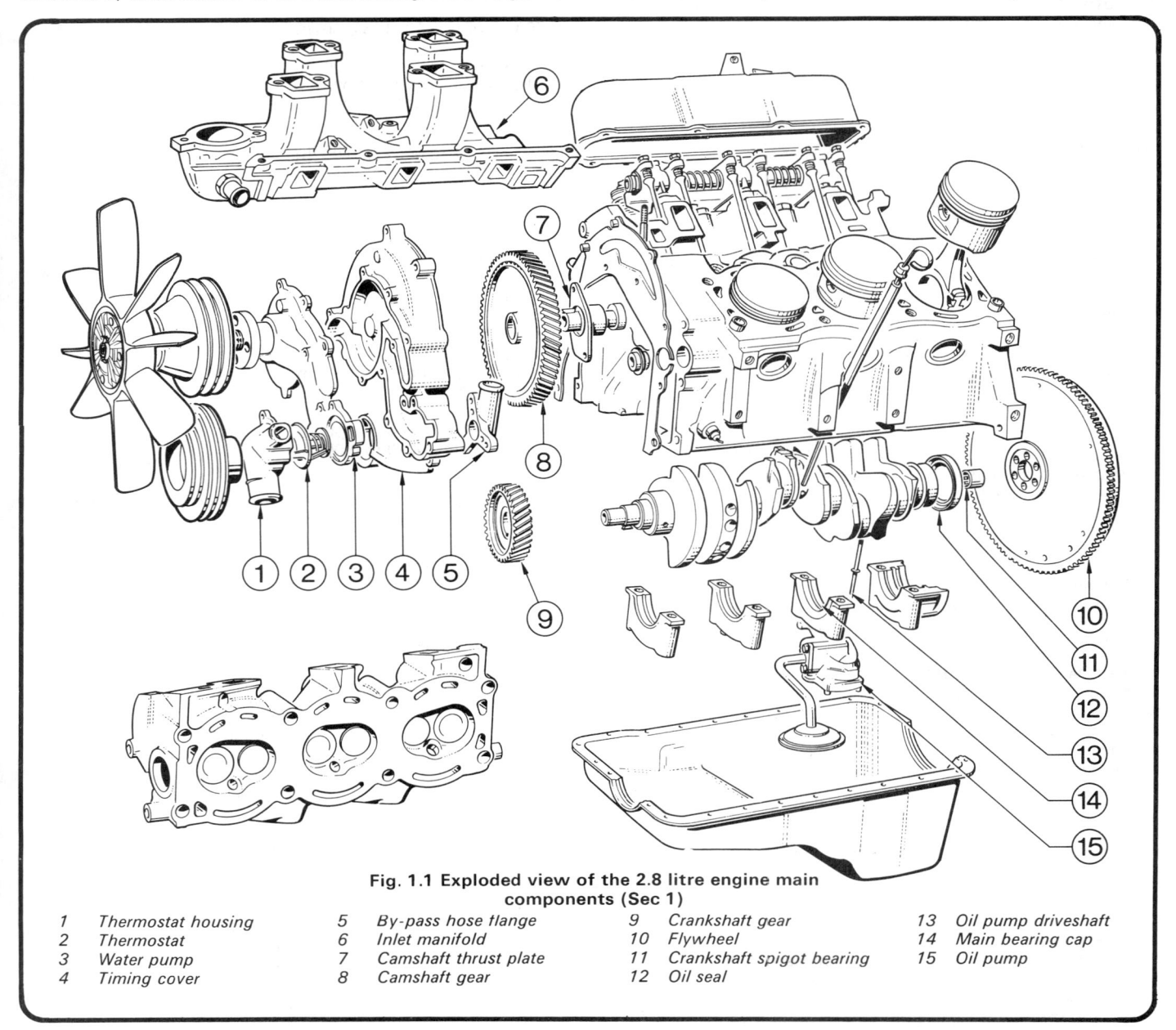

Fig. 1.1 Exploded view of the 2.8 litre engine main components (Sec 1)

1	Thermostat housing	5	By-pass hose flange	9	Crankshaft gear
2	Thermostat	6	Inlet manifold	10	Flywheel
3	Water pump	7	Camshaft thrust plate	11	Crankshaft spigot bearing
4	Timing cover	8	Camshaft gear	12	Oil seal
				13	Oil pump driveshaft
				14	Main bearing cap
				15	Oil pump

Chapter 1 Engine

point tighten it a further 3/4 turn.
7 Remove the filler cap on the left-hand side of the engine and fill the engine with the correct quantity and grade of oil (photo).
8 Remove the oil level dipstick and wipe it clean, then re-insert it and remove it again. The oil level should be up to the maximum mark (photo), but note that this check is only accurate if the car is on level ground.
9 If necessary, top up the level noting that the distance between the minimum and maximum level marks represents approximately 1.0 litre (1.76 pint).
10 Clean the oil filler cap then refit it.

Check and adjust valve clearances
11 Refer to Section 35.

Tighten inlet manifold bolts
12 Remove the air chamber and tighten the inlet manifold bolts to the specified torque with reference to Chapter 3. On completion refit the air chamber.

Renew the crankcase emission valve
13 Pull the valve and hose from the rear of the right-hand rocker cover (photo).
14 Release the clip and remove the valve from the hose.
15 Fit the new valve in reverse order.

3 Major operations possible with engine in car

The following major operations can be carried out without removing the engine from the car.

(a) Removal and refitting of the cylinder heads
(b) Removal and refitting of rocker shaft and pushrods
(c) Removal and refitting of front engine mountings
(d) Removal and refitting of the flywheel (gearbox removed)
(e) Removal and refitting of the sump and oil pump
(f) Removal and refitting of the timing gears
(g) Removal and refitting of the big-end bearings
(h) Removal and refitting of the pistons and connecting rods
(i) Renewal of the crankshaft front seal in timing cover

4 Major operations requiring engine removal

The following major operations may be carried out after removal of the engine:

(a) Removal and refitting of the crankshaft and main bearings
(b) Renewal of the crankshaft rear seal

5 Methods of engine removal

1 Although it is possible to remove the engine and gearbox as an assembly it is recommended that the engine is removed as a separate unit. If the engine and gearbox are removed as an assembly it has to be lifted out at a very steep angle and for this a considerable lifting height is required.
2 If the work being undertaken requires the removal of the gearbox as well as the engine it is recommended that the gearbox is removed first (refer to Chapter 6).

6 Engine – removal

It is essential to have a good hoist, and two axle-stands if an inspection pit is not available.
The sequence of operations listed in this Section is not critical as the position of the person undertaking the work, or the tool in his hand, will determine to a certain extent the order in which the necessary operations are performed. Obviously the engine cannot be removed until everything is disconnected from it and the following sequence will ensure that nothing is forgotten.
1 Remove the bonnet as described in Chapter 11.
2 Disconnect the battery negative lead and the earth cable from the

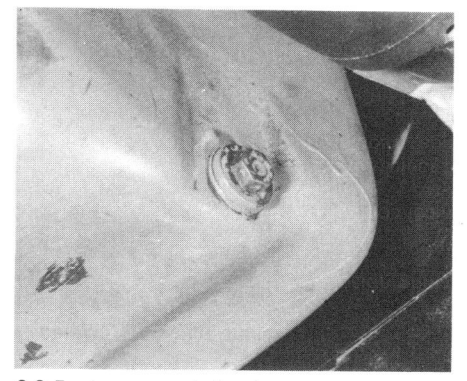

2.2 Engine sump drain plug

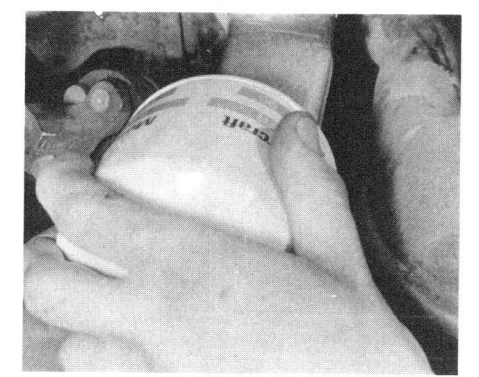

2.6 Fitting the oil filter cartridge

2.7 Filling the engine with oil

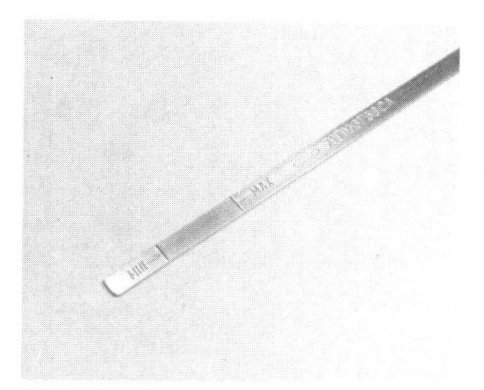

2.8 Engine oil level dipstick markings

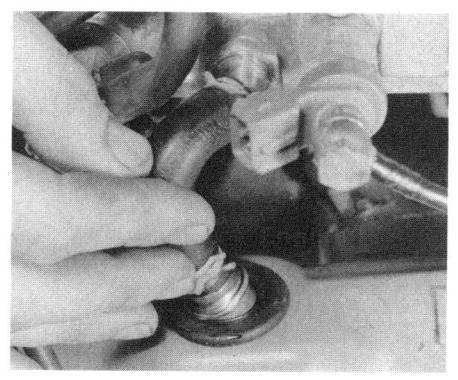

2.13 Removing the crankcase emission valve

6.2 Earth cable on the alternator adjustment bracket (arrowed)

alternator adjustment bracket (photo).
3 Drain the cooling system and remove the radiator as described in Chapter 2.
4 Remove the expansion tank and bracket as described in Chapter 2 (photos).
5 Disconnect the hoses from the auxiliary air valve, thermostat housing, and the heater tubes on the bulkhead.
6 Disconnect the brake servo vacuum hose from the air chamber.
7 Disconnect the inlet duct from the throttle housing and mixture control unit.
8 Disconnect the throttle cable with reference to Chapter 3.
9 Pull the centre HT lead from the distributor cap and position to one side.
10 Disconnect the multi-plug from the distributor.
11 Disconnect the wiring from the temperature gauge sender unit (photo), thermotime switch, warm-up regulator, auxiliary air device and cold start valve.
12 Unscrew the union nut and disconnect the oil pressure gauge line (photo).
13 Disconnect the earth cable from the the bulkhead (photo).
14 Position a suitable container beneath the power steering pump, then disconnect the fluid hoses (Chapter 10) and drain the fluid. Place the hoses on one side.
15 Remove the alternator as described in Chapter 12.
16 Remove the power steering pump drivebelt as described in Chapter 10.
17 Remove the air chamber then remove the injectors, leaving them attached to the fuel lines, and position the lines to one side. Refer to Chapter 3 for the removal procedure.

18 Apply the handbrake then jack up the front of the car and support on axle stands.
19 Remove the starter motor as described in Chapter 12.
20 Remove the clutch cable from the gearbox with reference to Chapter 5.
21 Unscrew and remove the oil filter cartridge (Section 2).
22 Disconnect the hoses from the oil cooler.
23 Using a 27 mm spanner unscrew the special nut securing the extension bracket to the oil cooler. Remove the bracket, together with the washers and O-ring (photo).
24 Unscrew the nuts from the two engine mountings (photo).
25 Unbolt the exhaust pipes from the left and right exhaust manifolds. Tie them to one side.
26 Unbolt the gearbox from the engine adaptor plate.
27 Lower the car to the ground.
28 Support the gearbox with a trolley jack.
29 Attach a hoist to the engine and just take its weight. To maintain engine balance, a suitable place to connect the hoist chains is around the front of the exhaust manifolds.
30 To prevent damage to the distributor cap and rotor arm, remove them with reference to Chapter 4.
31 Unscrew the gearbox-to-engine bolts.
32 Raise the engine while guiding the engine mountings from the slotted brackets (photo). It may be helpful to completely remove the right-hand mounting. With the engine raised clear of the brackets, pull it forwards to disengage it from the gearbox and location dowels.
33 When the clutch is clear of the gearbox input shaft lift the engine from the engine compartment while guiding it past the surrounding components (photo).

6.4A Removing the expansion tank mounting bracket upper ...

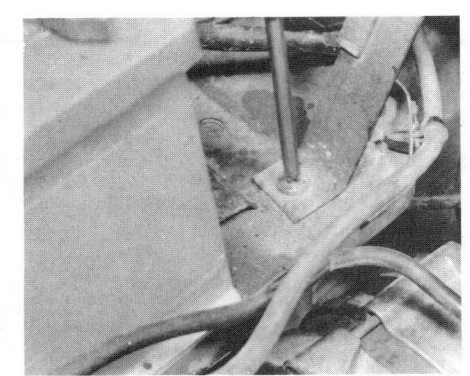

6.4B ... and lower screws

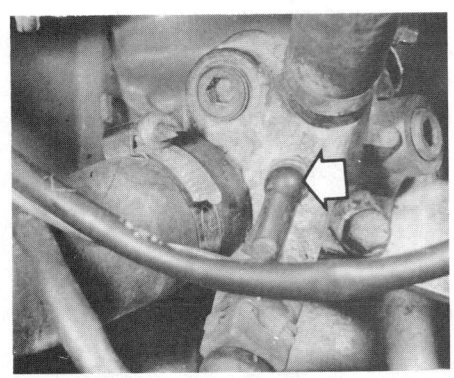

6.11 Temperature gauge sender unit wiring (arrowed)

6.12 Oil pressure gauge line (arrowed)

6.13 Removing the earth cable from the bulkhead

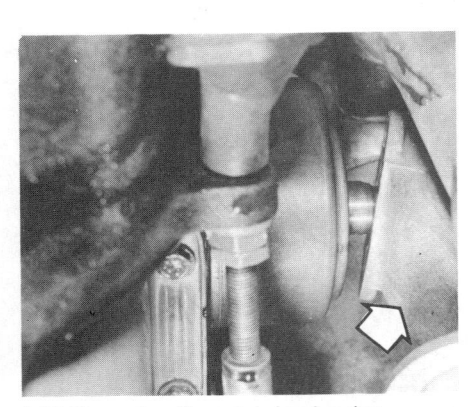

6.23 Removing the extension bracket (arrowed) from the oil cooler

Chapter 1 Engine

6.24 Left-hand engine mounting nut

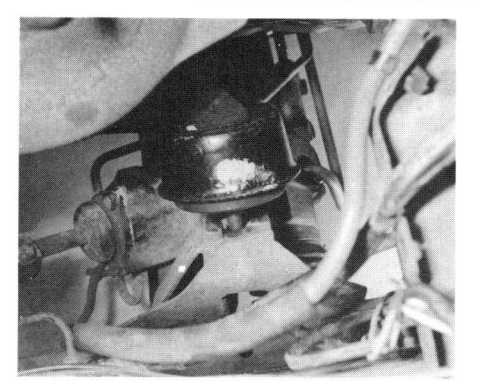

6.32 Lifting the engine clear of the mounting brackets

6.33 Lifting the engine from the engine compartment

7 Engine dismantling – general

1 When the engine is removed from the car, it, and particularly its accessories, are vulnerable to damage. If possible mount the engine on a stand, or failing this, make sure it is supported in such a manner that it will not be damaged whilst undoing tight nuts and bolts.
2 Cleanliness is important when dismantling the engine to prevent exposed parts from contamination. Before starting the dismantling operations, clean the outside of the engine with paraffin, or a good grease solvent if it is very dirty. Carry out this cleaning away from the area in which the dismantling is to take place.
3 If an engine stand is not available carry out the work on a bench or wooden platform. Avoid working with the engine directly on a concrete floor, as grit presents a real source of trouble.
4 As parts are removed, clean them in a paraffin bath. Never immerse parts with internal oilways in paraffin but wipe down carefully with a petrol-dampened rag. Clean oilways with wire.
5 It is advisable to have suitable containers to hold small items in their groups as this will help when reassembling the engine and also prevent possible loss.
6 Always obtain complete sets of gaskets when the engine is being dismantled. It is a good policy to always fit new gaskets in view of the relatively small cost involved. Retain the old gaskets when dismantling the engine with a view to using them as a pattern to make a replacement gasket if a new one is not available.
7 When possible, refit nuts, bolts and washers in their locations as this helps to protect the threads from damage and will also be helpful when the engine is being reassembled, as it establishes their location.
8 Retain unserviceable items until the new parts are obtained, so that the new part can be checked against the old part to ensure that the correct item has been supplied.

8 Engine ancillaries – removal

Although the items listed may be removed separately with the engine installed (as described in the relevant Chapters) it is more appropriate to take them off after the engine has been removed from the car when extensive dismantling is being carried out. The items are:

Fuel injection system components (Chapter 3)
Distributor (Chapter 4)
Fan and water pump (Chapter 2)
Power steering pump (Chapter 10)
Inlet and exhaust manifolds (Chapter 3)
Clutch (Chapter 5)
Spark plugs (Chapter 4)

9 Engine – dismantling

1 Remove the oil level dipstick.
2 Unscrew the centre sleeve and remove the oil cooler and gasket (photo).
3 If not done previously, unscrew the sump drain plug and drain the oil into a suitable container.
4 Unbolt the engine mountings (photos).
5 Unbolt the power steering pump idler pulley and bracket (photo).
6 Disconnect the bypass hose from the timing cover rear elbow (photo).
7 Remove the rocker cover securing bolts and lift off the covers (photo). Disconnect the breather hose where applicable.
8 Undo the rocker shaft securing bolts and remove the rocker shafts and oil splash shields (photos). Note which way the splash shields are

9.2 Oil cooler showing centre sleeve

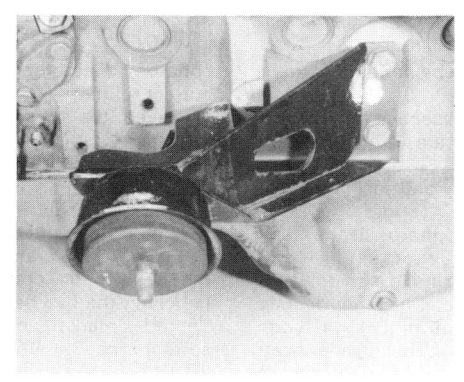

9.4A Left-hand side engine mounting

9.4B Right-hand side engine mounting

Chapter 1 Engine

9.5 Power steering pump idler pulley and bracket (arrowed)

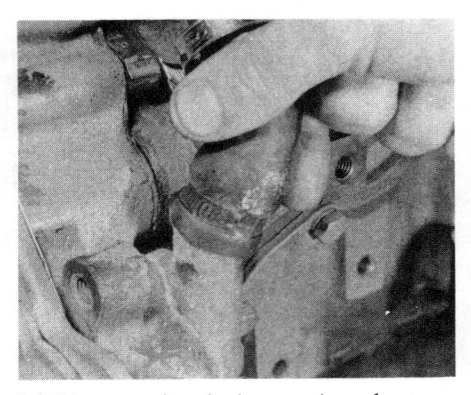

9.6 Disconnecting the by-pass hose from the timing cover rear elbow

9.7 Unbolting the rocker covers

9.8A Unscrew the bolts ...

9.8B .. and remove the rocker shafts ...

9.8C ... and oil splash shields

9.9 Remove the pushrods

9.10 Removing the cylinder head bolts

9.11 Removing the right-hand side cylinder head

fitted. Mark the rocker shafts so that they can be refitted in their original positions.
9 Lift out the pushrods and keep them in their respective positions in relation to the rocker shafts to ensure that they are refitted in their original locations (photo).
10 Unscrew the cylinder head bolts progressively and in the reverse order to the tightening sequence shown in Fig. 1.7 (photo).
11 Each head can now be lifted from the block (photo). If the head sticks to the cylinder block try to break the seal by rocking it. If this does not free it a soft-faced hammer can be used to strike it sharply and break the cylinder head joint seal. Never use a metal hammer directly on the head as this may fracture the casting. Also never try to prise the head free by forcing a screwdriver or chisel between the cylinder head and cylinder block as this will damage the mating surfaces.
12 Remove the cylinder head gaskets.
13 Unscrew the oil pressure switch.
14 Using a piece of bent wire, prise the valve tappets from the block

(photos). Place the tappets with their respective pushrods to ensure correct refitting.
15 Invert the engine and unbolt the sump (photo). Remove the gaskets.
16 Unbolt the pick-up pipe from the oil pump, and remove the gasket (photos).
17 Unbolt the oil pump and withdraw the driveshaft, noting which way round it is fitted (photos).
18 Remove the crankshaft pulley securing bolt and take the pulley off the crankshaft (photos). Restrain the crankshaft from turning by chocking the flywheel.
19 Index mark the position of the flywheel in relation to the crankshaft so that it can be refitted in the same position.
20 Chock the flywheel to restrain it from turning, then remove the six bolts securing the flywheel to the crankshaft and lift off the flywheel (photo).
21 Remove the engine adaptor plate (photo).

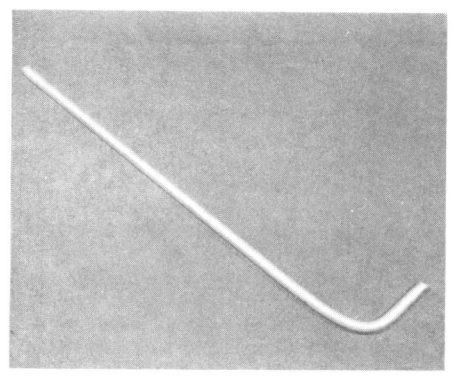

9.14A Using a piece of bent wire ...

9.14B ... prise up the tappets ...

9.14C ... and remove them from the block

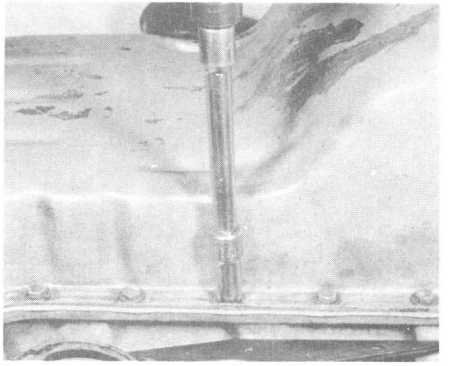

9.15 Unbolting the sump

9.16A Unbolt the oil pump pick-up pipe ...

9.16B ... and remove the gasket

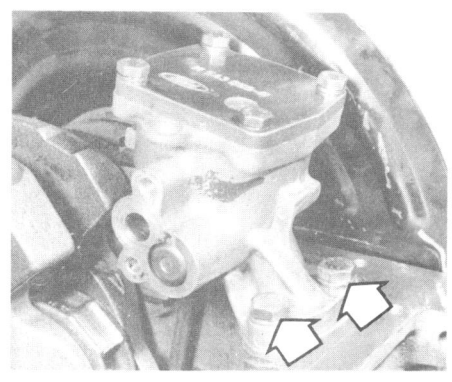

9.17A Oil pump mounting bolts (arrowed)

9.17B Removing the oil pump driveshaft (arrowed)

9.18A Unscrew the crankshaft pulley bolt ...

9.18B ... remove the bolt ...

9.18C ... and slide off the pulley

9.20 Removing the flywheel bolts

Chapter 1 Engine

22 Unbolt the thermostat rear water elbow. Remove the gasket.
23 Unbolt the timing cover and remove the gasket (photos).
24 Unscrew the bolt securing the camshaft gear to the camshaft and pull off the gear (photos).
25 Unscrew the front intermediate plate attaching bolts and remove the intermediate plate. Remove the gasket and spacer (photos).
26 If the crankshaft gear needs to be removed use a standard puller to draw it off the crankshaft (photo).
27 Unscrew the camshaft thrust plate securing bolts, remove the thrust plate and withdraw the camshaft (photos).
28 Check that the big-end bearing caps and connecting rods have identification marks. This is to ensure that the correct caps are fitted to the correct connecting rods and at reassembly are fitted in their correct cylinder bores. Note that the pistons have an arrow (or notch) marked on the crown to indicate the forward facing side (photos).
29 Remove the No 1 cylinder big-end nuts, then tap off the big-end cap. Keep the shell bearings with the cap and connecting rod from which they are removed. To remove the shell bearings, press the bearing on the side opposite the groove in both the connecting rod and the cap, and the bearing will slide out (photos).
30 Withdraw the piston and connecting rod upwards out of the cylinder bore by tapping with the handle of a hammer from underneath.
31 Repeat the procedure given in paragraphs 29 and 30 to remove the remaining pistons and connecting rods.
32 Make sure that the identification marks are visible on the main bearing caps so that they can be refitted in their original positions at reassembly. The caps are numbered and an arrow indicates the front of the engine (photo). Note that as from February 1982 Nos 2 and 3 caps are retained with longer bolts incorporating a domed head (photo). Keep these identified for position.
33 Undo the securing bolts and lift off the main bearing caps and the

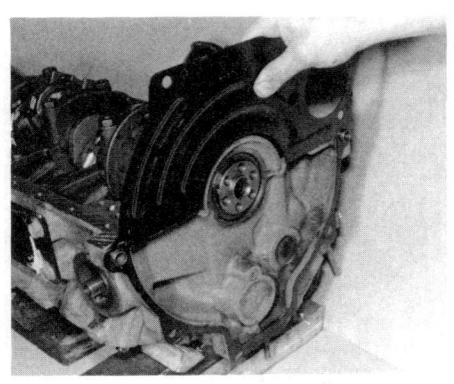

9.21 Removing the engine adaptor plate

9.23A Unbolt the timing cover ...

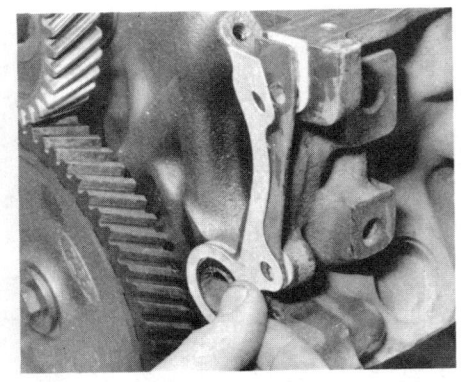

9.23B ... and remove the gasket

9.24A Remove the securing bolt ...

9.24B ... and slide off the camshaft gear

9.25A Unscrew the bolts ...

9.25B ... and remove the intermediate plate ...

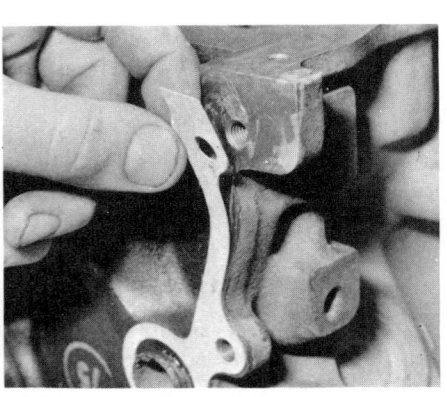

9.25C ... and gasket

Are your plugs trying to tell you something?

Normal.
Grey-brown deposits, lightly coated core nose. Plugs ideally suited to engine, and engine in good condition.

Heavy Deposits.
A build up of crusty deposits, light-grey sandy colour in appearance.
Fault: Often caused by worn valve guides, excessive use of upper cylinder lubricant, or idling for long periods.

Lead Glazing.
Plug insulator firing tip appears yellow or green/yellow and shiny in appearance.
Fault: Often caused by incorrect carburation, excessive idling followed by sharp acceleration. Also check ignition timing.

Carbon fouling.
Dry, black, sooty deposits.
Fault: over-rich fuel mixture.
Check: carburettor mixture settings, float level, choke operation, air filter.

Oil fouling.
Wet, oily deposits. Fault: worn bores/piston rings or valve guides; sometimes occurs (temporarily) during running-in period.

Overheating.
Electrodes have glazed appearance, core nose very white – few deposits. Fault: plug overheating. Check: plug value, ignition timing, fuel octane rating (too low) and fuel mixture (too weak).

Electrode damage.
Electrodes burned away; core nose has burned, glazed appearance. Fault: pre-ignition. Check: for correct heat range and as for 'overheating'.

Split core nose.
(May appear initially as a crack). Fault: detonation or wrong gap-setting technique. Check: ignition timing, cooling system, fuel mixture (too weak).

WHY DOUBLE COPPER IS BETTER FOR YOUR ENGINE.

Unique Trapezoidal Copper Cored Earth Electrode
50% Larger Spark Area
Copper Cored Centre Electrode

Champion Double Copper plugs are the first in the world to have copper core in both centre *and* earth electrode. This innovative design means that they run cooler by up to 100°C – giving greater efficiency and longer life. These double copper cores transfer heat away from the tip of the plug faster and more efficiently. Therefore, Double Copper runs at cooler temperatures than conventional plugs giving improved acceleration response and high speed performance with no fear of pre-ignition.

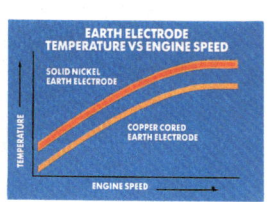

TRAPEZOIDAL COPPER CORED EARTH ELECTRODE
NEW TRAPEZOIDAL COPPER CORED EARTH ELECTRODE
CONVENTIONAL SOLID NICKEL ALLOY EARTH ELECTRODE
50% INCREASE IN SPARK AREA

EARTH ELECTRODE TEMPERATURE VS ENGINE SPEED
SOLID NICKEL EARTH ELECTRODE
COPPER CORED EARTH ELECTRODE
TEMPERATURE / ENGINE SPEED

Champion Double Copper plugs also feature a unique trapezoidal earth electrode giving a 50% increase in spark area. This, together with the double copper cores, offers greatly reduced electrode wear, so the spark stays stronger for longer.

- FASTER COLD STARTING
- FOR UNLEADED OR LEADED FUEL
- ELECTRODES UP TO 100°C COOLER
- BETTER ACCELERATION RESPONSE
- LOWER EMISSIONS
- 50% BIGGER SPARK AREA
- THE LONGER LIFE PLUG

Plug Tips/Hot and Cold.
Spark plugs must operate within well-defined temperature limits to avoid cold fouling at one extreme and overheating at the other.
Champion and the car manufacturers work out the best plugs for an engine to give optimum performance under all conditions, from freezing cold starts to sustained high speed motorway cruising.
Plugs are often referred to as hot or cold. With Champion, the higher the number on its body, the hotter the plug, and the lower the number the cooler the plug.

Plug Cleaning
Modern plug design and materials mean that Champion no longer recommends periodic plug cleaning. Certainly don't clean your plugs with a wire brush as this can cause metal conductive paths across the nose of the insulator so impairing its performance and resulting in loss of acceleration and reduced m.p.g.
However, if plugs are removed, always carefully clean the area where the plug seats in the cylinder head as grit and dirt can sometimes cause gas leakage.
Also wipe any traces of oil or grease from plug leads as this may lead to arcing.

CHAMPION

DOUBLE COPPER

1 This photographic sequence shows the steps taken to repair the dent and paintwork damage shown above. In general, the procedure for repairing a hole will be similar; where there are substantial differences, the procedure is clearly described and shown in a separate photograph.

2 First remove any trim around the dent, then hammer out the dent where access is possible. This will minimise filling. Here, after the large dent has been hammered out, the damaged area is being made slightly concave.

3 Next, remove all paint from the damaged area by rubbing with coarse abrasive paper or using a power drill fitted with a wire brush or abrasive pad. 'Feather' the edge of the boundary with good paintwork using a finer grade of abrasive paper.

4 Where there are holes or other damage, the sheet metal should be cut away before proceeding further. The damaged area and any signs of rust should be treated with Turtle Wax Hi-Tech Rust Eater, which will also inhibit further rust formation.

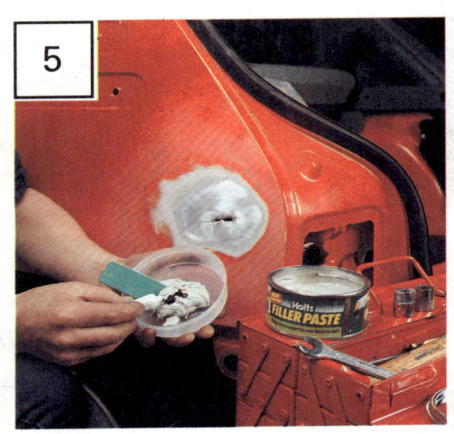

5 *For a large dent or hole* mix Holts Body Plus Resin and Hardener according to the manufacturer's instructions and apply around the edge of the repair. Press Glass Fibre Matting over the repair area and leave for 20-30 minutes to harden. Then ...

5A ... brush more Holts Body Plus Resin and Hardener onto the matting and leave to harden. Repeat the sequence with two or three layers of matting, checking that the final layer is lower than the surrounding area. Apply Holts Body Plus Filler Paste as shown in Step 5B.

5B *For a medium dent,* mix Holts Body Plus Filler Paste and Hardener according to the manufacturer's instructions and apply it with a flexible applicator. Apply thin layers of filler at 20-minute intervals, until the filler surface is slightly proud of the surrounding bodywork.

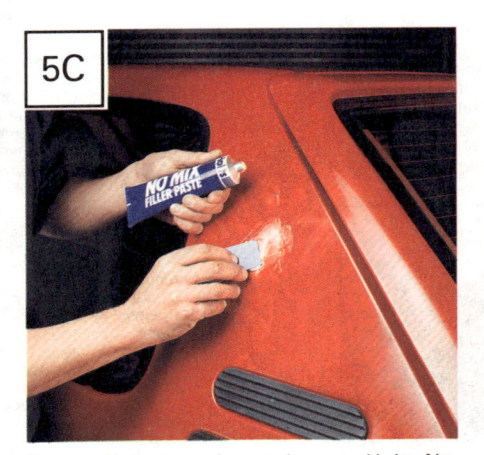

5C *For small dents and scratches* use Holts No Mix Filler Paste straight from the tube. Apply it according to the instructions in thin layers, using the spatula provided. It will harden in minutes if applied outdoors and may then be used as its own knifing putty.

6 Use a plane or file for initial shaping. Then, using progressively finer grades of wet-and-dry paper, wrapped round a sanding block, and copious amounts of clean water, rub down the filler until glass smooth. 'Feather' the edges of adjoining paintwork.

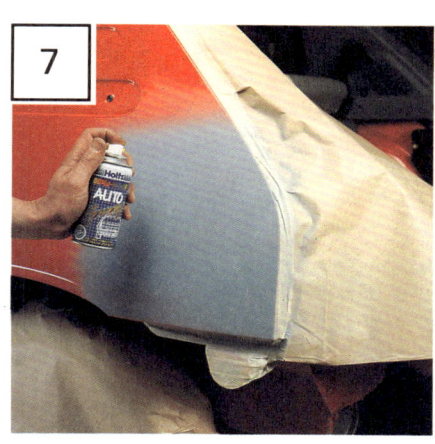

7 Protect adjoining areas before spraying the whole repair area and at least one inch of the surrounding sound paintwork with Holts Dupli-Color primer.

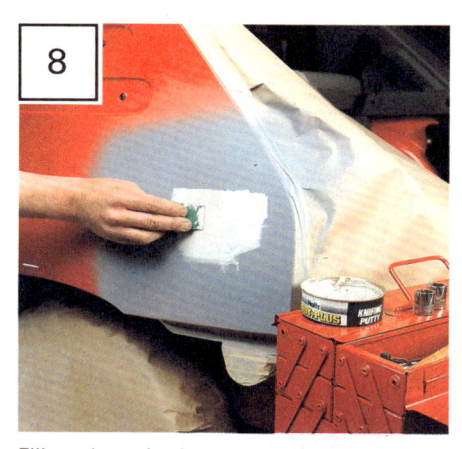

8 Fill any imperfections in the filler surface with a small amount of Holts Body Plus Knifing Putty. Using plenty of clean water, rub down the surface with a fine grade wet-and-dry paper – 400 grade is recommended – until it is really smooth.

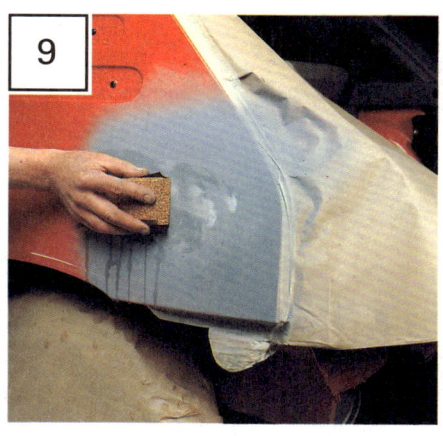

9 Carefully fill any remaining imperfections with knifing putty before applying the last coat of primer. Then rub down the surface with Holts Body Plus Rubbing Compound to ensure a really smooth surface.

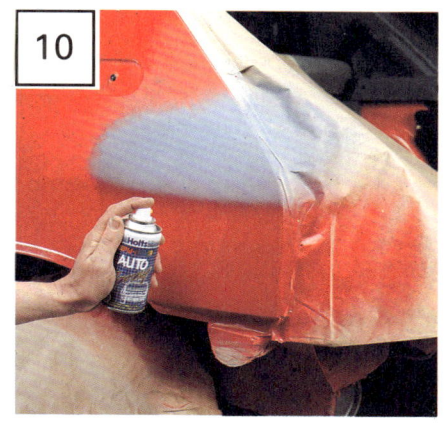

10 Protect surrounding areas from overspray before applying the topcoat in several thin layers. Agitate Holts Dupli-Color aerosol thoroughly. Start at the repair centre, spraying outwards with a side-to-side motion.

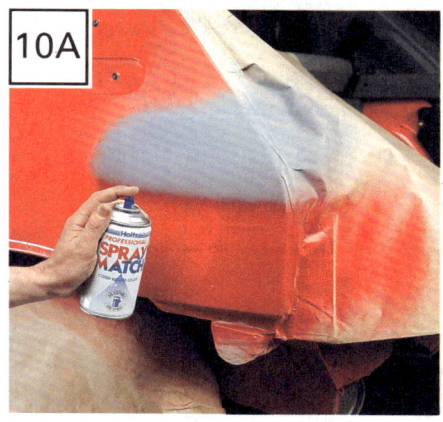

10A If the exact colour is not available off the shelf, local Holts Professional Spraymatch Centres will custom fill an aerosol to match perfectly.

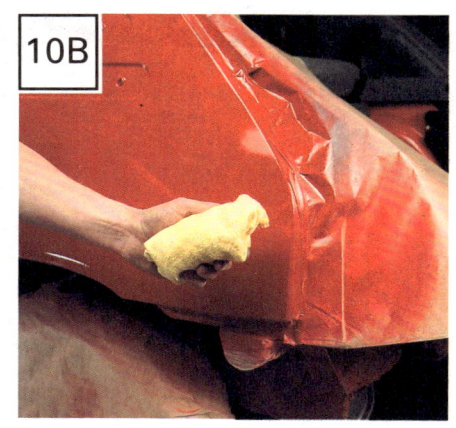

10B To identify whether a lacquer finish is required, rub a painted unrepaired part of the body with wax and a clean cloth.

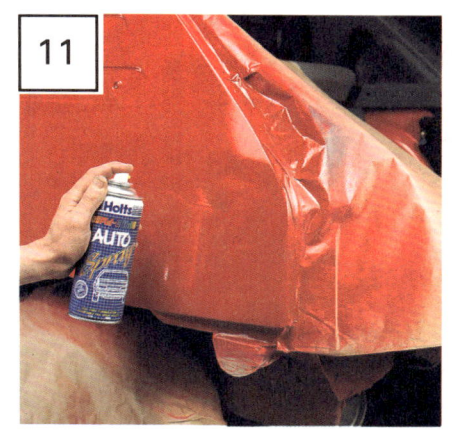

11 If *no* traces of paint appear on the cloth, spray Holts Dupli-Color clear lacquer over the repaired area to achieve the correct gloss level.

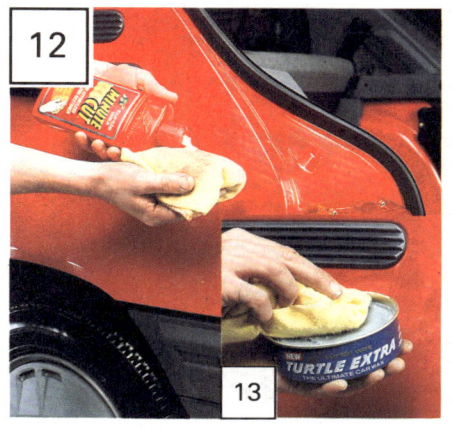

12 The paint will take about two weeks to harden fully. After this time it can be 'cut' with a mild cutting compound such as Turtle Wax Minute Cut prior to polishing with a final coating of Turtle Wax Extra.

14 When carrying out bodywork repairs, remember that the quality of the finished job is proportional to the time and effort expended.

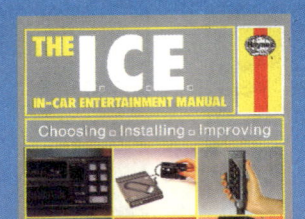

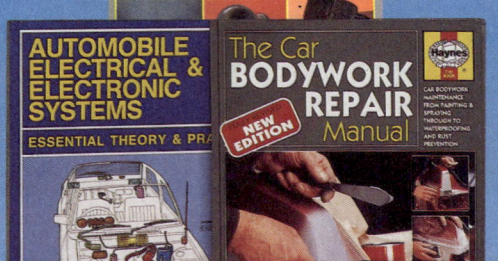

HAYNES No1 for DIY

Haynes publish a wide variety of books besides the world famous range of *Haynes Owners Workshop Manuals*. They cover all sorts of DIY jobs. Specialist books such as the *Improve and Modify* series and the *Purchase and DIY Restoration Guides* give you all the information you require to carry out everything from minor modifications to complete restoration on a number of popular cars. In addition there are the publications dealing with specific tasks, such as the *Car Bodywork Repair Manual* and the *In-Car Entertainment Manual*. The *Household DIY* series gives clear step-by-step instructions on how to repair everyday household objects ranging from toasters to washing machines.

Whether it is under the bonnet or around the home there is a Haynes Manual that can help you save money. Available from motor accessory stores and bookshops or direct from the publisher.

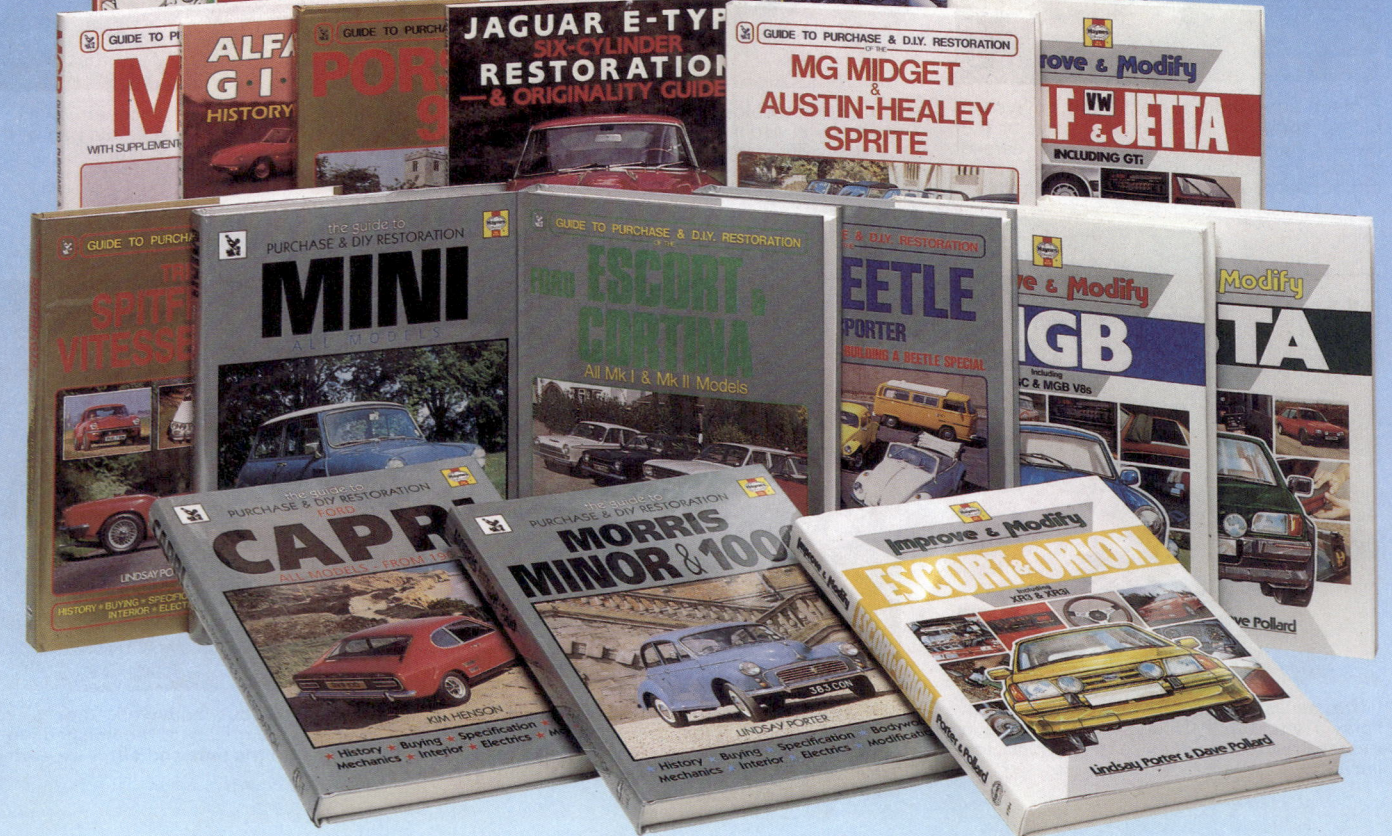

9.26 Pulling off the crankshaft gear

9.27A Unscrew the bolts ...

9.27B ... and remove the camshaft thrust plate

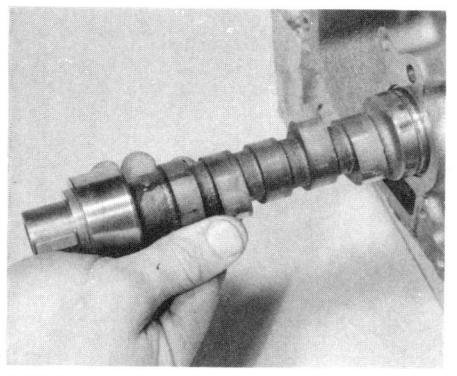

9.27C Withdrawing the camshaft

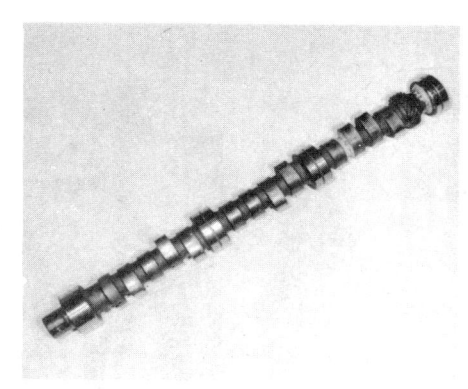

9.27D Camshaft removed from the engine

9.28A Check that the big-end bearing caps and connecting rods are marked for position (arrows)

9.28B The arrow on the piston crown faces the front of the engine

9.29A Removing the big-end nuts

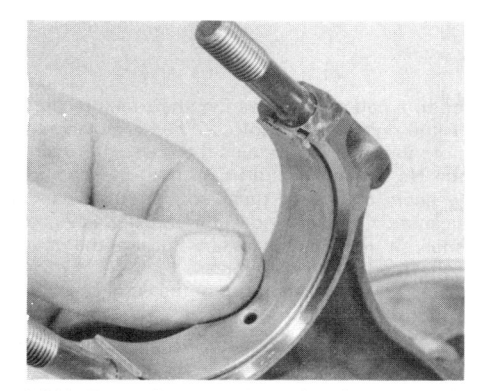

9.29B Removing a shell bearing from the connecting rod

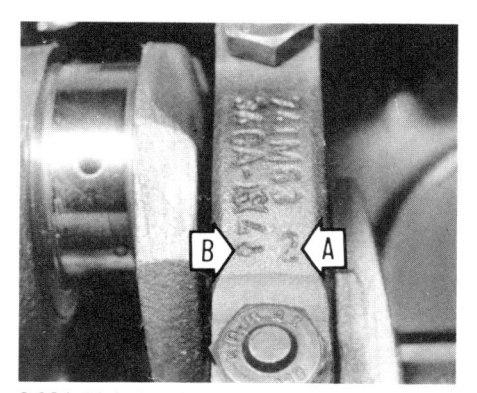

9.32A Main bearing cap number (A) and directional arrow (B)

9.32B The domed (top) and standard (bottom) main bearing bolts

34 Chapter 1 Engine

9.33A Removing a main bearing bolt

9.34 Removing the crankshaft

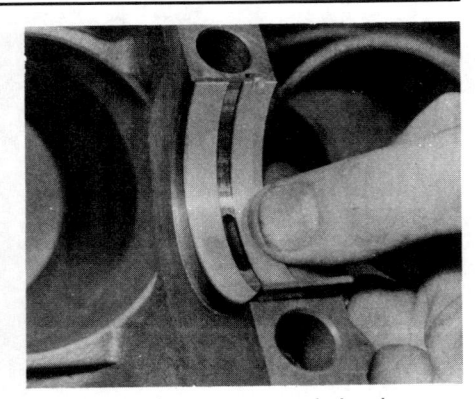

9.35A Removing an upper main bearing shell

9.35B Removing the upper half of No 3 main bearing shell

12.1 The rocker shaft has a retaining pin at each end (arrowed)

bottom half of each bearing shell, taking care to keep the bearing shell in the right caps (photo). When removing the rear main bearing cap note that this also retains the crankshaft rear oil seal. When removing the No 3 main bearing cap, note the position of the two half thrust washers and mark them so that they can be refitted in the same position. On later models the thrust washers are integral with the main shell. It may be necessary to tap the main bearing caps with a soft-faced hammer to release them.
34 Lift the crankshaft out of the crankcase and remove the rear oil seal (photo).
35 Remove the upper halves of the main bearing shells and the upper half of the No 3 bearing thrust washers (where applicable) from the crankcase and place them with the respective main bearing caps (photos).

10 Sump – removal and refitting with engine in car

1 Disconnect the battery negative lead.
2 Remove the distributor cap (Chapter 4) and the radiator shroud (Chapter 2).
3 Apply the handbrake then jack up the front of the car and support on axle stands.
4 Unscrew the sump drain plug and drain the oil into a suitable container. When completed refit the plug and tighten.
5 Remove the starter motor as described in Chapter 12.
6 Unscrew the nuts from both engine mountings.
7 Remove the adaptor plate bolt from the gearbox.
8 Unscrew the sump bolts and prise the sump from the crankcase.
9 Raise the engine as far as possible using a hoist or trolley jack, and support with stands beneath the gearbox and front of the engine.
10 Lower the sump from the engine.
11 Scrape off all the gaskets and clean the mating faces.
12 Refitting is a reversal of removal, but fit the new gaskets with reference to Section 33. Finally fill the engine with oil.

11 Engine components – examination for wear

When the engine has been stripped down and all the parts have been cleaned, they should be examined for wear. In cases where no definite wear limit is given, it must be a matter of judgment on whether or not a part is to be renewed or refitted, taking into consideration the further life expected from the engine, the degree of reliability required, the cost of the new part and the amount of dismantling which will be necessary to renew the part later.

12 Rocker assembly – dismantling, examination and reassembly

1 Tap out the roll pin from one end of the rocker shaft and remove the spring washer (photo).
2 Slide the rocker arms, rocker supports and springs off the rocker shaft. Keep them in the correct order so that they can be reassembled in the same position. If a rocker support sticks it can be removed by tapping it with a soft-faced hammer.

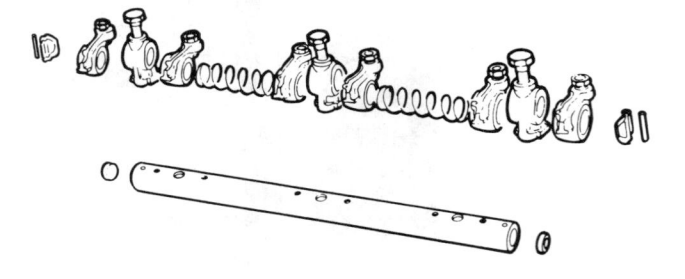

Fig. 1.2 Exploded view of rocker shaft assembly (Sec 12)

Chapter 1 Engine

3 Examine the rocker shaft and rocker arms for wear. If the rocker arm surface that contacts the valve stem is considerably worn, renew the rocker arm. If it is worn slightly step-shaped it may be cleaned up with a fine oil stone.

4 Commencing January 1984, modified rocker shaft springs have been fitted. Where tappet noise is evident on earlier models between engine speeds of 1500 and 2000 rpm, the later type of springs may be fitted.

5 Oil the parts and reassemble them on their shafts in the original order. With both rocker shafts fitted the oil holes must face downwards to the cylinder heads. This position is indicated by a notch on one end face of the rocker shaft.

13 Tappets and pushrods – examination and renovation

1 Examine the valve tappets for wear and damage, renew if suspect.
2 Check the pushrods for signs of bending or wear. Correct or renew as necessary.

14 Camshaft and camshaft bearings – examination and renovation

1 If there is excessive wear in the camshaft they will have to be renewed. As the fitting of new bearings requires special tools this should be left to your local Ford dealer.
2 The camshaft may show signs of wear on the bearing journals or cam lobes. The main decision to take is what degree of wear necessitates renewing the camshaft, which is expensive. Scoring or damage to the bearing journals cannot be removed by regrinding; renewal of the camshaft is the only solution.
3 The cam lobes may show signs of ridging or pitting on the high points. If ridging is slight then it may be possible to remove it with a fine oil stone or emery cloth. The cam lobes, however, are surface hardened and once the hard skin is penetrated wear will be very rapid.

15 Cylinder heads – dismantling, renovation and reassembly

1 Clean the dirt and oil off the cylinder heads. Remove the carbon deposits from the combustion chambers and valve heads with a scraper or rotary wire brush.
2 Remove the valves by compressing the valve springs with a suitable valve spring compressor and lifting out the collets. Release the valve spring compressor and remove the valve spring retainer, spring and valve (photos). Mark each valve so that they can be fitted in the same location.
3 With the valves removed, clean out the carbon from the ports.
4 Examine the heads of the valves and the valve seats for pitting and burning. If the pitting on valve and seat is slight it can be removed by grinding the valves and seats together with coarse, and then fine, valve grinding paste. If the pitting is deep the valves will have to be reground on a valve grinding machine and the seats will have to be recut with a valve seat cutter. Both these operations are a job for your local Ford dealer or motor engineering specialist.
5 Check the valves guides for wear by inserting the valve in the guide, the valve stem should move easily in the guide without side play. Renewal of worn guides requires special tools and should be left to your local Ford dealer.
6 When grinding slightly pitted valves and valve seats with carborundum paste proceed as follows: Apply a little coarse grinding paste to the valve seat and using a suction type valve grinding tool, grind the valve into its seat with a semi-rotary movement, lifting the valve from time to time. A light spring under the valve head will assist in this operation. When a dull matt even surface finish appears on both the valve and the valve seat, clean off the coarse paste and repeat the grinding operation with a fine grinding paste until a continuous ring of light grey matt finish appears on both valve and valve seat. Carefully clean off all traces of grinding paste. Blow through the gas passages with compressed air.
7 Check the valve springs for damage and also check the free length, refer to the Specifications at the beginning of this Chapter. Renew if defective.

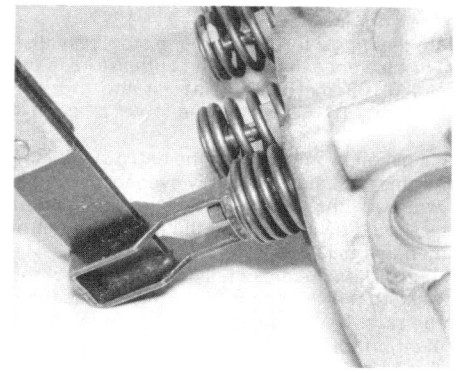

15.2A Compress the spring with a valve spring compressor ...

15.2B ... then remove the collets (magnetic probe shown) ...

15.2C ... followed by the retainer ...

15.2D ... and valve spring

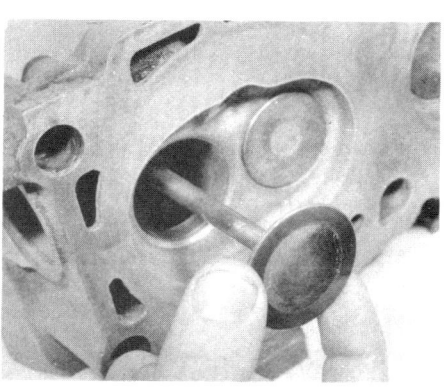

15.2E Removing an inlet valve

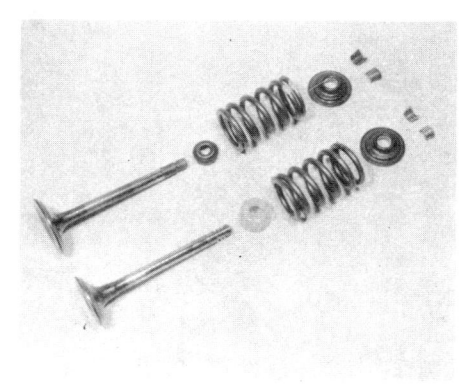

15.2F Inlet valve (top) and exhaust valve (bottom) components

15.8 Inlet valve seal (left) and exhaust valve seal (right)

8 Lubricate the valve stem with engine oil and insert it in the valve guide. Slide on a new oil seal. The light plastic seals go on the exhaust valves (photo).
9 Fit the valve spring and valve spring retainer.
10 Use a suitable valve spring compressor to compress the valve spring until the collets can be fitted in position in the slots in the valve stem. Release the valve spring compressor.
11 After fitting all the parts, tap the top of the valve springs lightly with a plastic hammer to ensure correct seating of the collets.

16 Cylinder bores – examination and renovation

1 A new cylinder is perfectly round and the walls parallel throughout its length. The action of the piston tends to wear the walls at right angles to the gudgeon pin due to side thrust. This wear takes place principally on that section of the cylinder swept by the piston rings.
2 It is possible to get an indication of bore wear by removing the cylinder heads with the engine still in the car. With the piston down in the bore first signs of wear can be seen and felt just below the top of the bore where the top piston ring reaches and there will be a noticeable lip (other than normal carbon build-up). If there is no lip it is fairly reasonable to expect that bore wear is not severe and any lack of compression or excessive oil consumption is due to worn or broken piston rings or pistons.
3 If it is possible to obtain a bore measuring micrometer, measure the bore in the thrust plane below the lip and again at the bottom of the cylinder in the same plane. If the difference is more than 0.076 mm (0.003 in) then a rebore is necessary. Similarly a difference of 0.076 mm (0.003 in) or more across the bore diameter is a sign of ovality calling for a rebore.

4 Any bore which is significantly scratched or scored will need reboring. This symptom usually indicates that the piston or rings are damaged. Even if only one cylinder is in need of reboring it will still be necessary for all cylinders to be bored and fitted with new oversize pistons and rings. Your Ford agent or local motor engineering specialist will be able to rebore and obtain the necessary matched pistons. If the crankshaft is undergoing regrinding also, it is a good idea to let the same firm renovate and reassemble the crankshaft and pistons to the block. A reputable firm normally gives a guarantee for such work. In cases where engines have been rebored already to their maximum, cylinder liners are available which may be fitted. In such cases the same reboring processes have to be followed and the services of a specialist engineering firm are required.

17 Pistons and piston rings – examination and renovation

1 Worn pistons and rings can usually be diagnosed when the symptoms of excessive oil consumption and low compression occur and are sometimes, though not always, associated with worn cylinder bores. Compression testers that fit onto the spark plug holes are available and these can indicate where low pressure is occurring. Wear usually accelerates the more it is left so when the symptoms occur early action can possibly save the expense of a rebore.
2 Another symptom of piston wear is piston slap – a knocking noise from the crankcase, not to be confused with big-end bearing failure. It can be heard clearly at low engine speed when there is no load (idling for example) and is much less audible when the engine speed increases. Piston wear usually occurs in the skirt or lower end of the piston and is indicated by vertical streaks in the worn area which is always on the thrust side. It can also be seen where the skirt thickness is different.
3 Piston ring wear can be checked by first removing the rings from the pistons as described later in this Section. Then place the rings in the cylinder bores from the top, pushing them down about 40 mm (1.5 in) with the head of the piston (from which the rings have been removed) so that they rest square in the cylinder. Then measure the gap at the ends of the ring with a feeler gauge (photo). If it exceeds the specified maximum then the ring needs renewal.
4 The grooves in which the rings locate in the piston can also become enlarged in use. The clearance between ring and piston, in the groove, should not exceed 0.102 mm (0.004 in) for the top two compression rings and 0.076 mm (0.003 in) for the lower oil control ring.
5 However, it is rare that a piston is only worn in the ring grooves and the need to renew them for this fault alone is hardly ever encountered. Wherever pistons are renewed the weight of the four piston/connecting rod assemblies should be kept within the limit variation of 8 g (0.28 oz) to maintain engine balance.
6 To remove the piston rings, slide them carefully over the top of the piston, taking care not to scratch the aluminium alloy; never slide them off the bottom of the piston skirt. It is very easy to break the cast iron piston rings if they are pulled off roughly, so this operation should be done with extreme care. It is helpful to make use of old feeler gauges.
7 Lift one end of the piston ring to be removed out of its groove and insert under it the end of the feeler gauge.
8 Turn the feeler gauge slowly round the piston and, as the ring

17.3 Checking the piston ring end gap with a feeler gauge

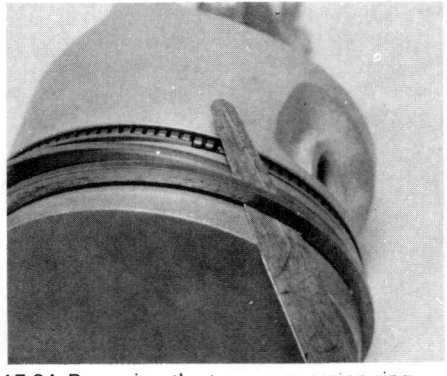

17.8A Removing the top compression ring

17.8B Removing the second compression ring

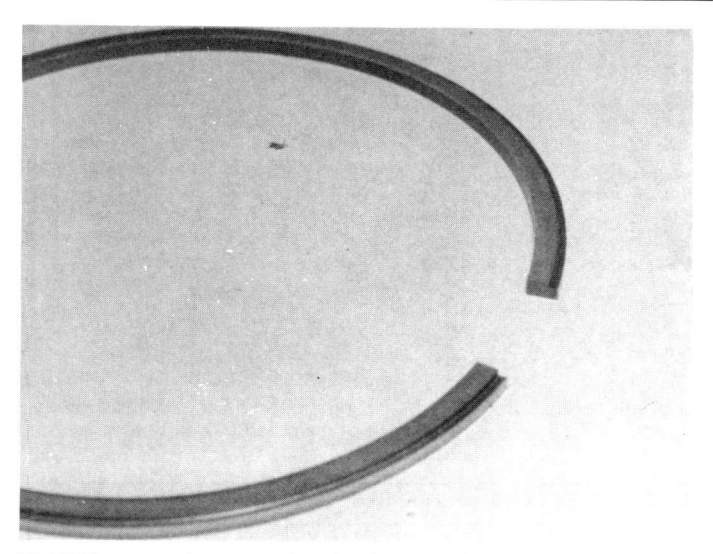

Chapter 1 Engine

17.8C The second compression ring is stepped

19.1 Checking the crankpin for ovality

comes out of its groove, apply slight upward pressure so that it rests on the land above. It can then be eased off the piston, with the feeler gauge stopping it from slipping into an empty groove if it is any but the top piston ring that it being removed (photos).
9 The piston rings must always be removed from the top of the piston.
10 Check that the piston ring grooves and oilways are thoroughly clean and unblocked. Piston rings must always be fitted over the head of the piston and never from the bottom.
11 The easiest method to use when fitting rings is to wrap a feeler gauge round the top of the piston and place the rings on one at a time, starting with the bottom oil control ring, over the feeler gauge. The feeler gauge, complete with ring, can then be slid down the piston over the other piston ring grooves until the correct groove is reached. The piston ring is then slid off the feeler gauge into the groove.
12 An alternative method is to fit the rings by holding them slightly open with the thumbs and both of the index fingers. This method requires a steady hand and great care as it is easy to open the ring too much and break it.

18 Connecting rods and gudgeon pins – examination and renovation

1 Gudgeon pins are a shrink fit into the connecting rods. Neither of these would normally need renewal unless the pistons were being changed, in which case the new pistons would automatically be supplied with new gudgeon pins.
2 Connecting rods are not subject to wear but in extreme circumstances such as engine seizure they could be distorted. Such conditions may be visually apparent but where doubt exists they should be checked for alignment and if necessary renewed or straightened. The bearing caps should also be examined for indications of filing down which may have been attempted in the mistaken idea that bearing slackness could be remedied in this way. If there are such signs then the connecting rods and caps should be renewed by your Ford agent or local engineering specialist.

19 Crankshaft – examination and renovation

1 Examine the main bearing journals and crankpins for score marks or scratches. Also check them for ovality using a micrometer (photo). Ovality in excess of 0.025 mm (0.001 in) should be regarded as excessive.
2 Contrary to normal practice, Ford stipulate that under no circumstances should the crankshaft be reground; so if damage or excessive wear is evident it will be necessary to obtain a new crankshaft.

19.4 The spigot bearing in the rear of the crankshaft

3 If the old crankshaft is being refitted, check that the oilways are clear. This can be done by probing with wire or by blowing through with an air line. Then insert the nozzle of an oil gun into the respective oilways, each time blanking off the previous hole, and squirt oil through the shaft. It should emerge through the next hole. Any oilway blockage must obviously be cleaned out prior to refitting the crankshaft.
4 Check the spigot bearing in the rear of the crankshaft and if necessary renew it (photo).

20 Main and big-end bearings – examination and renovation

1 With careful servicing and regular oil and filter changes, bearings will last for a very long time, but they can still fail for unforeseen reasons. With big-end bearings the indication is a regular rythmic loud knocking from the crankcase. The frequency depends on engine speed and is particularly noticeable when the engine is under load. This symptom is accompanied by a fall in oil pressure, although this is not normally noticeable unless an oil pressure gauge is fitted. Main bearing failure is usually indicated by serious vibration, particularly at higher engine revolutions, accompanied by a more significant drop in oil pressure and a 'rumbling' noise.

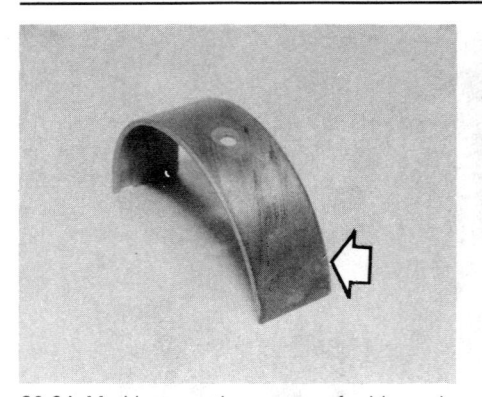

20.6A Markings on the reverse of a big-end bearing shell (arrowed)

20.6B 'Standard' markings on the reverse of a main bearing shell (arrowed)

20.7A Plastigage used to check the running clearance of main and big-end bearings (arrowed)

2 Bearing shells in good condition have bearing surfaces with a smooth, even matt silver/grey colour all over. Worn bearings will show patches of a different colour when the bearing metal has worn away and exposed the underlay. Damaged bearings will be pitted or scored. If the crankshaft is in good condition it is merely a question of obtaining another set of bearings the same size. A new crankshaft will need new bearing shells as a matter of course.

3 The original bore in the cylinder block may have been standard or the first grade oversize and in the latter instance this will be indicated by the bearing caps which will be marked with white paint.

4 The original size of the crankshaft main bearing journals may have been standard or 0.25 mm (0.010 in) undersize. If the journals were undersize originally, this will be indicated by a green stripe on the first balance weight.

5 The original size of the big-end journal was either standard or 0.25 mm (0.010 in) undersize. If undersize, the corresponding journal web will be marked with a green spot.

6 If the bearing shells are to be renewed, take the old ones along to your supplier and this will act as a check that you are getting the correct size bearings. Undersize bearings are marked as such on the reverse face (photos).

7 The running clearance of the bearings may be checked using Plastigage. The plastic thread is positioned on the journal then the bearing shell and cap assembled, and the bolts/nuts tightened to the specified torque. When the cap is removed, the width of the thread is measured with a card gauge which indicates the running clearance (photos).

21 Timing gears – examination and renovation

1 Inspect the gear teeth for damage, or signs of excessive wear which will cause noisy operation.

2 The backlash between the camshaft gear and the crankshaft gear must not exceed 0.27 mm (0.011 in) and the backlash should be checked at four different points around the periphery of the gear (photo).

22 Flywheel – examination and renovation

1 Inspect the flywheel for damage and check the ring gear for wear and damage.

2 If the ring gear is badly worn or has missing teeth it should be renewed. The old ring can be removed from the flywheel by cutting a notch between two teeth with a hacksaw and then splitting it with a cold chisel.

3 To fit a new ring gear requires heating the ring to 204°C (400°F). This can be done by polishing four equal spaced sections of the gear laying it on a suitable heat resistant surface (such as fire bricks) and heating it evenly with a blow lamp or torch until the polished areas turn a light yellow tint. Do not overheat or the hard wearing properties will be lost. The gear has a chamfered inner edge which should go against the shoulder when put on the flywheel. When hot enough place the gear in position quickly, tapping it home is necessary, and let it cool naturally without quenching in any way.

23 Oil pump – dismantling, examination and reassembly

1 If oil pump wear is suspected it is possible to obtain a repair kit. Check for wear first as described later in this Section and if confirmed, obtain an overhaul kit or a new pump. The two rotors are a matched pair and form a single replacement unit. Where the rotor assembly is to be re-used and the outer rotor, prior to dismantling, must be marked on its front face in order to ensure correct reassembly.

2 Remove the intake pipe and oil strainer if not already removed.

3 Note the relative position of the oil pump cover and body and then undo and remove the bolts and spring washers (photo). Lift away the cover.

4 Carefully remove the rotors from the housing.

5 Using a centre punch, tap a hole in the centre of the pressure relief valve sealing plug, (make a note to obtain a new one).

6 Screw in a self-tapping screw and, using an open-ended spanner, withdraw the sealing plug.

20.7B Card gauge for checking width of Plastigage

21.2 Checking the timing gear backlash

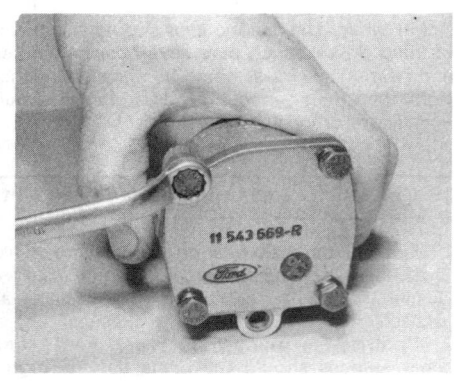

23.3 Unbolting the oil pump cover

Chapter 1 Engine 39

23.8 Checking the rotor lobe clearance

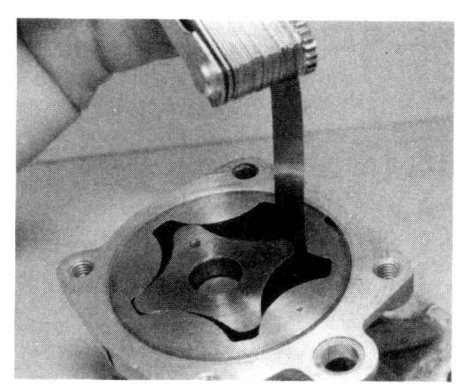

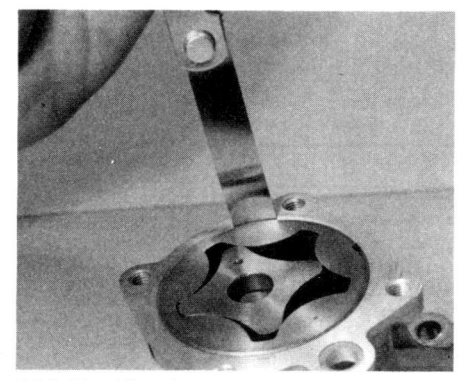

23.9 Checking the outer rotor clearance

23.10 Checking the endfloat

7 Thoroughly clean all parts in petrol or paraffin and wipe dry using a non-fluffy rag. The necessary clearances may now be checked using a machined straight-edge (a good steel rule) and a set of feeler gauges. The critical clearances are between the lobes of the centre rotor and convex faces of the outer rotor; between the rotor and the pump body; and between both rotors and the end cover plate.

8 The rotor lobe clearance may be checked using feeler gauges and should be within the limits 0.05 to 0.20 mm (0.002 to 0.008 in) (photo).

9 The clearance between the outer rotor and pump body should be within the limits 0.15 to 0.30 mm (0.006 to 0.012 in) (photo).

10 The endfloat clearance may be measured by placing a steel straight-edge across the end of the pump and measuring the gap between the rotors and the straight-edge (photo). The gap in either rotor should be within the limits 0.03 to 0.10 mm (0.0012 to 0.004 in).

11 If the only excessive clearances are endfloat it is possible to reduce them by removing the rotors and lapping the face of the body on a flat bed until the necessary clearances are obtained. It must be emphasised, however, that the face of the body must remain perfectly flat and square to the axis of the rotor spindle, otherwise the clearances will not be equal and the end cover will not be a pressure tight fit to the body. It is worth trying, of course, if the pump is in need of renewal anyway but unless done properly, it could seriously jeopardise the rest of the overhaul. Any variations in the other two clearances should be overcome with a new unit.

12 With all parts scrupulously clean, first refit the relief valve and spring and lightly lubricate with engine oil.

13 Using a suitable diameter drift, drive in a new sealing plug, flat side outwards until it is flush with the intake pipe mating face.

14 Lubricate both rotors with engine oil and fit them in the body. Fit the oil pump cover and secure with the bolts tightened in a diagonal and progressive manner to the torque wrench setting given in the Specifications.

15 Fit the driveshaft into the rotor driveshaft and ensure that the rotor turns freely.

16 Fit the intake pipe and oil strainer to the pump body together with a new gasket.

24 Lubrication system – description

The pressed steel oil sump is attached to the underside of the crankcase and acts as a reservoir for the engine oil. The oil pump draws oil from the sump via an oil strainer and intake pipe, then passes it into the full-flow oil filter. The filtered oil flows from the centre of the oil filter element and passes through a short drilling, on the right-hand side, to the oil pressure switch and to the main gallery (on the left-hand side of the crankcase) through a transverse drilling.

Four drillings connect the main gallery to the four main bearings and the camshaft bearings in their turn are connected to all the main bearings. The big-end bearings are supplied with oil through diagonal drillings from the nearest main bearing.

When the crankshaft is rotating, oil is thrown from the hole in each big-end bearing and ensures splash lubrication of the gudgeon pins and thrust side of the cylinders. The timing gears are also splash lubricated through an oil drilling.

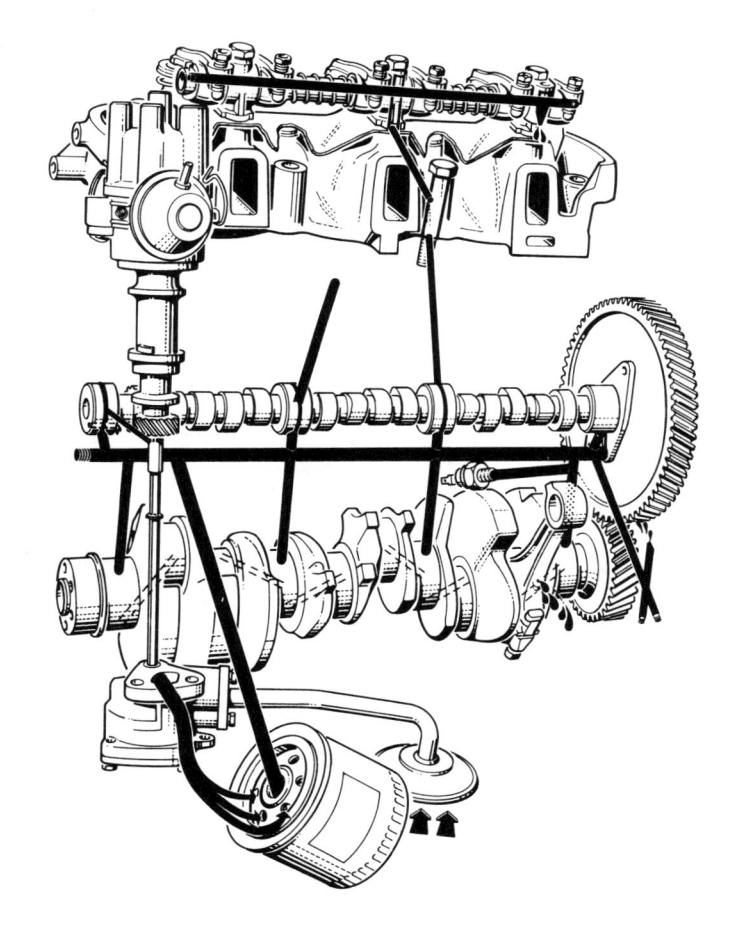

Fig. 1.3 Engine lubrication system for the 2.8 litre engine (Sec 24)

The crankshaft third bearing journal intermittently feeds oil under pressure, to the rocker shafts through a drilling in the cylinder block and cylinder head. The oil then passes back to the engine sump, via large drillings in the cylinder block and cylinder head.

25 Crankcase ventilation system – description

The closed crankcase ventilation system is used to control the emission of crankcase vapour. It is controlled by the amount of air drawn in by the engine when it is running and the throughput of the vent valve.

The system is known as the PCV system (positive crankcase ventilation) and the advantage of the system is that should the 'blow by' exceed the capacity of the PCV valve, excess fumes are fed into the engine through the air cleaner.

26 Engine reassembly – general

1 To ensure maximum life with minimum trouble from an overhauled engine, not only must every part be correctly assembled but everything must be spotlessly clean, all oilways must be clean, locking washers and spring washers must always be fitted where needed and all bearings and other sliding surfaces must be thoroughly lubricated during assembly.
2 Before assembly, renew any bolts, studs and nuts whose threads are in any way damaged and whenever possible use new spring washers. Obtain a complete set of new gaskets and all new parts as necessary.
3 When refitting parts ensure that they are refitted in their original positions and directions. Oil seal lips should be smeared with grease before fitting. A liquid gasket sealant should be used where specified to prevent leakage.
4 Apart from your normal tools, a supply of clean rags, an oil can filled with clean engine oil and a torque wrench are essential.

27 Crankshaft – refitting

1 Wipe the bearing shell locations in the crankcase with a clean rag and fit the main bearing upper half shells in position.
2 Clean the main bearing shell locations and fit the half shells in the caps (photo). If the old bearings are being refitted (although this is a false economy unless they are practically new) make sure they are fitted in their original positions.

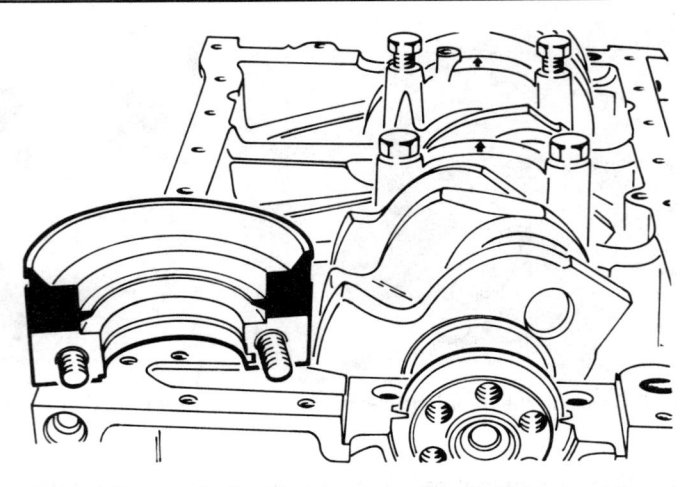

Fig. 1.4 Rear main bearing cap area (shaded) for applying sealing compound (Sec 27)

3 Apply a little grease to each side of the No. 3 main bearing so as to retain the thrust washers (early models only).
4 Fit the upper halves of the thrust washers into their grooves each side of the main bearing. The slots must face outwards with the tag located in the groove (early models only).
5 Lubricate the crankshaft main bearing journals and the main bearing shells with engine oil (photo).
6 Place the rear main bearing oil seal in position on the end of the crankshaft. Alternatively fit it later.
7 Carefully lower the crankshaft into the crankcase.
8 Apply a thin coating of sealing compound to the mating faces of the crankcase and the rear main bearing cap (photo).

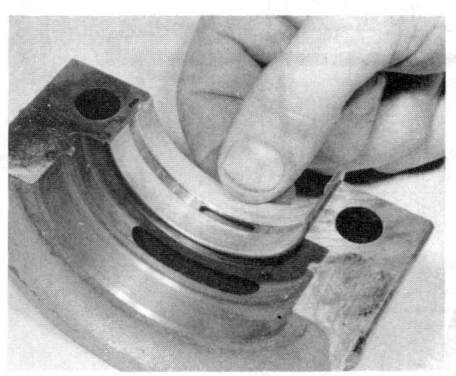

27.2 Fitting a bearing shell to the rear main bearing cap

27.5 Oiling No 3 main bearing upper shell

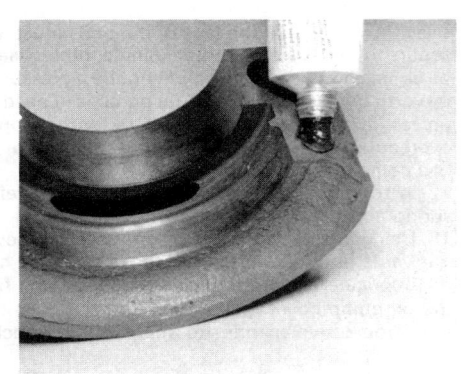

27.8 Applying sealing compound to the rear main bearing cap

27.10 Fitting the rear main bearing cap

27.11 Tightening the main bearing cap bolts

27.12 Temporarily stick adhesive tape on the rear of the crankshaft when fitting the rear oil seal to protect the seal lips

Chapter 1 Engine

27.13A Crankshaft endfloat can be checked with feeler gauges ...

27.13B ... or a dial gauge

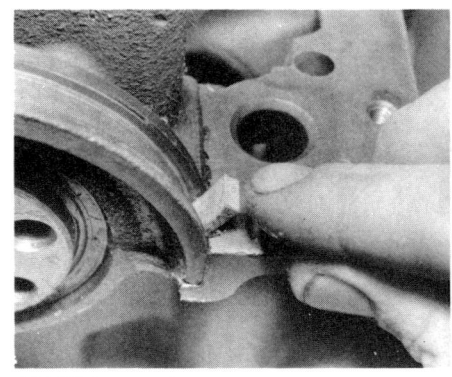

27.15A Locate the dowel seals on the main bearing cap ...

27.15B ... and press fully in with a screwdriver

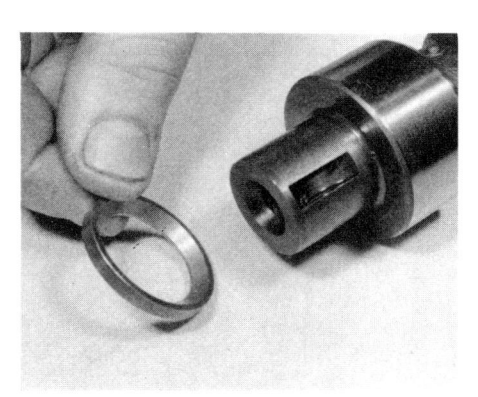

28.1A Fit the chamfered spacer ...

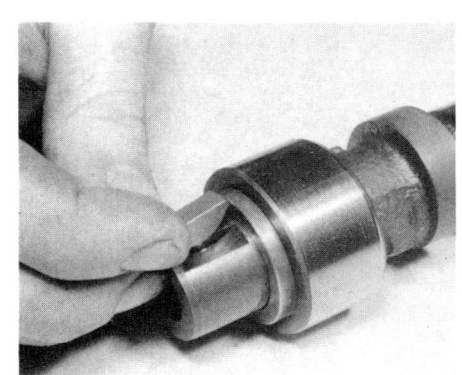

28.1B ... and key on the camshaft

9 Apply a smear of grease to both sides of the No 3 main bearing cap so as to retain the thrust washers. Fit the thrust washers with the tag located in the groove and the slots facing outwards (early models only).
10 Fit the main bearing caps with the arrows on the caps pointing to the front of the engine (photo).
11 Progressively tighten the main bearing securing bolts to the specified torque, except No 3 bearing bolts which should only be finger-tight. Now press the crankshaft fully to the rear then slowly press it fully forward, hold it in this position and tighten the No 3 main bearing cap securing bolts to the specified torque (photo). This ensures that the thrust washers are correctly located.
12 Press the rear main bearing oil seal firmly against the rear main bearing. If fully fitting it, protect the seal lips with adhesive tape (photo).
13 Using feeler gauges, check the crankshaft endfloat by inserting the feeler gauge between the crankshaft journal side and the thrust washers. The clearance must not exceed the figure given in the Specifications. The check can also be made using a dial gauge (photos).
14 Rotate the crankshaft to ensure that it is not binding or sticking. Should it be very stiff to turn or have high spots it must be removed and thoroughly checked.
15 Coat the rear main bearing cap dowel seals with sealing compound and fit in position. Use a blunt screwdriver or similar tool to press them fully in (photos). They should be fitted with the rounded face pointing towards the bearing cap.

28 Camshaft and front intermediate plate – refitting

1 Slide the spacer onto the camshaft with the chamfered side first and fit the key (photos).
2 Lubricate the camshaft bearings, the camshaft and the thrust plate.
3 Carefully insert the camshaft from the front and fit the thrust plate and self-locking securing bolts. Tighten the bolts to the specified torque.
4 Fit the timing cover guide sleeves and O-ring seals onto the crankcase. The chamfered end of the guide sleeves must face outward towards the timing cover.
5 Ensure the mating faces of the crankcase and the front intermediate plate are clean and then apply sealing compound to both faces. Position the gasket on the crankcase and then fit the intermediate plate (photo).

28.5 Fitting the intermediate plate

42   Chapter 1 Engine

28.6 Temporary bolt for locating the intermediate plate

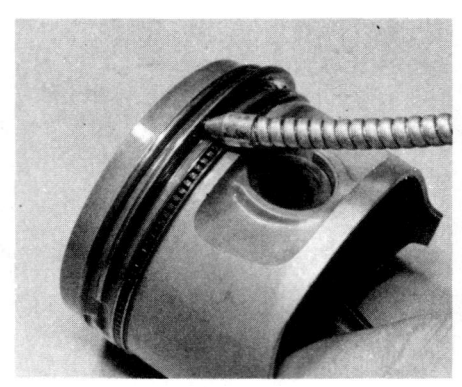

29.5 Oiling the piston rings

29.7 Oiling the cylinder bore

29.8 Using a piston ring compressor when inserting the piston into the bore

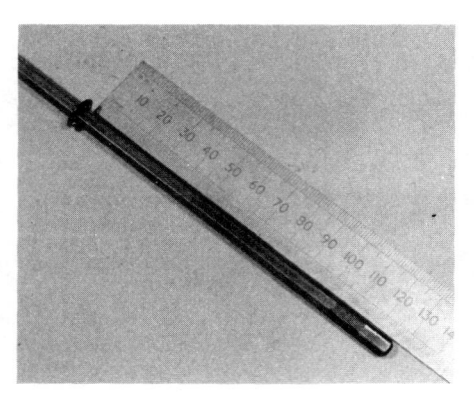

30.2 Checking the stop washer position on the oil pump driveshaft

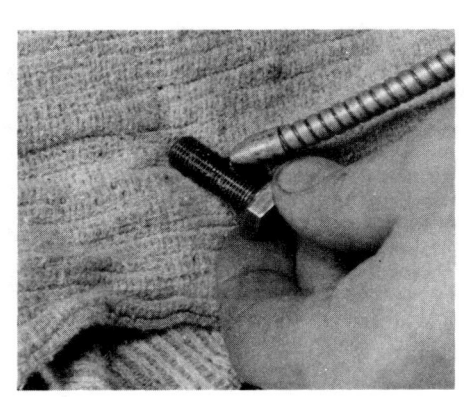

31.3 Oil the flywheel bolts before fitting

6   Fit the two centre bolts finger-tight then fit another two bolts temporarily for locating purposes. Tighten the centre securing bolts then remove the temporary fitted locating bolts (photo).

29 Pistons and connecting rods – refitting

1   Wipe clean the connecting rod half of the big-end bearing and the underside of the shell bearing and fit the shell bearing in position with its locating tongue engaged with the corresponding cut-out in the rod.
2   If the old bearings are nearly new and are being refitted then ensure they are refitted in their correct locations on the correct rods.
3   The pistons, complete with connecting rods, are fitted to their bores from the top of the block.
4   Locate the piston ring gaps in the following manner:

   Top: 150° from one side of the oil control ring helical expander gap
   Centre: 150° from the opposite side of the oil control ring helical expander gap
   Bottom: oil control ring Helical expander: opposite the marked piston front side
   Oil control ring, intermediate rings: 25 mm (1.0 in) each side of the helical expander gap

5   Well lubricate the piston and rings with engine oil (photo).
6   Fit a universal ring compressor and prepare to insert the first piston into the bore. Make sure it is the correct piston-connecting rod assembly for that particular bore, that the connecting rod is the correct way round and that the front of the piston (marked with an arrow or a notch) is to the front of the engine.
7   Lubricate the cylinder bore (photo) and insert the connecting rod and piston assembly into the cylinder bore up to the bottom of the piston ring compressor.
8   Gently but firmly tap the piston through the piston ring compressor and into the cylinder bore, using the shaft of a hammer (photo).
9   Generously lubricate the crankpin journals with engine oil and turn the crankshaft so that the crankpin is in the most advantageous position for the connecting rods to be drawn onto it.
10 Wipe clean the connecting rod bearing cap and back of the shell bearing, and fit the shell bearing in position ensuring that the locating tongue at the back of the bearing engages with the locating groove in the connecting rod cap.
11 Generously lubricate the shell bearing and offer up the connecting rod bearing cap to the connecting rod.
12 Refitting the connecting rod nuts.
13 Tighten the nuts with a torque wrench to the specified setting.
14 When all the pistons and connecting rods have been fitted, rotate the crankshaft to check that everything is free, and that there are no high spots causing binding.

30 Oil pump – refitting

1   Ensure the mating faces of the oil pump and the crankcase are clean.
2   Insert the hexagonal oil pump driveshaft with the pointed end towards the distributor location. The stop washer (photo) must be positioned 127.5 mm (5.020 in) from the rounded end of the driveshaft.
3   Fit the oil pump to the crankcase and secure with two bolts tightened to the specified torque. **Note:** *When a new or overhauled oil pump is being fitted, it should be filled with engine oil and turned by hand for one complete revolution before being fitted, this will prime the pump.*

31 Flywheel and clutch – refitting

1   Fit the engine adaptor plate on the two locating dowels.
2   Ensure the mating faces of the flywheel and crankshaft are clean and fit the flywheel to the crankshaft, aligning the marks made at dismantling, unless new parts are being fitted.

Chapter 1 Engine

3 Fit the six flywheel securing bolts (oiled) and lightly tighten (photo).
4 Chock the flywheel to restrain it from turning, and tighten the securing bolts in a diagonal and progressive manner to the specified torque.
5 Refit the clutch disc and pressure plate assembly to the flywheel making sure the disc is the correct way round (see Chapter 5).
6 Secure the pressure plate assembly with the retaining bolts, centralise the clutch disc using an old gearbox input shaft or suitable mandrel and then tighten the retaining bolts to the specified torque in a diagonal and progressive manner.

32 Timing gears and timing cover – refitting

1 Check that the keyways in the end of the crankshaft are clean and the keys are free from burrs. Fit the keys into the keyways.
2 If the crankshaft gear was removed refit it to the crankshaft using a suitable diameter tube to drive it fully home. Alternatively temporarily use the pulley and bolt to press it on (photos).
3 Fit the camshaft gear on the camshaft with the punch mark in alignment with the mark on the crankshaft gear (photo). Note that if there are two punch marks on the crankshaft gear, ensure that the correct mark is aligned (Fig. 1.5).
4 Fit the camshaft gear retaining washer and bolt. Tighten the bolt to the specified torque.
5 If available use a dial gauge to check the camshaft endfloat (photo).
6 Determine the position of the oil seal in the timing cover, then drive out the old seal (photos).
7 Clean the front face of the intermediate plate and the face of the timing cover, then coat both mating faces with sealing compound.
8 Position a new gasket on the intermediate plate and fit the timing cover to the cylinder block. Insert the bolts finger tight.
9 Centre the timing cover around the nose of the crankshaft using vernier calipers at four diametric points, then fully tighten the bolts (photo).
10 Fit the front oil seal to the timing cover and draw it into the position determined in paragraph 6 using a socket and bolt (photos).
11 Trim any protruding gasket from the sump mating face (photo).

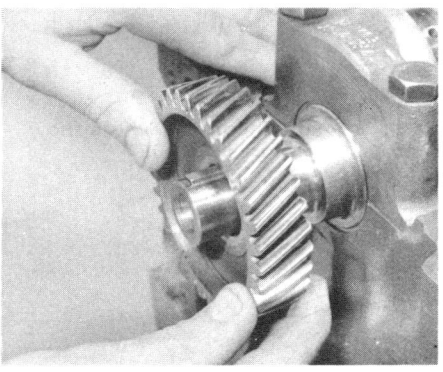

32.2A Locate the crankshaft gear on the crankshaft ...

32.2B ... and use the pulley to press it on

32.3 Alignment dots on the timing gears

32.5 Checking the camshaft endfloat with a dial gauge

32.6A Checking the position of the old oil seal in the timing cover

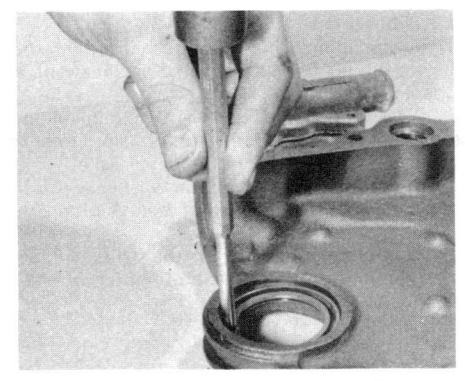

32.6B Driving out the timing cover oil seal

32.9 Using vernier calipers to check that the timing cover is centred on the crankshaft nose

32.10A Fitting the crankshaft front oil seal using a socket and bolt

32.10B Fitted position of the crankshaft front oil seal in the timing cover

32.11 Trimming the gasket from the sump mating face

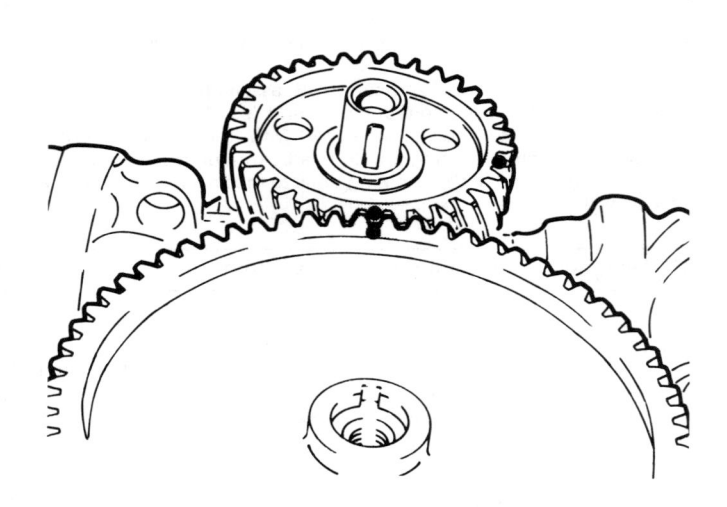

Fig. 1.5 Correct timing gear mark alignment where the crankshaft gear has two marks (Sec 32)

33 Sump, water pump and crankshaft pulley – refitting

1 Clean the mating faces of the crankcase and sump. Ensure that the grooves in the seal carriers are clean.
2 Fit the rubber seals in the grooves of the seal carriers. Apply sealing compound beneath the ends of the rear seal (photos).
3 Locate the side gaskets with their tabs under the cut-outs in the rubber seals (photo). Retain the gaskets with a little sealing compound if necessary.
4 Locate the sump on the gaskets, making sure that the bolt holes are still aligned.
5 Fit the sump securing bolts and tighten them to the correct torque in two stages as specified in Specifications at the beginning of this Chapter.
6 Place a new gasket on the timing cover and fit the water pump. Fit the thermostat and thermostat housing, refer to Chapter 2. Fit the rear water elbow together with a new gasket.
7 Coat the crankshaft pulley washer with sealing compound. Fit the pulley, washer and securing bolt. Tighten the bolt to the specified torque (photo).
8 Locate the oil cooler on the cylinder block, together with a new seal, then insert the sleeve and tighten it with the cooler pipes angled as shown in Fig. 1.6.

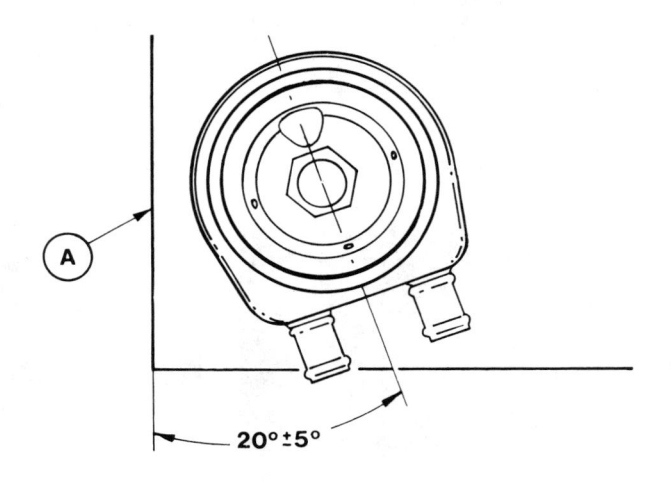

Fig. 1.6 Oil cooler fitting angle (Sec 33)

A Rear face of cylinder block

32.2A Fitting the rubber seal to the timing cover

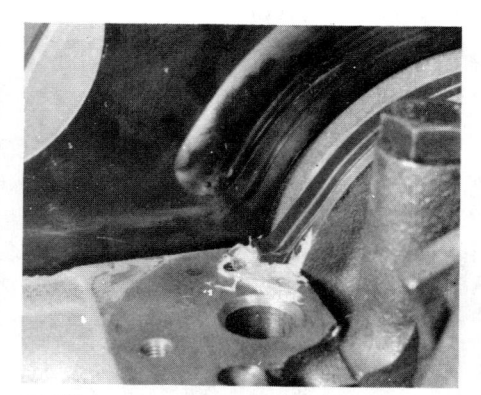

33.2B Apply sealing compound beneath the rear rubber seal

33.3 Locating the side gasket tabs beneath the seals

Chapter 1 Engine

33.7 Applying sealing compound to the crankshaft pulley washer

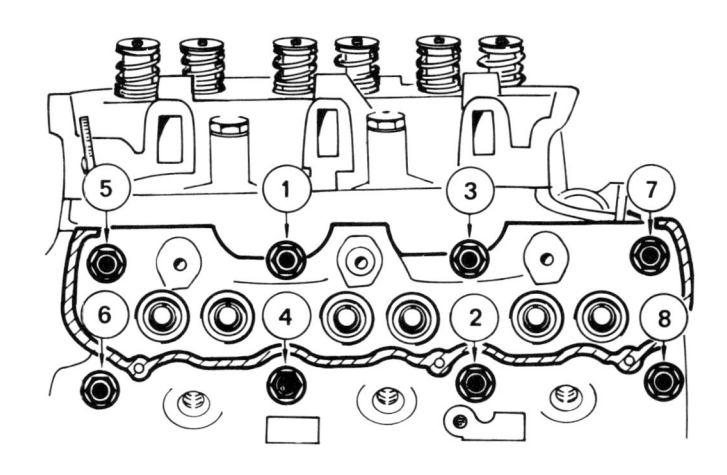

Fig. 1.7 Tightening sequence for cylinder head bolts on the 2.8 litre engine (Sec 34)

34 Cylinder heads, rocker gear and inlet manifold – refitting

1 Lubricate the valve tappets with clean engine oil and insert them in the cylinder block. Ensure they are fitted in their original locations.
2 Ensure that the mating faces of the cylinder block and the cylinder heads are clean.
3 Position the new cylinder head gaskets over the guide bushes on the cylinder block. Check that they are correctly located. The right and left-hand gaskets are different. The gaskets are marked FRONT TOP (photos).
4 Carefully lower the cylinder heads onto the cylinder block and fit the holding down bolts which should be first lightly oiled.
5 Tighten the bolts, in the sequence shown in Fig. 1.7, in the stages as specified in Specifications (photo).
6 Lubricate the pushrods with engine oil and insert them in the cylinder block.
7 Place the oil splash shields in position on the cylinder heads and fit the rocker shaft assemblies. Guide the rocker arm adjusting screws into the pushrod sockets (photo).
8 Tighten the rocker shaft securing bolts progressively to the specified torque (photo).
9 Refit the inlet manifold as described in Chapter 3.
10 Adjust the valve clearances as described in Section 35.

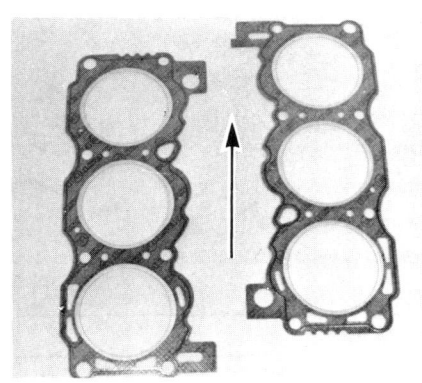

34.3A The difference between the left and right head gaskets. The arrow indicates front

34.3B Cylinder head gasket position markings

34.3C Fitting the head gaskets on the cylinder block

34.5 Tightening the cylinder head bolts

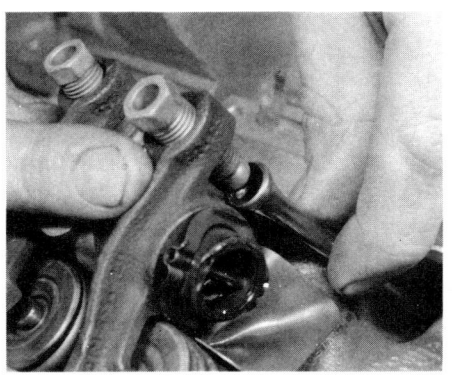

34.7 Guiding the rocker arm adjusting screws onto the pushrods

34.8 Tightening the rocker shaft bolts

35 Valve clearances – checking and adjustment

1 Adjust the inlet and exhaust valve clearances when the engine is cold, between 20 and 40°C (68 and 104°F). Clearances are important. If the clearance is too great the valves will not open as fully as they should. They will also open late and close early, this will affect the engine performance. If the clearances are too small the valves may not close completely which could result in lack of compression and very soon, burnt out valves and valve seats.

2 When turning the engine during valve clearance adjustments always turn the engine in the direction of normal rotation by means of a spanner on the pulley nut.

3 Turn the engine and align the crankshaft pulley mark with the O-mark on the timing cover.

4 If the crankshaft pulley is rotated backwards and forwards slightly, the valves of No 1 or 5 cylinder will be seen to be rocking (the two rocker arms moving in opposite directions). If the valves of No 1 cylinder are rocking, rotate the crankshaft through 360° so that those on No 5 cylinder are rocking.

5 When the valves of No 5 cylinder are in this position, check the valve clearances of No 1 cylinder by inserting a feeler gauge of the specified thickness between the rocker arm and the valve stem. Adjust the clearance, if necessary, by turning the rocker arm adjusting screw until the specified clearance is obtained (photo). Inlet and exhaust valve clearances are different.

6 If the engine is now rotated $1/3$ of a turn the valves of No 3 cylinder will be rocking and the valves of No 4 cylinder can be checked and adjusted.

7 Proceed to adjust the clearances according to the firing order as follows. Fig. 1.8 shows the cylinder numbers and valves listed in their correct order, working from the front of the engine.

Valves rocking	Valves to adjust
No 5 cylinder	No 1 cylinder
No 3 cylinder	No 4 cylinder
No 6 cylinder	No 2 cylinder
No 1 cylinder	No 5 cylinder
No 4 cylinder	No 3 cylinder
No 2 cylinder	No 6 cylinder

8 Fit the rocker cover gaskets and rocker covers (photo). Tighten the securing bolts to the specified torque.

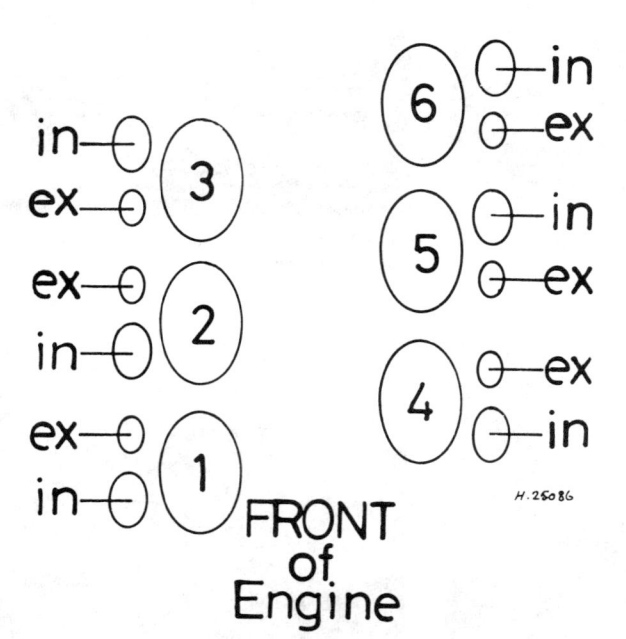

35.5 Adjusting the valve clearances

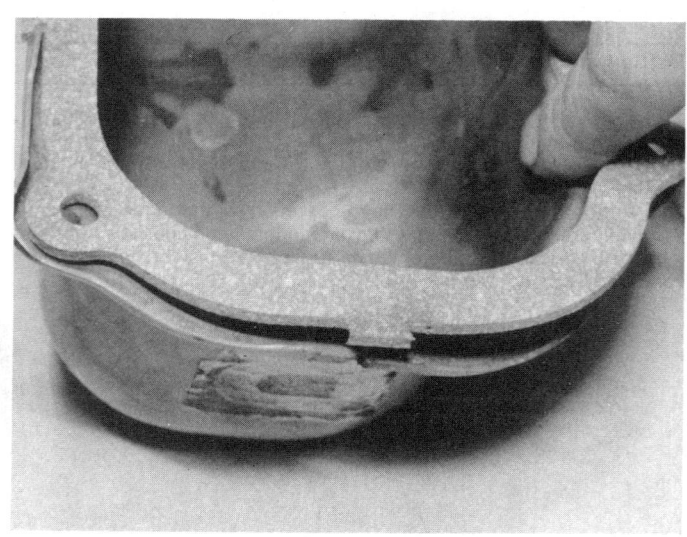

35.8 Fitting the gasket to a rocker cover

Fig. 1.8 Location of inlet and exhaust valves (Sec 35)
Note that this illustration depicts 2.8 litre engines only

36 Engine ancillaries – refitting

Refer to Section 8 and refit the listed components with references to the Chapters indicated.

37 Engine – refitting

Refitting the engine is a reversal of the removal procedure given in Section 6, but in addition note the following points:

(a) Lightly grease the gearbox input shaft
(b) Check that the clutch release arm is correctly located
(c) Adjust the clutch cable as described in Chapter 5
(d) Adjust the tension of the power steering pump and alternator drivebelt(s) as described in Chapters 10 and 2 respectively
(e) Fill the cooling system (Chapter 2) and power steering system (Chapter 10), then fill the engine with oil (Section 2).

38 Engine – initial start-up after major overhaul or repair

1 Make sure that the battery is fully charged and that all lubricants, coolant and fuel are replenished.

Chapter 1 Engine 47

2   Double check all fittings and electrical connections. Ensure that the distributor is correctly fitted and that the ignition timing static setting is correct. If in doubt refer to Chapter 4.
3   Remove the spark plugs and the '–' connection from the ignition coil. Turn the engine over on the starter motor until the oil pressure warning light is extinguished or until oil pressure is recorded on the gauge. This will ensure that the engine is not starved of oil during the critical few minutes running after initial start-up. The fuel system will also be primed during this operation.
4   Reconnect the '–' connection on the ignition coil and refit the spark plugs and leads. Start the engine.
5   As soon as the engine fires and runs, keep it going at a fast tickover only (no faster) and bring it up to normal working temperature.
6   As the engine warms up there will be odd smells and some smoke from parts getting hot and burning off oil deposits. The signs to look for are leaks of water or oil, which will be obvious if serious. Check also the exhaust pipe and manifold connections as these do not always find their exact gas tight position until the warmth and vibration have acted on them, and it is almost certain that they will need tightening further. This should be done, of course, with the engine stopped.
7   When normal running temperature has been reached, adjust the engine idle speed as described in Chapter 3.
8   Stop the engine and wait a few minutes to see if any lubricant or coolant is dripping out when the engine is stationary.
9   After the engine has run for 20 minutes remove the engine rocker covers and rocker shafts and recheck the tightness of the cylinder head bolts (photo). Also check the tightness of the sump bolts. In both cases use a torque wrench.
10   Road test the car to check that the timing is correct and that the engine is giving the necessary smoothness and power. Do not race the engine – if new bearings and/or pistons have been fitted it should be treated as a new engine and run in at a reduced speed for the first 1000 miles (1600 km).

PART B: 3.0 LITRE ENGINE

39 General description

The general configuration of the 3.0 litre engine is similar to the 2.8

38.9 Rechecking the tightness of the cylinder head bolts after running the engine for 20 minutes

litre engine described in Section 1. The significant differences are as follows:

(a) Flat cylinder heads with the combustion chambers in the pistons
(b) Individually mounted rocker arms instead of a rocker shaft
(c) Water pump located on the right of the engine
(d) Distributor and oil pump driven from the front of the camshaft
(e) Mechanical fuel pump operated by an eccentric on the front of the camshaft
(f) Rear oil seal located in a separate carrier
(g) Oil filter located on the front left-hand side of the engine

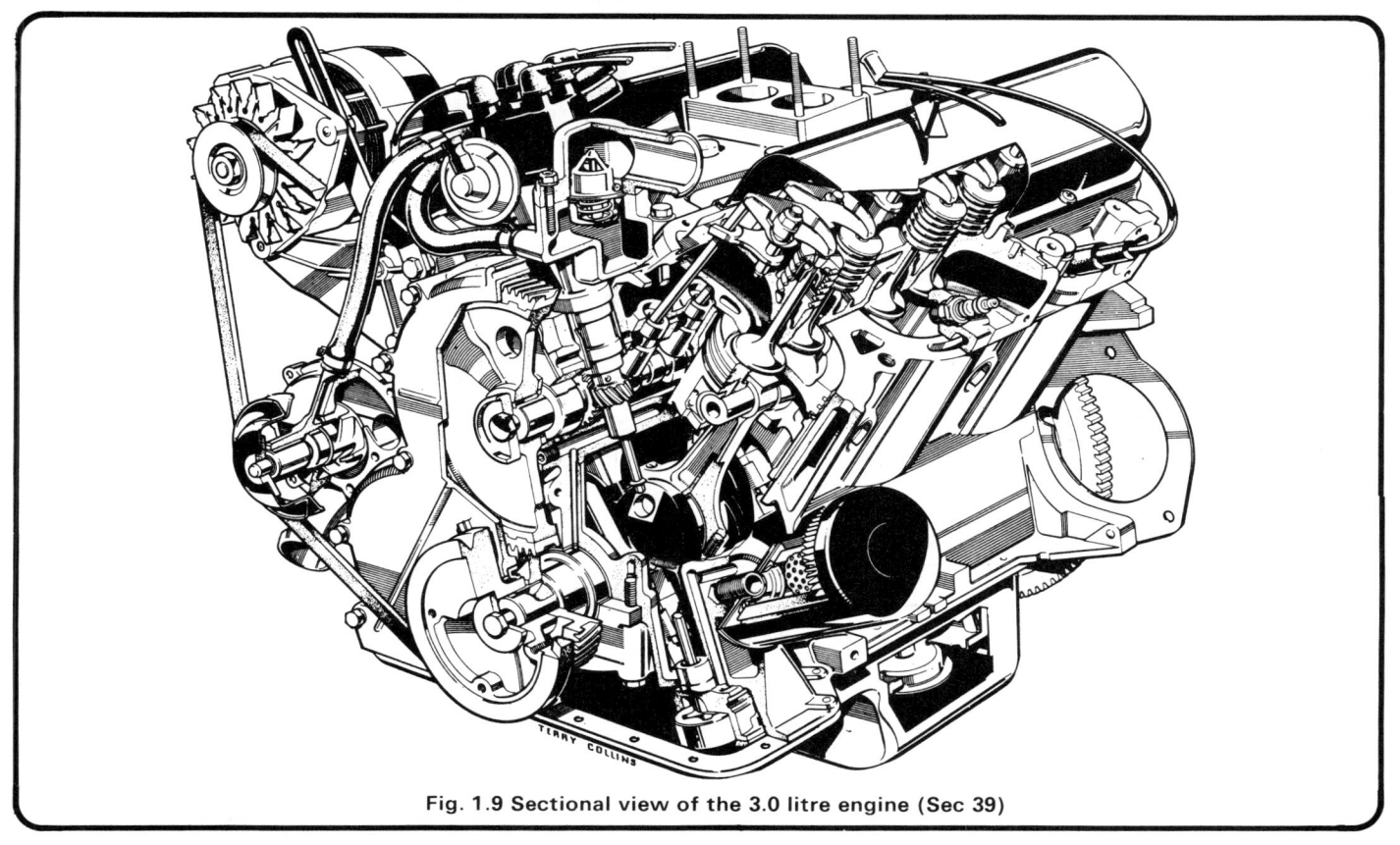

Fig. 1.9 Sectional view of the 3.0 litre engine (Sec 39)

40 Routine maintenance

At the intervals specified in the Routine Maintenance section in the front of the manual carry out the following procedures.

Change engine oil, renew oil filter and clean oil filler cap
1 Refer to Section 2.

Check and adjust valve clearances
2 Refer to Section 89.

Tighten inlet manifold bolts
3 Remove the air cleaner and tighten the inlet manifold bolts to the specified torque with reference to Chapter 3. On completion refit the air cleaner.

Renew the crankcase emission valve
4 Refer to Section 2.

41 Major operations possible with engine in car

The following major operations may be carried out without removing the engine:

(a) Removing and refitting the cylinder heads
(b) Removing and refitting the timing gears
(c) Removing and refitting the engine front mountings
(d) Removing and refitting the engine/gearbox rear mounting
(e) Removing and refitting the camshaft

42 Major operations requiring engine removal

The following major operations may be carried out after removal of the engine:

(a) Removing and refitting the flywheel
(b) Removing and refitting the rear main bearing oil seal
(c) Removing and refitting the crankcase sump*
(d) Removing and refitting the connecting rod bearing
(e) Removing and refitting the pistons and connecting rods
(f) Removing and refitting the oil pump
(g) Removing and refitting the crankshaft and main bearings
(h) Removing and refitting the camshaft bushes

* Although it is possible to remove the sump with the engine in the car, it is recommended that for any operations which require the removal of the sump, such as the removal of the crankshaft and big end bearings, the engine is removed.

43 Methods of engine removal

The engine may be lifted out either on its own or in unit with the transmission. On models fitted with automatic transmission it is recommended that the engine be lifted out on its own, unless a substantial crane or overhead hoist is available, because of the weight factor. If the engine and transmission are removed as a unit they have to be lifted out at a very steep angle, so make sure that there is sufficient lifting height available.

44 Engine – removal

1 Open and prop up the bonnet, then cover the wings with cloths, or cardboard, to protect them from scratches during the subsequent operations.
2 Mark the outline of the bonnet hinges and then with the help of an assistant, remove the four bonnet screws, release the bonnet stay and carefully lift the bonnet clear.
3 Disconnect the battery leads, remove the battery clamp bolt and lift the battery out.
4 Remove the air cleaner.

5 Remove the splash shield by removing the bolts and clips.
6 Place a container of 10 litre (17.6 pint) capacity underneath the radiator, remove the radiator bottom hose and drain the coolant.
7 Remove the upper hose from the water pump. Remove the bolts from the air deflector panel and from the radiator shroud and remove them.
8 Remove the four bolts securing the radiator to the front panel and lift the radiator out.
9 Disconnect the hot water hose from the water pump and automatic choke.
10 Remove the starter motor heat shield and then disconnect the leads from the alternator, temperature sender unit, ignition coil and starter motor.
11 Remove the two bolts securing the throttle linkage bracket assembly to the inlet manifold and remove it (photo).
12 Remove the fuel lines from the fuel pump (photo), the servo vacuum hose from the inlet manifold and the oil pressure line from the connector and/or lead from the oil pressure switch.
13 Remove the two bolts securing the starter motor and remove it.
14 Remove the two nuts to disconnect the rubber insulators from the engine mountings.
15 Remove the two nuts from each exhaust flange and disconnect the exhaust pipes from the exhaust manifold.

44.11 Throttle linkage and bracket

44.12 Fuel pump and fuel line

Chapter 1 Engine

16 On manual transmission models, remove four clips and remove the clutch housing cover. On automatic transmission models disconnect and plug the oil cooler hoses.
17 Support the transmission then remove the transmission-to-engine bolts.
18 On automatic transmission models, drain the torque converter and unscrew the driveplate bolts with reference to Chapter 6.
19 Take the weight of the engine on the hoist and then pull the engine forward, to separate it from the transmission. On automatic transmission models make sure that the torque converter remains fully in the transmission.
20 When the engine has been separated from the transmission, lift the engine slowly, checking frequently to see that the engine does not foul any part of the bodywork.
21 When the engine is clear of the body, lower it to the ground, or transfer it to a work bench.

45 Engine dismantling – general

Refer to Section 7.

46 Engine ancillaries – removal

With the engine separated from the transmission remove the externally mounted components as follows:

Carburettor and fuel pump (Chapter 3)
Inlet and exhaust manifolds (Chapter 3)
Spark plugs and distributor (Chapter 4)
Alternator (Chapter 12)
Fan and pulley (Chapter 2)
Water pump and thermostat (Chapter 2)
Crankcase ventilation tube
Oil filter (Section 40)
Oil level dipstick and filler cap
Engine mounting brackets
Clutch (Chapter 5)
Power steering pump (Chapter 10)

47 Cylinder heads – removal with engine in car

1 Disconnect the battery leads.
2 Remove the splash shield, by removing four bolts and four clips.
3 Place a container of 10 litre (17.6 pint) capacity underneath the radiator, remove the radiator bottom hose and drain the system.
4 Remove the radiator top hose.
5 Remove the air cleaner, disconnect the fuel pipe from the carburettor and the breather hose from the rocker cover.
6 Remove the servo vacuum hose from the inlet manifold.
7 Remove the three bolts and remove the throttle linkage bracket complete with linkage.
8 Disconnect the radiator hose from the automatic choke. Disconnect the vacuum line from carburettor.
9 Disconnect the lead from the temperature sender unit.
10 Disconnect the HT leads from the spark plugs and from the ignition coil and remove the LT lead to the distributor. Remove one fixing bolt and lift off the distributor assembly.
11 Disconnect the alternator cables. Slacken the alternator mounting bolts and remove the drivebelt, then remove the alternator fixing bolts and remove the alternator and its fixing bracket.
12 Remove the spark plugs. Remove the nuts securing the rocker covers and lift off the covers.
13 Remove the five bolts securing the inlet manifold and take off the inlet manifold complete with carburettor.
14 Loosen the rocker arm adjusting nuts until the rocker arms can be turned aside (Fig. 1.10) then remove the pushrods and lay them out in line, or push their ends through a sheet of paper on which their position in the engine can be noted. It is very important that the pushrods are refitted to the same valve as that from which they were removed.

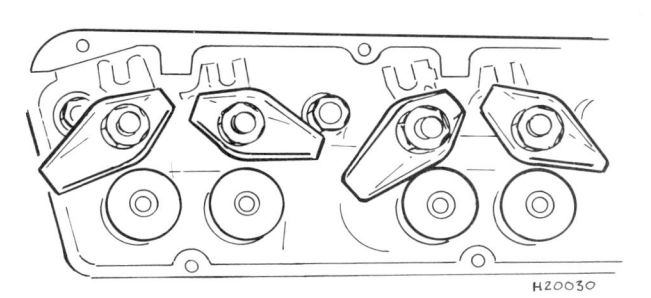

Fig. 1.10 Turn the rocker arms to one side to remove the pushrods (Sec 47)

15 Remove twelve bolts and detach the exhaust manifolds from the cylinder heads.
16 Slacken the cylinder head bolts progressively in the reverse order to that shown in Fig. 1.19. Remove the bolts and lift off the cylinder head. If the gasket has stuck and the cylinder head cannot be lifted off, tap the cylinder head with a hide-faced, or plastic-headed hammer. Do not try to prise them off with a screwdriver or damage will result.

48 Cylinder heads – removal with engine out of car

Follow the sequence given in Section 47, starting at paragraph 10, disregarding references to parts which have already been removed.

49 Valves – removal

1 Remove the deposits from the combustion chambers and valve heads by scraping and wire brushing. Take care not to scratch the cylinder head gasket faces.
2 Lay the cylinder head on its side and fit the forked end of a valve spring compressor over the spring retainer and the head of the compressor screw in the centre of the valve head.
3 Compress the spring about a quarter of an inch to free the split collet from the top of the valve stem. If the spring retainer and collet sticks, do not screw the compressor down further, because this may damage the valve stem, but tap the spring retainer sharply with a light hammer.
4 When the two halves of the collet have been removed, unscrew the spring compressor until the spring is no longer compressed and remove the compressor.
5 Remove the spring retainer, the spring and the oil seal from the valve stem, then withdraw the valve from the cylinder head.
6 Discard the valve stem oil seals and place the other components together, identifying them so that they are refitted in the correct place in the cylinder head.
7 Remove the spark plugs from the cylinder head, if not already removed.

50 Rocker arms – removal

1 Remove the stiff nut from the rocker arm mounting stud.
2 Lift off the rocker arm ball seating and the rocker arm and lay the three parts together. It is important that the rocker arm, its ball seating and nut are refitted to the same stud as the one from which they were removed.

51 Tappets – removal

1 Remove the tappets from their bores with a magnet and place them in a rack in sequence.
2 If any are difficult to remove, rotate the crankshaft, which will cause the camshaft to turn and move the tappets to release them from any gum in their bores.

52 Crankshaft pulley – removal

1 Prevent the crankshaft from rotating, by jamming the flywheel if the engine is out of the car, or by locking the transmission if the engine is still in the car.
2 Release the pulley bolt by fitting a ring spanner and tapping the spanner sharply, then remove the bolt and washer.
3 Use a sprocket puller to draw the pulley off the crankshaft. Do not attempt to lever it off, because the engine front cover is fragile and levering against it may distort, or even break it.
4 Withdraw the Woodruff key from the end of the crankshaft and place it with the pulley, the bolt and its washer.
5 The crankshaft pulley may be removed with the engine in the car, but it is first necessary to remove the radiator and the drivebelts.

53 Flywheel – removal

1 Prevent the flywheel from rotating by jamming the ring gear against an improvised stop.
2 On models with manual transmission, remove the clutch pressure plate and disc as described in Chapter 5.
3 Release the six bolts securing the flywheel to the crankshaft and remove five of them.
4 While supporting the flywheel in place, remove the sixth bolt and lift the flywheel off carefully, so that the mating surfaces of flywheel and crankshaft are not damaged.

54 Sump – removal

1 With the engine out of the car, ensure that the sump has been drained and then turn the engine over so that the sump is uppermost. Fit wooden blocks to support the engine in this position.
2 Remove the bolts securing the oil sump to the crankcase and remove the sump. If the sump is stuck to the crankcase, it should be released by prising one side with a screwdriver.
3 To remove the sump with the engine in the car, first remove the dipstick and disconnect the battery.
4 Remove the four bolts and four clips and detach the oil splash shield.
5 Place a container of at least 6 litre (10.6 pint) capacity beneath the sump, remove the drain plug and drain the sump. Refit the drain plug and tighten it to the torque wrench setting given in the Specifications.
6 Detach the engine rubber mountings, by removing their nuts and washers.
7 Remove the clamping bolt from the top and bottom ends of the steering intermediate shaft. Push the top end of the shaft upwards until the lower end can be disengaged.
8 Place a jack under the transmission and use it to raise the engine slightly.
9 Remove the four bolts securing the front anti-roll bar to the side-member and the four bolts securing each engine mounting bracket to the sidemember.
10 Remove the oil sump bolts and if the sump has stuck to the crankcase, release the sump by pressing it sideways with a screwdriver.

55 Front cover – removal

1 If the engine is in the car, it will first be necessary to disconnect the battery, drain the cooling system and remove the radiator.
2 Loosen the alternator mounting bolts, swing the alternator downwards towards the cylinder block and remove the drivebelt.
3 Remove the bolt from the centre of the cooling fan and pull off the fan and clutch assembly.
4 Remove the bolt and washer from the centre of the crankshaft pulley and then use a claw puller to draw the pulley off the shaft. Do not attempt to lever the pulley off, because this may result in damage to the front cover.
5 Remove the six bolts securing the sump to the front cover and then the eleven bolts securing the front cover to the cylinder block.
6 Lift the cover off and remove the gasket.

56 Timing gears – removal

1 Remove the front cover to expose the timing gears.
2 Remove the bolt and washer securing the large gear to the camshaft, then pull the gear off and remove the Woodruff key.
3 Draw the gear from the crankshaft and remove the Woodruff key.

57 Camshaft – removal

1 The camshaft may be removed with the engine in the car, if the following preparatory work is carried out.
2 Drain the cooling system and remove the radiator.
3 Disconnect the battery.
4 Remove the spark plug leads from the spark plugs.
5 Disconnect the leads from the alternator, remove the alternator and its drivebelt.
6 Undo and remove the screws securing the rocker covers and lift off the rocker covers and their gaskets.
7 Slacken the nuts securing the rocker arms until the rocker arms can be swung aside.
8 Remove the pushrods, identifying each one as it is removed, so that it can be refitted in the same position.
9 Remove the cylinder heads, as described in Section 47, then use a magnet to lift out the tappets. Place each tappet with the pushrod which it operates, so that they also will be refitted in their original positions.
10 Remove the front cover as described in Section 55.
11 The camshaft plate is secured with two cross-head countersunk screws. These cannot be removed with an ordinary screwdriver and any attempt to do so will result in the screws being damaged. If an impact screwdriver is not available, the screws may be removed with a suitable key and the brace of a socket set.
12 Remove the camshaft thrust plate and the spacer behind it.
13 Pull the camshaft forward to withdraw it, taking care to keep the shaft in line, so that the edges of the cams do not damage the camshaft bearings.

58 Oil pump – removal

1 Remove the crankcase sump as already described.
2 Remove the bolt attaching the oil screen to the main bearing cap.

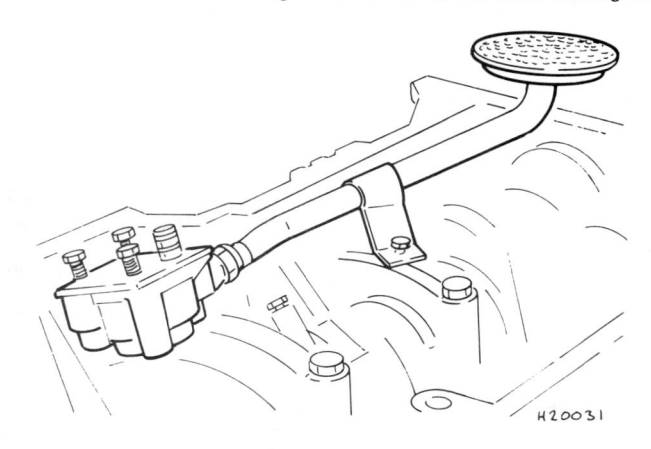

Fig. 1.11 Oil pump on the 3.0 litre engine (Sec 58)

3 Remove the two bolts securing the oil pump to the crankcase and withdraw the pump assembly and its driveshaft.

59 Pistons, connecting rods and bearings – removal

1 If the crankcase sump and the cylinder heads are removed, the pistons and connecting rods can be removed with the engine in the car. The bearing shells can be removed without removing the cylinder heads.
2 Before removing any connecting rod cap, make sure that the cylinder number is marked on both parts. If no marking is visible, put mating marks on the joint before removing the bolts.
3 Turn the crankshaft until the connecting rod being removed is at its lowest position.
4 Remove the connecting rod nuts and pull off the bearing cap.
5 Rotate the crankshaft until the connecting rod being removed is in its highest position then use the handle of a hammer to push the assembly out of the top of the cylinder.
6 Withdraw the assembly from the cylinder and place the bearing cap back on the connecting rod.
7 If the bearings are being renewed, remove the old bearings by sliding them round until the notch in the bearing shell is disengaged from the recess in the bearing seating and then lift the bearing out. If the bearings are removed and the same ones are being used again, it is vital that they are refitted in the same seating as the one from which they were removed.

60 Crankshaft rear oil seal – removal

1 To remove the oil seal with the engine in the car, first remove the gearbox, or automatic transmission as detailed in Chapter 6. On manual transmission models, it is then necessary to remove the clutch assembly as described in Chapter 5.
2 Release the six bolts securing the flywheel and remove five of them. While holding the flywheel, remove the sixth bolt and lift away the flywheel.
3 Remove the four bolts securing the oil seal carrier and lift off the carrier and seal assembly.
4 Drive the seal out of the carrier, taking care not to damage the bore of the carrier.

Fig. 1.12 Removing the rear oil seal from the carrier (Sec 60)

61 Crankshaft and main bearings – removal

1 Remove the engine from the car.
2 Remove the clutch assembly on manual transmission models.
3 Remove the flywheel and the rear oil seal carrier.
4 Make sure that the crankcase sump has been drained, then turn the engine upside down, chock it securely and remove the sump.
5 Remove the crankshaft pulley and the engine front cover.
6 Remove the oil pump and strainer.
7 Check that all the main and connecting rod bearing caps have their identifying numbers stamped on, so that they can be refitted in their original positions.
8 Remove the crankshaft gear.
9 Remove the two nuts and the bearing cap of each connecting rod in turn and lay out the bearing caps in order.
10 Remove the two bolts from each of the main bearing caps, lift the caps off and lay them out in order. Note that the centre main bearing cap has thrust washers. Mark their positions before removing them, to ensure that they are refitted exactly as removed.
11 Carefully lift the crankshaft out, taking care to keep it parallel with the crankcase, so that the thrust bearing surfaces of the upper bearing halves are not damaged.
12 Remove the bearing inserts from the cylinder block, being careful to note the position from which each of the four bearings and each of the two thrust washers were removed.

62 Engine components – examination for wear

Refer to Section 11.

63 Rocker arms – examination and renovation

1 The studs on which the rocker arms pivot are a press fit into the cylinder head, and if a straight edge is laid across all six, it can be seen whether any have begun to pull out. If this is the case, the stud will have to be removed, the hole reamed out and an oversize stud fitted.
2 Check that the threads on the studs are in good condition and that a torque of at least 5.9 Nm (4.4 lbf ft) is required to turn each adjuster on oiled threads. If this figure is not achieved, fit a new adjuster and if still deficient, fit a locknut to the adjuster.
3 Examine the rocker arms and fulcrum seats for signs of ridging, or excess wear on the bearing surfaces. If either part shows excessive wear, both the rocker arm and the fulcrum seat must be renewed.
4 When checking the rockers arms, remove them one at a time to eliminate any possibility of arms and seats becoming interchanged, or being refitted to the wrong stud.

64 Tappets and pushrods – examination and renovation

Refer to Section 13.

65 Camshaft and camshaft bearings – examination and renovation

1 Undersize camshaft bearings are available, so it is possible to salvage a worn camshaft by regrinding the journals. If the cams, or skew gear, are worn significantly, or are damaged, a new camshaft must be fitted and the bearings should then also be renewed.
2 Worn or damaged camshaft bearings can be replaced by new ones, but this requires special tools and should be entrusted to a Ford dealer, or to a workshop which undertakes camshaft reclaiming.

66 Cylinder bores – examination and renovation

Refer to Section 16.

67 Pistons and piston rings – examination and renovation

Refer to Section 17.

68 Connecting rods and gudgeon pins – examination and renovation

Refer to Section 18.

69 Crankshaft – examination and renovation

1 Examine the four main bearing journals and the six crankpins, which should all be smooth and highly polished. If there are any deep grooves, or deep scratches and there is a similar deterioration in the corresponding bearing shells, the crankshaft should be reground.
2 If the journals are in good condition and it is suspected that the crankshaft may be sufficiently worn to require regrinding, it is worth seeking the advice of a main Ford agent, or crankshaft regrinding service.
3 If the crankshaft is reground, the works will also supply the new bearings which will be required to match the journal diameters.
4 Check the spigot bearing in the rear of the crankshaft on manual transmission models and if necessary renew it.

70 Main and big-end bearings – examination and renovation

Refer to Section 20.

71 Timing gears – examination and renovation

Refer to Section 21.

72 Flywheel/driveplate – examination and renovation

1 Inspect the flywheel or driveplate for damage and check that the ring gear does not have any broken or badly worn teeth.
2 On manual transmission models it is possible to renew the ring gear as described in Section 22, but on automatic transmission models it is not possible to remove the ring gear from the driveplate.

73 Valves and valve seats – examination and renovation

1 Examine the valves for pits and grooves and the valve stems for excessive wear or distortion.
2 Examine the valve head for signs of burning of the valve seats and check the fit of the valves in the valve guides, to see if the valve guides are worn.
3 Worn valve guides can be reamed out and valves with oversize stems fitted. If a valve guide has been reamed out, the corresponding valve seat must be refaced.
4 Valves and valve seats which show little sign of wear, can be lapped in.
5 To lap the valves, fit a suction type grinding tool to the valve head, smear a little coarse carborundum grinding paste on to the seating face of the valve and insert the valve into the cylinder head. Rub the stem of the valve grinding tool between the palms of the hands, so that the valve is rotated backwards and forwards. When it is felt that cutting has ceased, lift the valve, rotate it through 90°, lower it and repeat the grinding operation, redistributing the grinding paste on the seat if necessary. Continue the process until a matt grey band is produced around the complete periphery of both the valve and the valve seat.
6 Repeat the operation using fine grinding paste, to obtain a finer finish, then clean both the valves and the valve seats to remove all traces of grinding paste.
7 If it is convenient to take the cylinder head and the valves to a workshop with the necessary equipment, it is far better to have all the valves and their seats refaced, so that all the valves and seats are at the correct angle and have the correct width of contact.

74 Decarbonising – general

1 When the cylinder heads are removed, either in the course of an overhaul or for inspection of bores or valve condition when the engine is in the car, it is normal to remove all carbon deposits from the piston crowns and head.
2 This is best done with a cup shaped wire brush and an electric drill and is fairly straightforward when the engine is dismantled and the pistons removed. Sometimes hard spots of carbon are not easily removed except by a scraper. When cleaning the pistons with a scraper, take care not to damage the surface of the piston in any way.
3 When the engine is in the car, certain precautions must be taken when decarbonising the pistons' crowns in order to prevent dislodged pieces of carbon falling into the interior of the engine which could cause damage to cylinder bores, piston and rings – or, if allowed into the water passages, damage to the water pump. Turn the engine so that the piston being worked on is at the top of its stroke and then mask off the adjacent cylinder bores and all surrounding water jacket orifices with paper and adhesive tape. Press grease into the gap all round the piston to keep carbon particles out and then scrape all carbon away by hand. Do not use a power drill and wire brush when the engine is in the car as it will virtually be impossible to keep all the carbon dust clear of the engine. When completed, carefully clear out the grease around the rim of the piston with a matchstick or something similar – bringing any carbon particles with it. Repeat the process on the other piston crowns. It is not recommended that a ring of carbon is left round the edge of the piston on the theory that it will aid oil consumption. This was valid in the earlier days of long stroke low revving engines but modern engines, fuels and lubricants cause less carbon deposits anyway and any left behind tend merely to cause hot spots.

75 Oil pump – dismantling, examination and reassembly

1 The oil pump maintains a pressure of about 2.8 kgf/cm^2 (39.8 lbf/in^2) and unless there is a significant drop in oil pressure which is proved to be due to the pump, rather than to worn bearings, it is better to leave the oil pump undisturbed.
2 To dismantle the pump, first remove it from the engine as described in Section 58.
3 Remove the two bolts securing the end cover to the body and remove the cover and the relief valve assembly, then the two piece rotor assembly.
4 Remove the two bolts and take the pick-up tube and screen assembly off the pump housing.
5 Wash all the parts in petrol, use a brush to clean the inside of the pump housing and the pressure relief valve chamber, and make sure that all particles of dirt and metal are removed. Allow the parts to dry naturally or blow them dry.
6 Check the inside of the pump housing, the outer race and the rotor for damage and excessive wear.
7 Examine the mating surface of the pump cover for wear. If the cover is scored, grooved, or shows any signs of wear, a new cover must be fitted.
8 With the rotor assembly fitted in the housing, place a straight-edge over the end of the housing and measure the rotor end play. If this exceeds 0.104 mm (0.004 in), the outer race, shaft and rotor must be renewed as an assembly.
9 Measure the rotor-to-housing clearance, which should not exceed 0.304 mm (0.012 in) and the inner to outer rotor gap, which should be less than 0.020 mm (0.008 in). If outside these dimensions, a new pump should be fitted.
10 Check that the relief valve and its seating are free of damage and that the relief valve spring has not collapsed, or is damaged, and fit new parts if necessary.
11 Reassemble the pump by first fitting the inner and outer rotors, then lubricating them with engine oil.
12 Apply engine oil to the relief valve plunger and spring and insert them into their housing. Refit the spring cap and the pump cover.
13 Refit the cover attachment screws and tighten them to the torque wrench setting given in the Specifications. After tightening the screws, check that the rotor moves freely.
14 Refit the pick-up tube and strainer assembly to the pump body, using a new gasket and then insert and tighten the two fixing screws.

Chapter 1 Engine   53

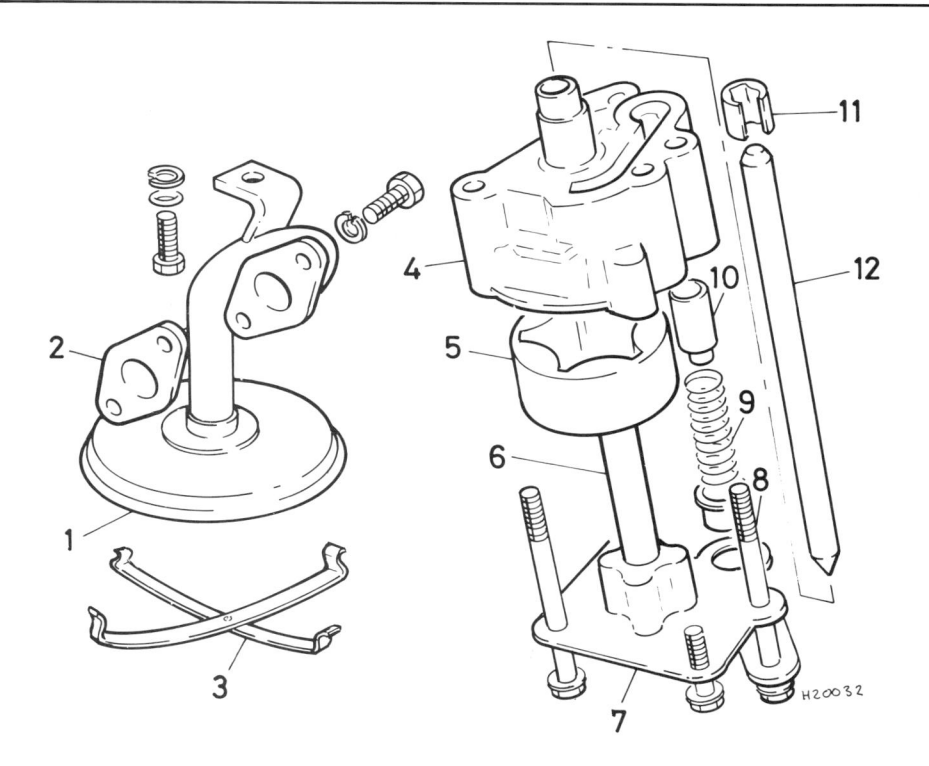

Fig. 1.13 Exploded view of the oil pump (Sec 75)

1 Pick-up tube and screen
2 Gasket
3 Clips
4 Pump body
5 Outer rotor
6 Inner rotor
7 End cover
8 Seat
9 Spring
10 Relief valve plunger
11 Clip
12 Shaft

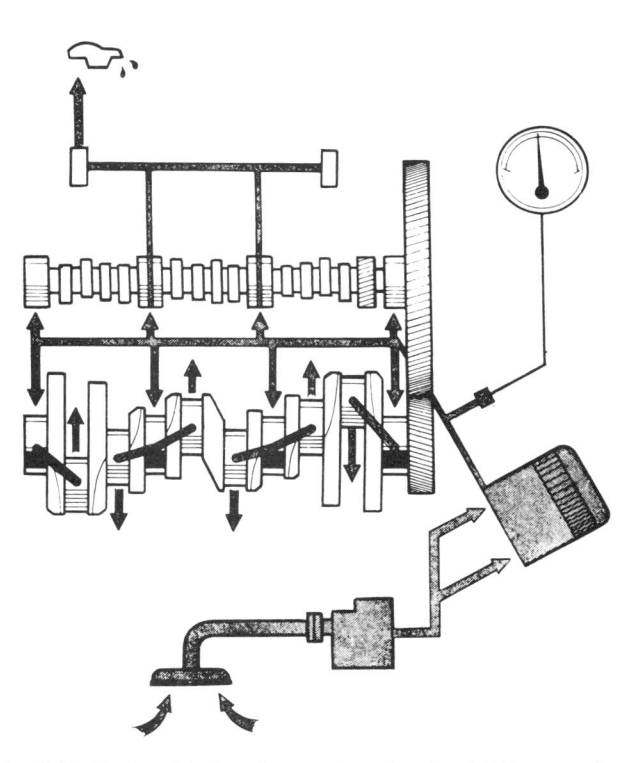

Fig. 1.14 Engine lubrication system for the 3.0 litre engine (Sec 76)

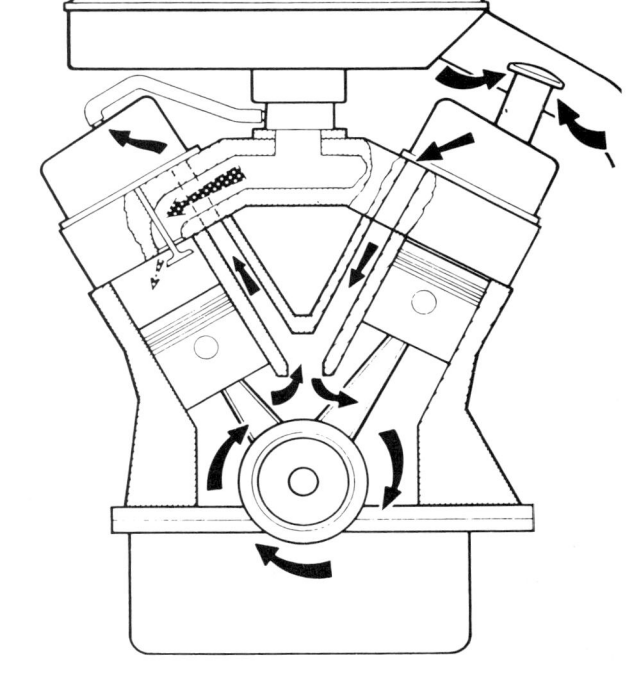

Fig. 1.15 Crankshaft ventilation system (Sec 77)

76 Lubrication system – description

The lubrication system is virtually the same as that for the 2.8 litre engine described in Section 24, except that the rocker arms are supplied with oil through hollow tappets and pushrods.

77 Crankcase ventilation system – description

A semi-sealed crankcase ventilation system is fitted. Fresh air is drawn through a breather in the oil filler cap and mixes with the blow-by gases in the crankcase. It is then drawn out of the vent valve in the right-hand rocker cover and into the engine through an adaptor beneath the carburettor.

The system ensures that unburnt gases are returned to the engine for further combustion.

78 Engine reassembly – general

Refer to Section 26.

79 Camshaft – refitting

1 Slide the spacer with its chamfered end first onto the camshaft and fit the key.
2 Lubricate the camshaft journals with engine oil and slide the camshaft in from the front, taking care to keep it straight, so as not to damage the camshaft bearings.
3 Fit the camshaft thrust plate without a gasket. Fit the two countersunk head retaining screws and tighten them.

80 Crankshaft – refitting

1 With the engine upside down, lay each of the upper bearing halves onto the appropriate crankcase web, making sure that the locking tang on the bearing is properly engaged in the corresponding slot in the crankcase web. If the old bearings are being refitted, make certain that each bearing half is fitted on the place from which it was removed. Fit the thrust washers to each side of the centre bearing, the grooved faces of the washers being outwards (Fig. 1.16). Retain the thrust washers in place with grease.
2 Fit the four bearing halves to the appropriate bearing caps. Fit the thrust washers on either side of the centre bearing cap, with the tab of the washer fitted into the slot in the bearing cap. Retain the thrust washers in place with grease.
3 Lubricate the crankcase bearing halves with engine oil and carefully lower the crankshaft onto them.
4 Take each bearing cap in turn, lubricate its bearing half with engine oil. Apply engine oil to the crankshaft journal and fit the bearing cap with the arrow pointing towards the front of the engine. Each bearing cap must be refitted in the position from which it was removed.
5 Insert the two retaining bolts and tighten them to the torque wrench setting given in the Specifications. Turn the crankshaft to ensure that it rotates freely, before fitting the next bearing cap. If the crankshaft does not rotate freely, remove the bearing cap and check that the bearing shells have seated properly and that the bearing cap is the correct one for that position.
6 Fit the crankshaft key. Place the crankshaft gear on the end of the crankshaft and position it by temporarily fitting the crankshaft pulley and the original washer and bolt. Tighten the bolt, then remove the bolt, washer and crankshaft pulley.
7 Measure the crankshaft endfloat (see Section 27) which should be as specified. If necessary, bring it between these limits by inserting thrust washers of the appropriate thickness.

81 Pistons and connecting rods – refitting

Refer to Section 29.

82 Oil pump – refitting

1 Prime the oil pump by filling the inlet port with engine oil and then rotating the driveshaft to distribute the oil within the pump body.
2 Insert the pointed end of the driveshaft into the pump body and fit the clip to retain it.
3 Place the pump in position, engage the driveshaft and then insert the two retaining bolts. Tighten the bolts to the torque wrench setting given in the Specifications.
4 Refit the bolt securing the oil pick-up tube to the main bearing cap and tighten it to the specified torque.

Fig. 1.16 Crankshaft thrust washers – arrowed (Sec 80)

83 Timing gears and front plate – refitting

1 Apply jointing compound to the outer edge of the front end of the cylinder block and to the rear face of the front adaptor plate.
2 Position the gasket and adaptor plate, securing them with two bolts inserted finger tight.
3 Fit the adaptor plate reinforcement (Fig. 1.17); insert its three bolts and tighten them, then remove the two previously inserted locating bolts.
4 Refit the camshaft gear, so that the keyways of both the crankshaft and the camshaft are in line and the marks on the gears are on the same line. The crankshaft gear has two marks on it and it is important that they are positioned as shown in Fig. 1.5.
5 Measure the endfloat of the camshaft after inserting the camshaft securing bolt and tightening it to the torque wrench setting given in the Specifications. The camshaft endfloat can be adjusted by fitting a different length spacer.
6 Measure the backlash of the camshaft gear at four equally spaced points around its periphery. This should be 0.22 to 0.32 mm (0.009 to 0.013 in).
7 Apply jointing compound to the circumference of the timing cover joint face and to the front of the adaptor plate and offer up the timing cover.
8 Refit the crankshaft pulley to position the timing cover, then insert the timing cover bolts and tighten them to the torque wrench setting given in the Specifications.
9 Apply jointing compound to one face of the crankshaft pulley washer, place this side towards the pulley, insert the pulley securing bolt and tighten it to the specified torque.

Fig. 1.17 Adapter plate reinforcement (Sec 83)

Fig. 1.18 Rear oil seal and adapter plate (Sec 84)

Fig. 1.19 Tightening sequence for cylinder head bolts on the 3.0 litre engine (Sec 87)

84 Rear oil seal and flywheel – refitting

1 Fit the engine rear adaptor plate and secure it with its five bolts.
2 Smear engine oil onto the crankshaft rim. Smear oil on the lip of the seal and position the seal on the crankshaft.
3 Align the seal carrier holes with those in the cylinder block, insert the four bolts and tighten them to the specified torque.
4 Fit the flywheel, or driveplate on automatic transmission models. Insert the six bolts and tighten them to the torque wrench setting given in the Specifications.

85 Sump – refitting

1 Ensure that the mating surfaces of both the sump and the crankcase are clean and that the sealing surface of the sump is not distorted.
2 Apply jointing compound to the points where the timing cover abuts the cylinder block and where the rear oil seal carrier adjoins it.
3 Place the gasket in position, fit the sump and insert the retaining bolts finger tight.
4 Progressively tighten the sump bolts in opposite parts until all of them are at the specified torque.
5 Fit a new washer to the sump drain plug, insert the drain plug and tighten it.

86 Cylinder heads – reassembly

1 Lay the cylinder head on its side, lubricate every valve guide with engine oil, and insert the valves into the guide from which they were removed.
2 To avoid damaging the lip of the stem oil seal, with consequent increased oil consumption, wrap a piece of adhesive tape, or foil, round the groove of every valve stem. Slide the seals over the valve stems and push them over the tops of the valve guides.
3 Remove the adhesive tape, or foil, from the valve stems.
4 Fit the valve spring over the valve, with the close coiled end towards the cylinder head, then fit the spring retainer on top of the spring.
5 Using a valve spring compressor, compress the spring just enough to insert the two split collets. If the valve spring is compressed too far, there is danger of the spring retainer damaging the valve stem oil seal.
6 After inserting the split collets, release the spring compressor and remove it.
7 Check that the split collets have seated properly and tap the top of the valve stem lightly with a plastic headed hammer to ensure this.

8 Repeat the procedure given in paragraphs 4 to 7 on the remaining valves.

87 Cylinder heads – refitting

1 If the tappets were removed, lubricate the tappets and their bores with engine oil and insert them with their dished ends uppermost. Ensure that every tappet is refitted into the bore from which it was taken.
2 Make sure that the mating surfaces of the cylinder block, cylinder heads and inlet manifold are clean, then place new cylinder head gaskets over the guides on the cylinder block. The cylinder head gaskets bear the marking TOP FRONT and need to be positioned correctly.
3 Lower the cylinder heads over their positioning studs. Although the cylinder heads are identical, they should have been marked when removed and should be refitted in their original positions.
4 Insert the cylinder head bolts finger tight then tighten them progressively in the order shown in Fig. 1.19 and in the stages as given in the Specifications.

88 Rocker arms and covers – refitting

1 Ensure that all the pushrods are inserted into their proper places and that their lower ends are engaged in the cups of the tappets.
2 Place the rocker arms over the pushrods, fit the fulcrum seats and adjuster nuts, and tighten the adjusting nuts finger tight.
3 Set No 1 cylinder at TDC and refit the distributor (see Chapter 4).
4 Adjust the valve clearances as described in the following Section.
5 Fit the rocker covers, using new gaskets, ensuring that the gaskets are seated evenly all round the cylinder head. Fit the cover attachment bolts and tighten them to the specified torque.

89 Valve clearances – checking and adjustment

Refer to Section 35, but use Fig. 1.20 for inlet and exhaust valve locations on 3.0 litre engines.

90 Engine ancillaries – refitting

Refer to Section 46 and refit the listed components with reference to the Chapters indicated.

Chapter 1 Engine

91 Engine – refitting

Refitting the engine is a reversal of the removal procedure given in Section 44, but in addition note the following points:

(a) Lightly grease the gearbox input shaft (manual transmission)
(b) Check that the clutch release arm is correctly located (manual transmission)
(c) Adjust the clutch cable as described in Chapter 5 (manual transmission)
(d) Before connecting the engine to the automatic transmission, check that the torque converter is fully engaged with the transmission pump by reference to Chapter 6
(e) Refill the automatic transmission with fluid as described in Chapter 6
(f) Adjust the tension of the power steering pump and alternator drivebelt(s) as described in Chapters 10 and 2 respectively
(g) Fill the cooling system (Chapter 2) and power steering system (Chapter 10), then fill the engine with oil (Section 40) (photo).

92 Engine – initial start-up after major overhaul or repair

Refer to Section 38.

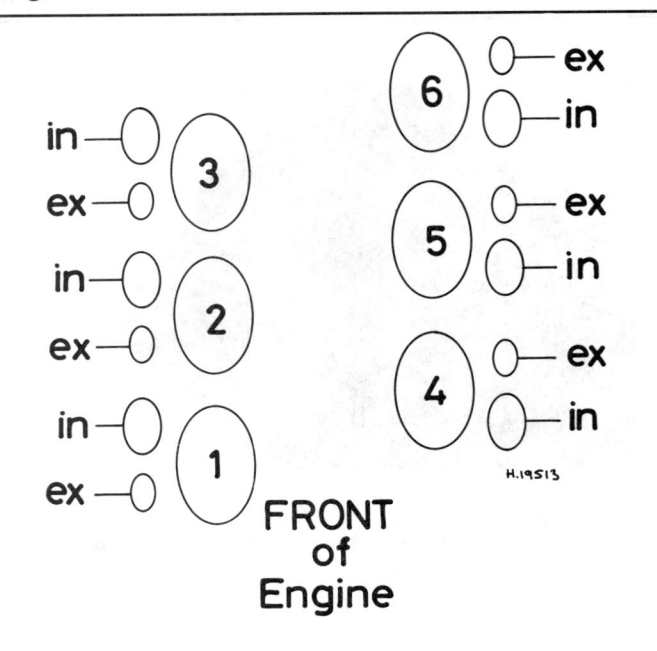

Fig. 1.20 Location of inlet and exhaust valves (Sec 89)
Note that this illustration depicts 3.0 litre engines only

PART C: FAULT DIAGNOSIS

93 Fault diagnosis – engine

Symptom	Reason(s)
Engine fails to turn over when starter is operated	Discharged or defective battery Dirty or loose battery leads Defective starter solenoid, or switch Defective starter motor Dirty or broken engine earthing strap
Engine turns at normal speed, but does not start	Ignition components wet or damp Defective low tension lead Defective or incorrectly set contact breaker (if applicable) Defective condenser (if applicable) Distributor cap defective, or central brush in distributor cap not making contact with the rotor arm Insufficient petrol in tank Vapour lock in fuel line (in hot conditions) Fuel pump failure Excessive choke, causing spark plugs to become wet
Engine stops and will not restart	Ignition failure resulting from breakdown, or presence of water under wet conditions Lack of fuel, or fuel blockage
Engine misfires, or runs unevenly	Loose ignition leads Faulty spark plug Tracking on distributor cap insulation Incorrect mixture Ignition too retarded Dirty contact breaker points (if applicable), or loose connection in distributor Faulty or loose condenser (if applicable) Leaking carburettor, or manifold joint
Lack of power and poor compression	Defective valves Blown cylinder head gasket Worn bores or damaged piston
Excessive oil consumption	Worn cylinder bores Defective valve stem seals Leaks

Chapter 2 Cooling system

Contents

Antifreeze mixture	6
Cooling system – draining	3
Cooling system – filling	5
Cooling system – flushing	4
Drivebelts – removal and refitting	10
Expansion tank – removal and refitting	14
Fan hub bearing (3.0 litre engine) – renewal	12
Fault diagnosis – cooling system	15
General description	1
Radiator – removal, inspection, cleaning and refitting	7
Routine maintenance	2
Temperature gauge sender unit – removal and refitting	13
Thermostat – removal, testing and refitting	8
Viscous drive fan – removal and refitting	11
Water pump – removal and refitting	9

Specifications

System type .... Pressurised, with belt-driven pump and fan (viscous drive type on later models), thermostat and expansion tank on later models

Thermostat
Opening temperature:
- 3.0 litre engines .... 85 to 98°C (185 to 208°F)
- 2.8 litre engines .... 79 to 83°C (174 to 181°F)

Fully open temperature:
- 3.0 litre engines .... 99 to 102°C (210 to 216°F)
- 2.8 litre engines .... 97°C (207°F)

Minimum thermostat travel .... 7.0 mm (0.28 in)

Drivebelt tension
Deflection midway between pulleys (longest run) .... 10.0 mm (0.4 in)

Pressure cap setting
- 3.0 litre engines .... 0.91 kgf/cm² (12.9 lbf/in²)
- 2.8 litre engines .... 0.85 to 1.10 kgf/cm² (12.1 to 15.6 lbf/in²)

Coolant
Type/specification .... Antifreeze to Ford spec SSM-978 9103-A (Duckhams Universal Antifreeze and Summer Coolant)

Capacity:
- 3.0 litre engines .... 9.3 litre (16.4 pint)
- 2.8 litre engines .... 8.7 litre (15.3 pint)

Torque wrench settings

	Nm	lbf ft
3.0 litre engines		
Fan	7 to 10	5 to 7
Water pump	7 to 10	5 to 7
Thermostat housing	16 to 20	12 to 15
2.8 litre engines		
Radiator	8 to 11	6 to 8
Thermostat housing	17 to 20	13 to 15
Water pump	9 to 13	7 to 10
Fan shroud	8 to 11	6 to 8
Fan pulley	20 to 25	15 to 18
Fan to viscous coupling	20 to 25	15 to 18

Chapter 2 Cooling system

1 General description

The cooling system is of the pressurised type consisting of a belt-driven pump, front mounted radiator and thermostat. The belt-driven fan is of standard type on early models, but of viscous drive type (photo) on later models, the latter incorporating a speed limitation device. On 3.0 litre engines the thermostat is located at the outlet to the top hose, but on 2.8 litre engines it is located on the inlet from the bottom hose. Later models are fitted with an expansion tank.

The system functions as follows: with the engine cold, the thermostat is closed and the coolant is pumped through the engine waterways and heater only. When the engine reaches a predetermined temperature, the thermostat opens and the coolant then circulates down through the radiator to provide additional cooling. Where an expansion tank is fitted, this ensures that the cooling system is always full by collecting coolant when the engine is hot and returning it as the engine cools.

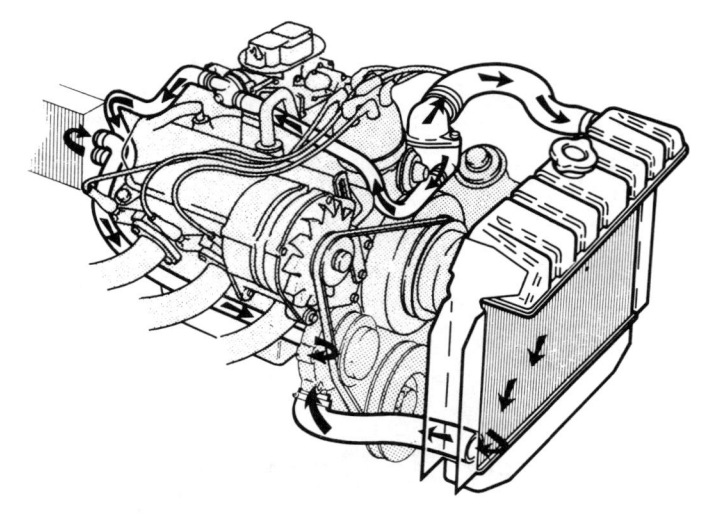

Fig. 2.1 Cooling system (3.0 litre engine) (Sec 1)

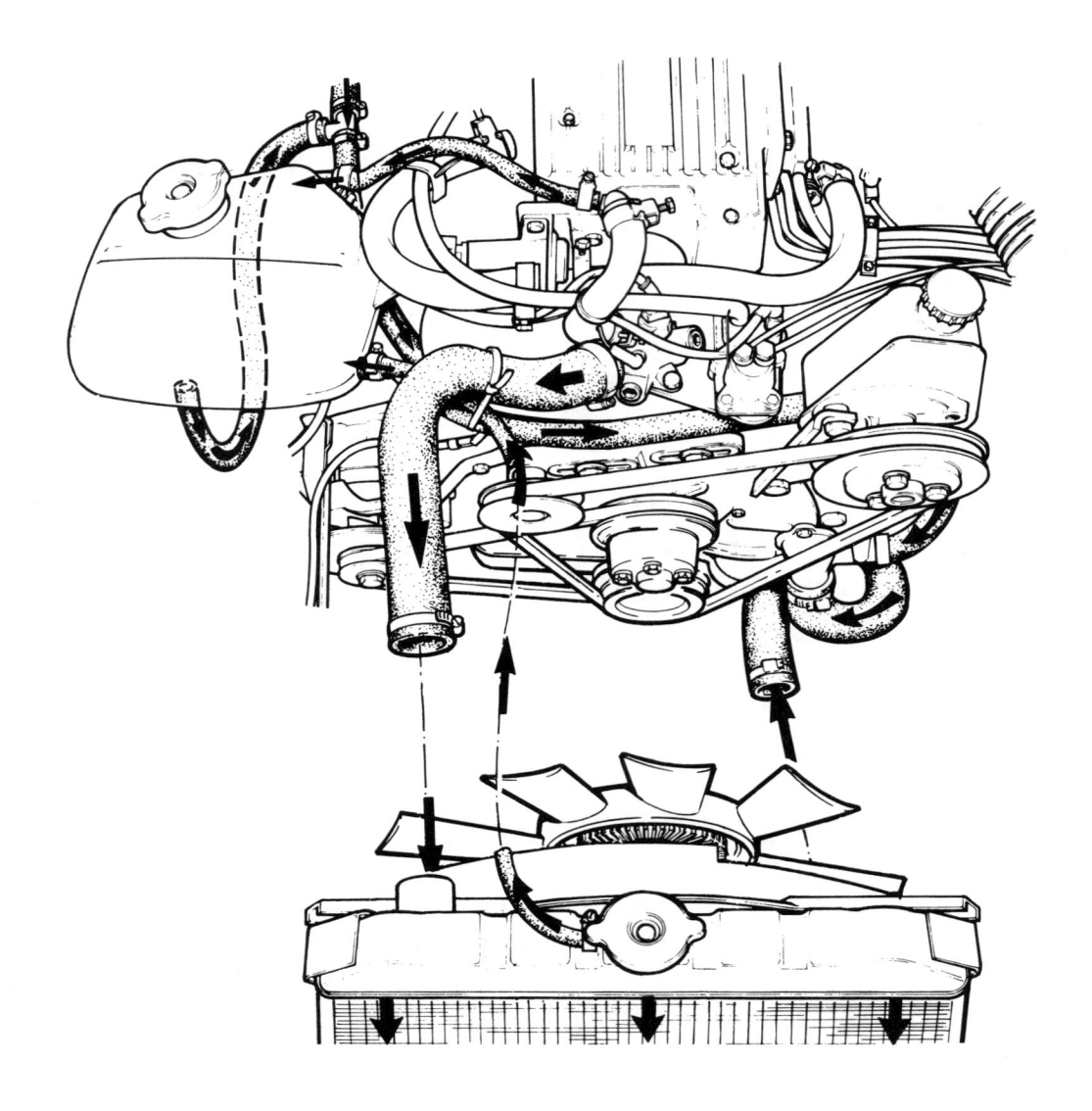

Fig. 2.2 Cooling system (2.8 litre engine) (Sec 1)

Chapter 2 Cooling system

1.1 Viscous drive fan coupling (temperature sensitive type shown)

2.5 Topping-up the coolant level in the expansion tank

2.7 Checking the drive-belt tension (2.8 litre engine)

2 Routine maintenance

At the intervals specified in the Routine Maintenance section in the front of the manual carry out the following procedures.

Check cooling system for leaks
1   Inspect the cooling system hoses including the radiator and heater hoses for damage, deterioration and security. If there is any indication of swelling or perishing renew the hoses.
2   Check the radiator for damage and leaks, and clean any debris from the cooling fins.

Check coolant level
3   Where an expansion tank is fitted (later models) the coolant level can be checked without removing the pressure cap as the tank is made of translucent plastic. The tank should be approximately half full with the coolant just covering the plastic level indicator. With a hot engine the level will be slightly higher.
4   On earlier models not fitted with an expansion tank the level must be checked with the engine cold. Unscrew the radiator pressure cap and check that the coolant level is between 25.0 mm (1.0 in) and 35.0 mm (1.4 in) below the base of the filler neck.
5   Top-up the fluid level using the correct solution of water and antifreeze with the engine cold (photo). If the engine is warm, the pressure cap must be removed slowly to allow the pressure to escape gradually, otherwise the pressure drop may cause the coolant to boil over causing personal injury.

Check drivebelt condition and tension
6   Inspect the complete length of the drivebelt(s) for deterioration which will appear as fraying or glazing to the pulley contact surfaces.
7   Check that the deflection of the drivebelt(s) midway between the water pump and alternator pulleys is approximately 10.0 mm (0.4 in) under moderate thumb pressure (photo).
8   Where adjustment is necessary, loosen the alternator pivot and adjustment bolts, reposition the alternator then tighten the bolts. A lever may be used at the pulley end of the alternator to obtain the correct tension if required.
9   Run the engine for several minutes then recheck the tension.

Renew antifreeze
10   Drain and flush the cooling system as described in Sections 3 and 4.
11   Fill the cooling system with the correct solution of water and antifreeze as described in Sections 5 and 6.

3 Cooling system – draining

1   If possible, ensure that the engine is cold before draining the coolant, to avoid the risk of scalding.
2   Remove the radiator or expansion tank filler cap. If the engine is hot, this must be done very cautiously, because a sudden release of

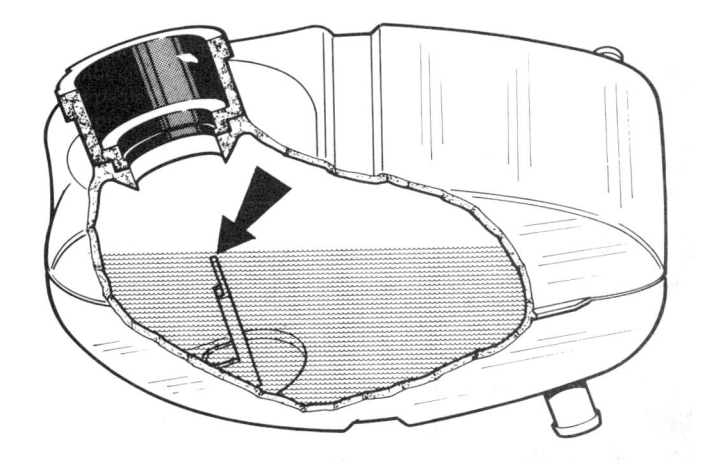

Fig. 2.3 Cutaway view of the expansion tank showing the plastic level indicator – arrowed (Sec 2)

pressure can result in the coolant boiling and blowing out. On a hot engine, place a cloth over the cap and turn it very gently, to release the pressure slowly.
3   Remove the splash shield from beneath the radiator, if applicable.
4   Unless the coolant is being discarded and the car is over a suitable drain, place a clean two and a half gallon container underneath the radiator.
5   Remove the radiator bottom hose and allow the coolant to drain.
6   If the system is to be completely drained, remove the cylinder block drain plug and drain out the remaining coolant from the engine. This drain plug is located on the side of the cylinder block to the rear of the oil filter.
7   On models with an expansion tank remove the radiator cap as well.

4 Cooling system – flushing

1   In time, the cooling system will lose its efficiency because of a build-up of rust and sediment in the radiator. To clean the radiator, remove the filler cap and the bottom hose and flush the radiator, by inserting a hose in the filler cap neck and running water through for about ten minutes.
2   If there is a heavy accumulation of sediment, it is better to reverse flush the radiator, by connecting the water supply to the bottom of the radiator and allowing it to flow out of the top. Special hose adapters are available to enable this to be done with the radiator in position, but it is preferable to remove the radiator and turn it upside down.
3   Remove the thermostat (Section 8) then reverse flush the engine with a water hose until clean water flows from the radiator bottom hose connection.

5 Cooling system – filling

1 Ensure that the cylinder block drain plug is screwed in firmly and that all the cooling system hoses are in position and secured with hose clips.
2 Using a mixture of 45% antifreeze and 55% water, fill the cooling system slowly, to minimise the risk of air locks. Ensure that the heater control is turned to the 'HOT' position otherwise an air lock may form in the heater.
3 On early models without an expansion tank, fill the radiator to the level given in Section 2 then refit the filler cap. Where an expansion tank is fitted, completely fill the radiator and refit the cap then half fill the expansion tank and refit the filler cap (photo). Note that the pressure cap with the double seal must always be fitted to the expansion tank.
4 On some engines an air bleed screw may be fitted in the heater hose. Loosen the screw before filling the system and retighten it when bubble free water flows.
5 Run the engine at a fast idle speed for several minutes then allow it to cool and top up the coolant level with reference to Section 2.

6 Antifreeze mixture

1 Apart from the protection against freezing conditions which the use of antifreeze provides, its use is essential to minimise corrosion of the cooling system.
2 The cooling system is initially filled with a solution of 45% antifreeze and it is recommended that this percentage is maintained.
3 With long-life types of antifreeze mixtures, renew the coolant every two years. With other types, drain and refill the system every twelve months. Whichever type is used, it must be of the ethylene glycol base.
4 The following table gives a guide to protection against frost but a mixture of less than 30% concentration will not give protection against corrosion:

Amount of antifreeze	Protection to
45%	$-32°C$ ($-26°F$)
40%	$-25°C$ ($-13°F$)
30%	$-16°C$ ($+3°F$)
25%	$-13°C$ ($+9°F$)
20%	$-9°C$ ($+16°F$)
15%	$-7°C$ ($+19°F$)

7 Radiator – removal, inspection, cleaning and refitting

1 Drain the cooling system as described in Section 3.
2 Remove the splash shield (if fitted) after removing its four bolts and four clips.
3 On automatic transmission models, place a tray beneath the oil cooler unions, wipe the unions clean and undo them. Quickly plug the pipe ends and the cooler unions, to prevent the loss of fluid.
4 Disconnect the top hose and where applicable the expansion tank hose from the radiator.
5 Where applicable, remove the screws securing the fan shroud to the radiator then position the shroud over the fan blades (photo).
6 For ease of removal also remove the front radiator cowl after removing the screws (photo).
7 Remove the four bolts, spring and flat washers which secure the radiator to the front panel and lift the radiator out carefully, so as not to damage the radiator matrix (photos).
8 With the radiator out of the car, reverse flush it to remove the sludge and clean the exterior with a jet of compressed air, or water, to remove any dirt or insects. If the radiator requires repair, it is better to fit an exchange radiator, or have the repair done by a radiator specialist.
9 If you are renewing an early radiator and you intend to fit a later slim top tank type, you will need to fit an expansion container. Consult your Ford dealer for the correct type and fitting details should this be the case, as a special service kit will also be required.
10 Refitting is a reversal of removal, but fill the cooling system with reference to Section 5. On automatic transmission models top up the transmission fluid with reference to Chapter 6.

5.3 Filling the radiator with coolant

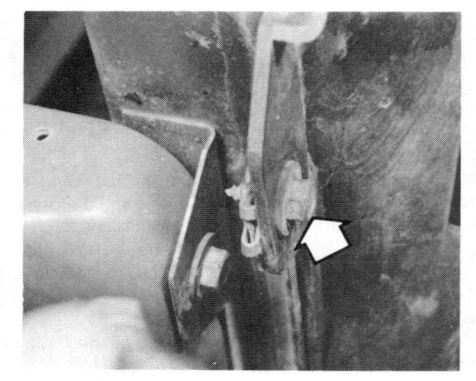

7.5 Fan shroud retaining screw (arrowed)

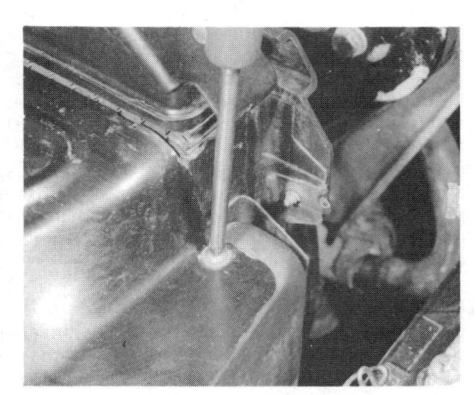

7.6 Removing the front radiator cowl

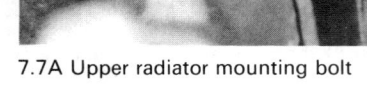

7.7A Upper radiator mounting bolt

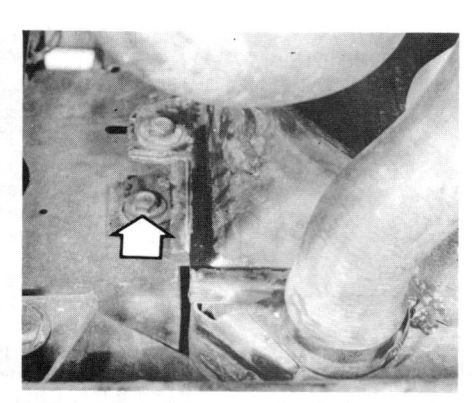

7.7B Lower radiator mounting bolt (arrowed)

7.7C Removing the radiator

Chapter 2 Cooling system

8 Thermostat – removal, testing and refitting

1 Drain the cooling system as described in Section 3.

3.0 litre engine

2 Slacken the clip of the top hose to the radiator and remove the hose (photo).
3 Undo and remove the two bolts and spring washers securing the thermostat housing to the cylinder head.
4 Tap the housing gently with a plastic-headed hammer, to break the gasket seal, then carefully lift the housing off. Remove the gasket in one piece if possible, otherwise scrape both joint faces to clean off the remains of the gasket.
5 Mark the thermostat to show which way round and which way up it is fitted, then use a screwdriver to remove it from the housing.

2.8 litre engine

6 Slacken the clips securing the radiator bottom hose and the heater hose to the thermostat housing and carefully remove the hoses (photo).
7 Undo and remove the three bolts and spring washers securing the thermostat housing.
8 Tap the thermostat housing with a soft-headed hammer to unstick it and lift the cover away (photo). Recover the front cover gasket if it is undamaged, otherwise scrape it off. Also remove the seal.
9 Carefully ease the thermostat out of its housing, being careful to note which way round it is fitted (photo).

All engines

10 Test the thermostat for correct functioning by suspending it on a string in a saucepan of cold water together with a thermometer. Heat the water and note the temperature at which the thermostat begins to open.
11 Compare the noted temperatures with those given in the Specifications at the beginning of this Chapter. If they differ considerably the thermostat must be renewed.
12 If the thermostat does not fully open in boiling water, or does not close down as the water cools, then it must be discarded and a new one fitted. Should the thermostat be stuck open when cold this will usually be apparent when removing it from the housing.
13 Refitting the thermostat is the reverse sequence to removal. Always ensure that the mating faces are clean and flat. If the thermostat housing is badly corroded fit a new housing. Always use a new gasket and seal if applicable (photo). Tighten the securing bolts to the specified torque wrench setting (photo). Refill the cooling system as described in Section 5.

9 Water pump – removal and refitting

1 Drain the cooling system as described in Section 3.

3.0 litre engine

2 Slacken the three alternator mounting bolts, push the alternator towards the engine, lift the drivebelt from the water pump pulley and remove it.
3 Slacken the lower radiator hose clip at the inlet connection to the pump. Carefully ease the hose from the water pump elbow.
4 Slacken the hose clip at the rear of the pump and carefully ease the hose from the pump.
5 Slacken the third hose clip situated at the top of the pump and carefully ease the hose from the top of the pump.
6 Undo the three bolts securing the water pump to the cylinder block. Note that each bolt is fitted with a spring washer and that, as the bolt is trapped by the fan pulley, it will be necessary to progressively slacken the bolts and pull the water pump away from the cylinder block. The location of the bolts should be noted, so that they are refitted in their original positions.
7 Once all three bolts have been unscrewed from the cylinder block the water pump may be removed and the gasket scraped off.

8.2 Thermostat housing and top hose (3.0 litre engine)

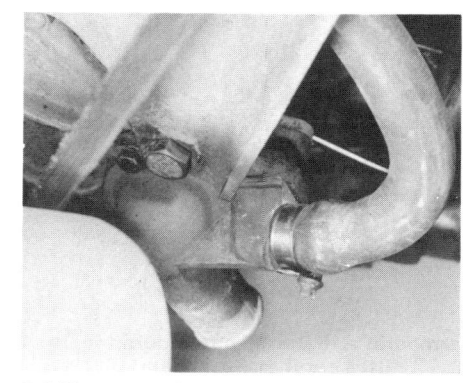

8.6 Thermostat housing with bottom hose disconnected and heater hose connected (2.8 litre engine)

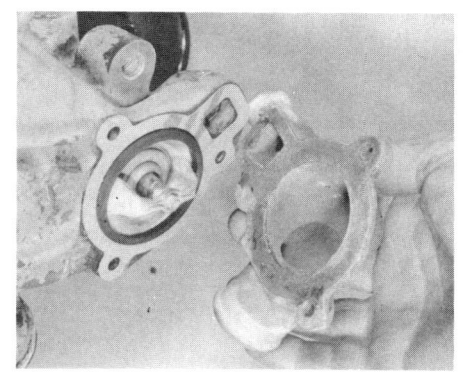

8.8 Removing the thermostat housing (2.8 litre engine)

8.9 Removing the thermostat (2.8 litre engine)

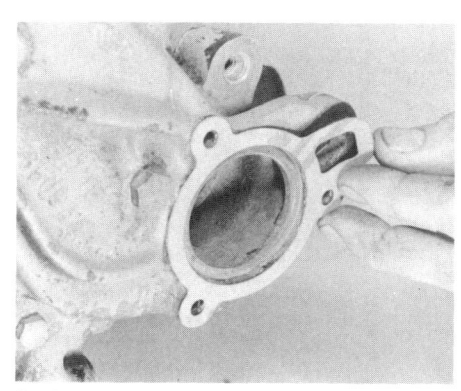

8.13A Fitting a new thermostat housing gasket (2.8 litre engine)

8.13B Tightening the thermostat housing bolts (2.8 litre engine)

Chapter 2 Cooling system

2.8 litre engine
8 Remove the radiator and shroud as described in Section 7.
9 Remove the drivebelt (Section 10).
10 Remove the fan and pulley (Section 11).
11 Remove the thermostat (Section 8).
12 Disconnect the hose from the thermostat rear elbow.
13 Unbolt the water pump from the engine, noting the location of each bolt as they are of various lengths (photos).
14 Scrape off the gasket.

All engines
15 Refitting is a reversal of removal, but clean the mating faces and always use a new gasket (photo). Tighten nuts and bolts to the specified torque. Refill the cooling system as described in Section 5.

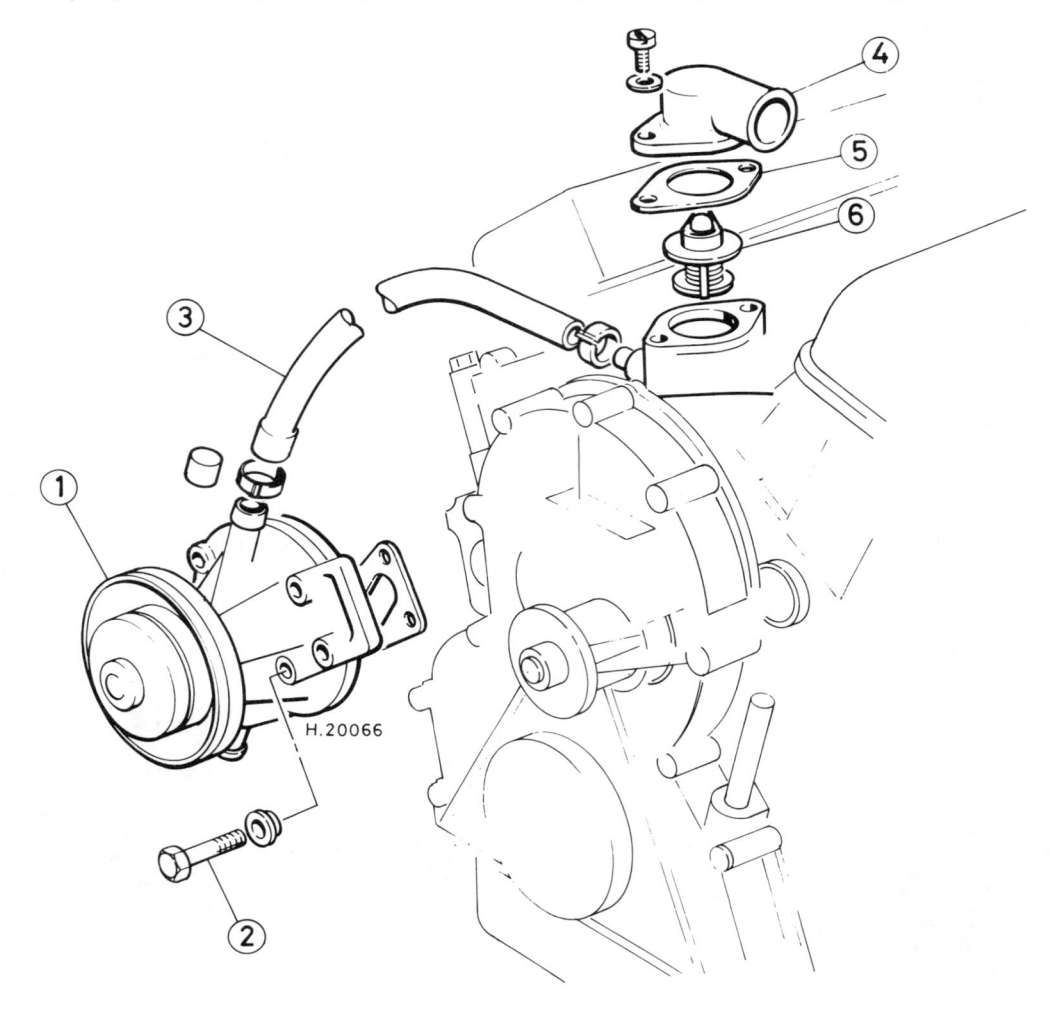

Fig. 2.4 Thermostat and water pump components on the 3.0 litre engine (Secs 8 and 9)

1 Water pump
2 Bolt
3 Transfer hose
4 Housing
5 Gasket
6 Thermostat

9.13A Unbolting the water pump (2.8 litre engine)

9.13B Water pump and impeller (2.8 litre engine)

9.15 Fitting the water pump gasket (2.8 litre engine)

Chapter 2 Cooling system

10 Drivebelts – removal and refitting

1 Where applicable, remove the power steering pump drivebelt with reference to Chapter 10.
2 Loosen the alternator pivot and adjustment bolts, swivel the alternator in to the engine, then slip the belt(s) from the pulleys.
3 Locate the new belt(s) on the pulleys, then tension with reference to Section 2.
4 Refer to Chapter 10 when refitting the power steering pump drivebelt.

11 Viscous drive fan – removal and refitting

1 For best access to the fan, first remove the radiator as described in Section 7.
2 Remove the drivebelt(s) (Section 10).
3 Where the fan has a central bolt, unscrew this and withdraw the fan from the extension shaft.
4 If the fan is secured by a concealed nut, ideally a 32 mm (1¼ in AF) spanner modified as shown in Fig. 2.5 should be used. The thickness of the spanner jaws must not exceed 5 mm (0.2 in), and it is important to realise that the nut has a **left-hand** thread. Alternatively, it is possible to unscrew two of the pulley bolts so that a spanner of normal thickness can be used to unscrew the nut.
5 With the fan assembly removed, unbolt the fan from the viscous coupling (photo).
6 If necessary unbolt the pulley (and extension shaft where applicable) from the hub flange (photo).
7 Refitting is a reversal of removal, but tighten the bolts/nut to the specified torque. Refit the radiator with reference to Section 7 and the drivebelt(s) with reference to Section 10.

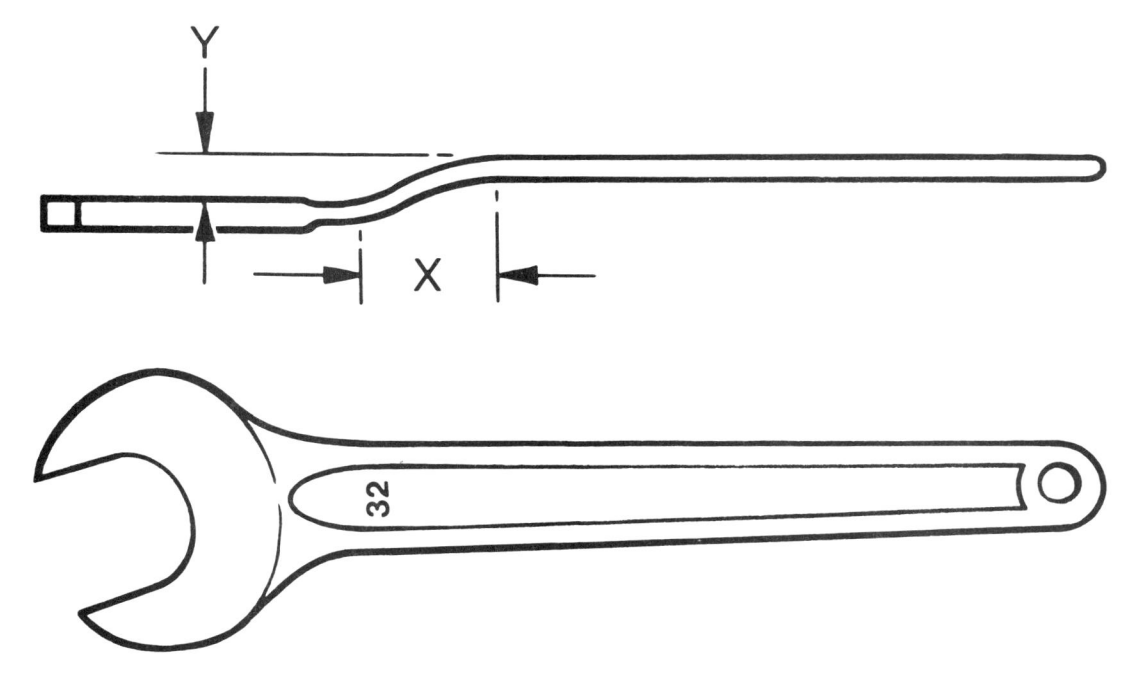

Fig. 2.5 Manifold spanner for removal of the viscous drive fan (Sec 11)

$X = 25.0$ mm (1.0 in) $Y = 12.0$ mm (0.5 in)

11.5 Bolts securing fan to viscous coupling

11.6 Unbolting the water pump pulley (2.8 litre engine)

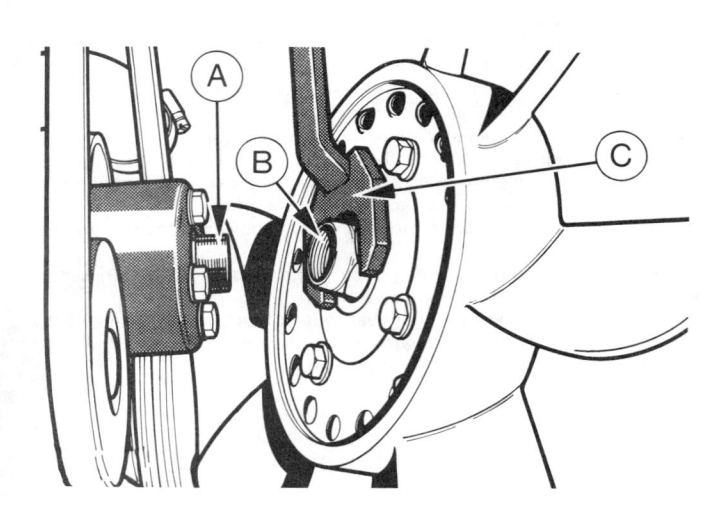

Fig. 2.6 Removing the viscous drive fan (Sec 11)
A Threaded hub
B Locknut (LH thread)
C Modified spanner

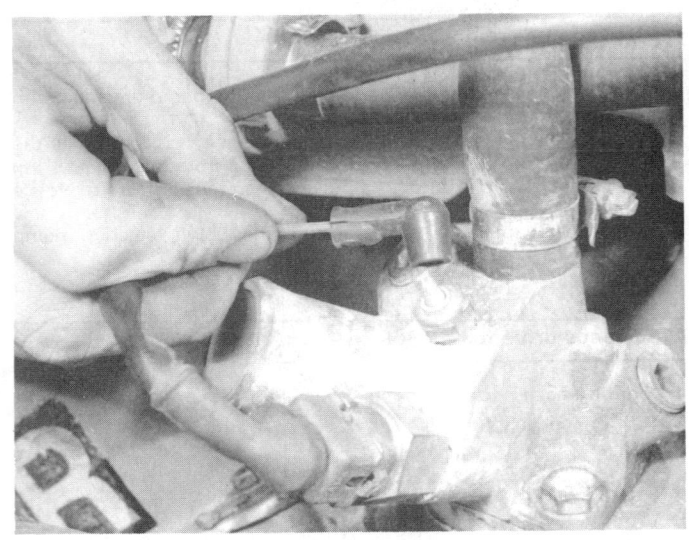

13.3 Disconnecting the wiring from the temperature gauge sender unit

12 Fan hub bearing (3.0 litre engine) – renewal

1 Remove the drivebelt (Section 10) then unbolt the fan and pulley. Where a viscous drive fan is fitted, refer to Section 11, but where a standard fan is fitted note the location of the spacers.
2 Remove the engine front cover (see Chapter 1).
3 Remove the circlip which retains the bearing in the housing, then press the bearing, expansion plug and hub out of the front cover.
4 Press the shaft out of the hub.
5 Press a new shaft and bearing assembly into the housing, so that the circlip grooves are in alignment and refit the circlip.
6 Press on the hub until its front face is 85.8 mm (3.38 in) from the rear face of the front cover.
7 Fit a new expansion plug to the shaft bore in the rear face of the bearing housing.
8 Refit the front cover, pulley and fan.
9 Refit the drivebelt with reference to Section 10.

13 Temperature gauge sender unit – removal and refitting

1 The temperature gauge sender unit is located on the outlet elbow at the front of the engine.
2 Refer to Section 3 and drain the cooling system sufficiently to bring the level below the outlet elbow.
3 Disconnect the wiring from the unit (photo).
4 Unscrew the sender unit from the elbow.
5 Refitting is a reversal of removal, but top up the system with reference to Section 5.

14 Expansion tank – removal and refitting

1 Refer to Section 3 and drain the cooling system sufficiently to bring the level below the radiator top hose.
2 Unscrew the cross-head mounting screws and lift the tank from the bracket (photo).
3 Disconect the hoses and remove the expansion tank (photos).
4 Refitting is a reversal of removal, but refill the cooling system as described in Section 5.

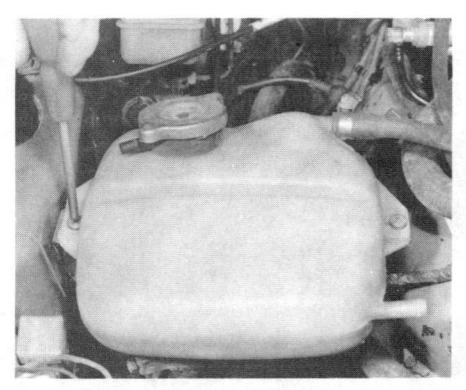

14.2 Removing the expansion tank mounting screws

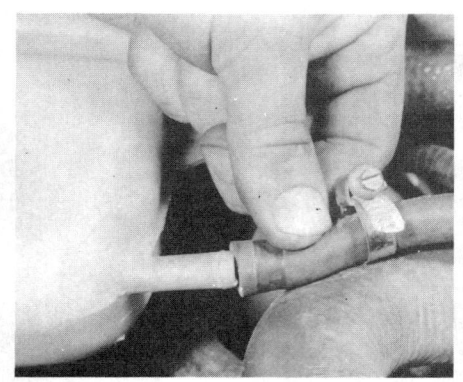

14.3A Disconnecting the side hose ...

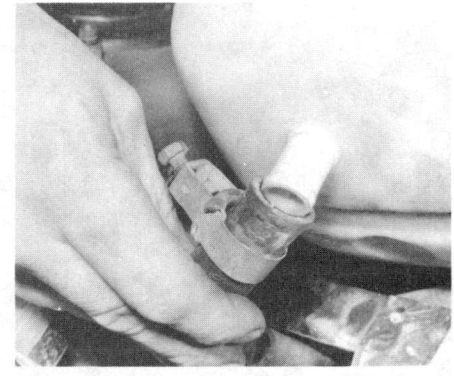

14.3B ... and bottom supply hose from the expansion tank

15 Fault diagnosis – cooling system

Symptom	Reason(s)
Overheating	Insufficient coolant
	Drivebelt slipping
	Radiator core blocked, or grille restricted
	Air lock in cooling system
	Water hose collapsed, or kinked, impeding flow
	Thermostat defective
	Ignition timing incorrect
	Viscous fan inoperative
	Incorrect, or defective pressure cap
Engine runs too cool	Thermostat defective, or no thermostat
	Wrong thermostat fitted
Loss of coolant	Defective pressure cap
	Leaking hoses
	Leak in radiator, or heater core
	Thermostat or pump gasket leak
	Blown cylinder head gasket
	Cracked cylinder block, or head

Chapter 3 Fuel and exhaust systems

Contents

Part A: Carburettor system

Accelerator cable – removal and refitting	17
Accelerator shaft and pedal – removal and refitting	18
Air cleaner – removal and refitting	3
Carburettor – general	7
Carburettor – removal and refitting	8
Exhaust manifold – removal and refitting	20
Exhaust system – removal and refitting	21
Fuel pump – cleaning	5
Fuel pump – removal and refitting	6
Fuel pump – testing	4
Fuel tank – cleaning and repair	14
Fuel tank – removal and refitting	13
Fuel tank filler pipe – removal and refitting	16
Fuel tank sender unit – removal and refitting	15
General description	1
Inlet manifold – removal and refitting	19
Routine maintenance	2
Weber carburettor – dismantling and reassembly	9
Weber carburettor – idling adjustment	12
Weber carburettor automatic choke – adjustment	11
Weber carburettor automatic choke – removal, overhaul and refitting	10

Part B: Fuel injection system

Accelerator cable – removal and refitting	28
Air cleaner – removal and refitting	24
Exhaust manifold – removal and refitting	30
Exhaust system – removal and refitting	31
Fuel tank, sender unit and filler pipe – general	27
General description	22
Idle mixture – adjustment	26
Inlet manifold – removal and refitting	29
Main system components – removal and refitting	25
Routine maintenance	23

Part C: Fault diagnosis

Fault diagnosis – fuel system (carburettor)	32
Fault diagnosis – fuel system (fuel injection)	33

Specifications

Part A: Carburettor system

General

System type	Dual venturi downdraught carburettor, mechanically operated diaphragm fuel pump, temperature controlled air cleaner on same models
Fuel pump delivery pressure	0.25 to 0.35 kgf/cm² (3.7 to 5.0 lbf/in²)
Fuel tank capacity	58.0 litre (12.7 gal)

Carburettor

Type	Weber dual venturi with automatic choke
Throttle barrel diameter	38 mm
Venturi diameter	27 mm
Main jet:	
1974/1975	145 mm or 142 mm
1976 onwards	142 mm
Idling speed:	
1974/1975	800 ± 25 rpm
1976	825 ± 25 rpm
1977 onwards	850 ± 25 rpm
Fast idle speed:	
1976	2200 ± 100 rpm
1974/1975 and 1977 onwards	2000 ± 100 rpm
Float level:	
Brass float	40.0 mm (1.575 in)
Plastic float	34.3 mm (1.350 in)
Float travel	12.5 ± 0.49 mm (0.492 ± 0.02 in)
CO % at idle:	
1974/1975	2.8 ± 0.2
1976	1.75
1977 onwards	1.0
Maximum choke vacuum pull-down:	
1974/1975	5.0 ± 0.25 mm (0.197 ± 0.01 in)
1976	7.0 ± 0.25 mm (0.276 ± 0.01 in)
1977 onwards	5.5 ± 0.25 mm (0.217 ± 0.01 in)
Choke phasing:	
1974/1975	2.3 ± 0.25 mm (0.091 ± 0.01 in)
1976	3.0 ± 0.25 mm (0.118 ± 0.01 in)
1977 onwards	2.8 ± 0.25 mm (0.110 ± 0.01 in)

Chapter 3 Fuel and exhaust systems

Torque wrench settings

	Nm	lbf ft
Fuel pump	16 to 20	12 to 15
Inlet manifold:		
Stage 1	4 to 8	3 to 6
Stage 2	8 to 15	6 to 11
Stage 3	18 to 22	13 to 16
Stage 4 (after running engine for 15 minutes at 1000 rpm)	18 to 22	13 to 16
Exhaust manifold to downpipe	20 to 27	15 to 20
Exhaust U-bolts and clamps	38 to 45	28 to 33
Carburettor	20 to 24	15 to 18

Part B: Fuel injection system

General

System type	Bosch K-Jetronic continuous injection
Idle speed	900 ± 25 rpm
CO % at idle	1.25 ± 0.2
Fuel pump delivery (minimum)	930.0 cc every 30 seconds
System pressure	5.55 ± 0.25 kgf/cm² (78.9 ± 3.6 lbf/in²)
Fuel tank capacity	58.0 litre (12.7 gal)

Torque wrench settings

	Nm	lbf ft
Fuel distributor	32 to 38	24 to 28
Warm-up regulator	7 to 10	5 to 7
Start valve	7 to 10	5 to 7
Auxiliary air device	7 to 10	5 to 7
Air chamber	7 to 9	5 to 7
Banjo bolts		
Fuel distributor	18 to 20	13 to 15
Warm-up regulator (M10)	11 to 15	8 to 11
Warm-up regulator (M8)	5 to 8	4 to 6
Injection pipes	5 to 8	4 to 6
Fuel pump, filter and accumulator	18 to 20	13 to 15

PART A: CARBURETTOR SYSTEM

1 General description

The fuel system consists of a rear mounted fuel tank, a diaphragm type fuel pump driven from the camshaft and a dual venturi downdraught carburettor. There are two separate exhaust assemblies, each consisting of a front pipe and resonator, intermediate pipe and silencer, and a tail pipe and silencer. The air cleaner on later models incorporates a temperature control flap.

Some models may be fitted with a parted vacuum switch (PVS) to prevent flat spots and misfiring occurring when the engine is cold and running with full or partial choke. When the engine is cold (below 71°C (160°F)) the PVS is closed. In this condition, carburettor vacuum is applied directly to the distributor via the non-return valve, which allows the vacuum advance to be maintained under acceleration, preventing flat spots or misfiring.

When the engine has reached its normal operating temperature the PVS is open. In this condition the carburettor is connected to the distributor via the PVS.

2 Routine maintenance

At the intervals specified in the Routine Maintenance section in the front of the manual carry out the following procedures.

Check and adjust the idling speed and mixture
1 Refer to Section 12.

Check condition of fuel hoses
2 Examine all the fuel hoses in the engine compartment and on the fuel tank, for signs of leakage or deterioration.
3 Check the security of all fuel hoses. Tighten the inlet manifold bolts.
4 Refer to Section 19 and tighten the inlet manifold bolts in the sequence given.

Renew the air filter element
5 Disconnect the battery.
6 Remove the two nuts securing the air cleaner lid and remove the lid. On later models the lid may be secured with spring fasteners.
7 Lift out the paper element and discard it.
8 Carefully clean the inside of the air cleaner body, ensuring that no dirt is allowed to drop into the carburettor.
9 Fit a new element and refit the air cleaner lid.
10 Reconnect the battery.

Check air cleaner temperature control (where fitted)
11 By looking through the main air cleaner spout, note the position of the flap valve. With the engine stationary the flap should be fully closed.
12 Start the engine and check the operation of the flap valve. With the engine idling the flap should open and stay open to allow hot air into the air cleaner. **Note:** *It is essential that this test is carried out with the engine cold.* If the flap remains closed when the engine is started then there is a fault in either the diaphragm unit or heat sensor and they should be tested as follows.
13 Check the vacuum supply pipe for leaks and renew it if suspect.

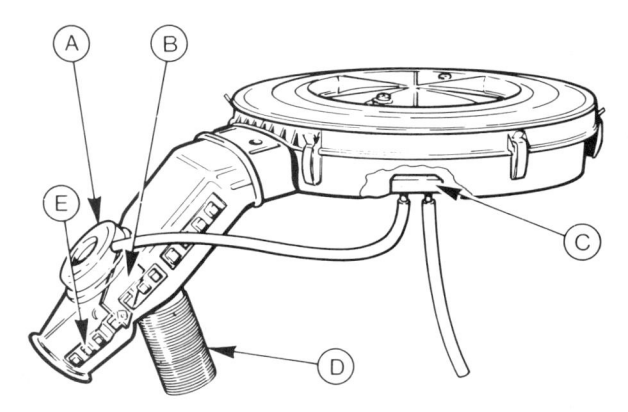

Fig. 3.1 Air temperature controlled air cleaner (Sec 2)

  A Vacuum motor   D Hot air intake
  B Flap valve   E Cold air intake
  C Heat sensor

14 Disconnect the vacuum supply pipe from the heat sensor. Observe the operation of the flap valve while at the same time applying suction to the end of the supply pipe. If the flap valve remains closed then either the diaphragm or the flap valve is faulty. If the flap now opens, the fault lies in the heat sensor.

3 Air clear – removal and refitting

1 Disconnect the battery.
2 Remove the two nuts securing the air cleaner lid and remove the lid, then take out the air cleaner element.
3 Knock back the four lock tabs and remove the four nuts securing the air cleaner to the carburettor body.
4 Lift the air cleaner body off the carburettor and disconnect the hot air tube. Where applicable also disconnect the vacuum hose.
5 Refitting is the reverse of removal. Use new locktabs if the original ones are suspect and fit a new paper element unless the existing one is very clean.

4 Fuel pump – testing

1 If there are no obvious leaks, a preliminary check of fuel pump operation can be made by disconnecting the outlet pipe of the fuel pump and connecting the pump outlet by a length of tube to a glass jar.
2 Crank the engine and observe whether a spurt of petrol emerges from the pump outlet about every second.
3 To check that the pump is within its specification, connect a pressure gauge to the pump outlet. Disconnect the HT lead from the coil, crank the engine and observe whether the outlet pressure is as specified.

5 Fuel pump – cleaning

1 Detach the fuel pipe from the pump inlet tube.
2 Undo and remove the centre screw and O-ring and lift off the sediment cap, filter and seal.
3 Thoroughly clean the sediment cap, filter and pumping chamber using a paintbrush and clean petrol to remove any sediment.
4 To reassemble is the reverse sequence to dismantling. Do not overtighten the centre screw as it could distort the sediment cap.

6 Fuel pump – removal and refitting

1 Loosen the hose clips from the two connections on the pump and remove the hoses. If crimped type clips are fitted cut them free and replace them with screw type clips.
2 Undo and remove the two nuts and spring washers securing the pump to the timing cover and lift the pump off. Remove the gasket.
3 Refitting the pump is the reverse of removal, but clean the mating faces and fit a new gasket.

7 Carburettor – general

The carburettor is of the Weber dual venturi downdraught type with an automatic choke to ensure easy starting from cold. The main and idling systems are duplicated, each barrel having separate systems supplied from a common float chamber.
A single accelerator pump of the diaphragm type supplies both barrels.
The main body of the carburettor incorporates the float chamber, throttle barrels, main venturi and accelerator pump body.
There are two throttle spindles, which are interconnected by gear quadrants to operate the throttle plates simultaneously. Each bank of cylinders is fed from a separate throttle barrel through ducts cast in the inlet manifold.
Incorporated in the float chamber cover is the fuel feed inlet connection and the twin air intakes for the two barrels.

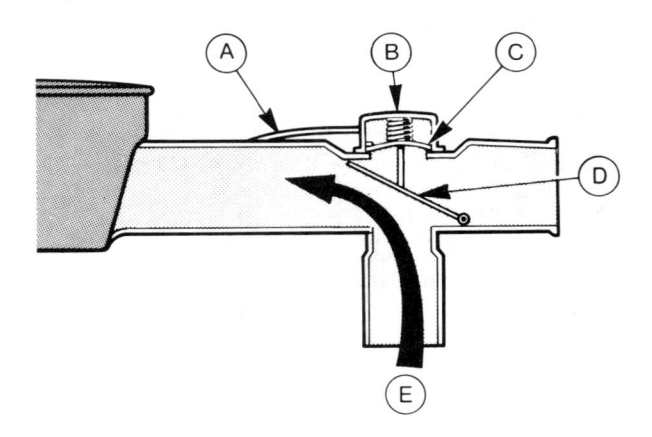

Fig. 3.2 Diagram of air temperature controlled air cleaner (Sec 2)

A Vacuum hose D Flap valve
B Diaphragm unit E Hot air flow
C Diaphragm

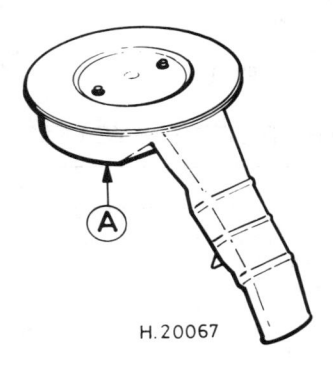

Fig. 3.3 Air cleaner (A) (Sec 3)

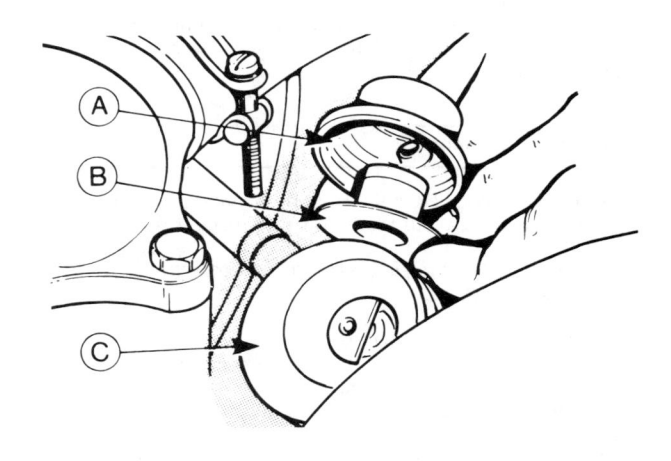

Fig. 3.4 Fuel pump filter removal (Sec 5)

A Cap B Filter C Pump body

Chapter 3 Fuel and exhaust systems

The fully automatic choke system is located on the right-hand side of the carburettor body and comprises a bi-metallic spring and a linkage to two offset choke plates mounted on a common spindle. The choke housing is heated by engine coolant.

8 Carburettor – removal and refitting

1 Disconnect the battery.
2 Remove the air cleaner assembly as detailed in Section 3.
3 Make sure that the cooling system is not under pressure, by removing the radiator cap and refitting it. With the radiator cap in position, disconnect the two hoses from the automatic choke and secure them with their open ends upwards. This will ensure only a very small quantity of coolant will be lost.
4 Disconnect the throttle link.
5 Disconnect the fuel feed and return pipes from the carburettor. If a crimped type hose clamp is fitted, cut the clamp to remove it and renew it with a screw type hose clamp.
6 Disconnect the vacuum pipe from the carburettor to the distributor.
7 Remove the four nuts and detach the carburettor assembly and gasket from the intake manifold. Note the position of the deflector plate.
8 Refitting is the reverse of the removal procedure, but first ensure that the joint faces are clean and undamaged. Use a new gasket, tighten the securing nuts to the torque wrench setting given in the Specifications and refit all the hoses. Top up the cooling system if necessary. Adjust the idle speed and mixture as described in Section 12.

9 Weber carburettor – dismantling and reassembly

Note: *On 1977 models onwards the carburettor has a 'by pass' idle adjusting screw which is sealed in production with a white or grey plastic cap. Under no circumstances should this screw be tampered with (Fig. 3.22).*

1 Remove the carburettor from the engine, as detailed in Section 8, and clean the carburettor exterior with petrol or a water soluble proprietary cleaner.
2 Carefully prise out the U-circlip with a screwdriver and disconnect the choke plate operating link.
3 Remove the six screws and detach the carburettor upper body.
4 Unscrew the brass nut located at the fuel intake and detach the fuel filter.
5 Tap out the float retaining pin, and detach the float and needle valve.
6 Remove the three screws and detach the power valve diaphragm assembly.
7 Unscrew the needle valve housing.
8 Unscrew the jet and jet plugs from the carburettor body, noting the positions in which they are fitted.
9 Remove the two emulsion tubes after removing the two main air correction jets.
10 Remove four screws, take off the accelerator pump cover and remove the diaphragm and spring. Similarly on automatic transmission models remove the anti-stall diaphragm.
11 Clean all the components and then ensure that the jets are unobstructed by blowing through them.
12 Carefully screw in the mixture screws until each of them contacts

Fig. 3.5 U-circlip on the choke operating link (Sec 9)

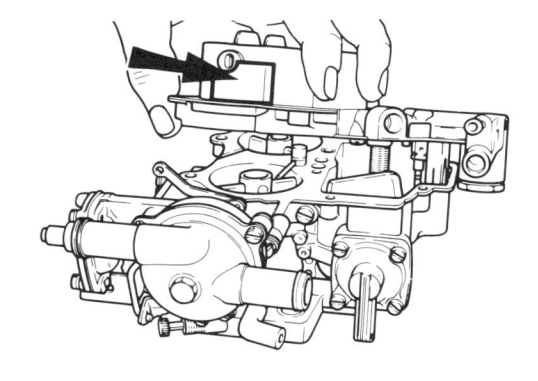

Fig. 3.6 Removing the upper body – arrowed (Sec 9)

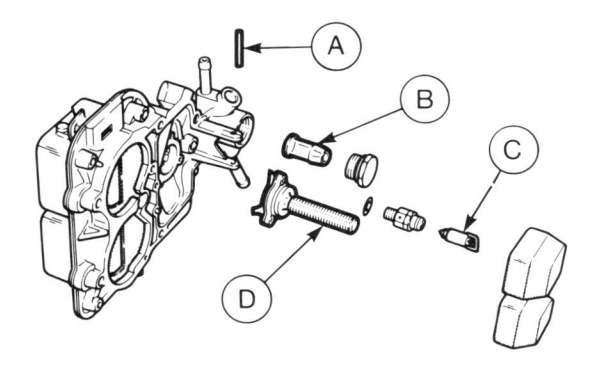

Fig. 3.7 Upper body components (Sec 9)

| A | Float retaining pin | C | Needle valve |
| B | Filter | D | Power valve |

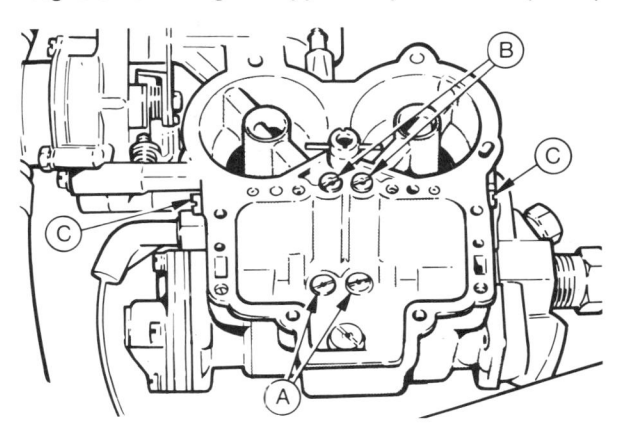

Fig. 3.8 Jet locations (Sec 9)

| A | Main jets | C | Idle jets |
| B | Air correction jets | | |

its sealing. Note and record the number of turns, so that the screws can be refitted to their original settings. Unscrew each screw fully and remove it and its spring.

13 Check the float assembly for signs of damage and ensure that the floats do not leak. Leaking floats can be detected by submerging the float in a pan of water. When heating the water, leaks will be indicated by bubbles rising from the float. Inspect the power valve, pump diaphragm and gaskets for splits. Check that the throttle plates and mixture screw needles are not worn, or damaged. Fit new parts where necessary.

14 Start reassembly by refitting the parts of the accelerator pump in the order shown in Fig. 3.12. Also refit the anti-stall diaphragm assembly.

15 Fit the two mixture screws and springs, setting each screw to its original position and refit the jets, jet plugs, emulsion tubes and air correction jets.

16 Refit the power valve diaphragm assembly as follows. Loosely fit the three diaphragm retaining screws, compress the return spring ensuring that the diaphragm is not twisted or distorted then tighten the screws and release the spring. Check for correct functioning by pushing the diaphragm down and then blocking the air bleed hole with a finger. Release the diaphragm while still keeping the air hole blocked. If the diaphragm stays down, it has sealed properly to its housing.

17 Refit the needle valve housing, needle valve and float assembly to the upper body.

18 Adjust the float level settings, firstly with the upper body vertical and the needle valve shut off. If the measurement is outside the limits given in the Specifications, bend the tag until the setting is correct and recheck.

19 Check the float travel by measuring from the body face when the upper body is horizontal. If the measurement is outside the limits given in the Specifications, bend the tag until the setting is correct and recheck. For both the float level adjustments the upper body gasket must be removed.

20 Refit the fuel intake filter and locating nut.

21 Position the gasket and refit the carburettor upper body. Make sure that the choke link locates through the upper body correctly.

22 Reconnect the choke link and secure it with the U-clip.

23 Adjust the throttle plates for synchronization. Wind back the idle screw until it is clear of the throttle mechanism and then loosen the synchronization adjusting screw. Hold the auto-choke plates in an open position, flick the throttle and allow both plates to close fully. Tap both plates to ensure that they have closed completely and then tighten the synchronization screw. To check, partially open the throttle using the idle screw, so that there is a clearance of 0.5 mm (0.02 in) between one of the throttle plates and its bore. Check that the other plate has the same opening.

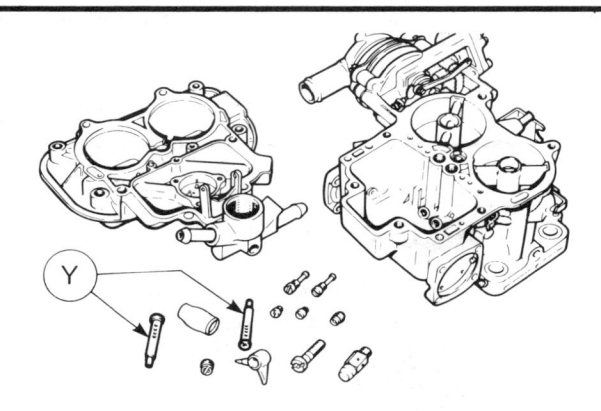

Fig. 3.9 Emulsion tubes (Y) (Sec 9)

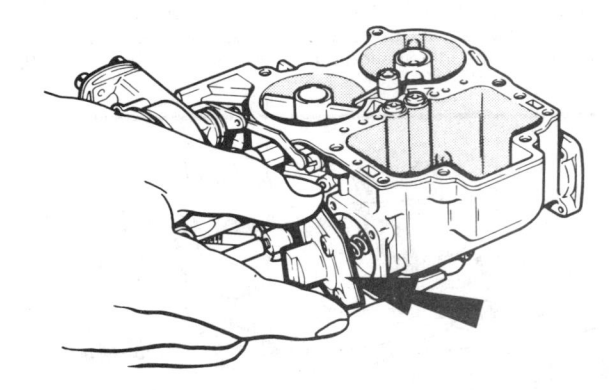

Fig. 3.10 Removing the accelerator pump cover – arrowed (Sec 9)

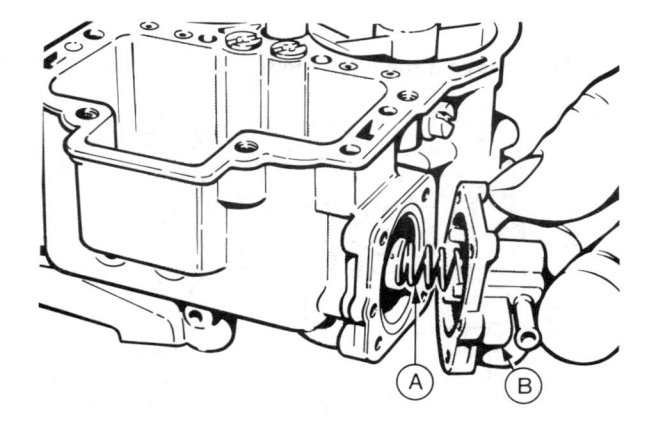

Fig. 3.11 Removing the anti-stall diaphragm cover (Sec 9)

A   Return spring   B   Cover

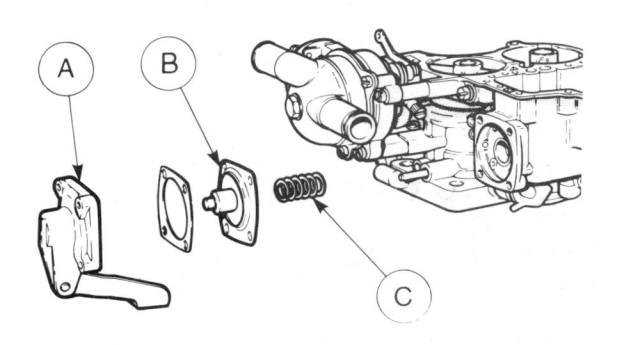

Fig. 3.12 Accelerator pump components (Sec 9)

A   Cover   C   Return spring
B   Diaphragm

Chapter 3 Fuel and exhaust systems

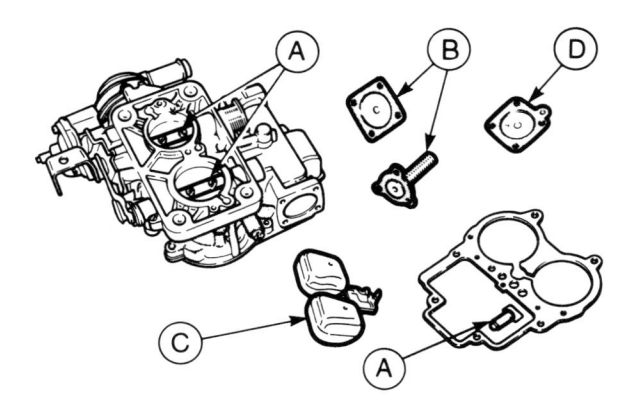

Fig. 3.13 Components to be checked (Sec 9)

A Check for wear and damage
B Check for splitting
C Check for leaking
D Check for splitting

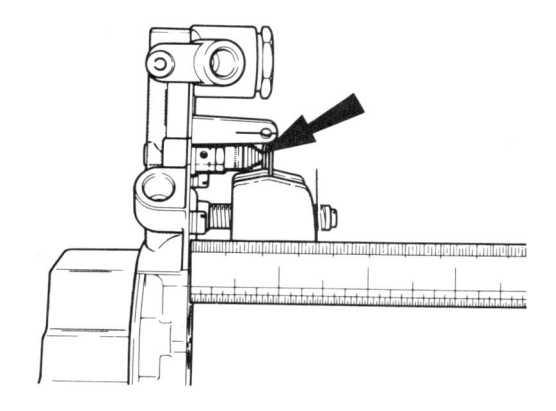

Fig. 3.14 Float level adjustment tag – arrowed (Sec 9)

10 Weber carburettor automatic choke – removal, overhaul and refitting

1 Disconnect the battery negative lead.
2 Remove the air cleaner as described in Section 3.
3 Unscrew the centre bolt and remove the cover.
4 Remove the three screws, detach the housing and unhook the bi-metallic spring.
5 Detach the internal heat shield.
6 Remove the single U-clip and disconnect the choke plate operating link.
7 Remove the three screws, disconnect the choke link at the operating lever and detach the choke assembly.
8 Remove the three screws and detach the vacuum diaphragm assembly.
9 Dismantle the remaining parts of the choke mechanism.
10 Clean all the components, inspect them for wear and damage and wipe them dry with a lint-free cloth. Do not use any lubricants during reassembly.
11 Reassemble the choke mechanism.
12 Refit the vacuum diaphragm and housing, ensuring that the diaphragm is flat before the housing is fitted.
13 Ensure that the O-ring is correctly located in the choke housing then reconnect the lower choke link. Position the assembly and secure it with the three screws; ensure that the upper choke link locates correctly through the carburettor body.
14 Reconnect the upper choke link to the choke spindle.
15 Check the vacuum pull-down and choke phasing, as described in Section 11.
16 Refit the internal heat shield ensuring that the hole in the cover locates correctly onto the peg cast in the housing.
17 Connect the bi-metal spring to the choke lever, position the housing and loosely fit the three retaining screws.
18 Rotate the housing until the marks are aligned, then tighten the three screws.
19 Refit the cover together with a new gasket and tighten the centre bolt. Top up the cooling system if necessary.
20 Reconnect the battery, run the engine and adjust the fast idle speed, as described in Section 11.
21 Refit the air cleaner.

Fig. 3.15 Automatic choke housing retaining screws – arrowed (Sec 10)

11 Weber carburettor automatic choke – adjustment

Note: *The procedure is described for a carburettor which is fitted in the car, but with the exception of fast idle speed adjustment, all tasks can be carried out on the bench if required where the carburettor has been removed.*

1 Disconnect the battery earth lead.
2 Remove the air cleaner, as described in Section 3.

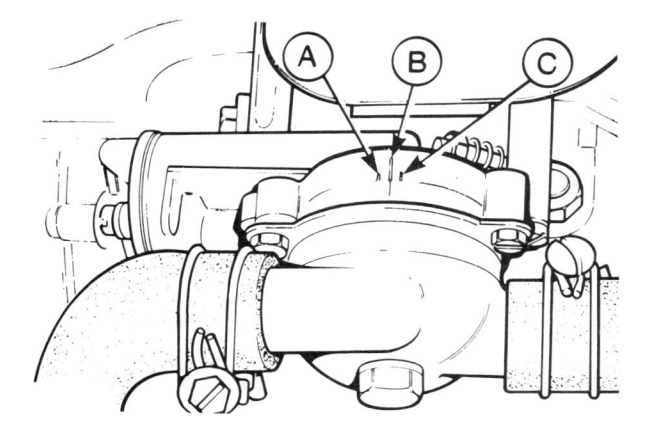

Fig. 3.16 Automatic choke housing alignment marks (Sec 10)

A Rich
B Normal setting
C Weak

Chapter 3 Fuel and exhaust systems

3 Unscrew the centre bolt and remove the cover. To prevent loss of coolant first remove and refit the pressure cap to release any pressure but do this only with the engine cold.
4 Remove the three screws, detach the housing and move it clear of the carburettor. For access to the lower screw it will be necessary to make up a suitable cranked screwdriver.
5 Detach the internal heat sheild.

Vacuum pull-down
6 Fit an elastic band to the choke plate lever and position it so that the choke plates are held closed. Open, then release, the throttle to ensure that the choke plates close fully. Unscrew the plug from the diaphragm unit then manually push open the diaphragm up to its stop from inside the choke housing. Do not push on the rod as it is spring-loaded, but push on the diaphragm plug body. The choke plate pull-down should now be measured, using an unmarked twist drill shank between the edge of the choke plate and the air horn wall, and compared with the specified figure. Adjust, if necessary, by screwing the adjusting screw in or out, using a short bladed screwdriver. Refit the end plug and detach the elastic band on completion.

Choke phasing
7 Hold the throttle partly open and position the fast idle cam so that the fast idle adjusting screw locates on the upper section of the cam. Release the throttle to hold the cam in this position, then push the choke plates down until the step on the cam jams against the adjusting screw. Measure the clearance betwen the edge of the choke plate and the air horn wall using a specified size drill. Adjust, if necessary, by bending the tag.
8 Refit the internal heat shield ensuring that the hole in the cover locates correctly onto the peg cast in the housing.
9 Connect the bi-metal spring to the choke lever, position the choke housing and loosely fit the three retaining screws.
10 Rotate the housing until the marks are aligned then tighten the three screws.
11 Refit the cover, together with a new gasket, and tighten the centre bolt. Top up the cooling system if necessary.
12 Reconnect the battery, run the engine and adjust the fast idle speed as described in the following paragraph.

Fast idle speed
13 Ideally a tachometer will be required in order to set the fast idle rpm to the specified value. Run the engine up to normal operating temperature, then switch off and connect the tachometer (where available). Open the throttle partially, hold the choke plates fully closed then release the throttle so that the choke mechanism is held in the fast idle position. Release the choke plates, checking that they remain fully open (if they are not open, the assembly is faulty or the engine is not at operating temperature). Without touching the accelerator pedal, start

Fig. 3.17 Checking the vacuum pull-down with a drill (Sec 11)

A Elastic band
B Drill
C Vacuum operated rod

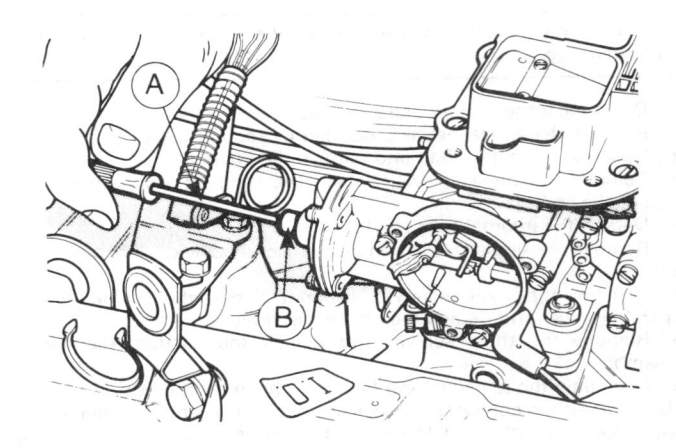

Fig. 3.18 Adjusting the vacuum pull-down (Sec 11)

A Screwdriver   B Diaphragm housing

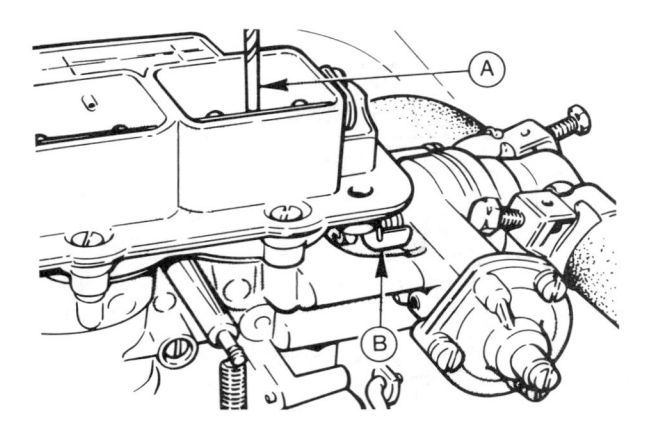

Fig. 3.19 Choke phasing adjustment (Sec 11)

A Drill   B Adjusting tag

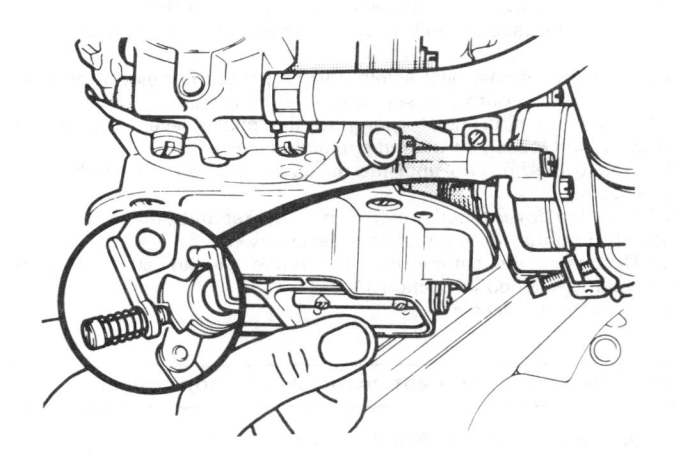

Fig. 3.20 Linkage set for checking fast idle adjustment (Sec 11)

Chapter 3 Fuel and exhaust systems

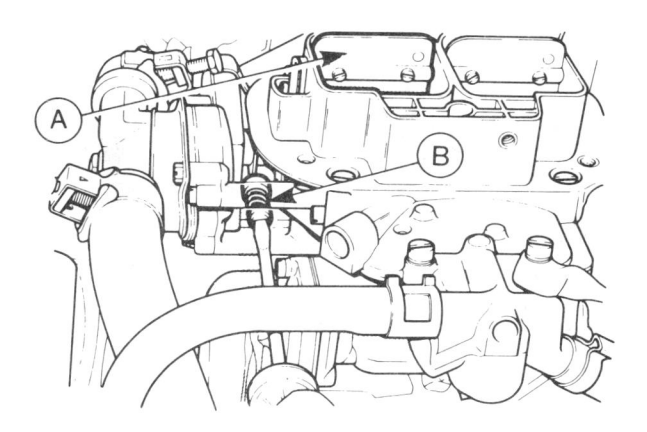

Fig. 3.21 Adjusting the fast idle speed (Sec 11)

A Choke fully open B Adjustment screw

the engine and adjust the fast idle screw as necessary to obtain the correct fast idle rpm.
14 Finally refit the air cleaner.

12 Weber carburettor – idling adjustment

Note: *In view of the increasing awareness of the dangers of exhaust pollution and the very low levels of carbon monoxide (CO) emission for which this carburettor is designed, the slow running mixture setting and the basic idle setting on Weber carburettors should not be adjusted without the use of a proper CO meter (exhaust gas analyser). Before commencing work also read the note at the beginning of Section 9.*

1 Warm up the engine to its normal operating temperature.
2 Connect a CO meter and a tachometer, if the latter is not already fitted to the car, according to manufacturer's instructions.
3 Clear the engine exhaust gases by running the engine at 3000 rpm for approximately 30 seconds and allow the engine to idle.
4 Wait for the meter to stabilise and compare the CO and idle speed readings against those given in the Specifications at the beginning of this Chapter.
5 Adjust the idle speed screw to give the correct rpm.
6 During normal routine maintenance servicing, normally no adjustment of the mixture (CO level) will be required. If however the CO level is found to be incorrect the following procedure should be adopted.
7 Remove the plastic cap covering each mixture screw (where applicable) by punching a small hole in the centre of each cap and prising out with a screwdriver. Fully wind in the two mixture screws until they stop and then wind out three full turns on each screw. Turn each screw both in and out by increments of a quarter of a turn at a time until the best running conditions are obtained. It is important that both screws are turned in the same direction and by the same amount each time. This will give an approximate setting for the mixture. Now clear the engine exhaust gases by running the engine at 3000 rpm for approximately 30 seconds and allow the engine to idle.
8 Adjust the mixture screws and the idle speed screw until the correct idle speed and CO reading are obtained. If the correct readings are not obtained within 10 to 30 seconds clear the engine exhaust gases as described in paragraph 3 and repeat the adjustment procedure until correct readings are obtained.
9 Where applicable fit new plastic caps to the mixture screws. Remove the tachometer and CO meter.

13 Fuel tank – removal and refitting

1 Great care is necessary when working on any part of the fuel system and it is safer to do the work outside if possible. Avoid having the car over a pit, where explosive vapour can accumulate and take care to avoid naked flames and any other sources of ignition.

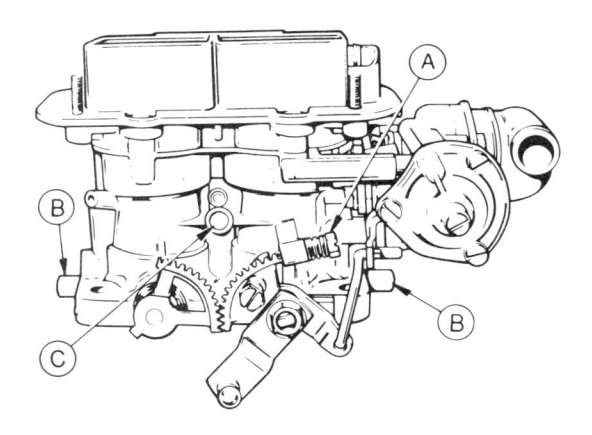

Fig. 3.22 Carburettor adjustment screws (Sec 12)

A Idle speed screw C Sealed plug (do not remove)
B Mixture screws

2 Disconnect the battery leads.
3 Syphon any petrol remaining in the tank into a metal container suitable for storing petrol. Label the container clearly and store it in a well ventilated place.
4 Choke the front wheels securely. Jack up the rear of the car and support it on blocks, or firmly based stands.
5 Disconnect the fuel feed pipe at the tank and free it from the clips along the front edge of the tank. Also disconnect the return pipe where fitted.
6 Remove the wiring connectors from the tank sender unit, noting their positions so that they can be reconnected correctly.
7 Unclip the vent pipe from the chassis and disconnect the breather pipe at the T-connection.
8 Loosen the tank straps (photo), support the tank in position and, while still supporting the tank, remove the straps. Carefully remove the tank assembly and tank guard, if fitted, leaving the fuel filler pipe in position.
9 Before refitting the tank, ensure that the four insulator pads (B in Fig. 3.23) are stuck to the tank in the positions shown.
10 Smear grease round the exterior of the filler pipe base, to facilitate its entry into the fuel tank seal. Position the tank and support it, then refit the two tank clips after reconnecting the fuel and vent pipes.
11 Ensure that the vent pipe is clipped into position and is not kinked or trapped.
12 Tighten the tank strap nuts until 35 to 40 mm (1.4 to 1.6 in) of thread is protruding through the nut.

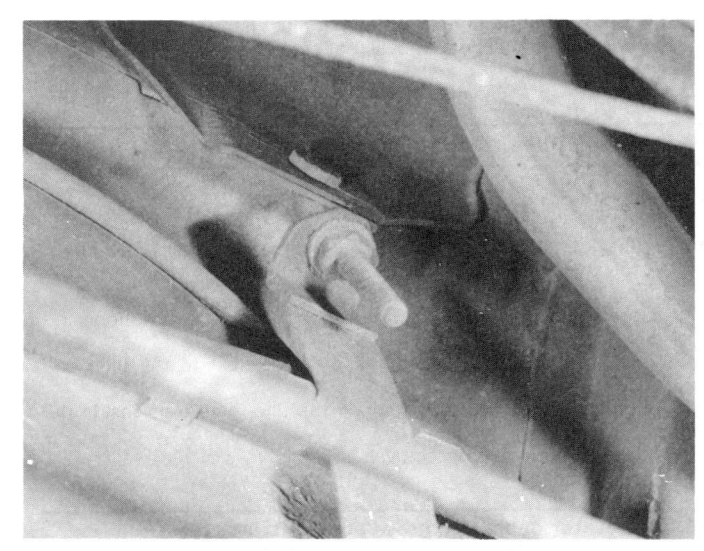

13.8 Fuel tank strap and bolt

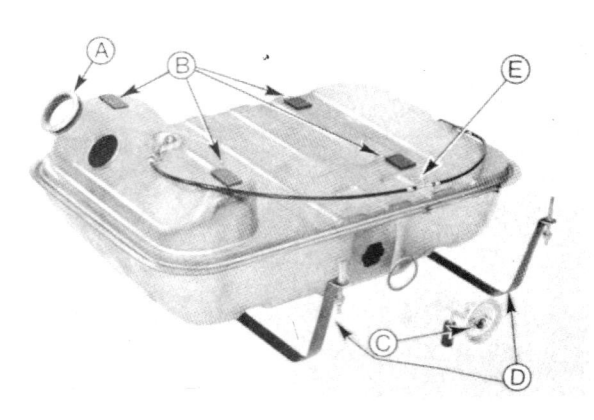

Fig. 3.23 Fuel tank components (Sec 13)

A Seal
B Rubber insulators
C Sender unit
D Straps
E T connection

16.6 Rubber gaiter at bottom of fuel tank filler pipe

14 Fuel tank – cleaning and repair

1 With time it is likely that sediment will collect in the bottom of the fuel tank. Condensation, resulting in rust and other impurities, will usually be found in the fuel tank of any car more than three or four years old.
2 When the tank is removed it should be vigorously flushed out with paraffin and turned upside down to drain. If facilities are available at the local garage the tank may be steam cleaned and the exterior repainted with a lead based paint.
3 Never weld or bring a naked light close to an empty fuel tank until it has been steam cleaned out for at least two hours or, washed internally with boiling water and detergent and allowed to stand for at least three hours.

15 Fuel tank sender unit – removal and refitting

1 Check the front wheels then jack up the rear of the car and support on axle stands.
2 Syphon the fuel from the fuel tank.
3 Where applicable disconnect the fuel line from the sender unit.
4 Disconnect the wiring.
5 Using two screwdrivers in the slots in the sender unit retaining ring, unscrew the sender unit from the fuel tank. Lift away the sealing ring and the sender unit, noting that the float must hang downwards.
6 Refitting is a reversal of removal, but clean the mating faces and fit a new sealing ring.

16.7 Fuel tank filler pipe upper mounting screw (arrowed)

16 Fuel tank filler pipe – removal and refitting

1 Remove the cap from the filler pipe.
2 Remove the right-hand side trim panel.
3 Peel back the insulation clear of the pipe cover panel (Ghia only).
4 Lift out the panel covering the spare wheel.
5 Unbolt the filler pipe cover panel.
6 Loosen the clips and release the rubber gaiter from the floor and filler pipe (photo).
7 Remove the screw at the top of the pipe next to the filler cap (photo).
8 Disconnect the hose and withdraw the filler pipe from inside the car.
9 Refitting is a reversal of removal.

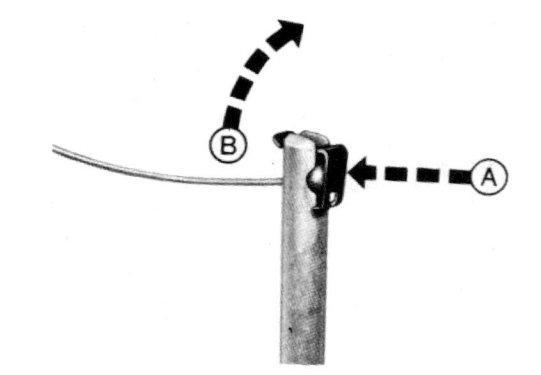

Fig. 3.24 Disconnecting the inner cable from the pedal (Sec 17)

Push at A and release at B

17 Accelerator cable – removal and refitting

1 Remove the lower facia panel inside the car.

2 Disconnect the cable from the pedal by unhooking the clip and releasing the inner cable.
3 Remove the brake servo unit as described in Chapter 9, then remove the screw securing the outer cable to the bulkhead.

Chapter 3 Fuel and exhaust systems

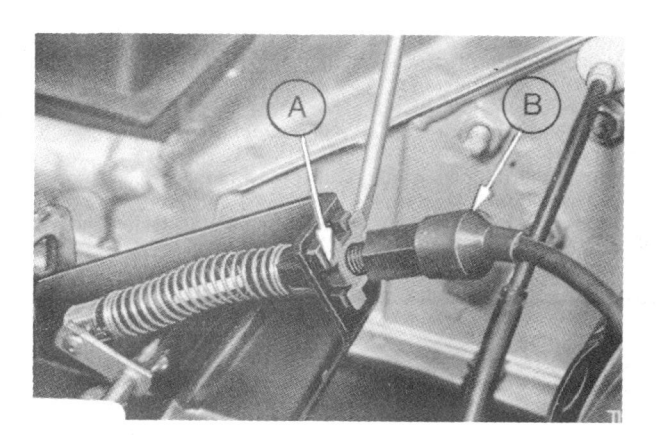

Fig. 3.25 Prising out the retaining clip (A) when removing the accelerator cable (B) (Sec 17)

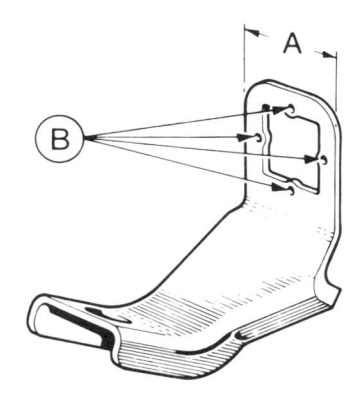

Fig. 3.26 Accelerator cable removal tool (Sec 17)

A = 25.0 mm (1.0 in)
B = Centre punch to provide raised shoulder

4 Remove the air cleaner as described in Section 3.
5 Pull off the sleeve and disconnect the plastic socket from the control linkage.
6 Prise the metal retaining clip from the bracket.
7 Using a screwdriver, depress the plastic tabs on the retainer while twisting the retainer, then pull from the mounting bracket. If this proves difficult make a removal tool to the dimensions given in Fig. 3.26, and press this onto the tabs to release them.
8 Refitting is a reversal of removal, but adjust it as follows. Place a weight on the accelerator pedal to hold it in its fully depressed position. On automatic transmission models check that the downshift cable is not preventing full movement. Back off the adjustment sleeve and adjust it so that the throttle plates are just fully open. Remove the weight and check the operation of the accelerator pedal.

18 Accelerator shaft and pedal – removal and refitting

Note: *If the pedal only is to be removed refer to paragraph 12.*
1 Disconnect the battery leads.
2 From inside the car remove the facia lower panel. It is retained by five screws (RHD) or three screws (LHD) along the rear edge and can be unclipped from the front edge.
3 Remove the accelerator cable from the pedal shaft as described in Section 16.

RHD variants
4 Disconnect the brake operating rod at the brake pedal, then remove the master cylinder and servo unit. Refer to Chapter 9 for further information.
5 Working through the rear bulkhead in the engine compartment, pull out the shaft end securing clip.
6 Rotate the right-hand shaft mounting bush through 45° in either direction and pull it out.
7 Detach the accelerator shaft assembly.

LHD variants
8 Loosen the clamp and detach the shaft extension rod.
9 Carefully drive out the right-hand mounting bush retaining clip from the shaft, then slide out the shaft until it touches the heater box.
10 Detach the right-hand mounting bush from the pedal box by rotating through 45° in either direction, then pulling it out. The accelerator shaft assembly can now be removed.

All models
11 Detach the remaining bush and clip from the shaft.
12 To remove the pedal, prise the flange away from the spigot on the shaft, then remove the pedal and spring.
13 When refitting the pedal, locate the spring on the spigot shaft, then clip the flanges onto the spigots and check that the pedal pivots correctly.

14 Refitting the pedal shaft is the reverse of the removal procedure, following which it will be necessary to adjust the cable, as described in Section 16. On RHD variants, check that the pedal has 6 to 14 mm (0.24 to 0.55 in) lift from the idle position. If necessary adjust the pedal lift-up stop to achieve this.

19 Inlet manifold – removal and refitting

1 Disconnect the battery negative lead.
2 Drain the cooling system (Chapter 2), remove the distributor (Chapter 4), and remove the carburettor (Section 8).
3 Disconnect the remaining hoses and wiring.
4 Unscrew the inlet manifold bolts progressively in the reverse order to that given in Fig. 3.27.
5 Lift the inlet manifold from the cylinder heads and remove the combined gasket.
6 To refit the inlet manifold, first clean the mating faces.
7 Apply sealing compound where the lower corners of the cylinder heads meet the cylinder block.
8 Locate a new gasket on the heads and block.
9 Fit the inlet manifold then insert the bolts and tighten them progressively in the sequence shown in Fig. 3.27 and in the stages given in the Specifications.
10 The remaining procedure is a reversal of removal. Refill the cooling system with reference to Chapter 2.

Fig. 3.27 Inlet manifold bolt tightening sequence for 3.0 litre engines (Sec 19)

20 Exhaust manifold – removal and refitting

1 Disconnect the battery negative lead.
2 Apply the handbrake then jack up the front of the car and support on axle stands.
3 Unscrew the nuts and disconnect the exhaust pipe and flange from the exhaust manifold. Remove the sealing ring.
4 Unscrew the nuts progressively and remove the manifold from the cylinder head. Remove the gaskets.
5 Refitting is a reversal of removal, but clean the mating faces and fit new gaskets. Tighten the nuts progressively.

21 Exhaust system – removal and refitting

1 Position the car over an inspection pit, or jack the car and support it securely on blocks or stands.
2 Disconnect the battery.
3 Lift up the resonator box and pull the rubber insulator off each side bracket.
4 Saturate all the clamp nuts with penetrating oil, or a proprietary corrosion inhibitor.
5 Disconnect the front pipe from the manifold, detach the sealing ring and lower the front section of the exhaust.

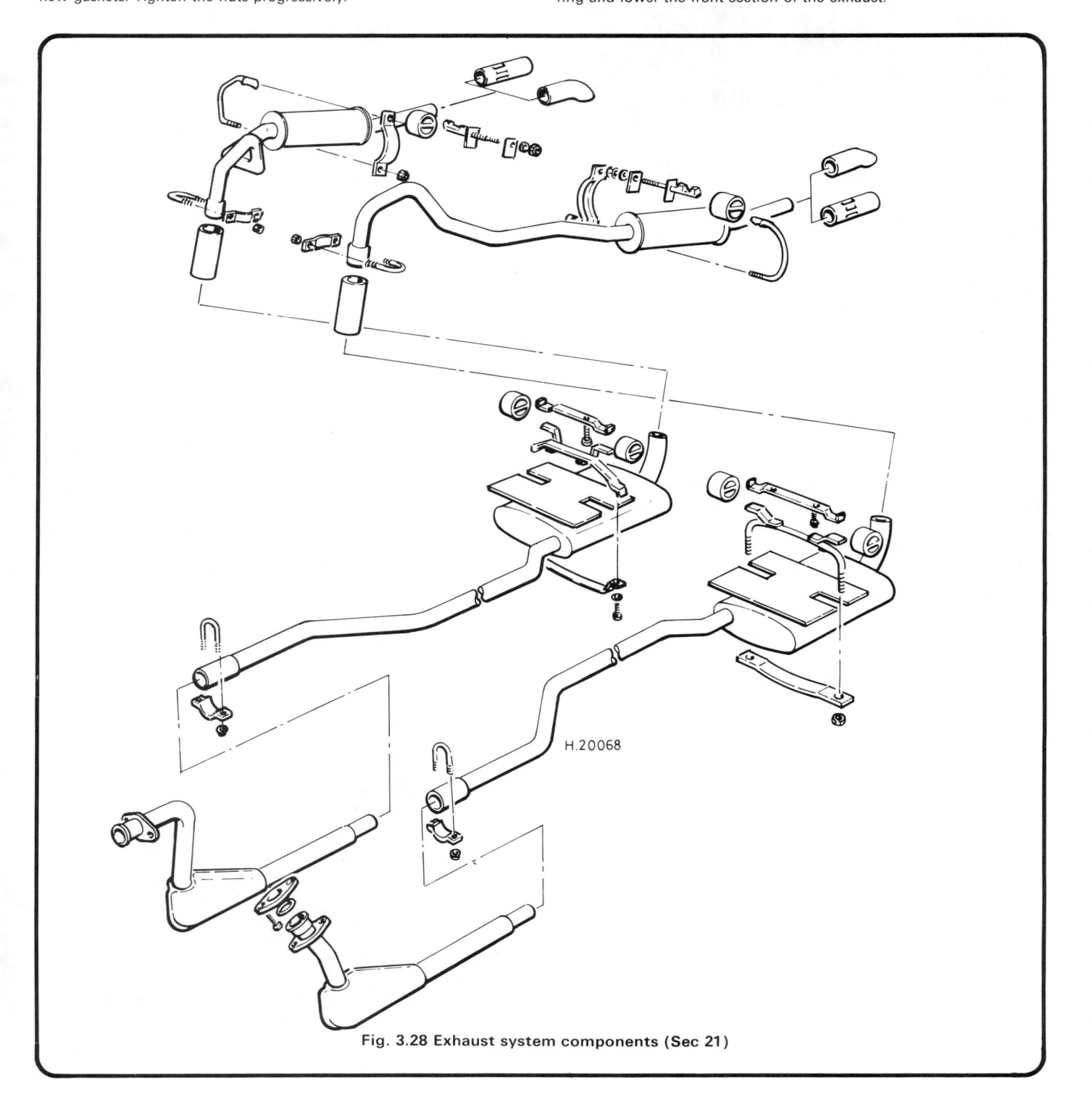

Fig. 3.28 Exhaust system components (Sec 21)

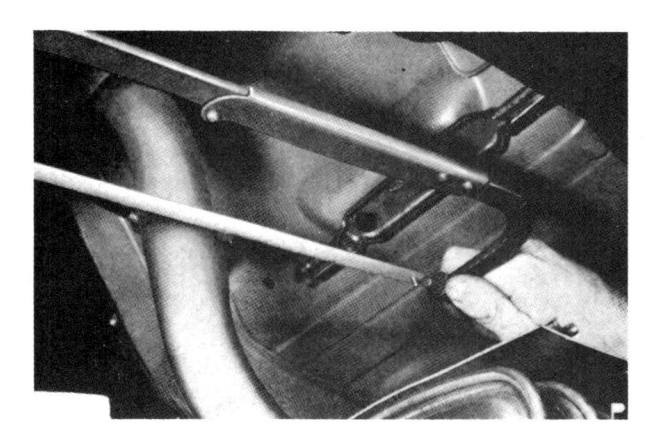

Chapter 3 Fuel and exhaust systems

Fig. 3.29 Cutting the rear of the exhaust system to facilitate its removal (Sec 21)

Fig. 3.30 Fitting the service sleeve to the exhaust system – arrowed (Sec 21)

6 Use a hacksaw to cut through the rear exhaust pipe about 240 mm (9.4 in) from the rear of the front resonator (Fig. 3.29).
7 Remove the nut securing the rear silencer bracket clamp, swing the bracket clear of the silencer and detach the rear exhaust section.
8 Detach and remove the rear mounting rubber and bracket clamp.
9 Remove the front pipe U-clamp and drift off the front pipe. Detach the resonator mounting bracket and the tail pipe trim, if fitted.
10 Start refitting by scribing a mark 45 mm (1.8 in) from the cut ends of the resonator to silencer pipe.
11 Position the resonator and front pipe assembly, loosely secure it to the bracket and manifold connection.
12 Slide a service sleeve (Fig. 3.30) onto the resonator pipe, positioning its end in line with the scribed mark made previously.
13 Fit the silencer pipe into the sleeve up to its scribed mark and refit the rear silencer clamp loosely. When fitting the clamp, ensure that its angled end is uppermost.
14 Align the exhaust system so that at no point is it nearer than 25 mm (1 in) to any component, or part of the bodywork. Tighten the manifold connection and then screw on the nuts of the resonator clamp until 13 mm (0.5 in) of thread protrudes through the nuts.
15 Position the two U-clamps on the service sleeve and tighten the nuts to the torque wrench setting given in the Specifications.
16 Refit the tail pipe trim if appropriate then reconnect the battery, start the engine and check for leaks.
17 Lower the car to the ground.

PART B: FUEL INJECTION SYSTEM

22 General description

The fuel injection system fitted to the 2.8 litre engine is of the continuous injection type and supplies a precisely controlled quantity of atomized fuel to each cylinder under all operating conditions.

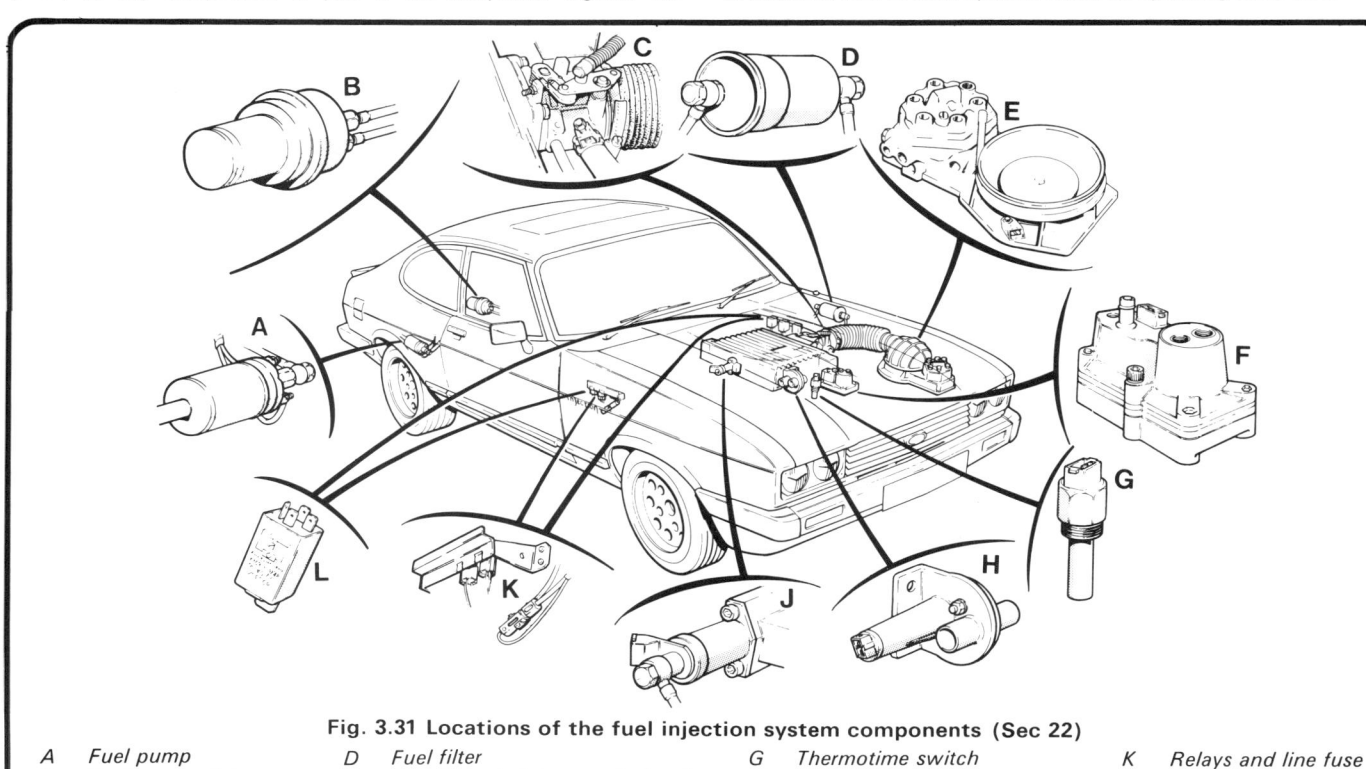

Fig. 3.31 Locations of the fuel injection system components (Sec 22)

A Fuel pump	D Fuel filter	G Thermotime switch	K Relays and line fuse
B Fuel accumulator	E Distributor/mixture control unit	H Auxiliary air device	L Impulse module
C Throttle housing	F Warm-up regulator	J Start valve	

This system, when compared with conventional carburettor arrangements, achieves a more accurate control of the air/fuel mixture resulting in reduced emission levels and improved prerformance.

The main components of the fuel injection system fall into three groups:

A *Fuel tank*
Fuel pump
Fuel accumulator
Fuel filter
Fuel distributor and mixture control
Throttle and injector valves
Air chamber

B *Warm-up regulator*
Auxiliary air device
Starter valve
Thermotime switch

C *Wiring*
Relays
Safety switch

The fuel pump is of electrically operated, roller cell type. A pressure relief valve is incorporated in the pump to prevent excessive pressure build up in the event of a restriction in the pipelines.

The fuel accumulator has two functions, (i) to dampen the pulsation of the fuel flow, generated by the pump and (ii) to maintain fuel pressure after the engine has been switched off. This prevents a vapour lock developing with consequent hot starting problems.

The fuel filter incorporates two paper filter elements to ensure that the fuel reaching the injection system components is completely free from dirt.

The fuel distributor/mixture control assembly. The fuel distributor controls the quantity of fuel being delivered to the engine, ensuring that each cylinder receives the same amount. The mixture control assembly incorporates an air sensor plate and control plunger.

The air sensor plate is located in the main air stream between the air cleaner and the throttle butterfly. During idling, the airflow lifts the sensor plate which in turn raises a control plunger which allows fuel to flow past the plunger and out of the injector valves. Increases in engine speed cause increased air flow which raises the control plunger and so admits more fuel.

The throttle valve assembly is mounted in the main air intake between the mixture control assembly and the air chamber. The throttle valve plate is controlled by a cable connected to the accelerator pedal.

The injector valves are located in the inlet manifold and are designed to open at a fuel pressure of 3.3 kgf/cm² (46.9 lbf/in²).

The air chamber is mounted on the top of the engine and functions as an auxiliary inlet manifold directing air from the sensor plate to each individual cylinder.

The warm-up regulator is located on the front of the engine and incorporates a bi-metal strip, vacuum diaphragm and control valve. The function of the regulator is to enrich the fuel/air mixture during the warm up period and also at full throttle operation.

The auxiliary air device is located on the front face of the air chamber. It consists of a pivoted plate, bi-metal strip and heater coil. The purpose of this device is to supply an increased volume of fuel/air mixture during cold idling rather similar to the fast idle system on carburettor layouts.

The fuel start valve is located on the main air chamber and is an electrically operated injector. Its purpose is to spray fuel into the air box at cold starting. This fuel comes from the fuel distributor and is atomized in the air chamber with air from the auxiliary air device.

The Thermotime switch is screwed into the engine adjacent to the warm up regulator. The switch incorporates a contact set, bi-metal strip and two heated elements. The switch controls the start valve by limiting the period of fuel injection from the valve and also preventing the valve from injecting any fuel at all when starting a hot engine.

Electrical relays and fuses. These are located under the facia panel on the right-hand side in the interior of the car. The relays comprise the main control relay and the power supply relay.

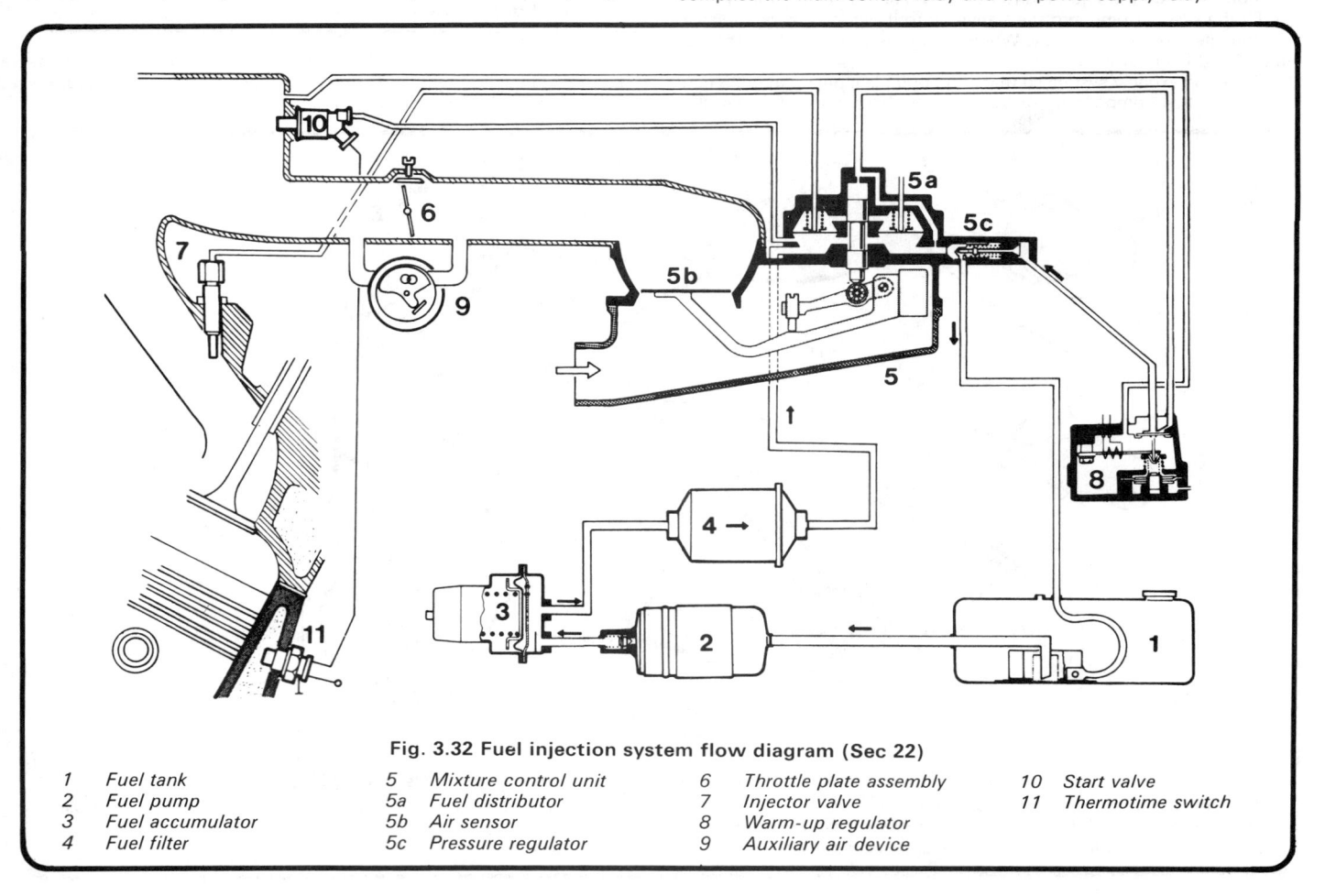

Fig. 3.32 Fuel injection system flow diagram (Sec 22)

1 Fuel tank	5 Mixture control unit	6 Throttle plate assembly	10 Start valve
2 Fuel pump	5a Fuel distributor	7 Injector valve	11 Thermotime switch
3 Fuel accumulator	5b Air sensor	8 Warm-up regulator	
4 Fuel filter	5c Pressure regulator	9 Auxiliary air device	

Chapter 3 Fuel and exhaust systems

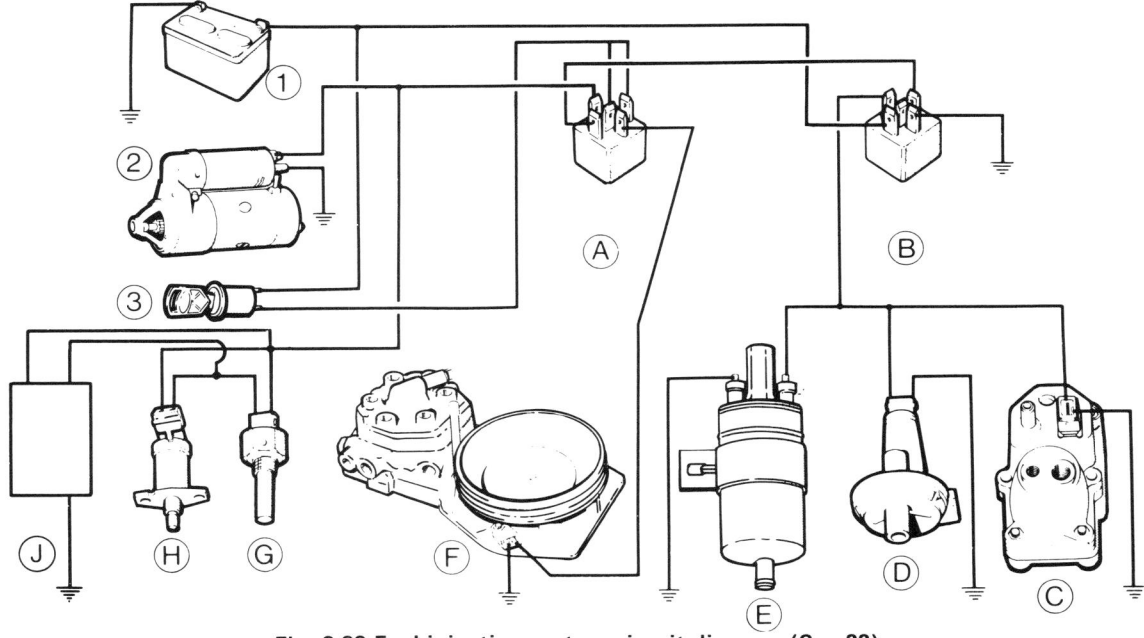

Fig. 3.33 Fuel injection system circuit diagram (Sec 22)

1 Battery
2 Starter motor
3 Ignition switch
A Main control relay
B Power supply relay
C Warm-up regulator
D Axiliary air device
E Fuel pump
F Safety switch on sensor plate
G Thermotime switch
H Start valve
J Impulse module

The safety switch is provided to cut off all power to the fuel injection system if the air sensor plate (mixture control assembly) is in the rest position even if the ignition switch is on. The safety switch would cut off the power to the fuel pump for example if a fuel line was damaged in an accident or burst through deterioration. The switch is located close to the mixture control unit. On later models the safety switch has been replaced by a safety control module. This module supplies power to the fuel pump provided that it is receiving pulses from the ignition coil, these pulses being evidence that the engine is running. When the engine stops for whatever reason, the safety control module cuts off power to the fuel pump.

23 Routine maintenance

At the intervals specified in the Routine Maintenance section in the front of the manual carry out the following procedures.

Check and adjust idling speed
1 Connect a tachometer to the engine then run the engine to normal operating temperature.
2 Increase the engine speed to 3000 rpm for 30 seconds then allow the engine to idle.
3 Check that the idle speed is as given in the Specifications and if necessary adjust the idle speed screw located beneath the throttle housing (photo).

Check condition of fuel hoses
4 Refer to Section 2.

Tighten the inlet manifold bolts and nuts
5 Refer to Section 29 and tighten the inlet manifold bolts and nuts in the sequence given.

Renew the fuel filter
6 Refer to Section 25.

Renew the air filter element
7 Disconnect the battery.
8 Release the spring clips on the air cleaner (photo).
9 Lift the air cleaner cover, together with the fuel distributor assembly, and lift out the air filter element (photo).
10 Clean the inside of the air cleaner body and cover.
11 Fit a new element then refit the cover.
12 Reconnect the battery.

23.3 Idle speed adjustment screw (arrowed)

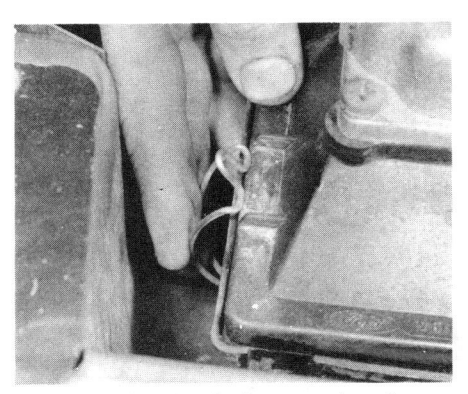

23.8 Releasing the air cleaner spring clip

23.9 Removing the air filter element

Chapter 3 Fuel and exhaust systems

24 Air cleaner – removal and refitting

1 Remove the element as described in Section 23.
2 Working under the left-hand front wheel arch, unbolt the air cleaner body from the side of the engine compartment.
3 Refitting is a reversal of the removal procedure with reference to Section 23.

25 Main system components – removal and refitting

Fuel pump

1 The fuel pump is located in a rubber mounting attached to the right-hand side of the fuel tank (photo).
2 Disconnect the battery.
3 Release the pressure in the system by temporarily loosening the feed line union on the fuel distributor. Catch the small amount of fuel which comes out in a piece of cloth.
4 With the rear of the car over an inspection pit or suitably raised, clamp the fuel inlet hose between the fuel tank and the pump using a brake hose clamp or similar. If there is only very little fuel in the tank, an alternative to clamping the hose is to drain the tank once the inlet hose is disconnected.
5 Disconnect the fuel inlet and outlet pipes from the pump.
6 Carefully identify the electrical connections to the pump and disconnect them.
7 Loosen the right-hand fuel tank supporting strap then unbolt the rubber mounting and fuel pump.
8 Prise the pump from the mounting.
9 Refitting is a reversal of removal.

Fuel accumulator

10 The fuel accumulator is bolted to the rear floor above the rear axle (photo).
11 Disconnect the battery.
12 Release the pressure in the system as described in paragraph 3.
13 Place the rear of the car over an inspection pit or raise it on ramps or axle-stands.
14 Disconnect the fuel pipes from the fuel accumulator and catch the small quantity of fuel which will be released.
15 Remove the two mounting screws and remove the accumulator.
16 Refitting is a reversal of removal, but remove the floor panel from inside the luggage compartment and refit the protective caps over the screw threads.

Fuel filter

17 The fuel filter is located on the wheel arch in the left-hand side of the engine compartment (photo).

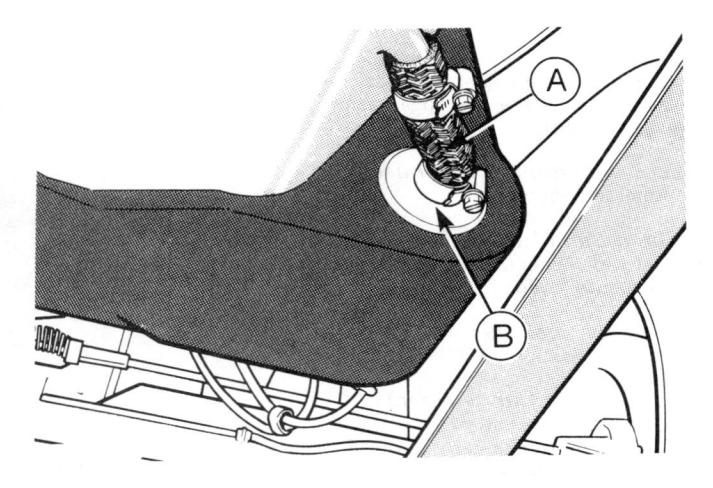

Fig. 3.34 Rear view of fuel pump (Sec 25)

A   Inlet pipe   B   Fuel pump

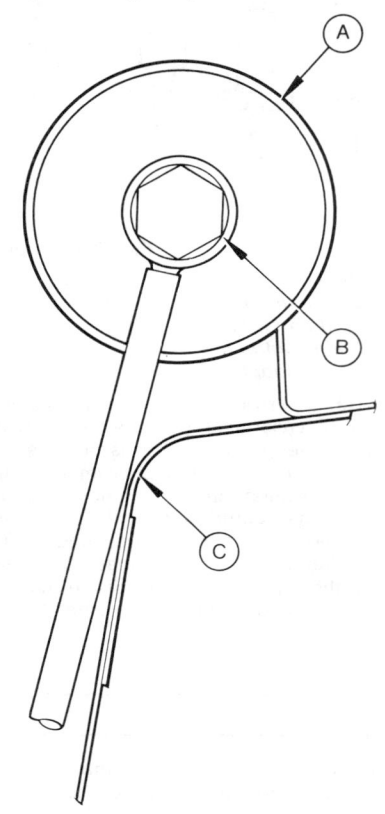

Fig. 3.35 Correct location of fuel filter outlet pipe (Sec 25)

A   Fuel filter   C   Engine compartment side
B   Outlet pipe   panel

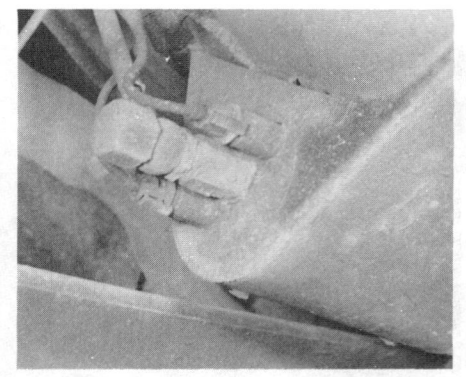

25.1 Fuel pump showing wiring and outlet pipe

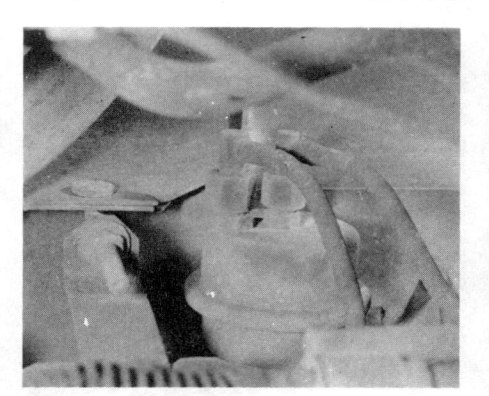

25.10 Fuel accumulator

25.17 Fuel filter

Chapter 3 Fuel and exhaust systems

18 Disconnect the battery.
19 Release the pressure in the system as described in paragraph 3.
22 Place some cloth beneath the fuel filter then disconnect the inlet and outlet unions.
21 Remove the mounting screws and withdraw the fuel filter. Remove the bracket if necessary.
22 Refitting is a reversal of removal, but make sure that the filter is positioned the correct way round, and locate the outlet pipe as shown in Fig. 3.35.

Fuel distributor
23 The fuel distributor is located on top of the air filter.
24 Disconnect the battery.
25 Release the pressure in the system as described in paragraph 3 but completely remove the feed line union.
26 Disconnect the remaining fuel lines from the distributor.
27 Unscrew the three mounting screws and lift the distributor away.
28 Refitting is a reversal of removal but use a new copper washer each side of the banjo unions and do not overtighten the union bolts.
29 Have the system pressure checked by your dealer.
30 Adjust the idle speed and mixture as described in Sections 23 and 26.

Warm-up regulator
31 The warm-up regulator is located on the front of the engine next to the cooling system outlet elbow.
32 Disconnect the battery.
33 Release the pressure in the system as described in paragraph 3.
34 Disconnect the vacuum hose and wiring plug (photo).
35 Disconnect the fuel lines (photo).
36 Unscrew the two socket-headed bolts and withdraw the regulator.
37 Refitting is a reversal of removal, but use a new copper washer each side of the banjo unions (photo). Do not overtighten the union centre bolts.

Fuel start valve
38 The fuel start valve is located on the right-hand side of the air chamber.
39 Disconnect the battery.
40 Disconnect the electric plug and fuel supply pipe from the valve (photo).
41 Unscrew the two socket-headed mounting bolts using an Allen key.
42 Refitting is a reversal of removal, but use a new mounting flange gasket. Use a new copper washer on each side of the banjo union and do not overtighten the centre bolt.

Auxiliary air device
43 The auxiliary air device is located on the right-hand front of the air chamber.
44 Disconnect the battery.
45 Disconnect the electric plug and two air hoses from the device (photos).
46 Unscrew the two mounting bolts and lift the assembly away.
47 Refitting is a reversal of removal.

Fuel injectors
48 Disconnect the battery.
49 Disconnect the accelerator cable and bracket from the throttle assembly.
50 Disconnect the main air supply trunking from the mixture control unit and throttle housing (photos).
51 Disconnect the electric plugs from the warm-up regulator, the thermotime switch, the auxiliary air device and the fuel start valve.
52 Disconnect the fuel supply pipe from the start valve also the vacuum pipe from the warm up regulator.
53 Unbolt the air chamber (eight bolts) and move it clear of the inlet manifold. To completely remove it, disconnect the remaining hoses. Remove the gasket (photo).
54 Unscrew the mounting bolts and pull the injectors from the inlet manifold (photo).
55 Using two spanners, loosen the union nuts and disconnect the injectors from the fuel lines.

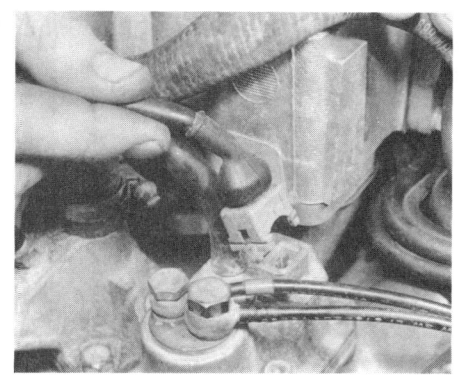

25.34 Disconnecting the wiring plug ...

25.35 ... and fuel lines from the warm-up regulator

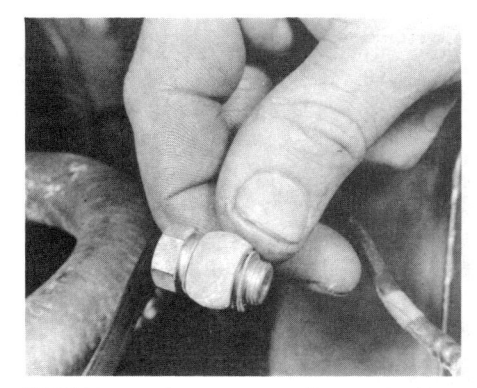

25.37 Renew the copper washers each side of the banjo union

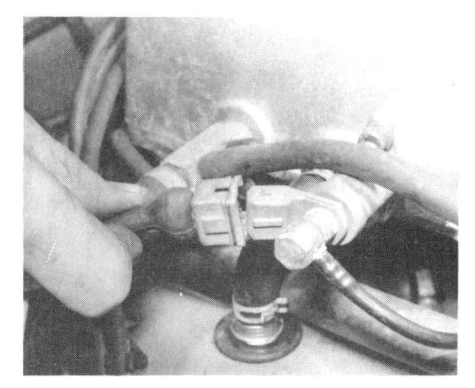

25.40A Disconnecting the fuel start valve wiring plug ...

25.40B ... and fuel supply pipe

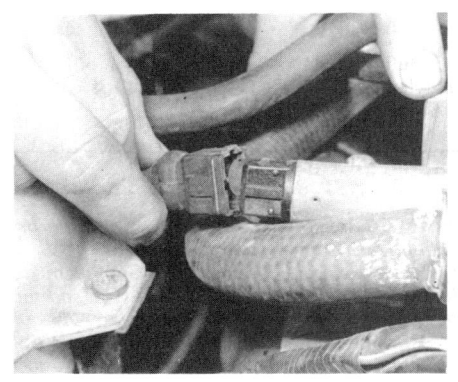

25.45A Auxiliary air device wiring plug ...

Chapter 3 Fuel and exhaust systems

25.45B ... and air hoses

25.50A Disconnecting the main air supply trunking from the mixture control unit ...

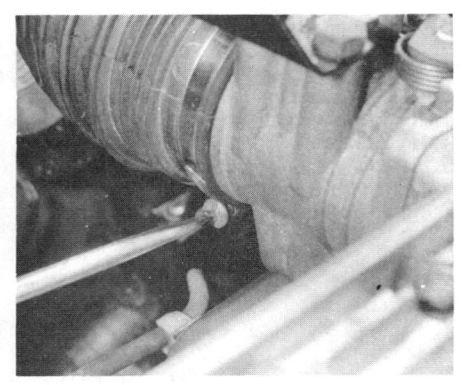

25.50B .. and throttle housing

25.53A Unscrew the bolts ...

25.53B ... remove the air chamber ...

25.53C ... and remove the gasket

25.54A Unscrew the mounting bolts ...

25.54B ... and remove the injectors

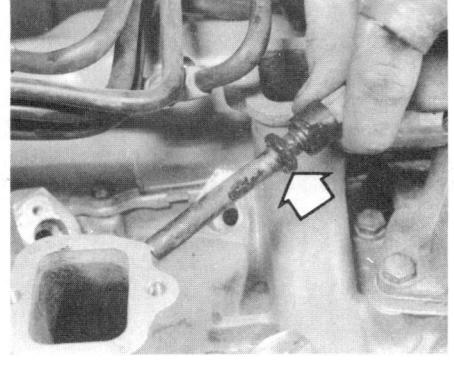

25.56 The injector O-ring seal (arrowed)

56 Refitting is a reversal of removal, but make sure that the injectors are clean and fit new O-ring seals (photo). Also renew the air chamber gasket.

26 Idle mixture – adjustment

Note: *The mixture setting is preset during production of the car and should not normally require adjustment. If new components of the system have been fitted, however, the mixture can be adjusted as follows using an exhaust gas analyser and tachometer. Adjustment is not possible without an exhaust gas analyser.*

1 Run the engine until it is at the normal operating temperature.
2 Connect an exhaust gas analyser and a tachometer (if not fitted as standard equipment) in accordance with the manufacturer's instructions.
3 Increase the engine speed to 3000 rpm and hold it there for 30 seconds to stabilise the exhaust gases and then allow the engine to return to idling.
4 Check the readings on the test instruments with those specified. If adjustment is required first turn the idle speed screw to give the correct idle speed.
5 Break off the tamperproof cap from the mixture control screw on top of the fuel distributor.
6 Stabilise the exhaust gases as described in paragraph 3.
7 Insert a 3 mm Allen key into the head of the mixture screw and turn the screw until the correct CO reading is obtained (photo). Readjust the idle speed screw.
8 If the mixture adjustment cannot be finalised within 30 seconds from the moment of stabilising the exhaust gases, repeat the operations described in paragraph 3 before continuing the adjustment procedure.

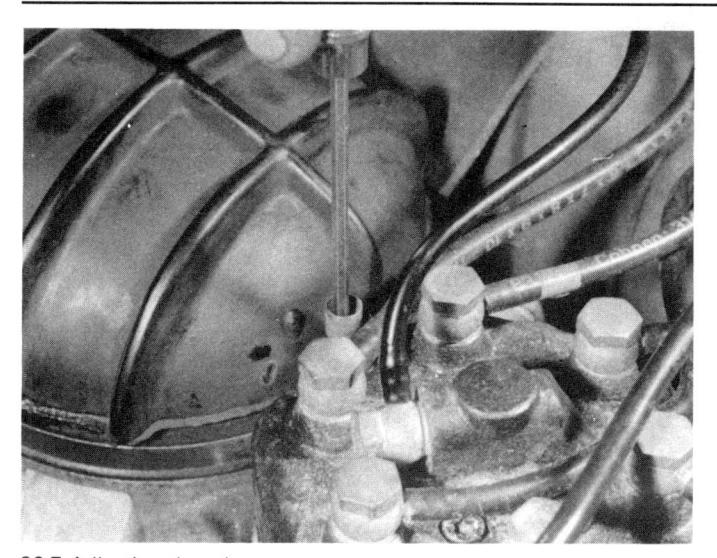

Chapter 3 Fuel and exhaust systems

26.7 Adjusting the mixture screw

27.1 Fuel tank sender unit

27 Fuel tank, sender unit and filler pipe – general

The removal and refitting procedures are similar to those given in Sections 13 to 16. However, the sender unit is located on the left-hand side of the fuel tank (photo) and the fuel pump is secured to the right-hand side.

28 Accelerator cable – removal and refitting

Refer to Section 17 (photo).

29 Inlet manifold – removal and refitting

1 Refer to Section 19, but instead of removing the carburettor, remove the injectors and air chamber as described in Section 25. Note that there is one nut at each end of the manifold, and that the bolts are of different lengths (photo).
2 Also unbolt the outlet elbow and warm-up regulator mounting bracket from the front of the inlet manifold (photos).
3 Follow the tightening sequence given in Fig. 3.36 which differs from the 3.0 litre engine (photo).

28.1A Removing the sleeve from the end of the accelerator cable

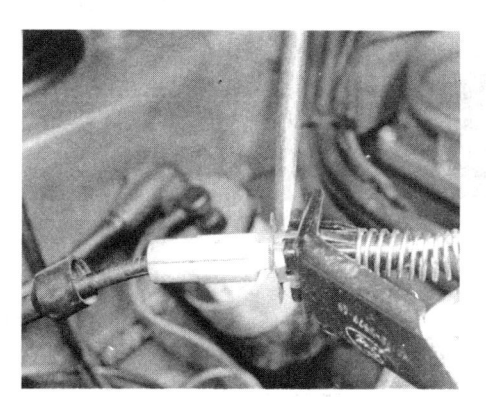

28.1B Prising out the retaining clip

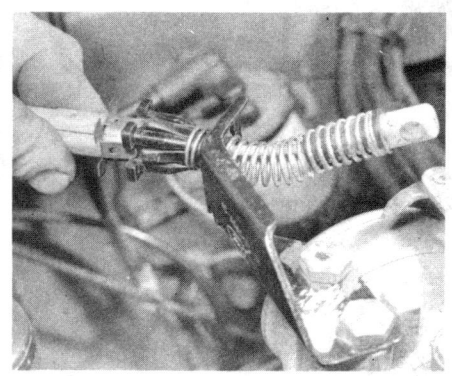

28.1C Removing the accelerator cable from the mounting bracket

29.1A Unscrew the inlet manifold bolts ...

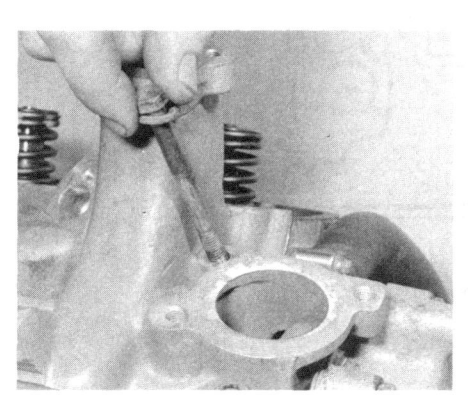

29.1B ... remove the bolts ...

29.1C ... and lift off the manifold

29.1D The inlet manifold gasket

29.1E Sealing compound location before fitting the gasket

29.2A Unscrew the bolts ...

29.2B ... and remove the outlet elbow and warm-up regulator mounting bracket

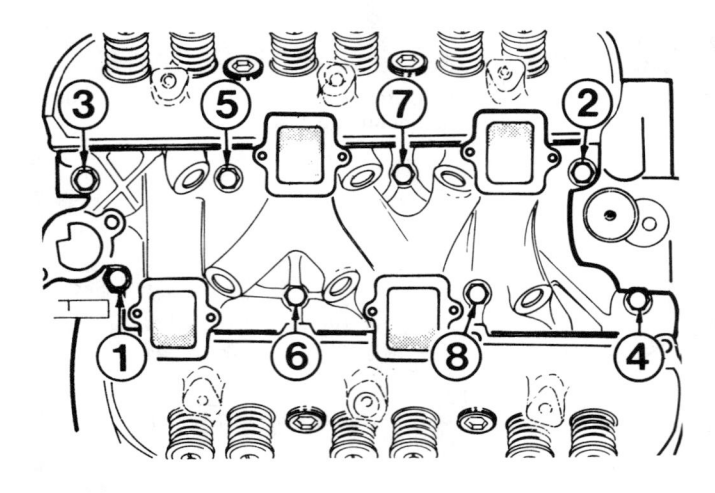

Fig. 3.36 Inlet manifold nut and bolt tightening sequence for 2.8 litre engines (Sec 29)

29.3 Tightening the inlet manifold nuts and bolts

Chapter 3 Fuel and exhaust systems

30 Exhaust manifold – removal and refitting

Refer to Section 20, but not that later models do not have a downpipe sealing ring (photos).

31 Exhaust system – removal and refitting

Refer to Section 21 (photo).

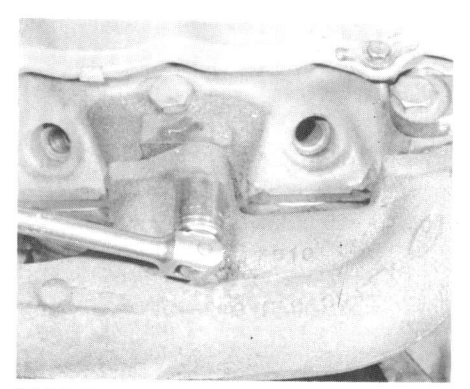

30.1A Tightening the exhaust manifold nuts

30.1B Right-hand exhaust manifold-to-downpipe connection

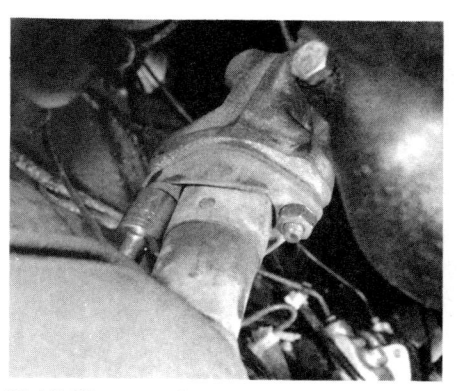

30.1C Disconnecting the left-hand exhaust downpipe from the manifold

30.1D Right-hand exhaust downpipe and flange

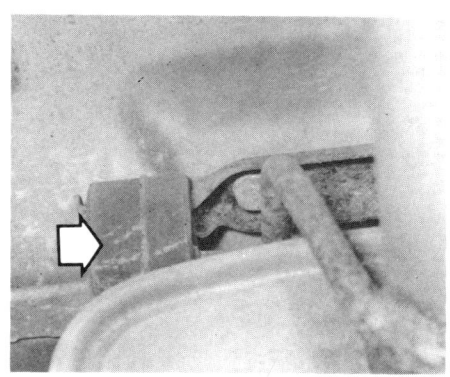

31.1A Exhaust system rubber mounting (arrowed)

31.1B Intermediate exhaust system-to-front pipe clamp

PART C: FAULT DIAGNOSIS

32 Fault diagnosis – fuel system (carburettor)

Symptom	Reason(s)
Fuel consumption excessive	Air filter element choked Fuel leak from carburettor, fuel pump or fuel lines Generally worn carburettor Needle valve sticking open Incorrect idling speed and/or mixture
Insufficient fuel delivery or weak mixture	Clogged fuel line or filter Faulty fuel pump Leaking inlet manifold gasket Weak mixture setting

33 Fault diagnosis – fuel system (fuel injection)

Symptom	Reason(s)
Engine will not start (cold or hot)	Air flow sensor plate incorrectly set Faulty fuel pump Fuel tank empty

Symptom	Reason(s)
Engine will not start (HOT)	Auxiliary air device not closing
Engine will not start (COLD)	Auxiliary air device not opening Start valve faulty Thermotime switch not closing
Engine misfires on road	Loose fuel pump electrical connections
Unsatisfactory road performance	System fuel pressure incorrect
Rough idling (always)	Mixture adjustment incorrect
Rough idling (during warm-up)	Auxiliary air device not operating correctly
Excessive fuel consumption	Mixture adjsutment incorrect Leak in system fuel lines
Engine runs on	Airflow sensor plate or control plunger not moving freely Injector valves leaking or their opening pressure too low

Chapter 4 Ignition system

Contents

Condenser (3.0 litre engine) – testing, removal and refitting	5	General description	1
Contact breaker points (3.0 litre engine) – adjustment	3	Ignition amplifier module (2.8 litre engine) – removal and refitting	10
Contact breaker points (3.0 litre engine) – removal and refitting	4	Ignition timing – adjustment	9
Distributor (all engines) – removal and refitting	6	Routine maintenance	2
Distributor (2.8 litre engine) – overhaul	8	Spark plugs – removal, servicing and refitting	11
Distributor (3.0 litre engine) – overhaul	7		
Fault diagnosis – ignition system	12		

Specifications

System type
3.0 litre engine .... Conventional, with contact breaker
2.8 litre engine .... Breakerless, electronic

Conventional ignition
Distributor
Rotation .... Clockwise
Condenser capacity .... 0.21 to 0.25 mfd
Contact breaker gap .... 0.64 mm (0.025 in)
Dwell angle .... 36° to 40°
HT lead resistance .... 30 000 ohm maximum
Firing order 1–4–2–5–3–6

Ignition timing
Pre 1976 model year .... 10° BTDC
1976 model year onwards .... 14° BTDC

Coil
Primary resistance .... 0.95 to 1.60 ohm
Secondary resistance .... 5000 to 9300 ohm

Spark plugs
Type .... Motorcraft AGR 22 or Champion RN9YC
Electrode gap .... 0.60 mm (0.024 in)

Breakerless ignition
Distributor
Rotation .... Clockwise
HT lead resistance .... 30 000 ohm maximum
Firing order 1–4–2–5–3–6

Ignition timing .... 12° BTDC

Coil
Primary resistance .... 0.75 to 0.85 ohm
Secondary resistance .... 5000 to 6000 ohm

Spark plugs
Type .... Motorcraft Super AGR 22C or Champion RN7YC
Electrode gap .... 0.60 mm (0.024 in)

All systems
Torque wrench settings
	Nm	lbf ft
Spark plugs (3.0 litre engine)	30 to 60	22 to 30
Spark plugs (2.8 litre engine)	25 to 38	18 to 28

Chapter 4 Ignition system

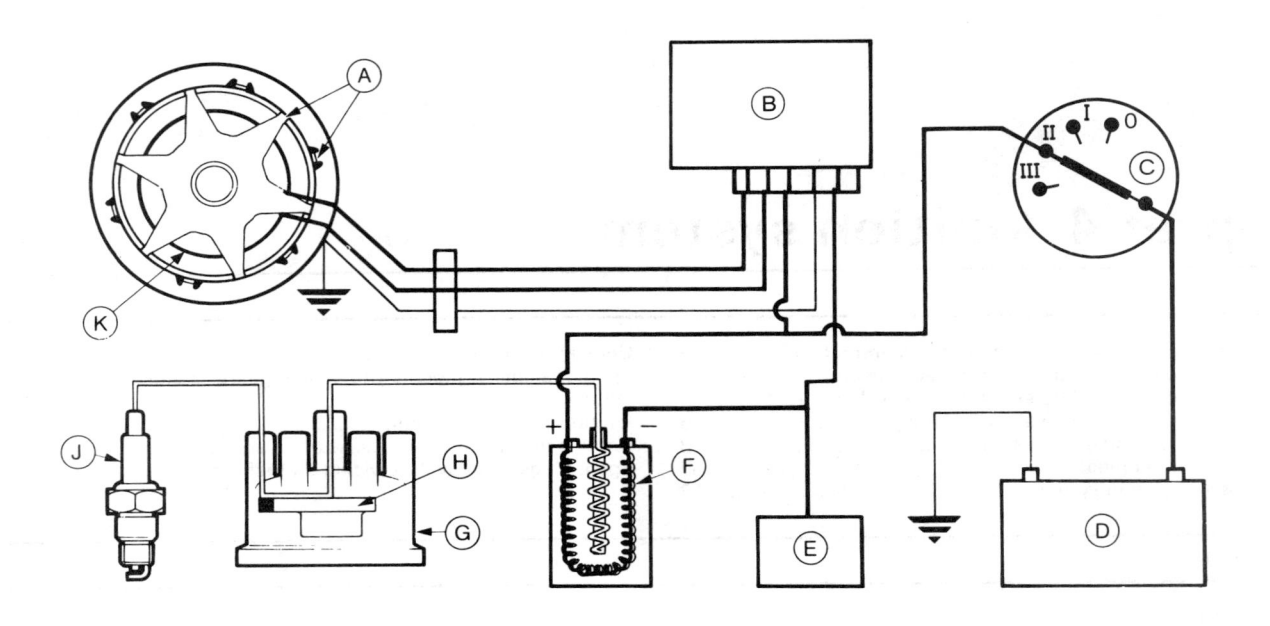

Fig. 4.1 Diagram of the breakerless ignition system on the 2.8 litre engine (Sec 1)

A Trigger and stator arms
B Amplifier module
C Ignition switch
D Battery
E Tachometer
F Coil
G Distributor cap
H Rotor arm
J Spark plug
K Tripper coil

1 General description

To achieve optimum performance from an engine and to meet stringent exhaust emission requirements, it is essential that the fuel/air mixture in the combustion chamber is ignited at exactly the right time relative to engine speed and load. The ignition system provides the spark necessary to start the mixture burning and the instant at which ignition occurs is varied automatically as engine operating conditions change.

The ignition system consists of a primary (low tension) circuit and a secondary (high tension) circuit. When the ignition is switched on, current is fed to the coil primary windings and a magnetic field is established. At the required point of ignition, the primary circuit is interrupted by the contact breaker points opening (3.0 litre engine) or the trigger arm passing the stator arm (2.8 litre engine). The magnetic field collapses and a secondary high voltage in induced in the secondary windings. This HT voltage is fed via the distributor rotor arm to the relevant spark plug. After delivering the spark the primary circuit is re-energised and the cycle is repeated.

The ignition timing is controlled centrifugally and by a vacuum unit to compensate for engine speed and load.

On the 3.0 litre engine the ignition system incorporates a ballast resistor wire which effectively boosts the HT voltage during starting.

2 Routine maintenance

At the intervals specified in the Routine Maintenance section in the front of the manual carry out the following procedures.

Clean and adjust the spark plugs, or renew them
1 Refer to Section 11.

Clean and check the distributor cap, rotor, HT leads and coil
2 Release the two clips securing the distributor cap to the distributor body and lift away the cap. Clean the cap inside and out with a dry cloth. Closely inspect the inside of the cap and the six segments. If there are any signs of cracking or if the segments are burned or scored excessively the cap will have to be renewed.

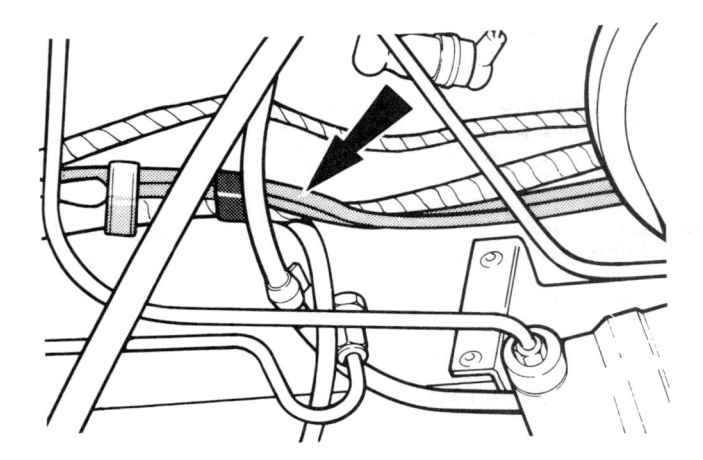

Fig. 4.2 Ballast resistor wire located on the left-hand side of the engine compartment – arrowed (Sec 1)

3 Also check the centre segment (3.0 litre engine) or carbon brush (2.8 litre engine) for wear.
4 Remove the rotor arm and check it for cracking and burning of the metal segment (photo). On the 2.8 litre engine check the spring-tensioned engine speed limiter incorporated in the rotor arm for free movement (photo). On the 3.0 litre engine check that the HT conductor spring on the rotor arm has sufficient tension to contact the centre segment in the cap.
5 With the rotor arm removed, lubricate the felt pad in the centre of the distributor shaft with two drops of engine oil.
6 Refit the rotor arm and distributor cap (photo).

Chapter 4 Ignition system

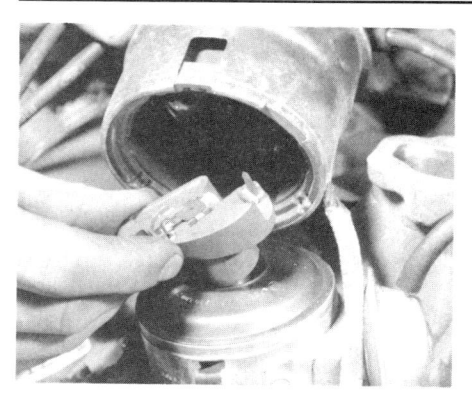

2.4A Removing the rotor arm (2.8 litre engine)

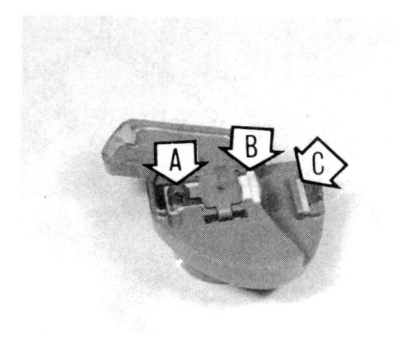

2.4B The engine speed limiter on the rotor arm (2.8 litre engine)
A Spring
B Moving contact
C Earthing contact

2.6 Securing the distributor cap with the spring clips (2.8 litre engine)

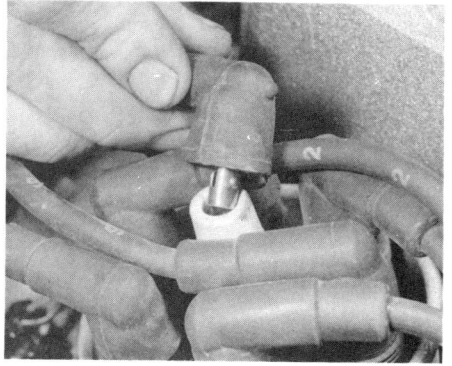

2.7 Checking the HT lead terminals on the distributor cap

2.8 Ignition coil and wiring

4.2 Removing the contact breaker low tension leads

7 Wipe clean the HT leads and check them for condition. Pull the leads from the distributor cap one at a time and make sure that no water has found its way into the socket terminals. Clean any corrosion away then refit the leads (photo).
8 Clean the ignition coil tower and check the HT lead as previously described (photo).

Check and adjust the contact breaker dwell angle (3.0 litre engine)
9 Refer to Section 3.

Check and adjust the ignition timing
10 Refer to Section 9.

Renew the contact breaker points
11 Refer to Section 4.

3 Contact breaker points (3.0 litre engine) – adjustment

1 Remove the distributor cap and rotor arm (Section 2)
2 Gently prise the contact breaker points open and examine the condition of their faces. If they are rough, pitted or dirty it will be necessary for new points to be fitted.
3 If the points are satisfactory, or have been renewed, measure the gap between the points with feeler gauges, by turning the crankshaft until the heel of the breaker arm is on a high point of the cam. Consult the Specifications at the beginning of this Chapter for the correct setting.
4 If the gap is incorrect, slacken the contact plate securing screw(s), then insert a screwdriver into the slot provided at the edge of the contact plate and move the plate until the gap is correct.
5 Tighten the screw(s) and recheck the gap.

6 Refit the rotor arm ensuring that it is correctly located in the slot at the top of the cam. Refit the distributor cap and retain in position with the two clips.
7 On modern engines a more accurate method of setting the points gap is by means of a dwell meter. Not only does this method give a precise points gap but it also evens out any variations in gap caused by wear in the distributor shaft or bushes or differences in any of the heights of the cam peaks.
8 The dwell angle is the number of degrees through which the distributor cam turns during the period between the instants of closure and opening of the contact breaker points. It can only be checked with a dwell meter connected in accordance with the maker's instructions.
9 If the angle is not as given in the Specifications, adjust the points gap. If the dwell angle is too large increase the points gap, if the dwell angle is too small, reduce the points gap.

4 Contact breaker points (3.0 litre engine) – removal and refitting

1 Remove the distributor cap and rotor arm (Section 2).
2 Slacken the self-tapping screw which secures the condenser and low tension leads to the contact breaker assembly and slide out the forked cable terminations (photo).
3 Undo and remove the two screws which secure the contact breaker base plate to the distributor and lift off the contact breaker assembly (photo).
4 Before refitting the contact breaker assembly, smear the distributor cam with a trace of grease.
5 Fit the contact breaker and its clamping screws, but do not fully tighten the screws until the contact breaker gap has been adjusted as described in Section 3.
6 Refit the rotor arm and distributor cap, refitting the two clips to secure the cap.

4.3 Contact breaker mounting screws (arrowed)

6.3 Disconnecting the vacuum advance pipe

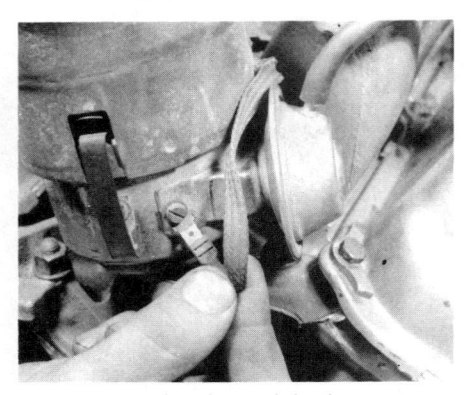

6.4 Disconnecting the earth lead

6.5 Pull the wires to disconnect the multi-plug connector

6.6 Rotor arm aligned with the slot on the distributor body rim (arrowed)

6.7 Crankshaft pulley notch and timing cover gradation (2.8 litre engine)

5 Condenser (3.0 litre engine) – testing, removal and refitting

1 The purpose of the condenser (sometimes know as a capacitor) is to ensure that when the contact breaker points open there is no sparking across them which would cause rapid wear of their faces and engine misfire.
2 The condenser is fitted in parallel with the contact breaker points. If it becomes faulty, it will cause ignition failure as the contact breaker points will be prevented from correctly interrupting the low tension circuit.
3 If the engine becomes very difficult to start or begins to miss after several miles of running, and the breaker points show signs of excessive burning, then the condition of the condenser must be suspect. One further test can be made by separating the points by hand with the ignition switched on. If this is accompanied by a bright flash, it is indicative that the condenser has failed.
4 Without special test equipment the only safe way to diagnose condenser trouble is to replace a suspect unit with a new one and note if there is any improvement.
5 To remove the condenser from the distributor take off the distributor cap and rotor arm.
6 Slacken the self-tapping screw holding the condenser lead and low tension lead to the contact breaker points. Slide out the forked terminal on the end of the condenser lead. Undo and remove the condenser retaining screw and remove the condenser from the breaker plate.
7 To refit the condenser, simply reverse the order of removal.

6 Distributor (all engines) – removal and refitting

1 Where applicable remove the air cleaner, as described in Chapter 3.
2 Remove the distributor cap only (Section 2).
3 Disconnect the vacuum advance pipe (photo).
4 Where applicable disconnect the earth lead from the terminal on the distributor body (photo).
5 Disconnect the distributor multi-plug connector. To disconnect the plug, the wires must be held and pulled, not the plug itself (photo).
6 Turn the engine until the rotor arm is pointing to the No 1 cylinder segment in the distributor cap. On later models this is also indicated by a slot on the distributor body rim (photo).
7 Check also that the notch in the crankshaft pulley is in line with the correct mark on the timing cover (photo). On the 3.0 litre engine the marks are in 2° intervals, but on the 2.8 litre engine they are in 3° intervals. See the Specifications for the correct setting. Mark the position of the distributor body in relation to the cylinder head.
8 Remove the single bolt at the base of the distributor and slowly lift out the distributor without turning its body. As the distributor is removed the skew gear will turn the rotor arm clockwise. Mark the new position of the rotor arm on the rim if there is not already a mark there (photos).
9 To refit the distributor ensure that the timing marks are still aligned as detailed in paragraph 7 of this Section. Hold the distributor over the cylinder head so that the body to the cylinder head marks are in alignment.
10 Position the rotor arm towards the mark on the distributor body and slide the assembly into position on the engine. As the gears mesh, the rotor will turn and align with No 1 segment in the distributor cap.
11 Refit the securing bolt to the base of the distributor but do not fully tighten at this stage.
12 Check that on the 3.0 litre engine the distributor vacuum unit is facing forwards, parallel to the centre line of the engine, or on the 2.8 litre engine the vacuum unit is facing in the direction shown in Fig. 4.4.
13 Reconnect the multi-plug connector. The manufacturers recommend that the multi-plug is fitted with a lithium based grease.
14 Reconnect the earth lead (if applicable) and the vacuum advance pipe.
15 Refer to Section 9 and set the ignition timing.
16 Refit the air cleaner (if applicable).

Chapter 4 Ignition system

6.8A Unscrew the clamp bolt ...

6.8B ... and remove the distributor

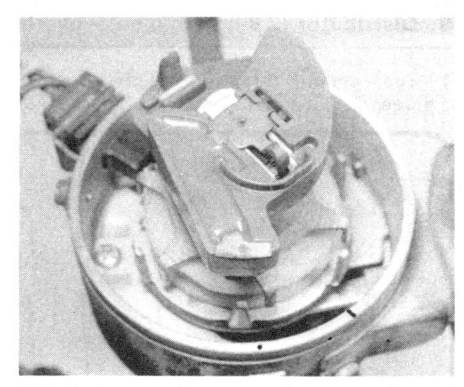

6.8C Rotor arm alignment after removal from engine

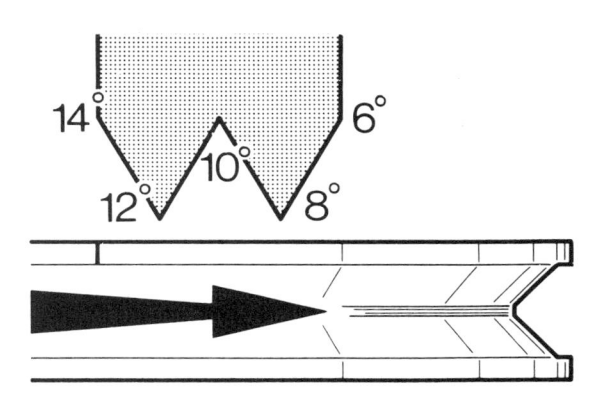

Fig. 4.3 Ignition timing marks on the 3.0 litre engine (Sec 6)

Arrow indicates direction of rotation

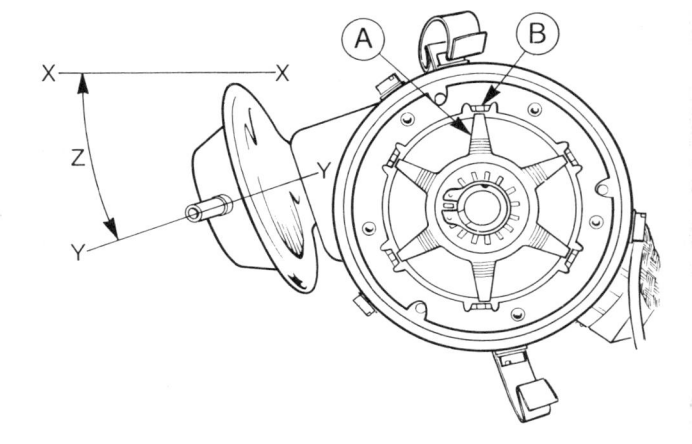

Fig. 4.4 Distributor fitting position on the 2.8 litre engine (Sec 6)

XX Rear face of engine
YY Vacuum unit axis
Z = 0° to 14°
A Trigger arm
B Stator arm

7 Distributor (3.0 litre engine) – overhaul

1 With the distributor removed from the engine, remove the rotor arm and the contact breaker assembly.
2 Prise the small circlip from the vacuum unit pivot post.
3 Remove the two screws which secure the breaker plate to the distributor body and lift the plate off.
4 Slacken the self-tapping screw which secures the condenser and low tension leads to the contact breaker assembly and slide out the forked cable terminations.
5 Undo and remove the condenser retaining screw and remove the condenser.
6 Remove the circlip, flat washer and wavy washer from the pivot post. Separate the two plates by bringing the holding down screw through the keyhole slot in the lower plate. Take care not to lose the earth spring on the pivot post.
7 Remove the two screws securing the vacuum unit to the distributor body and remove the vacuum unit.
8 To dismantle the vacuum unit, remove the plug from the end of the unit and withdraw the spring, vacuum stop and shims.
9 Before dismantling the centrifugal advance mechanism, note that the two springs are different and mark them so that they are refitted in their original places. Remove the springs, prise off the circlips securing the centrifugal weights, mark the weights to identify their original positions and then remove them.
10 Mark which of the end slots in the mechanical advance plate engages with the stop in the action plate. If this is not done, it is possible to have the rotor arm 180° from its correct position.
11 Remove the felt pad from the centre of the cam, expand the circlip which is then exposed and remove it. Lift off the cam and advance plate assembly.
12 Do not remove the distributor spindle unless it is necessary to fit a new gear, or spindle. To remove the gear, use a punch and hammer to drive out the lock pin and then pull the gear from the spindle. Take care to retain any shims, or washers which may be fitted. With the gear removed, the spindle assembly can be withdrawn from the distributor.
13 Before reassembling, carefully clean the body of the distributor and all its component parts. If the spindle has been removed, lubricate it with engine oil before reassembly.
14 Reassembly is a straightforward reversal of dismantling but there are several points which should be noted.
15 Lubricate the centrifugal weights and other parts of the centrifugal advance mechanism, the distributor shaft and the part of the spindle which carries the cam assembly, using engine oil. Do not oil excessively.
16 If the driveshaft has been removed, first refit the thrust washers below the action plate, before inserting the shaft into the distributor body. Fit the wavy washer and thrust washer to the lower end of the shaft and then refit the gear, securing it with a new spring pin. If a new gear, or new driveshaft, has been fitted, a new hole should be drilled at 90° to the existing hole.
17 After assembling the centrifugal weights and springs, check that they move freely, without binding.
18 Before assembling the breaker plates, make sure that the three nylon bearing studs are located in the holes in the upper bearing plate and that the earthing spring is fitted to the pivot post.
19 On completion of assembly, adjust the contact breaker gap to the specified setting.

Chapter 4 Ignition system

8 Distributor (2.8 litre engine) – overhaul

1 There are only two overhaul jobs possible, these being the fitting of a trigger coil kit and the renewal of the vacuum advance unit. Excessive wear or damage to the distributor body or driveshaft will mean the renewal of the complete distributor.

2 Refer to Fig. 4.5 for the order of removal of the components and lightly oil the shaft and vacuum unit arm during reassembly (photos).

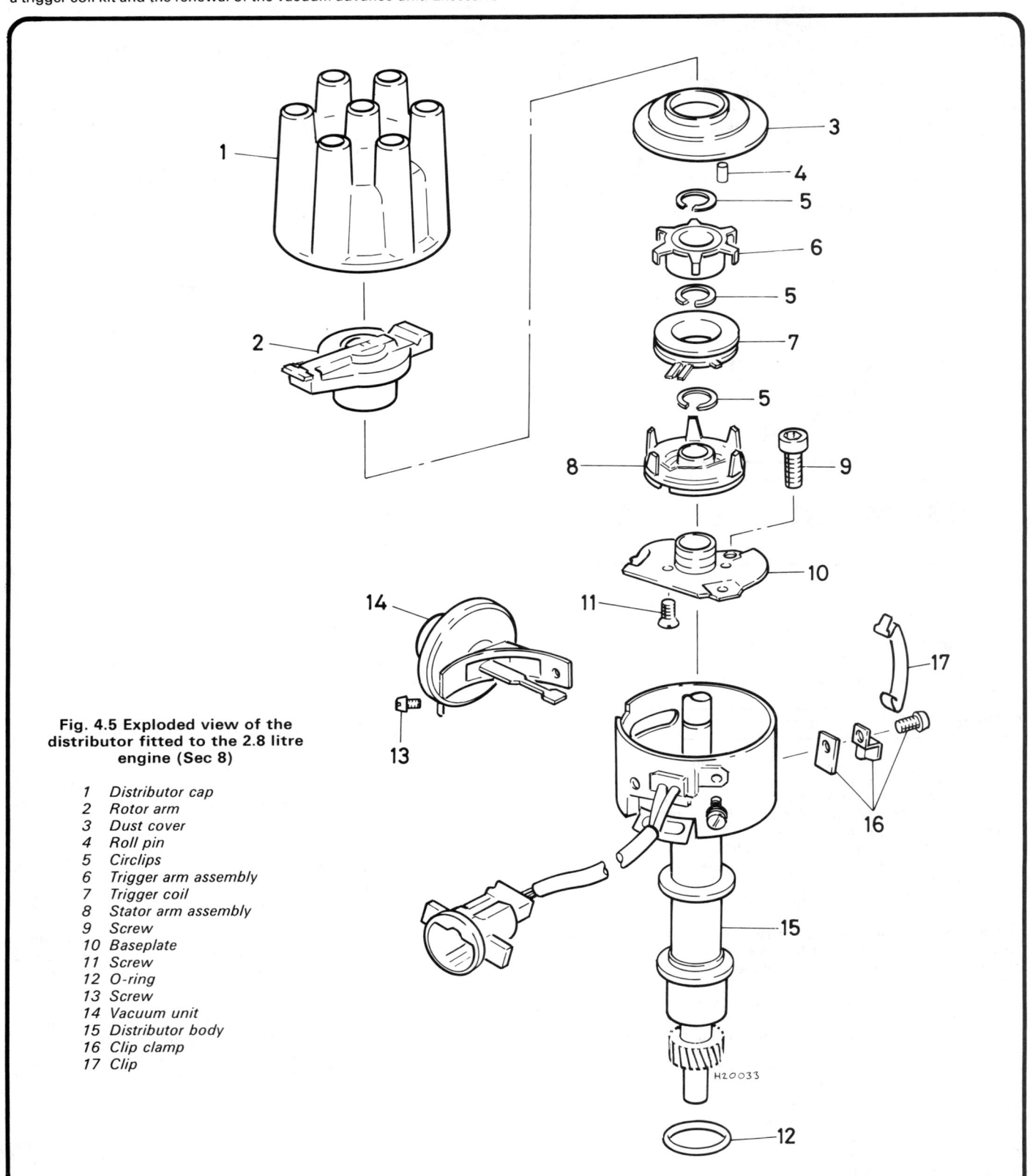

Fig. 4.5 Exploded view of the distributor fitted to the 2.8 litre engine (Sec 8)

1 Distributor cap
2 Rotor arm
3 Dust cover
4 Roll pin
5 Circlips
6 Trigger arm assembly
7 Trigger coil
8 Stator arm assembly
9 Screw
10 Baseplate
11 Screw
12 O-ring
13 Screw
14 Vacuum unit
15 Distributor body
16 Clip clamp
17 Clip

Chapter 4 Ignition system

8.2A Removing the plastic dust cover (2.8 litre engine)

8.2B Upper view of the trigger arm and coil (2.8 litre engine)

8.2C Wiring connector and securing screw (arrows)

8.2D Vacuum unit securing screw with earth lead terminal

9 Ignition timing – adjustment

1 With the distributor fitted as described in Section 6, check that the correct timing marks are aligned.
2 To check the initial (static) setting, remove the distributor cap and check that the rotor arm is pointing to the No 1 cylinder segment in the cap indicated by the slot on the rim.
3 On the 3.0 litre engine the contact points must be just separating. The exact instant can be checked by connecting a 12 volt test lamp between the contact breaker low tension terminal and earth. With the ignition on, the distributor should be turned anti-clockwise until the bulb just lights.
4 On the 2.8 litre engine the trigger arms and stator arms must be aligned. If not, turn the distributor as necessary.
5 Tighten the distributor clamp bolt and refit the cap.
6 Set in this way the timing will be approximately correct but a more accurate method is by using a stroboscopic timing light.
7 Highlight the specified timing marks on the timing cover and crankshaft pulley with white paint or chalk.
8 Connect a timing light to the No 1 spark plug lead.
9 Disconnect the vacuum advance pipe at the carburettor or air box (fuel injection) and plug the pipe then run the engine until normal operating temperatures are reached.
10 With the engine idling point the timing light at the crankshaft pulley. The white painted marks will appear stationary and if the timing is correct they will be in alignment.
11 To adjust the timing, stop the engine, slacken the clamp bolt and turn the distributor slightly. Tighten the clamp bolt, start the engine and recheck the timing. Repeat this procedure until the marks are in line.
12 To check the mechanical advance, increase the engine speed and note whether the white mark on the pulley moves away from the mark on the pointer. If it does the mechanical advance is functioning.
13 With the engine idling the vacuum advance can be checked by sucking the advance pipe that was removed from the carburettor or air box. This should also cause the white mark on the pulley to move away from the mark on the pointer.
14 On completion remove the timing light and reconnect the vacuum pipe.

10 Ignition amplifier module (2.8 litre engine) – removal and refitting

1 The ignition amplifier module is located on the left-hand side of the engine compartment, just behind the front suspension strut upper mounting.

Chapter 4 Ignition system

10.2 Ignition amplifier module and multi-plug connectors

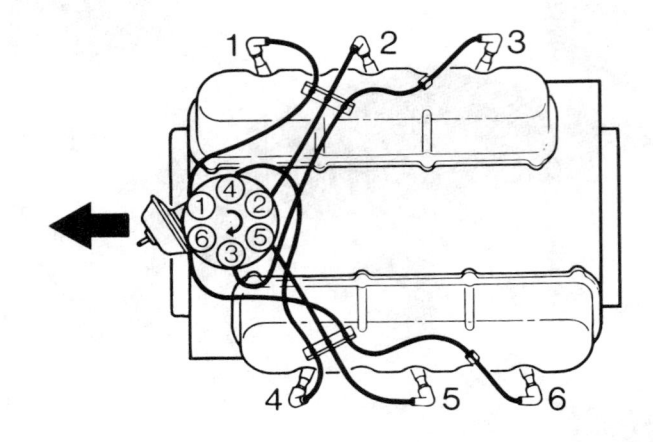

Fig. 4.6 HT lead positions on the 3.0 litre engine (Sec 11)

Arrow indicates front of engine

2 With the ignition switched off disconnect the two multi-plug connectors (photo). Pull the wires, not the plug itself.
3 Unscrew the mounting screws and remove the module.
4 Refitting is a reversal of removal, but the manufacturers recommend that the multi-plugs are filled with a lithium based grease.

11 Spark plugs – removal, servicing and refitting

1 Remove the air cleaner or inlet ducting as required.
2 Identify the HT leads for position then disconnect them from the spark plugs by pulling on the terminal ends, not the main leads (photo).
3 Using a spark plug spanner, unscrew the plugs from the cylinder heads (photos).
4 If the insulator nose of the spark plug is clean and white, with no deposits, this is indicative of a weak mixture, or too hot a plug (a hot plug transfers heat away from the electrode slowly – a cold plug transfers it away quickly).
5 The plugs fitted as standard are as listed in the Specifications at the beginning of this Chapter. If the tip and insulator nose are covered with hard black-looking deposits, then this is indicative that the mixture is too rich. Should the plug be black and oily, then it is likely that the engine is fairly worn, as well as the mixture being too rich.
6 If the insulator nose is covered with light tan to greyish brown deposits, then the mixture is correct and it is likely that the engine is in good condition.
7 If there are any traces of long brown tapering stains on the outside of the white portion of the plug, then the plug will have to be renewed, as this shows that there is a faulty joint between the plug and body and the insulator, and compression is being allowed to leak away.

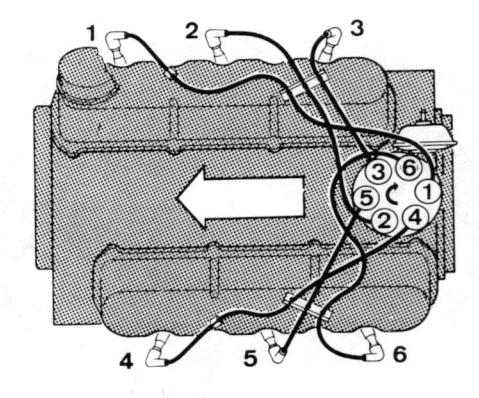

Fig. 4.7 HT lead positions on the 2.8 litre engine (Sec 11)

Arrow indicates front of engine

8 Plugs should be cleaned by a sand blasting machine which will free them from carbon more thoroughly than cleaning by hand. The machine will also test the condition of the plugs under compression. Any plug that fails to spark at the recommended pressure should be renewed.

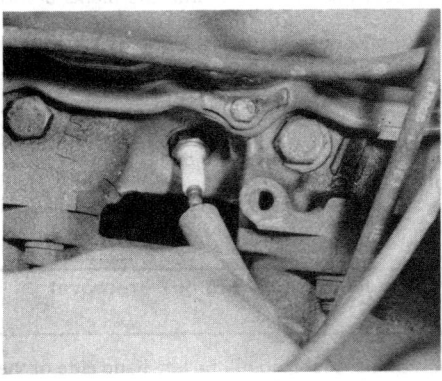

11.2 Disconnect the HT lead ...

11.3A ... unscrew the spark plug ...

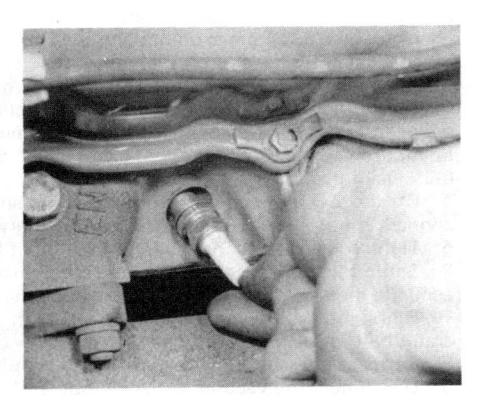

11.3B ... and remove it

Chapter 4 Ignition system

9 The spark plug gap is of considerable importance, as, if it is too large or too small, the size of the spark and its efficiency will be seriously impaired. The spark plug should be set to the figure given in the Specifications at the beginning of this Chapter:
10 To set it, measure the gap with a feeler gauge, and then bend open, or close, the outer plug electrode until the correct gap is achieved. The centre electrode should never be bent as this may crack the insulation and cause plug failure if nothing worse.
11 Refit the plugs, and refit the leads from the distributor in the correct firing order, which is given in the Specifications. Screw the plugs in by hand initially then tighten them to the specified torque using the plug spanner.

12 Fault diagnosis – ignition system

By far the majority of breakdown and running troubles are caused by faults in the ignition system either in the low tension or high tension circuits.

There are two main symptoms indicating ignition faults. Either the engine will not start or fire, or the engine is difficult to start and misfires. If it is a regular misfire the fault is almost sure to be in the secondary or high tension circuit. If the misfiring is intermittent, the fault could be in either the high or low tension circuits. If the car stops suddenly, or will not start at all, it is likely that the fault is in the low tension circuit. Loss of power and overheating, apart from faulty carburation or fuel injection settings, are normally due to faults in the distributor or to incorrect ignition timing.

Engine fails to start

1 If the engine fails to start and the car was running normally when it was last used, first check there is fuel in the petrol tank. If the engine turns over normally on the starter motor and the battery is evidently well charged, then the fault may be in either the high or low tension circuits. First check the HT circuit. Note: *If the battery is known to be fully charged, the ignition light comes on, and the starter motor fails to turn the engine* check the tightness of the leads on the battery terminals *and also the secureness of the earth lead to its connection to the body. It is quite common for the leads to have worked loose, even if they look and feel secure. If one of the battery terminal posts gets very hot when trying to work the starter motor this is a sure indication of a faulty connection to that terminal.*
2 One of the commonest reasons for bad starting is wet or damp spark plug leads and distributor. Remove the distributor cap. If condensation is visible internally dry the cap with a rag and also wipe over the leads. Refit the cap.
3 If the engine still fails to start, check that current is reaching the plugs, by disconnecting each plug lead in turn at the spark plug end, and holding the end of the cable about 5 mm (0.2 in) away from the cylinder block. Spin the engine on the starter motor.
4 Sparking between the end of the cable and the block should be fairly strong with a strong regular blue spark. (Hold the lead with rubber to avoid electric shocks.) If current is reaching the plugs, then remove them and clean and regap them. The engine should now start.
5 If there is no spark at the plug leads, take off the HT lead from the centre of the distributor cap and hold it to the block as before. Spin the engine on the starter once more. A rapid succession of blue sparks between the end of the lead and the block indicate that the coil is in order and that the distributor cap is cracked, the rotor arm faulty, or the carbon brush in the top of the distributor cap or rotor arm spring is not making good contact. Possibly, the points are in bad condition (mechanical contact breaker). Renew them as described in this Chapter, Section 4.

6 If there are no sparks from the end of the lead from the coil check the connections at the coil end of the lead. If it is in order start checking the low tension circuit.
7 Use a 12V voltmeter or a 12V bulb and two lengths of wire. On conventional distributors switch on the ignition and ensure that the points are open. On breakerless distributors ensure that the segments on the rotor are not adjacent to the trigger coil. Make a test between the low tension wire to the coil (+) terminal and earth. A reading of 7 to 8 volts should be obtained. No reading indicates a break in the supply from the ignition switch or a fault in the ballast resistance wire. A correct reading indicates a faulty coil or condenser, or a broken lead between the coil and the distributor. On breakerless ignition systems the trigger coil and amplifier module are suspect. Have them checked by a Ford dealer.
8 Take the condenser wire off the points assembly and with the points open test between the moving point and earth. If there is now a reading then the fault is in the condenser. Fit a new one as described in this Chapter, Section 5.
9 With no reading from the moving point to earth, take a reading between earth and the CB or negative (−) teminal of the coil. A reading here shows a broken wire which will need to be renewed between the coil and distributor. No reading confirms that the coil has failed and must be renewed, after which the engine will run once more. Remember to refit the condenser wire to the points assembly.

Engine misfires

10 If the engine misfires regularly, run it at a fast idling speed. Pull off each of the plug caps in turn and listen to the note of the engine. Hold the plug cap in a dry cloth or with a rubber glove as additional protection against a shock from the HT supply.
11 No difference in engine running will be noticed when the lead from the defective plug is removed. Removing the lead from one of the good cylinders will accentuate the misfire.
12 Remove the plug lead from the end of the defective plug and hold it about 5 mm (0.2 in) away from the block. Start the engine. If the sparking is fairly strong and regular the fault must lie in the spark plug.
13 The plug may be loose, the insulation may be cracked, or the points may have burnt away giving too wide a gap for the spark to jump. Worse still, one of the points may have broken off.
14 If there is no spark at the end of the plug lead, or if it is weak and intermittent, check the ignition lead from the distributor to the plug. If the insulation is cracked or perished, renew the lead. Check the connections at the distributor cap.
15 If there is still no spark, examine the distributor cap carefully for tracking. This can be recognised by a very thin black line running between two or more electrodes, or between an electrode and some other part of the distributor. These lines are paths which now conduct electricity across the cap thus letting it run to earth. The only answer is a new distributor cap.
16 Apart from the ignition timing being incorrect; other causes of misfiring have already been dealt with under the section dealing with the failure of the engine to start. To recap – these are that:

    (a) *The coil may be faulty giving an intermittent misfire*
    (b) *There may be a damaged wire or loose connection in the low tension circuit*
    (c) *The condenser may be short circuiting (if fitted)*
    (d) *There may be a mechanical fault in the distributor (broken driving spindle or contact breaker spring, if fitted)*

17 If the ignition timing is too far retarded, it should be noted that the engine will tend to overheat, and there will be a quite noticeable drop in power. If the engine is overheating and the power is down, and the ignition timing is correct, then the carburettor or fuel injection should be checked, as it is likely that this is where the fault lies.

Chapter 5 Clutch

Contents

Clutch assembly – inspection	4	Clutch release bearing – removal and refitting	6
Clutch assembly – refitting	5	Fault diagnosis – clutch	9
Clutch assembly – removal	3	General description	1
Clutch cable – renewal	7	Routine maintenance	2
Clutch pedal – removal and refitting	8		

Specifications

General
Type .... Single dry plate, diaphragm spring, cable-operated
Clutch disc diameter .... 241.3 mm (9.5 in)
Lining thickness .... 3.81 mm (0.15 in)
Clutch pedal free travel .... 27.0 ± 4.0 mm (1.06 ± 0.16 in)

Torque wrench setting
	Nm	lbf ft
Pressure plate assembly	16 to 20	12 to 15

1 General description

All manual transmission models covered by this manual are fitted with a single diaphragm spring clutch. The unit comprises a steel cover which is dowelled and bolted to the rear face of the flywheel and contains the pressure plate, diaphragm spring and fulcrum rings.

The clutch friction disc is free to slide along the splined gearbox input shaft and is held in position between the flywheel and the pressure plate by the pressure of the pressure plate spring. Friction lining material is riveted to the disc and it has a spring cushioned hub to absorb transmission shocks and to help ensure a smooth take off.

The circular diaphragm spring is mounted on shoulder pins and held in place in the cover by two fulcrum rings. The spring is also held to the pressure plate by three spring steel clips which are riveted in position.

The clutch is actuated by a cable controlled by the clutch pedal. The clutch release mechanism consists of a release lever and a bearing which contacts the release fingers on the pressure plate assembly. The effect of any wear of the friction material in the clutch is adjusted out by means of a cable adjuster at the lower end of the cable where it passes through the bellhousing.

Depressing the clutch pedal actuates the clutch release lever by means of the cable. The release lever pushes the release bearing forward to bear against the release fingers so moving the centre of the diaphragm spring within the annular rings which act as fulcrum points. As the centre of the spring is pushed in, the outside of the spring is pushed out, so moving the pressure plate backward and disengaging the pressure plate from the friction disc.

When the clutch pedal is released, the diaphragm spring forces the pressure plate into contact with the friction linings on the disc and at the same time pushes it a fraction of an inch forward on its splines. The disc is now firmly sandwiched between the pressure plate and the flywheel, so the drive is taken up.

Chapter 5 Clutch

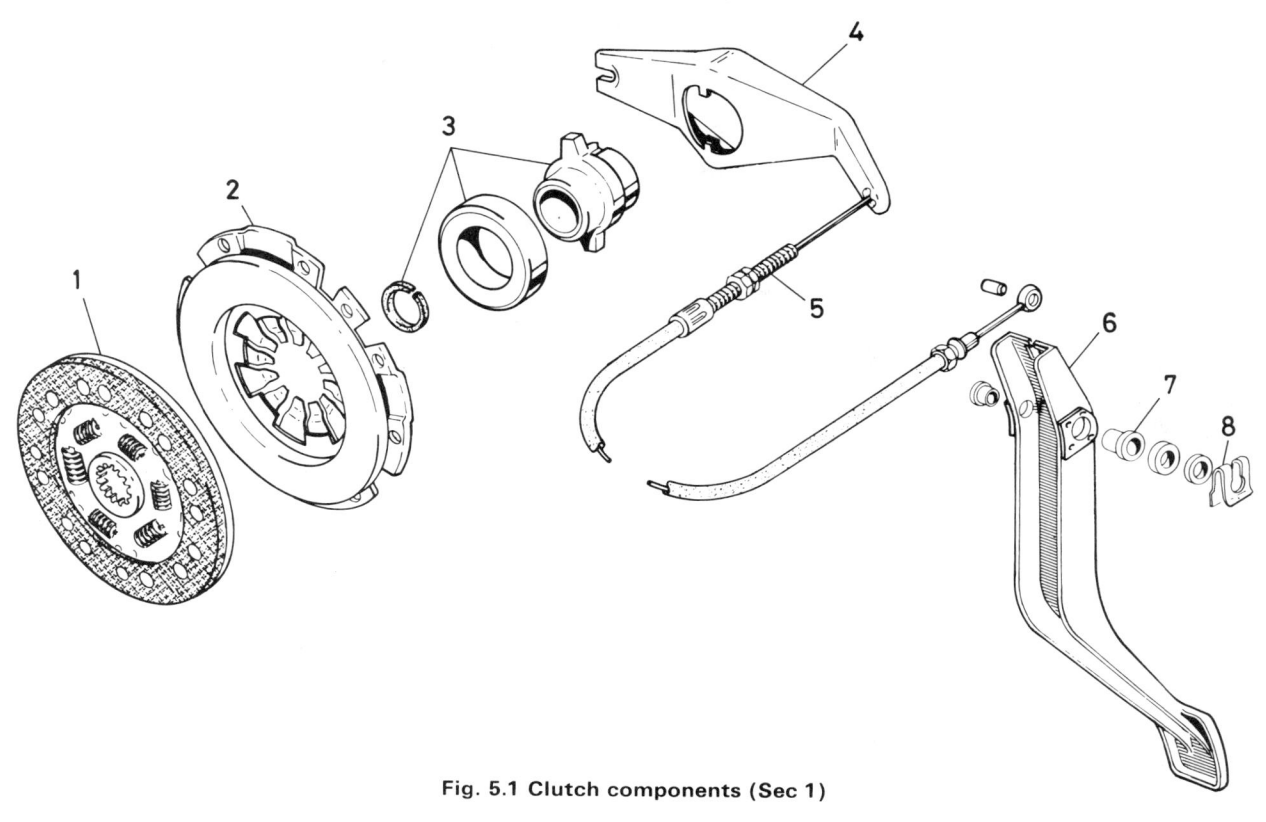

Fig. 5.1 Clutch components (Sec 1)

1. Friction disc
2. Pressure plate assembly
3. Release bearing and hub
4. Release lever
5. Cable
6. Clutch pedal
7. Bush
8. Spring clip

2 Routine maintenance

At the intervals specified in the Routine Maintenance Section in the front of the manual, carry out the following procedure.

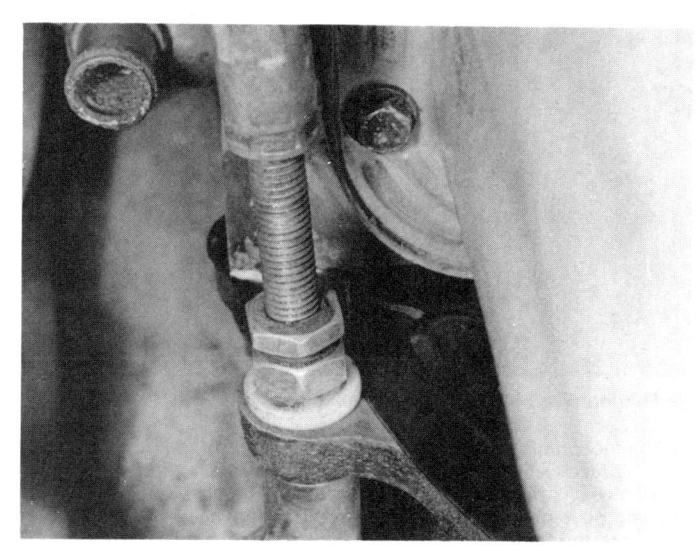

2.2 Clutch cable adjustment and locknuts

Clutch pedal and cable adjustment

1 Apply the handbrake then jack up the front of the car and support on axle stands.
2 Ensure that the clutch cable is not kinked, then pull the outer cable forward from the clutch housing to take up the free play. Release the locknut and turn the adjusting nut so that it is against the abutment face of the clutch housing (photo).
3 Fully depress the clutch pedal several times to ensure correct seating of the cable components, then check that the free travel measured at the end of the clutch pedal is as shown in Fig. 5.2

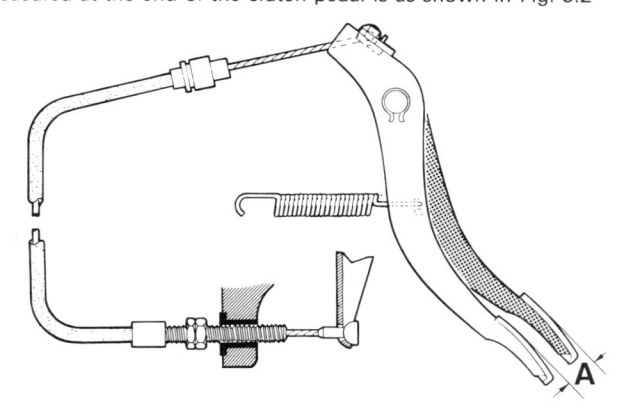

Fig. 5.2 Clutch pedal free travel adjustment (Sec 2)

$A = 27.0 \pm 4.0$ mm (1.06 ± 0.16 in)

Chapter 5 Clutch

3.1 View of the clutch with the gearbox removed

3.3 One method of locking the starter ring gear while unscrewing the clutch cover bolts

4 If necessary, turn the adjusting nut until the pedal free travel is correct. Tighten the locknut against the adjusting nut where fitted – some early models are not fitted with a locknut.
5 Lower the car to the ground.

3 Clutch assembly – removal

1 Remove the gearbox as described in Chapter 6 (photo).
2 Mark the clutch cover and flywheel in relation to each other to ensure identical positioning if the components are re-used.
3 Progressively unscrew the clutch cover bolts then withdraw the assembly from the dowels on the flywheel and at the same time recover the friction plate. If necessary, hold the flywheel stationary by locking the starter ring gear (photo).

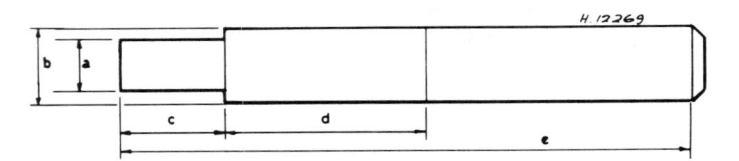

Fig. 5.3 Clutch centralising mandrel dimensions (Sec 5)

a = 16.26 mm (0.64 in)
b = 25.40 mm (1.00 in)
c = 31.75 mm (1.25 in)
d = 44.45 mm (1.75 in) minimum
e = 152.40 mm (6.00 in) approx

4 Clutch assembly – inspection

1 Examine the friction disc for wear of the friction material, for broken hub springs, distortion of the rim and wear on the splines. Unless the clutch plate is in very good condition it is a false economy not to fit a new one.
2 The friction disc should be renewed as as assembly. It is advised that this course is preferable to trying to fit a new friction lining.
3 Check the machined faces of the flywheel and pressure plate. If the flywheel is scored, it should be removed and machined. If the pressure plate is scored, a new assembly should be fitted.
4 Examine the clutch assembly for any signs of oil leakage into the clutch and if signs of leaks are found, rectify the leaks before fitting the clutch.
5 Examine the diaphragm spring for wear and damage. If the diaphragm spring is unserviceable, a new pressure plate assembly must be fitted.
6 Check the release bearing for smoothness of operation. It should be reasonably free, bearing in mind that it is pre-packed with grease and there should not be any roughness, or slackness, in it.
7 Check the condition of the clutch spigot bearing in the end of the crankshaft. Further information on this is given in Chapter 1.

5 Clutch assembly – refitting

Note: A clutch centralising tool is required in this Section.
1 It is important that no oil or grease gets onto the friction material of the friction disc or on to the faces of the flywheel and pressure plate. It is advisable to have clean hands when refitting the clutch and to wipe the flywheel and pressure plate surfaces with clean rag before reassembly is started.

5.2 Friction disc FLYWHEEL SIDE marking

2 The friction disc is marked *flywheel side* (photo) and it is important that this face, which can also be identified by its having the longer central boss, is placed against the flywheel. The cushion springs and retaining plate must be offset away from the flywheel.
3 Locate the friction disc and cover assembly on the flywheel with the cover correctly positioned on the locating dowels (photo). If the original assembly is being refitted align the previously made marks.
4 Insert the retaining bolts finger tight.

Chapter 5 Clutch

5.3 Locating the friction disc on the flywheel

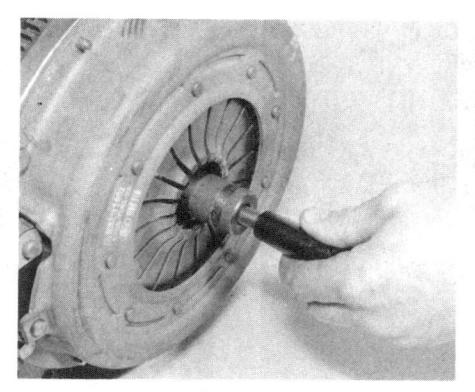

5.6 Centralising the friction disc

5.7 Tightening the clutch cover bolts

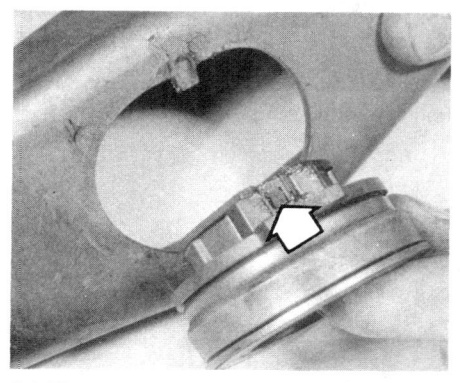

6.4 The plastic tabs which retain the release bearing in the lever (arrowed)

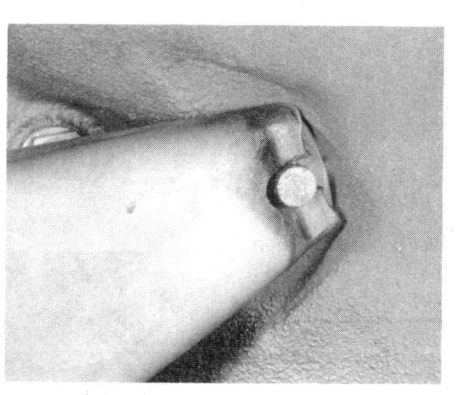

6.5 Release lever location on the pivot stud

6.6 View of the release bearing

5   The friction disc must now be centralised so that when the engine and gearbox are mated the input shaft splines will pass through the splines in the friction disc. Ideally a universal clutch centralising tool or an old gearbox input shaft should be used, but alternatively a wooden mandrel can be made to the dimensions shown in Fig. 5.3.
6   Insert the centralising tool through the friction disc and into the spigot bearing (photo). Rotate the tool or move it side to side until the friction disc is accurately centred.
7   Progressively tighten two opposite cover bolts until the friction disc is firmly held in position, then tighten all the bolts progressively and in diagonal sequence (photo).
8   Remove the centralising tool.
9   Wipe down the gearbox input shaft and apply a **little** molybdenum based grease to the splines and spigot bearing surface.
10 Refit the gearbox with reference to Chapter 6.

6 Clutch release bearing – removal and refitting

1   Remove the gearbox as described in Chapter 6. Alternatively, if the engine is to be removed for repairs, access to the release bearing is possible leaving the gearbox in position.
2   Loosen and back off the cable adjustment nuts.
3   Where applicable, prise the clutch cable rubber gaiter from the bellhousing and unhook the cable end fitting from the release lever.
4   Pull the release bearing from the guide tube. Where necessary use a screwdriver to prise the bearing from the lever and release the plastic tabs (photo).
5   Pull the release lever sideways to disengage it from the pivot retaining stud (photo), then withdraw it over the guide tube and input shaft.
6   The release bearing is supplied complete with the hub (photo). If required, the two components can be separated using a vice and soft metal drift.
7   Before refitting the release bearing, apply a little molybdenum grease to the guide tube and contact faces of the release lever.
8   Refitting is a reversal of removal, but finally check the clutch pedal and cable adjustment as described in Section 2.

7 Clutch cable – renewal

1   Apply the handbrake then jack up the front of the car and support on axle stands.
2   Prise the clutch cable rubber gaiter from the bellhousing, loosen the adjustment nut(s), and unhook the cable end fitting from the release lever. Pull off the gaiter (photos).
3   Remove the cable from the hole in the bellhousing, noting the position of the sleeve (photo).
4   Working inside the car, remove the lower facia panel for access to the clutch pedal.
5   Disconnect the cable from the clutch pedal by pushing the pin through the cable eye.

7.2A Prise off the rubber gaiter ...

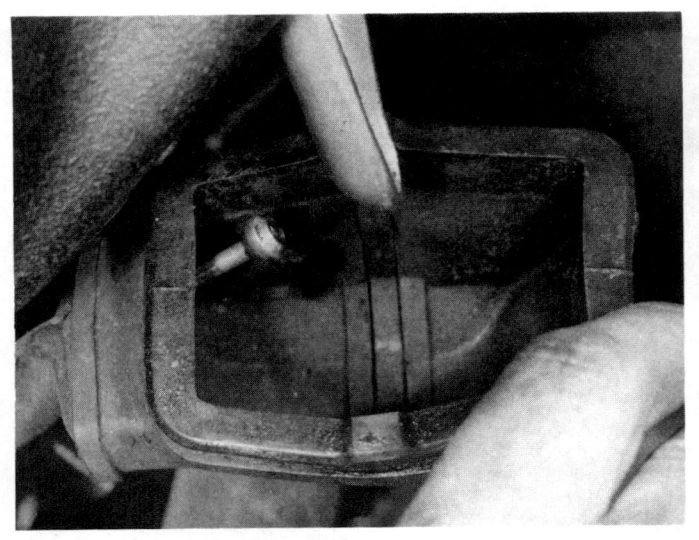

7.2B .. and unhook the inner cable

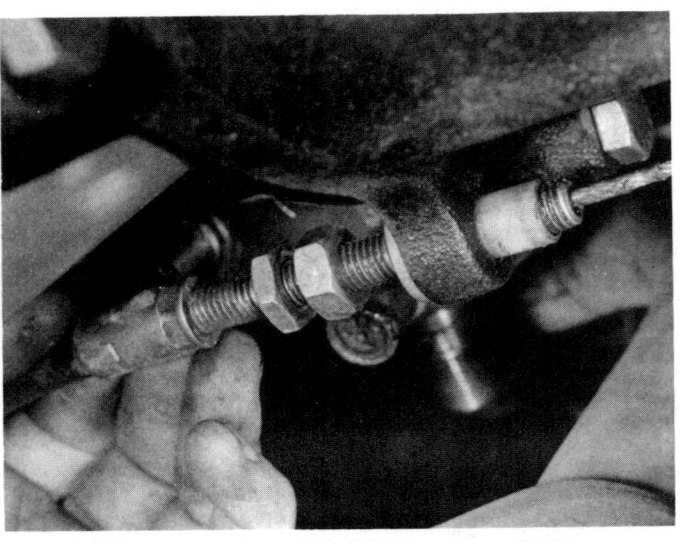

7.3 Removing the cable from the bellhousing

6 Working in the engine compartment pull the cable through the bulkhead and remove it from the car.
7 Fit the new cable using a reversal of the removal procedure, then adjust the pedal free travel as described in Section 2.

8 Clutch pedal – removal and refitting

1 Working inside the car, remove the lower facia panel below the steering column.
2 Remove the left-hand spring clip and its washer from the pedal pivot shaft.
3 Unhook the clutch pedal return spring.
4 Push the pedal shaft to the right until it is clear of the pedal and remove the pedal.
5 Push the pin out of the eye in the end of the clutch cable to disconnect the cable.
6 Remove the clutch pedal pivot bush.
7 Examine the pedal and bush for wear and if necessary renew them.
8 Refitting is a reversal of removal, but lubricate the pivot bush with a little grease and finally adjust the pedal free travel as described in Section 2.

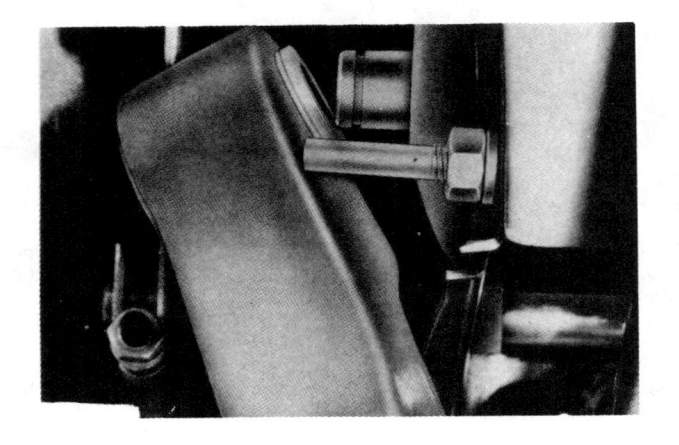

Fig. 5.4 Removing the clutch pedal (Sec 8)

9 Fault diagnosis – clutch

Symptom	Reason(s)
Judder when taking up drive	Loose or worn engine mountings Clutch disc contaminated with oil, or linings worn Clutch cable sticking or frayed Faulty pressure plate assembly
Clutch fails to disengage	Clutch disc sticking on input shaft splines Clutch cable broken Faulty pressure plate assembly Clutch pedal and cable adjustment incorrect
Clutch slips	Clutch pedal and cable adjustment incorrect Clutch disc contaminated with oil, or linings worn Faulty pressure plate assembly
Noise when depressing clutch pedal	Worn or dry release bearing Faulty pressure plate assembly Worn clutch disc or input shaft splines
Noise when releasing clutch pedal	Faulty pressure plate assembly Broken clutch disc cushioning springs

Chapter 6
Manual gearbox and automatic transmission

Contents

Part 1: Manual gearbox

Fault diagnosis – manual gearbox	17
Gearbox – removal and refitting	3
Gearbox (4-speed) – dismantling	4
Gearbox (4-speed) – examination and renovation	8
Gearbox (4-speed) – reassembly	9
Gearbox (5-speed) – disengagement of 5th gear	16
Gearbox (5-speed) – dismantling	11
Gearbox (5-speed) – examination and renovation	14
Gearbox (5-speed) – reassembly	15
Gearchange lever (4-speed) – modifications	10
General description	1
Input shaft (4-speed) – dismantling and reassembly	6
Input shaft (5-speed) – dismantling and reassembly	13
Laygear (4-speed) – dismantling and reassembly	7

Mainshaft (4-speed) – dismantling and reassembly	5
Mainshaft (5-speed) – dismantling and reassembly	12
Routine maintenance	2

Part 2: Automatic transmission

Automatic transmission – removal and refitting	20
Downshift cable – removal, refitting and adjustment	23
Extension housing oil seal – renewal	25
Fault diagnosis – automatic transmission	26
General description	18
Routine maintenance	19
Selector mechanism – adjustment	22
Selector mechanism – removal, overhaul and refitting	21
Starter inhibitor/reverse lamp switch – removal and refitting	24

Specifications

Part 1: Manual gearbox

Type .... Four or five forward speeds (all synchromesh) and reverse

Ratios
Four-speed gearbox:
1st	3.16 : 1
2nd	1.94 : 1
3rd	1.41 : 1
4th	1 : 1
Reverse	3.346 : 1

Five-speed gearbox:
    1st .... 3.36 : 1
    2nd .... 1.81 : 1
    3rd .... 1.26 : 1
    4th .... 1 : 1
    5th .... 0.825 : 1
    Reverse .... 3.365 : 1

Layshaft cluster gear (four-speed gearbox)

Endfloat .... 0.15 to 0.45 mm (0.006 to 0.018 in)
Thrust washer thickness .... 1.55 to 1.60 mm (0.061 to 0.063 in)

Lubrication

Lubricant type/specification:
    All models except 5-speed 2.8 Injection .... Hypoid gear oil, viscosity SAE 80EP to Ford spec SQM-2C 9008-A (Duckhams Hypoid 80)
    5-speed 2.8 Injection models .... Semi-synthetic gear oil to Ford spec ESD-M2C-175-A (Duckhams Hypoid 75W/90S)

Lubricant capacity
    Four-speed .... 2.0 litre (3.5 pint)
    Five-speed .... 1.9 litre (3.3 pint)

Torque wrench settings

	Nm	lbf ft
Four-speed gearbox		
Clutch housing to gearbox	39 to 48	29 to 35
Clutch housing to engine	30 to 37	22 to 27
Mainshaft nut	35 to 41	26 to 30
Input shaft bearing retainer	17 to 21	12 to 15
Extension housing	54 to 61	40 to 45
Selector shaft bracket	7 to 10	5 to 7
Selector housing cover	16 to 21	12 to 15
Extension housing cover	9 to 11	7 to 8
Gearbox case cover	17 to 21	12 to 15
Five-speed gearbox		
Guide sleeve	9 to 11	7 to 8
Extension housing	45 to 49	33 to 36
Gearbox top cover	9 to 11	7 to 8
Selector locking mechanism	17 to 19	13 to 14
Filler plug	23 to 27	17 to 20
5th gear collar nut	120 to 150	89 to 111
5th gear locking plate	21 to 26	16 to 19
Gear lever	21 to 26	16 to 19

Part 2: Automatic transmission

Type .... Ford C3

Torque converter

Type .... Trilock (hydraulic)
Ratio .... 2.22 : 1

Transmission ratios

1st .... 2.47 : 1
2nd .... 1.47 : 1
3rd .... 1 : 1
Reverse .... 2.11 : 1

Lubricant

Capacity (approx) .... 7.5 litre (13.2 pint)
Type:
    Early models (black dipstick) .... ATF to Ford specification SQM 2C 9007 AA (Duckhams Q-Matic)
    Later models (red dipstick) .... ATF to Ford specification SQM 2C 9010A (Duckhams D-Matic)

Torque wrench settings

	Nm	lbf ft
Torque converter to driveplate	36 to 41	27 to 30
Downshift cable bracket	16 to 24	12 to 18
Downshift lever nut:		
Outer	10 to 15	7 to 11
Inner	41 to 54	30 to 40
Inhibitor switch	16 to 20	12 to 15
Brake band adjusting screw locknut	47 to 61	35 to 45
Fluid line	9 to 14	7 to 10
Oil cooler line	16 to 20	12 to 15
Automatic transmission to engine	30 to 37	22 to 27
Torque converter drain plug	27 to 40	20 to 30

Chapter 6 Manual gearbox and automatic transmission

PART 1: MANUAL GEARBOX

1 General description

The manual gearbox is of four or five speed type with one reverse gear. All forward gears are engaged through baulk ring synchromesh units. The mainshaft gears rotate freely on the shaft and are in constant mesh with their corresponding gears on the laygear. They are helically cut to achieve quiet running.

The reverse gear on the laygear has straight-cut spur teeth which drive the 1st/2nd synchroniser hub via the reverse idler gear when reverse gear is engaged.

Gears are engaged by a three rail selector system on the four-speed gearbox, and a single rail selector system on the five-speed gearbox.

2 Routine maintenance

At the intervals specified in the Routine Maintenance section in the front of the manual carry out the following procedure.

Check gearbox oil level
1 Jack up the car and support on axle stands, making sure that it is level. Alternatively position the car over an inspection pit or on ramps.
2 Using a suitable square key, unscrew the filler plug from the left-hand side of the gearbox (photo). Note that on the four-speed gearbox the filler plug is located above the drain plug.
3 For accurate results the check should not be made immediately after a run, but if this is unavoidable allow five or ten minutes for the oil to settle.
4 Make up a piece of bent wire to the dimensions shown in Fig. 6.3 then insert it and rest the horizontal section on the threads of the plug hole. Remove the wire and check the oil level. On all four-speed gearboxes, and five-speed gearboxes manufactured after May 1985 (build code FB) the oil should be level with or a maximum of 5.0 mm (0.2 in) below the bottom edge of the plug hole. For five-speed gearboxes manufactured up to April 1984 (build code EG) the oil must be level with the bottom of the plug hole. On five-speed gearboxes manufactured between May 1984 and April 1985 (tag suffix ending in E) the oil must be between 10.0 and 15.0 mm (0.4 and 0.6 in) below the bottom edge of the plug hole.
5 Where necessary top up the level with the specified grade of oil (photo).
6 Refit and tighten the filler plug then lower the car to the ground.

3 Gearbox – removal and refitting

1 If the gearbox is to be removed from the car without removing the engine, the gearbox is removed from beneath the car and a large ground clearance is required. If an inspection pit or ramps are not available, jack the car as high as possible, support it securely on blocks or stands and chock the wheels.
2 Disconnect the battery.

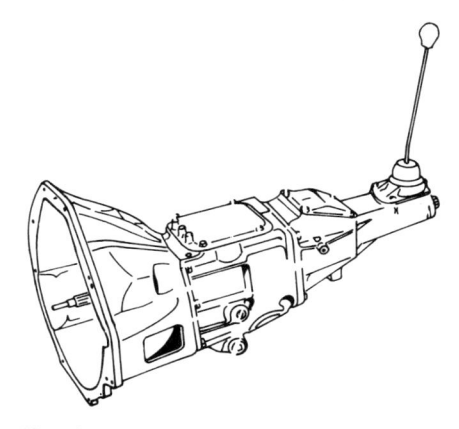

Fig. 6.1 Four-speed gearbox (Sec 1)

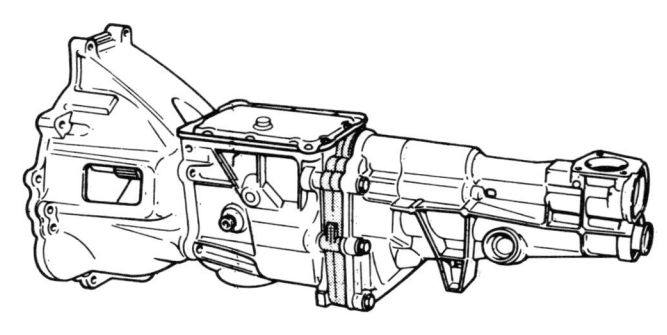

Fig. 6.2 Five-speed gearbox (Sec 1)

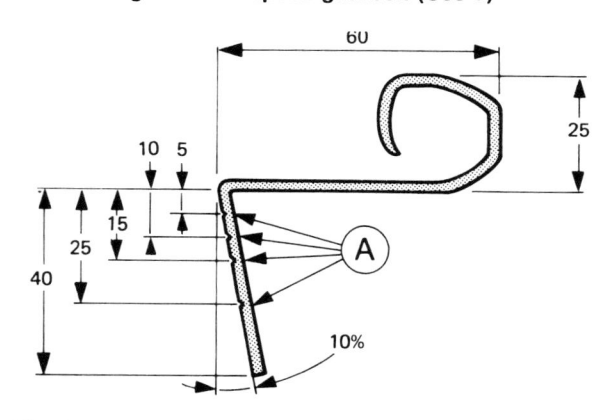

Fig. 6.3 Wire dimensions for making a gearbox oil level dipstick (Sec 2)

All dimensions in mm   A File notches

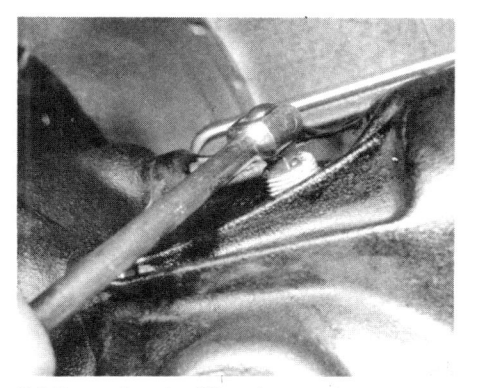

2.2 Removing the filler plug on the five-speed gearbox

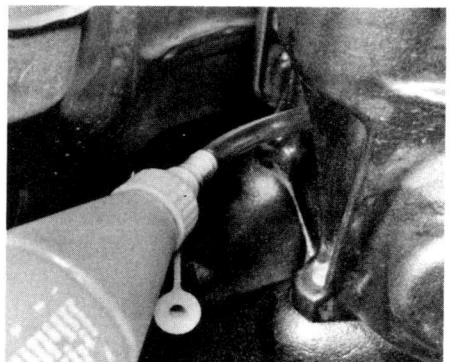

2.5 Topping-up the gearbox oil

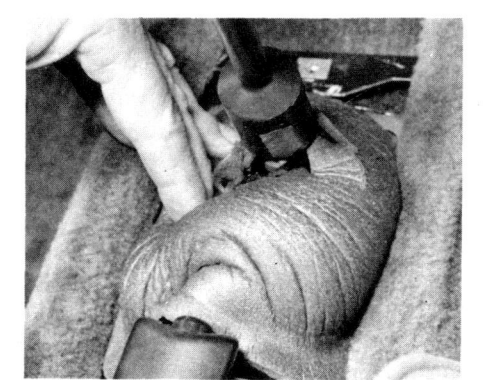

3.4A On the five-speed gearbox remove the packing ...

Fig. 6.4 Using a cranked spanner to unscrew the gear lever retainer (Sec 3)

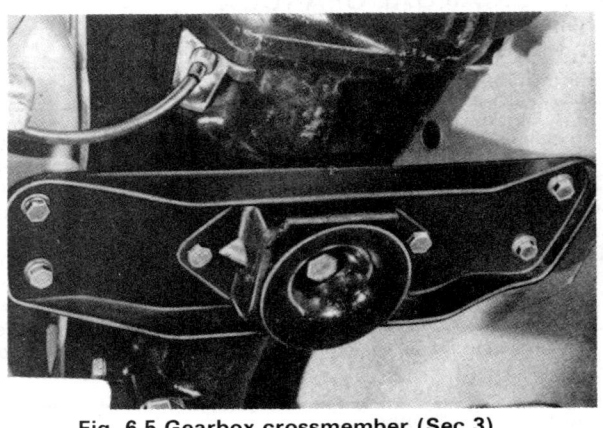

Fig. 6.5 Gearbox crossmember (Sec 3)

3 Unscrew the gear lever knob and remove the cover and rubber gaiter. If a package tray or centre console is fitted, these must be removed (refer to Chapter 11 if necessary).
4 On the four-speed gearbox, bend back the lock tab and unscrew the gear lever retainer using a suitably cranked spanner or careful use of a hammer and drift. On the five-speed type, remove the foam packing then unscrew the Torx screws holding the retainer. Lift out the gear lever (photos).
5 On the four-speed gearbox, place a drip tray beneath the gearbox, remove the drain plug and, when the oil has drained, refit the plug and tighten it. There is no drain plug on the five-speed gearbox.
6 Put mating marks on the two faces of the rear axle joint flange. Remove the four bolts from the rear axle pinion flange and the two bolts from the propeller shaft centre bearing support and withdraw the propeller shaft assembly from the extension housing. Tie a polythene bag round the end of the gearbox extension to keep out dirt to catch any oil which has not drained.

7 Disconnect the clutch cable with reference to Chapter 5, and where applicable the fuel injection fuel lines with reference to Chapter 3.
8 Remove the screw and lockplate securing the speedometer drive and disconnect the drive.
9 Remove the cover (if fitted) from the reversing light switch. Note which way the leads are fitted and then remove them (photo).
10 Detach the exhaust pipes from the exhaust manifold and tie them out of the way, so that the engine and transmission can be lowered.
11 Support the gearbox on a jack, or with blocks and remove the bolt securing the transmission to the rear engine mounting and the four bolts securing the gearbox crossmember to the floor (photo). Remove the crossmember.
12 Lower the transmission a little and insert a block of wood between the sump and the front engine mounting so that the engine does not drop too far when the transmission is removed.
13 Remove the bolts securing the clutch housing to the engine and backplate (photos). Pull the transmission assembly to the rear to

3.4B ... unscrew the Torx screws ...

3.4C ... and lift out the gear lever

3.9 Disconnecting the wiring from the reversing light switch

3.11 Unbolting the gearbox crossmember from the floor

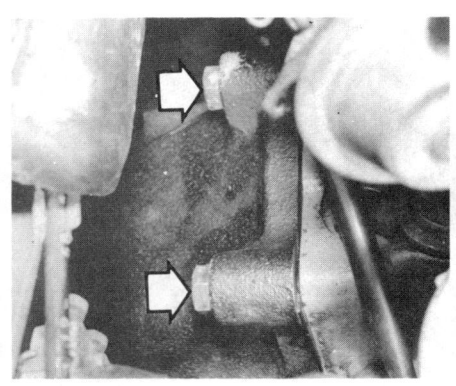

3.13A Two upper clutch housing-to-engine bolts (arrowed)

3.13B A lower clutch housing-to-backplate bolt (five-speed gearbox)

Chapter 6 Manual gearbox and automatic transmission

disengage the gearbox input shaft from the clutch pilot bearing and clutch friction disc. It is important that the engine and transmission are kept in line while this is being done.

14 When the gearbox shaft is clear of the clutch, remove the transmission assembly from beneath the car.

15 When refitting, ensure that the clutch pilot bearing in the end of the crankshaft is in place and is serviceable. Tie the clutch lever to the clutch housing, to prevent the release lever from slipping out while the transmission assembly is being fitted.

16 Smear some molybdenum-based grease onto the end and splines of the gearbox input shaft and refit the gearbox by reversing the removal procedure.

17 It is important when offering up the gearbox, that it is exactly in line with the crankshaft, otherwise the gearbox input shaft will not enter the clutch friction disc and the crankshaft. If there is difficulty in mating the splines of the gearbox shaft and the friction disc, select a gear to restrain the movement of the gearbox shaft and turn the gearbox slightly until the splines enter. Do not attempt to force the transmission onto the engine. This may damage the splines and make fitting impossible.

18 After refitting has been completed, check the oil level in the transmission and top up as necessary.

4 Gearbox (4-speed) – dismantling

1 Remove the four bolts from the top cover of the transmission extension housing and from the transmission case. Remove the covers, taking care not to lose the three springs under the transmission case cover.

2 Use a magnet to remove the three selector detent balls.

3 Remove the bolt from the selector finger and withdraw the finger from the selector shaft.

4 Remove the two bolts from the selector shaft bearing support and remove it.

5 Remove the three bolts from the selector housing cover. Remove the cover with the shaft, then rotate the selector shaft slightly and withdraw it from the selector housing cover.

6 Drive out the roll pin at the side of the extension housing and remove the reverse gear relay lever.

7 Using a pin punch and a hammer, drive out the three roll pins from the 1st/2nd, 3rd/top and reverse gears sufficiently to release the selector forks. Remove 3rd/top gear selector rail circlip and then withdraw the selector rails towards the extension housing. Note the plunger in the 3rd/top selector rail.

8 Remove the three selector forks.

9 Remove the plug from the side of the case and use a magnet to extract the plunger beneath it.

10 Bend back the tabs of the washers of the two bolts on the extension housing. Remove the banjo bolt, spring and ball and the reversing light switch, spring and plunger.

11 Use two screwdrivers and carefully push the speedometer out of the extension housing.

12 If there has been no sign of leaking from the extension housing oil seal, take care not to damage it. If the seal is defective, lever it out, taking care not to damage its seating.

13 Remove the four attachment bolts and detach the extension housing from the transmission case.

14 Using a long thin drift, tap out the layshaft from the front to the rear. Allow the laygear cluster to drop into the bottom of the box, out of mesh with the mainshaft gears.

15 Withdraw the mainshaft and extension assembly from the gearbox casing, pushing the 3rd/top synchroniser hub forward slightly to obtain the necessary clearance. A small roller bearing should be on the nose of the mainshaft, but if it is not there it will be found in its recess in the input shaft and should be removed.

16 Remove the four bolts securing the drive gear bearing retainer to the front of the gearbox and slide the retainer off the input shaft.

17 Remove the circlip from the outside of the front bearing and tap the outer race of the bearing to drive the input shaft and bearing assembly into the gearbox, then lift them out of the box.

18 Remove the layshaft gear cluster, the two thrust washers and any loose needle rollers.

19 Remove the reverse idler shaft by screwing a suitable bolt into the end of the shaft and then levering the shaft out using two open-ended spanners.

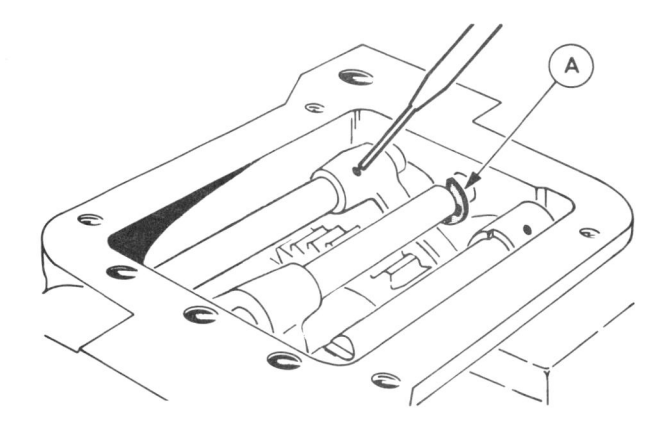

Fig. 6.6 Removing the selector fork roll pins (Sec 4)

A Circlip

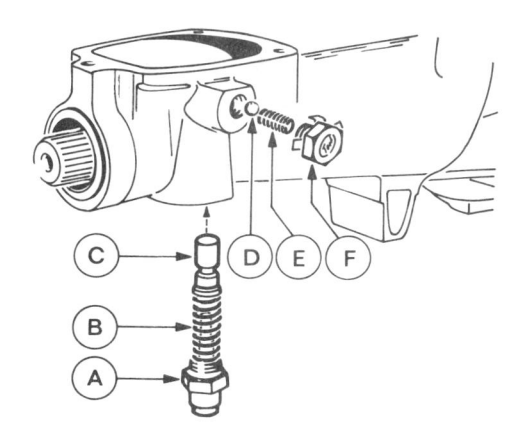

Fig. 6.7 Reverse gear interlock and light switch (Sec 4)

A Reverse light switch
B Spring
C Plunger
D Interlock ball
E Screw plug
F Retaining bolt

5 Mainshaft (4-speed) – dismantling and reassembly

1 Remove 3rd/top gear synchroniser hub circlip. Ease the hub and third gear forward by levering them gently with a pair of snipe-nosed pliers.

2 Remove the hub, the synchroniser ring and 3rd gear from the front of the mainshaft.

3 Mount the plain part of the mainshaft in a soft jawed vice, unlock the mainshaft nut and remove it.

4 Remove the lockplate, speedometer drivegear, its drive ball and then lever off the rear bearing, complete with bearing retainer. Separate the bearing and its retainer.

5 Remove the oil scoop ring and first gear.

6 Remove 1st/2nd gear synchroniser circlip and pull off the hub with the baulk rings and second gear.

7 Check that there is a mark to align the selector sleeve in relation to the hub. If no mark can be seen, make one with a dab of paint and then dismantle the synchroniser hubs.

8 Before starting reassembly, clean all parts thoroughly and check their condition. Smear all parts with transmission oil before fitting.

9 Reassemble the synchroniser sleeve and hub so that the marks on them align. Insert the sliding keys and fit the sliding key springs so that their open ends are staggered relative to each other.

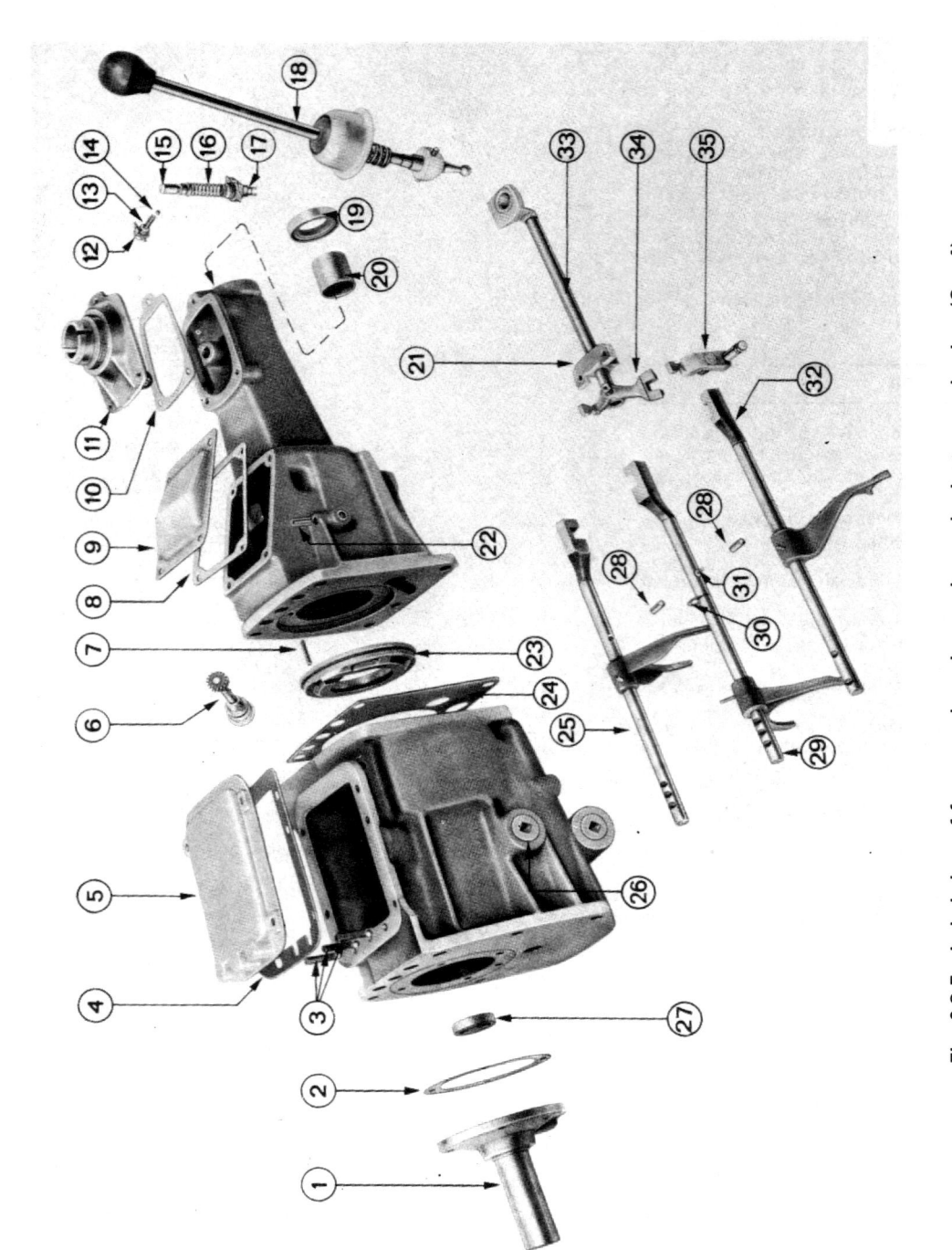

Fig. 6.8 Exploded view of four-speed gearbox casing and selector mechanism (Sec 4)

1 Main drive gear bearing retainer
2 Gasket
3 Detent balls with springs
4 Transmission case gasket
5 Transmission case cover
6 Speedometer pinion
7 Dowel
8 Extension housing cover gasket
9 Extension housing top cover
10 Selector housing gasket
11 Selector housing cover
12 Cap bolt with lockplate
13 Spring
14 Detent ball
15 Plunger
16 Spring
17 Reversing light switch
18 Gear lever assembly
19 Oil seal
20 Extension housing bush
21 Selector shaft bearing support
22 Reverse relay arm roll pin
23 Bearing support
24 Extension housing gasket
25 Selector rail with 1st/2nd gear selector fork
26 Filler plug
27 Drive gear, bearing retainer oil seal
28 Plunger
29 Selector rail with 3rd/4th gear selector fork
30 Circlip
31 Plunger
32 Selector rail with reverse gear selector fork
33 Selector shaft
34 Selector finger
35 Reverse gear relay lever

Fig. 6.9 Exploded view of four-speed gearbox internal components (Secs 4, 5, 6 and 7)

1 Circlip
2 Circlip
3 Grooved ball-bearing
4 Input shaft
5 Needle roller bearing
6 Circlip
7 3rd/4th gear synchroniser baulk ring
8 3rd/4th gear synchroniser hub/sleeve assembly
9 3rd gear
10 Transmission mainshaft
11 Speedometer drive locking ball
12 2nd gear
13 1st/2nd gear synchroniser baulk ring
14 1st/2nd gear synchroniser with reverse
15 Circlip
16 1st gear
17 Oil scoop ring
18 Intermediate bearing support
19 Bearing
20 Spacer
21 Speedometer wormdrive gear
22 Lockplate
23 Mainshaft nut
24 Layshaft thrust washer
25 Shim
26 Needle rollers (22)
27 Laygear
28 Layshaft thrust washer
29 Layshaft
30 Reverse idler gear
31 Idler shaft

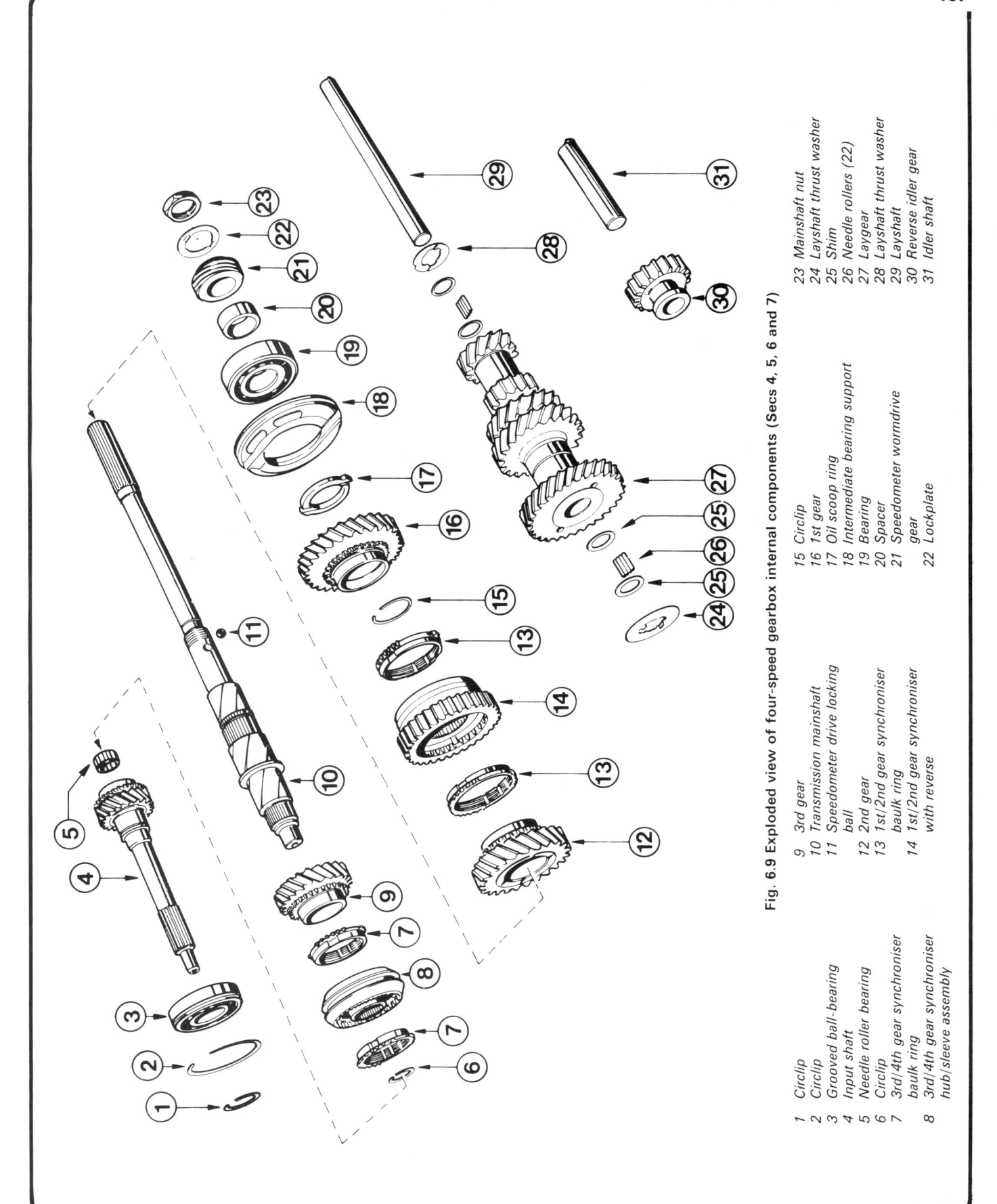

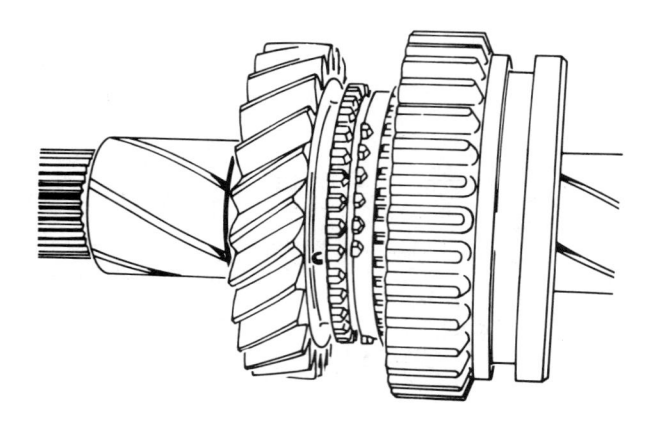

Fig. 6.10 Reverse gearing facing forwards (Sec 5)

10 Place the synchroniser baulk ring onto the cone of second gear and slide it onto the mainshaft together with 1st/2nd gear synchroniser hub. The gear on the hub should face forward.
11 Fit the synchroniser baulk ring and the circlip which holds the synchroniser hub in place.
12 Slide on first gear, so that the synchroniser cone portion lies inside the synchronising ring which has just been fitted.
13 Fit the oil scoop, with the oil groove facing rearwards.
14 Apply multi-purpose grease to the ball-bearing seat of the bearing retainer. Refit the retainer with the bearing inserted and then slide on the spacer.
15 Insert the drive ball of the speedometer gear into the shaft and slide the speedometer worm gear over it.
16 Fit the lockplate so that the two tabs fit into the speedometer worm gear recess. Screw on the nut and tighten it to the torque wrench setting given in the Specifications, then lock the nut by bending the tab washer.
17 Fit the synchroniser baulk ring to the cone of third gear and fit them to the front of the mainshaft. Fit the 3rd/top gear synchroniser hub, with its wide boss towards the rear and then retain it in place by fitting the circlip.

6 Input shaft (4-speed) – dismantling and reassembly

1 It is not necessary to dismantle the input shaft unless a new bearing or a new shaft is being fitted.
2 With a pair of circlip pliers, expand and remove the small circlip, which secures the bearing to the input shaft.
3 With a soft-headed hammer, gently tap the bearing forward until it can be pulled from the shaft.
4 When fitting the bearing, ensure that the groove in the periphery of the outer bearing track is away from the gear, otherwise it will not be possible to fit the large circlip which retains the bearing in place in the housing.
5 Either stand the input shaft upright on a bench and tap the bearing into place using a piece of tube of suitable diameter, or use the jaws of a vice to support the bearing and tap the rear of the input shaft with a soft faced hammer.
6 When the bearing is fully home, fit the circlip which retains the bearing on the shaft.

7 Laygear (4-speed) – dismantling and reassembly

1 Remove a shim from each end of the laygear, remove 22 needle rollers from each end of the gear.
2 When reassembling, insert a shim into the bottom of the bore. Daub the bore with general purpose grease and insert the 22 needle rollers. When all the rollers are in place, smear grease over them to retain them and then secure the shim over the ends of the rollers with grease. Due

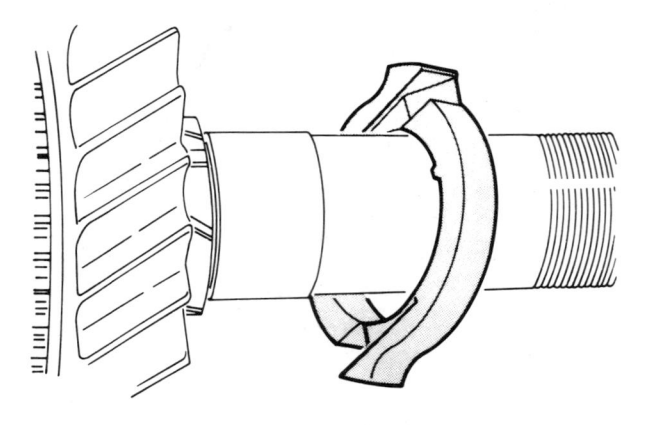

Fig. 6.11 Position of oil scoop ring with groove facing gear (Sec 5)

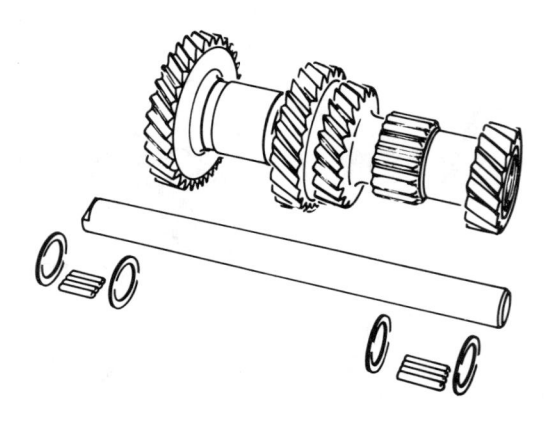

Fig. 6.12 Layshaft and gear cluster (Sec 7)

to the high levels of torque transmitted through the gearbox it is important that all the needle rollers have exactly the same diameter. If the needle rollers in a bearing have different diameters, this will cause uneven load distribution which can result in premature wear. If it is necessary to renew the needle rollers, make sure that the rollers used for any one location all come from the same pack. If more than one pack is purchased, do not mix the needle rollers from separate packs.

8 Gearbox (4-speed) – examination and renovation

1 Thoroughly clean the interior of the gearbox, and check for dropped needle rollers and spring pins.
2 Carefully clean and then examine all the component parts for general wear, distortion, slackness of fit, and damage to machined faces and threads.
3 Examine the gears for excessive wear and chipping of the teeth. Renew them as necessary.
4 Examine the layshaft for signs of wear, where the needle rollers bear. If a small ridge can be felt at either end of the shaft it will be necessary to renew it. Renew the thrust washers at each end.
5 The four synchroniser baulk rings are bound to be badly worn and it is false economy not to renew them. New rings will improve the smoothness and speed of the gearchange considerably.
6 The needle roller bearing and cage, located between the nose of the mainshaft and the annulus in the rear of the input shaft, is also liable to wear, and should be renewed as a matter of course.
7 Examine the condition of the two ball-bearing assemblies, one on the input shaft and one on the mainshaft. Check them for noisy

Chapter 6 Manual gearbox and automatic transmission

operation, looseness between the inner and outer races, and for general wear. Normally they should be renewed on a gearbox that is being rebuilt.

8 If either of the synchroniser units are worn it will be necessary to buy a complete assembly as the parts are not sold individually. Also check the sliding keys for wear.

9 Examine the ends of the selector forks where they rub against the channels in the periphery of the synchroniser units. If possible compare the selector forks with new units to help determine the wear that has occurred. Renew them if worn.

10 If the bearing bush in the extension is badly worn it is best to take the extension to your local Ford garage to have the bearing pulled out and a new one fitted. **Note:** *This is normally done with the mainshaft assembly still located in the extension housing.*

11 The oil seals in the extension housing and main drive gear bearing retainer should be renewed as a matter of course. Drive out the old seal with the aid of a drift or broad screwdriver. It will be found that the seal comes out quite easily.

12 With a piece of wood or suitable sized tube to spread the load evenly, carefully tap a new seal into place ensuring that it enters the bore squarely.

9 Gearbox (4-speed) – reassembly

1 When any bolt screws into a through bore, jointing compound should be applied to the threads of the bolt before it is inserted.

2 Fit the reverse idler gear with the groove on the gear towards the rear. Insert the idler shaft and drive it in with a plastic-headed hammer until the bottom of the step on the end of the shaft is flush with the end of the gearbox. The milled flats should be proud of the gearbox face and should be aligned with the layshaft bore.

3 Fit the large thrust washer to the layshaft bore at the front of the gearbox and the small thrust washer to the layshaft bore at the rear of the box. The tabs on the washers should be towards the case and the washers should be greased to retain them in place.

4 Check that the needle rollers and end washers of the layshaft gear are in place, then lower the gear cluster into the case carefully, making sure that the thrust washers are not displaced.

5 Push the input shaft, together with its ball-bearing, into the front of the gearbox until the circlip round the bearing bears against the front of the gearbox case.

6 Fit the drive bearing retainer, using a new gasket, making sure that the oil return bore of the gasket and bearing retainer are aligned as in Fig. 6.13. Apply jointing compound to the threads of the bolts, insert them and tighten to the torque wrench setting given in the Specifications.

7 Oil the new input shaft needle roller bearing and push it into the bore of the input shaft. Slide the top gear synchroniser baulk ring on to the cone of the input shaft.

8 Fit the extension housing gasket, then insert the assembled mainshaft into the rear of the box and drive it in until the nose of the mainshaft is fully engaged in the input shaft needle bearing.

9 Carefully turn the gearbox upside down, so that the layshaft gear cluster drops into engagement with the mainshaft and input shaft. Check that the gear and the two thrust washers are in line and then carefully push in the layshaft from the rear. If necessary tap the shaft in with the plastic-headed hammer, but take great care not to dislodge the thrust washers or needle rollers. When fully home, the bottom of the step on the idler shaft should be in line with the face of the box and the milled flats should be in line with the idler shaft bore.

10 Turn the gearbox the right way up and align the hole in the ball bearing support with the dowel on the extension housing. Fit the extension housing. Apply jointing compound to the threads of the bolts, insert them and tighten to the torque wrench setting given in the Specifications.

11 If a new extension bush has been fitted, check that the notch or oil groove in it aligns with the oil return groove in the extension housing. if the new extension housing oil seal has not already been fitted, slide it over the extension shaft and carefully tap it in with a hammer and a block of wood.

12 Fit a new O-ring to the speedometer pinion assembly, refit the assembly and fit the circlip to secure it.

13 Insert the reverse gear interlock plunger, ball and spring. Fit the bolt with a lock plate and insert the bolt. Tighten the bolt and bend the lock plates to secure it.

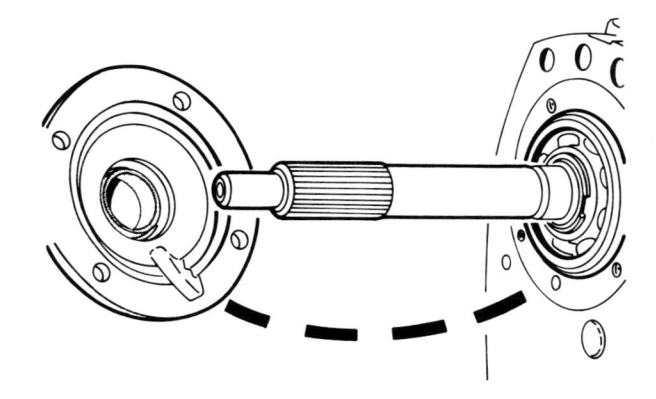

Fig. 6.13 Positioning of front bearing retainer oil return groove (Sec 9)

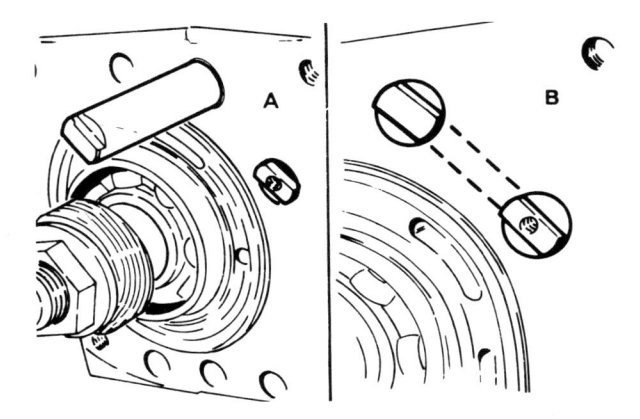

Fig. 6.14 Refitting the layshaft (Sec 9)

A Inserting layshaft
B Correct alignment of layshaft and reverse idler shaft

14 Insert all three selector forks into their positions in the gearbox.

15 Slide 1st/2nd gear selector rail through the right-hand bore in the case, then thread it through the 1st/2nd gear selector fork. Align the holes in the fork and rail and tap in the roll pin. Insert the plunger into the case.

16 Fit the plunger into the 3rd/top gear selector rail, insert the rail through the centre bore in the case and thread the rail through the 3rd/top gear selector fork. Fit the circlip to the groove in the selector rail. Align the holes in the selector fork and rail, insert the roll pin and tap it home.

17 Push reverse gear selector rail through the left-hand bore in the housing, and then thread it through the reverse gear selector fork. Align the holes in the fork and rail, insert a roll pin and tap it home.

18 Fit the interlock plunger, apply sealing compound to the plug and drive in the plug.

19 Insert reverse gear relay lever and secure it with a roll pin.

20 Insert the selector shaft through the selector housing cover and fit the cover using a new gasket. Insert the cover retaining bolts and screw them in finger tight.

21 Fit the selector shaft bracket.

22 Slide the selector finger onto the selector shaft. Engage the screw in the recess in the shaft and tighten the screw.

23 Drop the three detent balls into their bores in the case, grease the three springs and insert them into the recesses of the gearbox cover. Fit a new cover gasket and position the cover onto it. Insert the cover bolts and tighten them to the torque wrench setting given in the Specifications.

Chapter 6 Manual gearbox and automatic transmission

24 Check the play between the selector finger and 3rd/top gear selector rail (Fig. 6.16). This dimension should be 0.9 mm (0.035 in). To make the check, move the selector lever up to the reverse gear stop. Using a lever, push 3rd/top gear selector rail to the rear and then measure the clearance by inserting feeler gauges between 3rd/top gear selector rail and selector finger. If the clearance is not as specified, it should be corrected by changing the reverse interlock gear plunger for one which will give the appropriate clearance.

25 Fit a new gasket to the extension housing cover. Position the cover, insert the fixing bolts and tighten them to the torque wrench setting given in the Specifications.

26 Make sure that the gearbox drain plug is fitted and tightened where applicable, then fill the gearbox to the correct level with the specified oil.

10 Gearchange lever (4-speed) – modifications

1 To improve the gearchange movement between the gears a smaller gear lever spring is fitted, the new spring diameter being 2.0 mm (0.08 in) against the original spring diameter of 2.5 mm (0.10 in).

2 To renew this spring, first remove the gear lever unit from the transmission as given in Section 3.

3 Press or drive out the roll pin and remove the plastic cup (Fig. 6.17) and spring.

4 When reassembling, check that the pin diameter is less than the width of the slots (X in Fig. 6.17). If necessary grind down the pin and de-burr the slots by careful filing. When fitted the pin slots must face upwards.

5 If the gear lever was loose when in use, an O-ring can be fitted in place of the retainer seal as shown in Fig. 6.18.

6 When this ring is fitted and the gear lever relocated, screw the retainer cap into position so that it just touches the O-ring. Mark the relative positions of the cap and selector housing, then tighten the cap a further 15 mm (0.6 in).

11 Gearbox (5-speed) – dismantling

1 Clean the exterior of the gearbox with paraffin and wipe dry.
2 Remove the clutch release bearing and arm referring to Chapter 5.

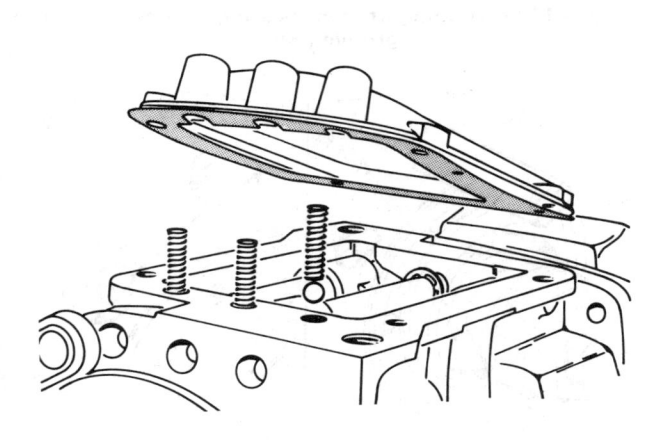

Fig. 6.15 Cover, detent balls and springs (Sec 9)

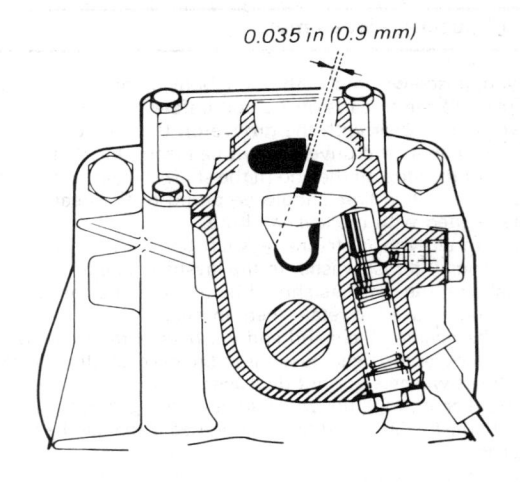

Fig. 6.16 Play between selector finger and 3rd/4th gear selector rail (Sec 9)

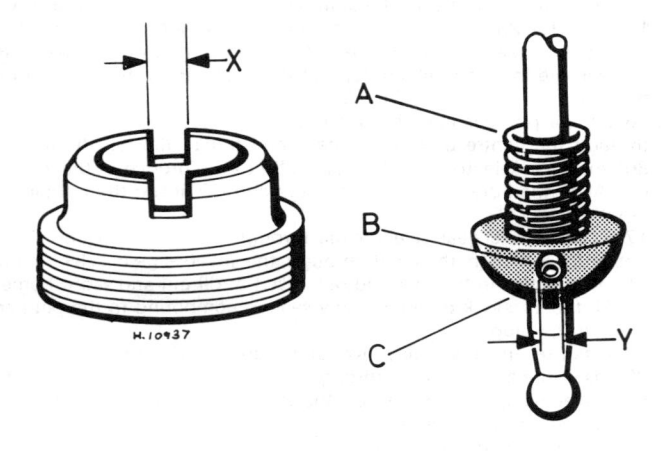

Fig. 6.17 Gearchange lever components (Sec 10)

A Spring
B Retaining pin
C Plastic cap

Dimension X must be greater than dimension Y

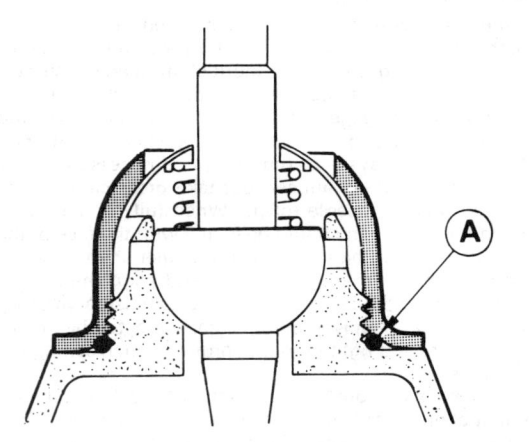

Fig. 6.18 Cross-section of gear lever showing O-ring (A) location (Sec 10)

Chapter 6 Manual gearbox and automatic transmission

3 Unscrew and remove the reversing light switch (photo).
4 Unbolt the clutch bellhousing from the front of the gearbox. Remove the gasket.
5 Unscrew the bolts and withdraw the clutch release bearing guide sleeve and gasket from the front of the gearbox (photo).
6 Unscrew the bolts and remove the top cover and gasket (photo).
7 Invert the gearbox and allow the oil to drain, then turn it upright again.
8 Unscrew the bolts and lift the 5th gear locking plate from the extension housing (photo).
9 Extract the 5th gear locking spring and pin from the extension housing (photos). Use a screw to remove the pin.
10 Working through the gear lever aperture, use a screwdriver or small drift to tap out the extension housing rear cover (photo).
11 Select reverse gear and pull the selector shaft fully to the rear. Support the shaft with a piece of wood, then drive out the roll pin and withdraw the connector from the rear of the selector shaft (photos).
12 Unbolt and remove the extension housing from the rear of the gearbox. If necessary tap the housing with a soft-faced mallet to release it from the dowels. Remove the gasket (photo).
13 Prise the cover from the extension housing and withdraw the speedometer drivegear (photo).
14 Select neutral then, using an Allen key, unscrew the selector locking mechanism plug from the side of the main casing. Extract the spring and locking pin, if necessary using a pen magnet (photos).
15 Drive the roll pin from the selector boss and selector shaft.
16 If necessary the selector shaft centralising spring and 5th gear locking control may be removed. Using a small screwdriver, push out the pin and plug and slide the control from the selector shaft (photos).
17 Note the location of the selector components, then withdraw the selector shaft from the rear of the gearbox and remove the selector boss and locking plate, 1st/2nd and 3rd/4th selector forks, and 5th gear selector fork and sleeve. Note that the roll pin hole in the selector boss is towards the front (photos).

11.3 Reversing lamp switch

11.5 Clutch release bearing guide sleeve

11.6 Top cover and gasket

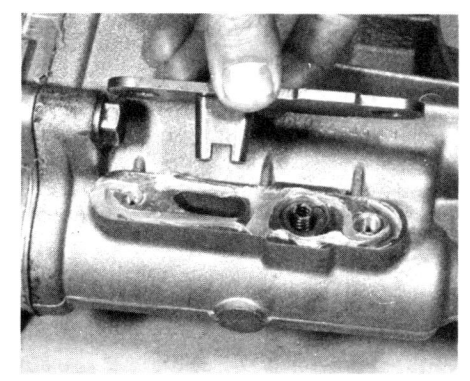

11.8 5th gear locking plate

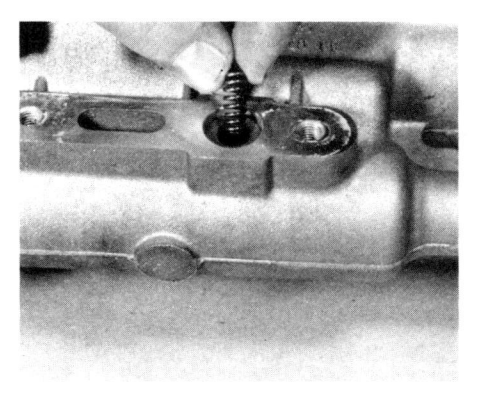

11.9A 5th gear locking spring

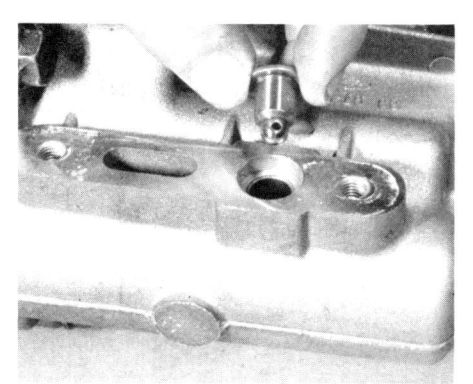

11.9B 5th gear locking pin

11.10 Extension housing rear cover

11.11A Driving out selector shaft roll pin

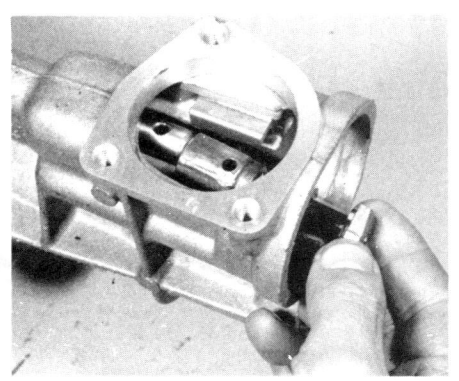

11.11B Removing selector shaft connector

11.12 Removing extension housing

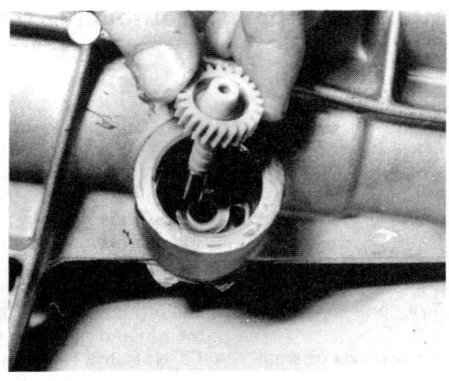

11.13 Removing speedometer drivegear

11.14A Unscrewing selector locking mechanism plug

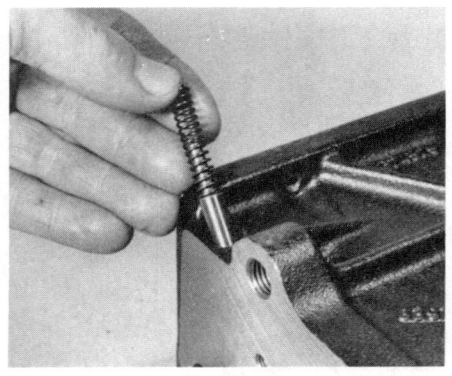

11.14B Selector locking spring and pin

11.16A 5th gear locking control plug

11.16B 5th gear locking control roll pin

11.17A Selector boss and lockplate

11.17B 1st/2nd selector fork

11.17C 3rd/4th selector fork

11.17D 5th gear interlock sleeve

11.17E 5th gear selector fork

Chapter 6 Manual gearbox and automatic transmission

18 Extract the circlip and pull the 5th gear synchroniser unit from the main casing, leaving it loose on the mainshaft (photos).
19 Slide 5th gear from the synchroniser unit hub (photo).
20 Select 3rd gear and either 1st or 2nd gear by pushing the respective synchronsier sleeves – this will lock the mainshaft and layshaft gear cluster.
21 Unscrew and remove 5th gear retaining nut while an assistant holds the gearbox stationary (photo). The nut is tightened to a high torque setting and an additional extension may be required.
22 Remove the washer and pull 5th gear from the layshaft gear cluster using a two-legged puller and socket in contact with the cluster. Remove the spacer ring (photos). Select neutral.
23 Extract the circlip retaining the layshaft gear cluster bearing in the intermediate housing (photo).
24 Using a soft-faced mallet, tap the intermediate housing free of the main casing and pull the intermediate housing rearwards as far as possible. Using a screwdriver inserted between the intermediate housing and main casing, prise the bearing from the shoulder on the layshaft gear cluster and remove it from the intermediate housing (photo).
25 Using a soft metal drift from the front of the main casing, drive the layshaft rearwards to allow the gear cluster to be lowered to the bottom of the casing.
26 Ease the input shaft from the front of the casing, if necessary using a small drift inside the gearbox to move the bearing slightly forwards, then using levers beneath the bearing circlip (photo).
27 Remove the 4th gear synchroniser ring. Remove the input shaft needle roller bearing from the end of the mainshaft or from the centre of the input shaft (photos).
28 Remove the mainshaft and intermediate housing from the main casing (photo). Remove the gasket.
29 Withdraw the layshaft and gear cluster from the main casing (photo).
30 Insert a suitable bolt (M8 x 60) into the reverse gear idler shaft, and

11.18A Extracting 5th gear synchro circlip

11.18B 5th gear synchro-hub

11.18C 5th gear synchro

11.19 Sliding 5th gear from mainshaft

11.21 Removing mainshaft nut

11.22A Withdrawing 5th gear from layshaft

11.22B Removing 5th gear spacer ring

11.23 Extracting layshaft bearing circlip

11.24 Prising off layshaft gear cluster bearing

Chapter 6 Manual gearbox and automatic transmission

11.26 Removing input (clutch) shaft

11.27A Removing 4th gear baulk ring

11.27B Removing input shaft needle roller bearing

11.28 Removing mainshaft and intermediate housing

11.29 Removing layshaft and laygear

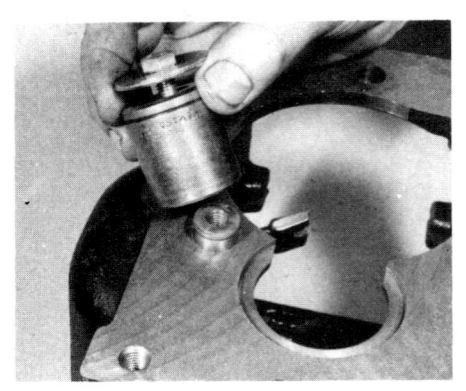

11.30A Removing reverse gear idler shaft

11.30B Reverse idler gear

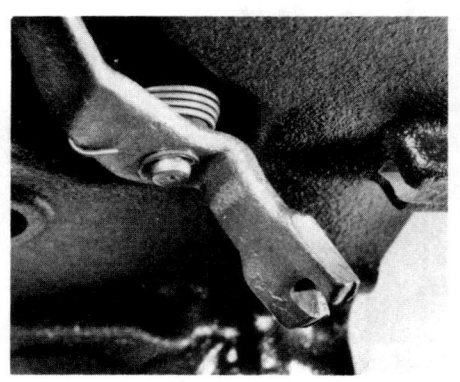

11.31 Reverse relay lever and pivot

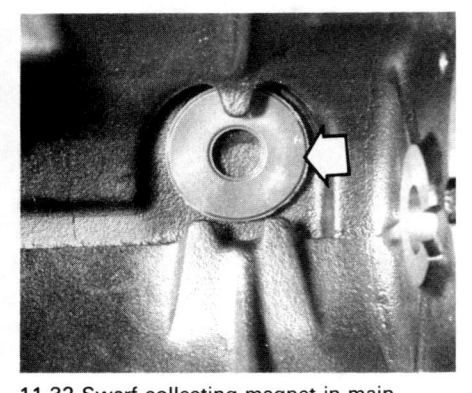

11.32 Swarf collecting magnet in main casing (arrowed)

using a nut, washer and socket pull out the idler shaft. Note the fitted position of the reverse idler gear, then remove it (photos).
31 Remove the guide from the reverse relay lever, then extract the circlip and remove the relay lever from the pivot (photo).
32 Remove the magnetic disc from the bottom of the main casing. Also remove any needle rollers which may have been displaced from the layshaft gear cluster (photo).

12 Mainshaft (5-speed) – dismantling and reassembly

1 Extract the circlip and slide the 3rd/4th synchroniser unit, together with 3rd gear, from the front of the mainshaft, using a two-legged puller where necessary. Separate the gear and unit, then remove the 3rd gear synchroniser ring (photos).
2 Remove the outer ring from 2nd gear, then extract the thrust washer halves (photos).
3 Slide 2nd gear from the front of the mainshaft and remove the 2nd gear synchroniser ring (photos).
4 Mark the 1st/2nd synchroniser unit hub and sleeve in relation to each other and note the location of the selector fork groove, then slide the sleeve forward from the hub and remove the sliding keys and springs. Note that the synchroniser hub cannot be removed from the mainshaft (photos).
5 Using a suitable puller, pull the speedometer drivegear off the rear of the mainshaft (photo).
6 Remove the 5th gear synchroniser unit and 5th gear from the mainshaft.
7 Extract the circlip retaining the mainshaft bearing, then support the intermediate housing on blocks of wood and drive the mainshaft through the bearing with a soft-faced mallet (photo).
8 Remove the oil scoop ring, 1st gear, and 1st gear synchroniser ring (photo).

12.1A Extracting 3rd/4th synchro circlip

12.1B Removing 3rd/4th synchro and baulk ring

12.1C Removing 3rd gear

12.2A 2nd gear thrust washer retaining ring

12.2B 2nd gear thrust washer segment

12.3A Removing 2nd gear from mainshaft

12.3B 2nd gear baulk ring

12.4A 1st/2nd synchro sleeve with reverse gear

12.4B Synchro sliding keys

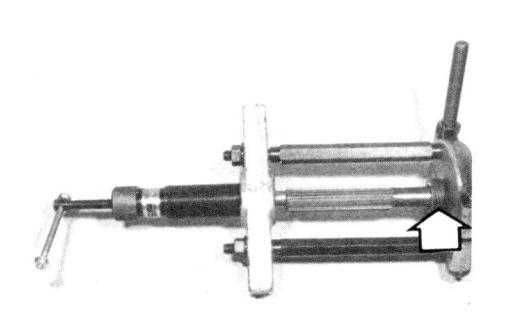

12.5 Removing speedometer drivegear (arrowed)

12.7 Extracting mainshaft bearing circlip

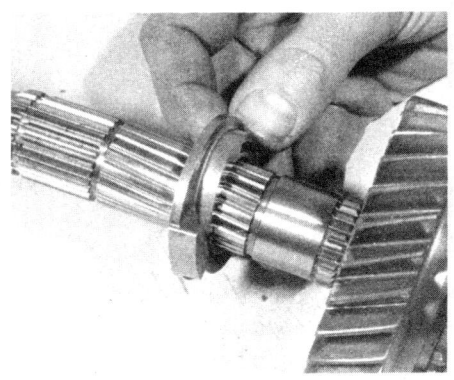

12.8 Oil scoop ring on mainshaft

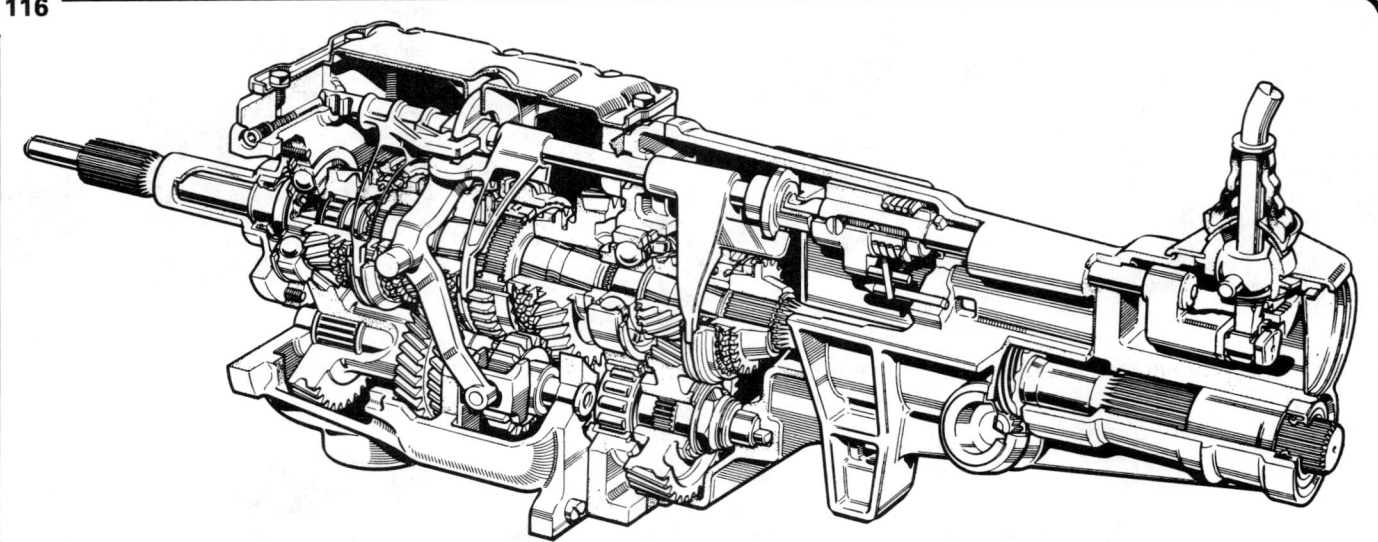

Fig. 6.19 Cutaway view of the five-speed gearbox (Secs 11 to 15)

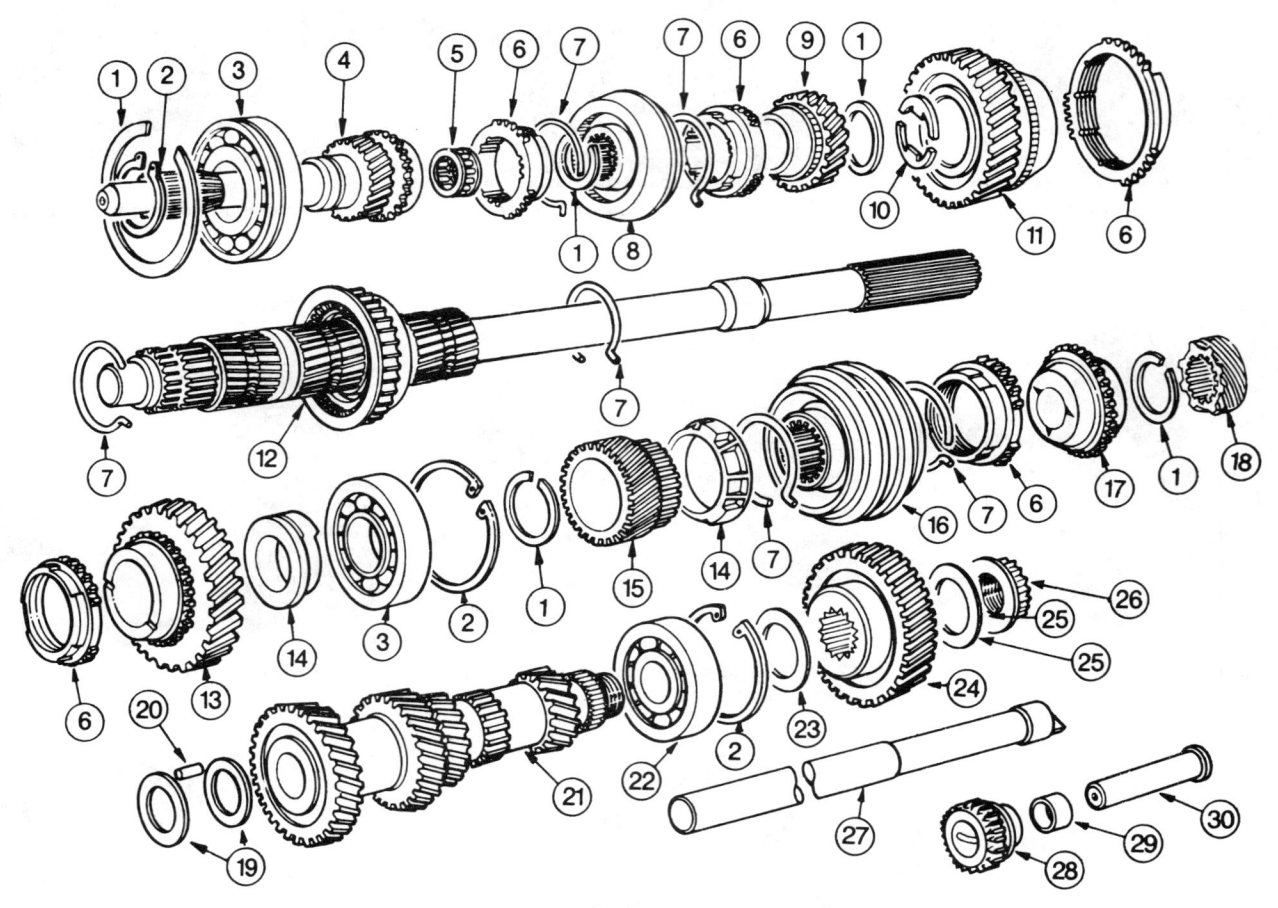

Fig. 6.20 Five-speed gearbox internal components (Secs 11 to 15)

1	Circlips	10	Thrust washers	17	5th gear synchro-hub	24	5th gear (layshaft)
2	Circlips	11	2nd gear	18	Speedometer drivegear	25	Washer
3	Ball-bearing	12	Mainshaft and 1st/2nd synchro hub	19	Spacers	26	Nut
4	Input shaft			20	Needle rollers	27	Layshaft
5	Needle roller bearing	13	1st gear	21	Layshaft gear	28	Reverse idler gear
6	Synchro baulk rings	14	Oil scoop rings	22	Roller bearing	29	Bush
7	Synchro springs	15	5th gear (mainshaft)	23	Spacer	30	Idler gear shaft
8	3rd/4th synchro unit	16	5th gear synchro unit				
9	3rd gear						

Chapter 6 Manual gearbox and automatic transmission

9 If necessary extract the circlip and drive the ball-bearing from the intermediate housing using a metal tube (photo). Also the synchroniser units may be dismantled, but first mark the hub and sleeve in relation to each other. Slide the sleeve from the hub and remove the sliding keys and springs.

10 Clean all the components in paraffin, wipe dry and examine them for wear and damage. Obtain new components as necessary. During reassembly lubricate the components with gearbox oil and where new parts are being fitted lightly grease contact surfaces.

11 Commence reassembly by assembling the synchroniser units. Slide the sleeves on the hubs in their previously noted positions, then insert the sliding keys and fit the springs as shown in Fig. 6.22.

12 Support the intermediate housing then, using a metal tube on the outer track, drive in the new bearing and fit the circlip.

13 Fit the spring to the rear of the 1st/2nd synchroniser hub followed by the 1st gear synchroniser ring (photo).

14 Slide the 1st gear and oil scoop ring (with the oil groove towards 1st gear) onto the mainshaft.

15 Using a metal tube on the mainshaft bearing inner track, drive the intermediate housing onto the mainshaft and fit the circlip. Make sure that the large circlip is towards the rear of the mainshaft.

16 Locate 5th gear and the 5th gear synchroniser with circlip loose on the mainshaft. Tap the speedometer drivegear lightly onto its shoulder – its final position will be determined later (photo).

17 Fit the 1st/2nd synchroniser sleeve to the hub in its previously noted position with the selector groove facing forward, then insert the sliding keys and fit the springs as shown in Fig. 6.22.

18 Fit the 2nd gear synchroniser ring to the 1st/2nd synchroniser unit with the sliding keys located in the slots.

19 Slide 2nd gear onto the front of the mainshaft and retain with the thrust washer halves and outer ring (photo).

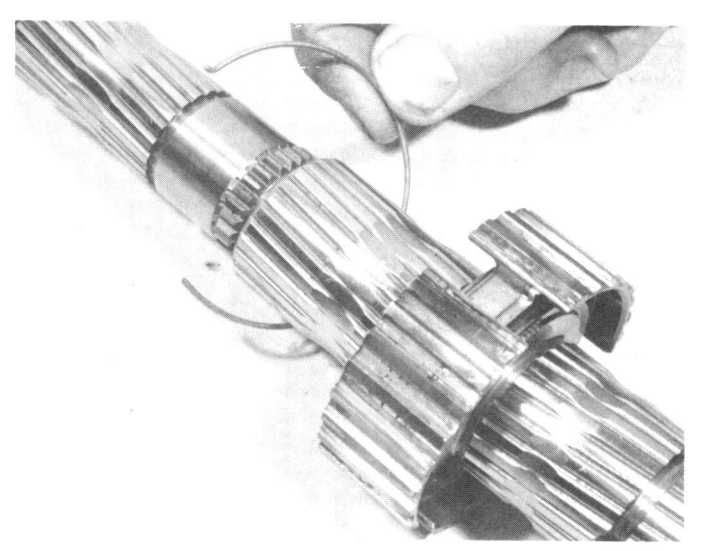

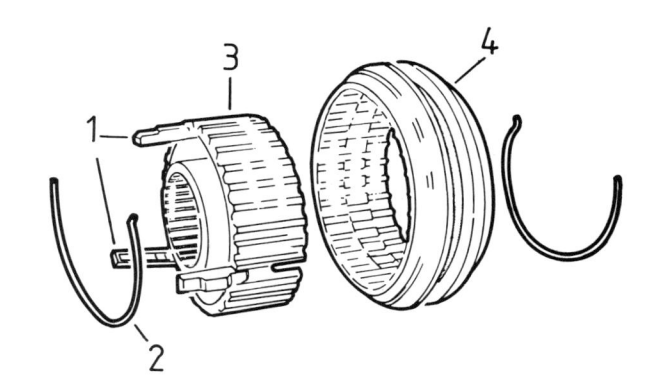

Fig. 6.21 Exploded view of synchroniser unit (Sec 12)

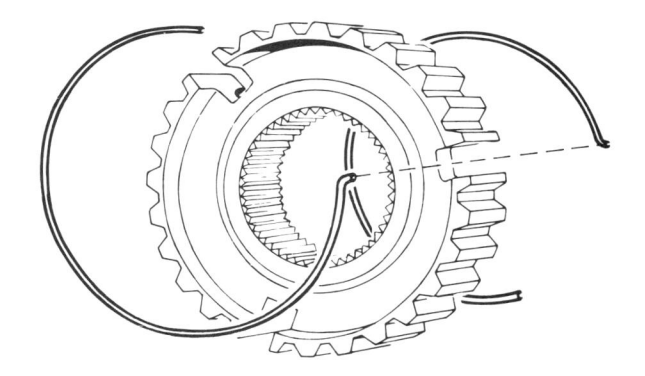

1 Sliding keys 3 Hub
2 Spring 4 Sleeve

Fig. 6.22 Alignment of synchroniser spring clips (Sec 12)

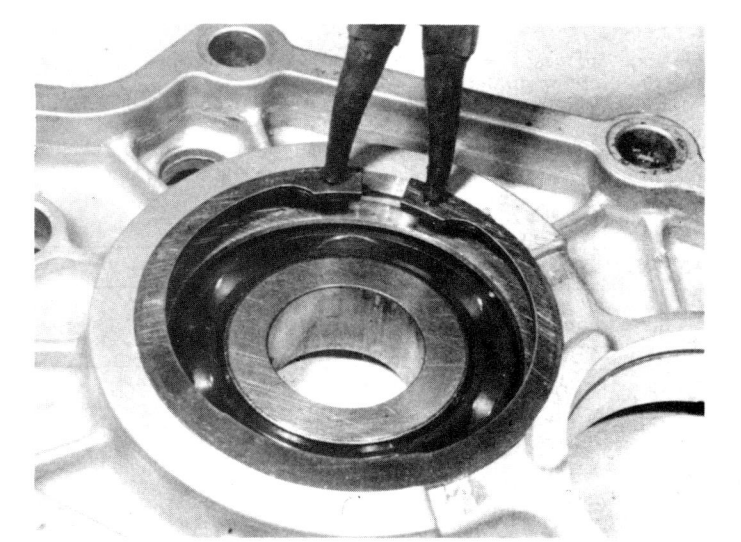

12.9 Intermediate housing bearing circlip

12.13 1st/2nd synchro spring

12.16 Locating the speedometer drivegear

12.19 2nd gear and thrust washer half, showing tab locating hole (arrowed)

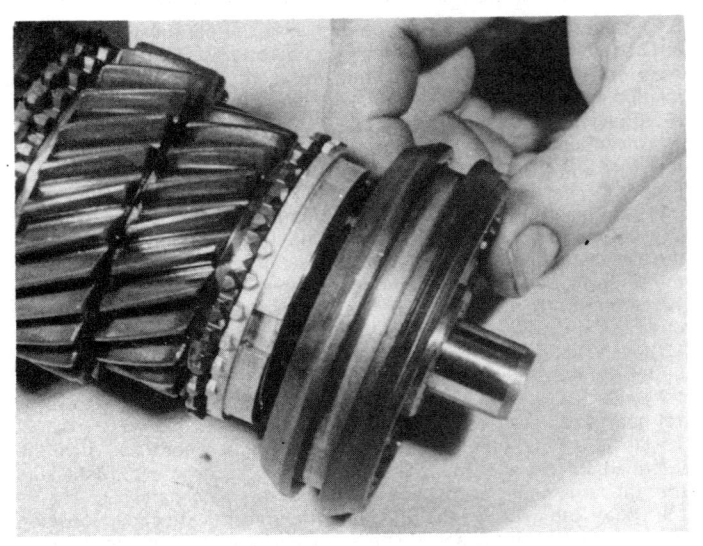

12.21 3rd/4th synchro unit

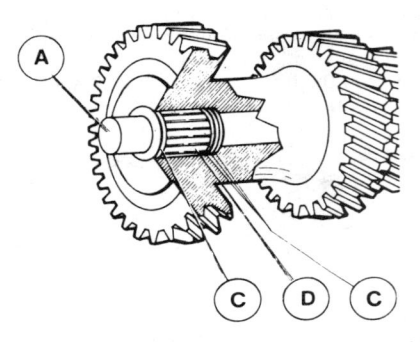

20 Slide 3rd gear onto the front of the mainshaft, then locate the synchroniser ring on the gear cone.
21 Locate the 3rd/4th synchroniser unit on the mainshaft splines with the long side of the hub facing the front (photo). Tap the unit fully home using a metal tube, then fit the circlip. Make sure that the slots in the 3rd gear synchroniser ring are aligned with the sliding keys as the synchroniser unit is being fitted.

13 Input shaft (5-speed) – dismantling and reassembly

Refer to Section 6.

14 Gearbox (5-speed) – examination and renovation

Refer to Section 8 but note the following points:

(a) There are five synchro rings
(b) The layshaft has no thrust washers
(c) There are two ball-bearings and one roller bearing
(d) When renewing the needle roller bearings in the layshaft, use rollers 25.75 mm (1.014 in) long. If the original rollers were 33 mm (1.299 in) long, two extra spacers will be required (Fig. 6.23)

15 Gearbox (5-speed) – reassembly

1 Locate the magnetic disc in the bottom of the main casing.
2 Fit the reverse relay lever onto the pivot and retain with the circlip. Fit the guide to the lever.
3 Position the reverse idler gear in the main casing with the long shoulder facing the rear and engaged with the relay lever. Slide in the idler shaft and tap fully home with a soft-faced mallet.
4 Smear grease inside the end of the layshaft gear cluster then fit the spacers and needle roller bearings – there are 21 needle rollers. If additional spacers are being used, place them behind the needle rollers (Fig. 6.23). Make sure that there is sufficient grease to hold the needle rollers in position while the layshaft is fitted. Due to the high levels of torque transmitted through the above transmissions, it is important that all the needle rollers of a countershaft needle roller bearing have exactly the same diameter. If the needle rollers in a bearing have different diameters, this will cause uneven load distribution which can result in premature wear. If it is necessary to renew the countershaft needle rollers, make sure that the rollers used for any one location all come from the same pack. If more than one pack is purchased, do not mix the needle rollers from separate packs.
5 Insert the layshaft in the gear cluster until the front of the shaft is flush with the front gear on the cluster.

Fig. 6.23 Layshaft needle roller bearing (Sec 14)

A Layshaft   C Spacers   D Needle rollers

6 Locate the layshaft and gear cluster in the bottom of the main casing.
7 Position a new gasket on the main casing, then fit the mainshaft and intermediate housing, and temporarily secure with two bolts.
8 Fit the input shaft needle roller bearing to the end of the mainshaft or in the centre of the input shaft (photo).

15.8 Input shaft needle roller bearing

Chapter 6 Manual gearbox and automatic transmission

9 Fit the 4th gear synchroniser ring to the 3rd/4th synchroniser unit with the cut-outs over the sliding keys, then fit the input shaft assembly and tap the bearing fully into the casing up to the retaining circlip (photo).
10 Invert the gearbox so that the layshaft gear cluster meshes with the input shaft and mainshaft gears.
11 Using a soft metal drift, drive the layshaft into the main casing until flush at the front face – the flat on the rear end of the layshaft must be horizontal (photo).
12 Using a metal tube, tap the layshaft gear cluster bearing into the intermediate housing and secure with the circlip (photo).
13 Fit the spacer ring then, using a metal tube tap 5th gear onto the splines of the layshaft gear cluster.
14 Fit the thrust washer and retaining nut. Select 3rd gear and either 1st or 2nd gear by pushing the respective synchroniser sleeve. While an assistant holds the gearbox stationary, tighten the nut to the specified torque, then lock it by peening the collar on the nut into the slot in the gear cluster (photos).

15 Select neutral, then slide 5th gear into mesh with the driving gear.
16 Slide the 5th gear synchroniser unit complete with spacer onto 5th gear. Then, using a metal tube, drive the dog hub and 5th synchroniser ring onto the mainshaft splines while guiding the synchroniser ring onto the sliding keys. Fit the circlip (photos).
17 Tap the speedometer drivegear into its correct position on the mainshaft – the distance between the gear and the 5th gear dog hub circlip should be 123.0 to 124.0 mm (4.843 to 4.882 in) (photo).
18 Locate the 5th gear selector fork in its synchroniser sleeve and locate the interlock sleeve in the groove (short shoulder to front), then insert the selector shaft through the sleeve and selector fork into the main casing (photo).
19 Locate the 1st/2nd and 3rd/4th selector forks in their respective synchroniser sleeves, position the selector boss and locking plate, and insert the selector shaft through the components into the front of the main casing. The roll pin hole in the selector boss must be towards the front.
20 If removed, refit the selector shaft centralising spring and 5th gear

15.9 Fitting input shaft

15.11 Layshaft correctly located before installation

15.12 Fitting the layshaft bearing

15.14A Tightening the mainshaft nut

15.14B Staking the mainshaft nut

15.16A 5th gear synchro spacer

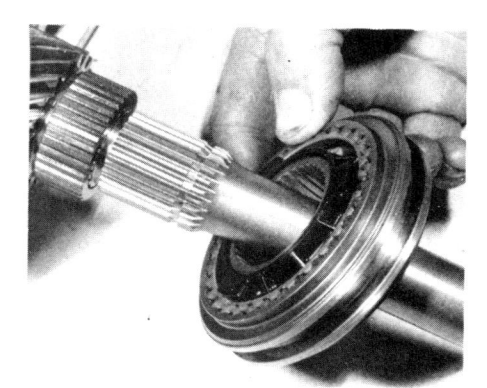

15.16B Fitting 5th gear synchro to mainshaft

15.16C 5th gear baulk ring and hub

15.16D 5th gear hub circlip

15.17 Measuring circlip-to-speedometer drivegear dimension

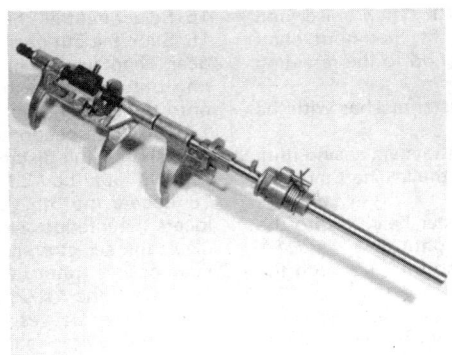

15.18 Selector shaft assembled

15.21 Fitting selector shaft/boss roll pin

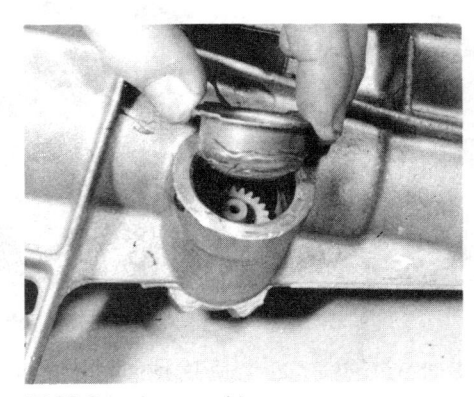

15.23 Speedometer drivegear cover

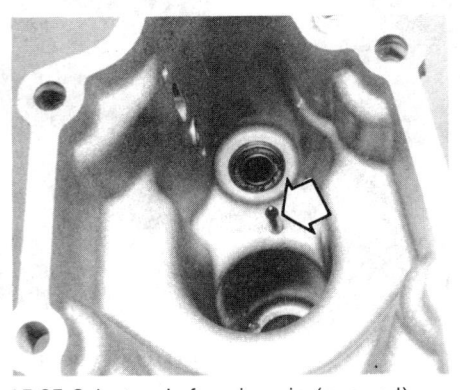

15.25 Selector shaft spring pin (arrowed)

locking control by inserting the pin and plug.
21 Align the holes, then drive the roll pin into the selector boss and selector shaft (photo).
22 Insert the selector locking pin and spring, apply sealer to the plug threads, then insert and tighten the plug using an Allen key.
23 Fit the speedometer drivegear to the rear extension housing. Apply a little sealant to the cover then press it into the housing (photo).
24 Remove the temporarily fitted bolts from the intermediate housing, then select 4th gear.
25 Stick a new gasket to the extension housing with grease, and fit the housing to the intermediate housing. Take care not to damage the rear oil seal, and make sure that the selector shaft centralising spring locates on the pin (photo).
26 Insert the bolts and tighten them to the specified torque in diagonal sequence. Before inserting the three bolts which go right through the main casing, apply sealant to their threads.
27 Select reverse gear and locate the connector on the rear of the selector rod. Support the rod with a piece of wood, then drive in the roll pin. Select neutral.
28 Press the rear cover into the extension housing.
29 Check that the 5th gear interlock sleeve is correctly aligned, then insert the 5th gear locking pin and spring.
30 Apply some sealant to the 5th gear locking plate, locate it on the extension housing, and insert and tighten the bolts to the specified torque.
31 Fit the gearbox top cover, together with a new gasket, and tighten the bolts to the specified torque in diagonal sequence.
32 Fit the clutch release bearing guide sleeve (oil slot downwards), together with a new gasket, and tighten the bolts to the specified torque in diagonal sequence. Where necessary apply sealant to the bolt threads.
33 Fit the clutch bellhousing to the front of the gearbox, together with a new gasket. Apply sealant to the bolt threads, then insert the bolts and tighten them to the specified torque in diagonal sequence.
34 Insert and tighten the reversing light switch in the extension housing.

35 Fit the clutch release bearing and arm with reference to Chapter 5.

16 Gearbox (5-speed) – disengagement of 5th gear

1 It is possible for vehicles built before November 1983 to have a problem with 5th gear jumping out of engagement during normal driving.
2 This trouble can be rectified by substituting modified 5th gear components to include the hub, baulk rings and selector fork.
3 Remove the gearbox as described in Section 3. This will necessitate removing the gear lever and turret cover which is held in position by Torx type screws.
4 Remove the clutch release bearing and lever from the bellhousing.
5 Unbolt and remove the top cover.
6 Carry out the operations described in paragraphs 8 to 17 of Section 11.
7 Mark the exact position of the speedometer wormdrive on the mainshaft and then draw it off using a puller.
8 Extract the circlip and remove the 5th speed synchro unit from the mainshaft.
9 Dismantle the synchro and reassemble using the new parts. When fitting the baulk rings, make sure that their bevelled side is towards the rear of the gearbox.
10 Carry out the operations described in paragraphs 16 to 22 and 25 to 27 in Section 15.
11 Insert the 5th gear locking pin and spring to the extension housing.
12 Smear the 5th gear locking plate with sealant and locate it with nuts hand tight.
13 Adjust the gearchange mechanism by pressing the gear lever fully to the right while in the neutral mode and then tighten the locking plate nuts to a torque of 25 Nm (18 lbf ft).
14 Fit the top cover and gasket and the clutch release components.

Chapter 6 Manual gearbox and automatic transmission

17 Fault diagnosis – manual gearbox

Symptom	Reason(s)
Weak or ineffective synchromesh	Synchronising cones worn, split or damaged
	Baulk ring synchromesh dogs worn, or damaged
Jumps out of gear	Broken gearchange fork rod spring
	Selector fork rod groove badly worn
Excessive noise	Incorrect grade of oil in gearbox or oil level too low
	Bush or needle roller bearings worn or damaged
	Gear teeth excessively worn or damaged
	Laygear thrust washers worn allowing excessive endplay (four-speed only)
Excessive difficulty in engaging gear	Clutch cable adjustment incorrect

PART 2: AUTOMATIC TRANSMISSION

18 General description

The automatic transmission takes the place of the clutch and gearbox, which are, of course, mounted behind the engine.

The unit has a large aluminium content which helps to reduce its overall weight and it is of compact dimensions. An oil cooler is fitted as standard and ensures cooler operation of the transmission under trailer towing and similar conditions. A vacuum connection to the inlet manifold provides smoother and more consistent downshifts under load than is the case with units not incorporating this facility.

The system comprises two main components:

(a) A three element hydrokinetic torque converter coupling capable of torque multiplication at an infinitely variable ratio
(b) A torque/speed responsive and hydraulically operated epicyclic gearbox comprising planetary gearsets providing three forward ratios and one reverse ratio

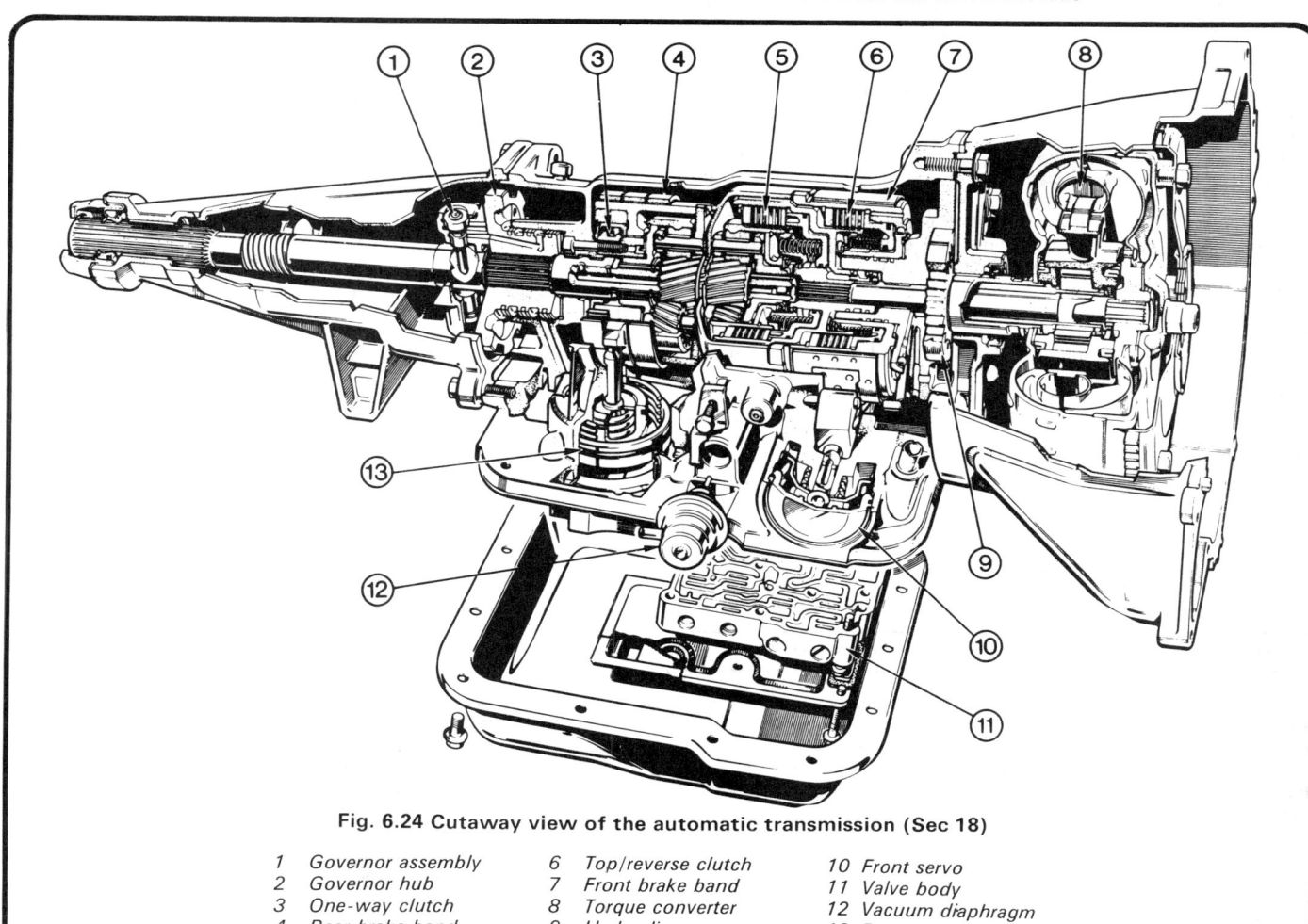

Fig. 6.24 Cutaway view of the automatic transmission (Sec 18)

1 Governor assembly
2 Governor hub
3 One-way clutch
4 Rear brake band
5 Forward clutch
6 Top/reverse clutch
7 Front brake band
8 Torque converter
9 Hydraulic pump
10 Front servo
11 Valve body
12 Vacuum diaphragm
13 Rear servo

Chapter 6 Manual gearbox and automatic transmission

When towing a vehicle with automatic transmission it is important to observe the precautions given in the 'Jacking and Towing' section at the front of the manual.

Due to the complexity of the automatic transmission unit, if performance is not up to standard, or overhaul is necessary, it is imperative that this be left to the local main agents who will have the special equipment for fault diagnosis and rectification.

The content of the following sections is therefore confined to supplying general information and any service information and instruction that can be used by the owner.

19 Routine maintenane

At the intervals specified in the Routine Maintenance section in the front of the manual carry out the following procedure.

Lubricate selector linkage
1 Apply the handbrake then jack up the front of the car and support on axle stands.
2 Oil the linkages at each end of the gearshift rod on the left-hand side of the transmission.
3 Lower the car to the ground.

Check transmission fluid level
4 Bring the engine and transmission to normal operating temperature and park the vehicle on level ground.
5 With the engine idling, move the transmission selector lever through all positions three times, then engage 'P' and wait for at least one minute.
6 With the engine still idling, withdraw the transmission dipstick. Wipe it clean, reinsert it, withdraw it for the second time and read off the level.
7 Two types of automatic transmission fluid are available (see Specifications for type numbers) and it is important that they are not mixed. The colour of the dipstick for fluid level checking denotes which fluid type to use.
8 If a transmission requiring later type fluid is fitted to a torque converter previously used with earlier type fluid, or *vice versa*, the components should be flushed to avoid mixing of the two fluids. Consult your Ford dealer if this condition arises.
9 If necessary top-up with the specified oil, through the dipstick tube, to the MAX mark. The quantity required to raise the level from MIN to MAX is approximately 0.3 litre (0.5 pint).

Adjust front brake band
10 Apply the handbrake then jack up the front of the car and support on axle stands.
11 Disconnect the downshift cable at the transmission.
12 Slacken the brake band adjuster screw locknut and unscrew the adjuster screw a few turns.
13 Tighten the adjuster screw to 14 Nm (10 lbf ft) using a suitable torque wrench. From this position slacken the adjuster screw by exactly 1$\frac{1}{2}$ turns (pre-1983 models) or 2 turns (1983 and later model), then tighten the locknut without disturbing the position of the adjuster screw.
14 Reconnect the downshift cable and lower the car to the ground.

20 Automatic transmission – removal and refitting

1 If possible, raise the car on a hoist or place it over an inspection pit. Alternatively, it will be necessary to jack up the car to obtain the maximum possible amount of working room underneath.
2 Place a large drain pan beneath the transmission sump (oil pan) then, working from the rear, loosen the attaching bolts and allow the fluid to drain. Remove all the bolts except the two front ones to drain as much fluid as possible, then temporarily refit two bolts at the rear to hold it in place.
3 Remove the starter motor as described in Chapter 12.
4 Where applicable, remove the access cover and adaptor plate bolts, rotate the engine until the torque converter drain plug is visible, then unscrew the plug and drain the fluid into a suitable container. Refit and tighten the plug.

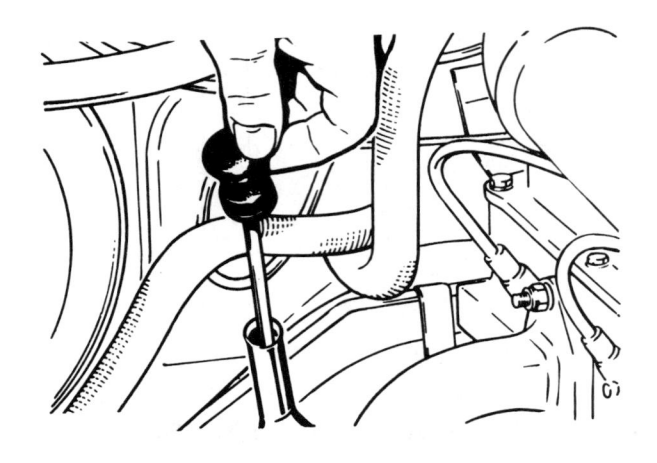

Fig. 6.25 Checking the automatic transmission fluid level (Sec 19)

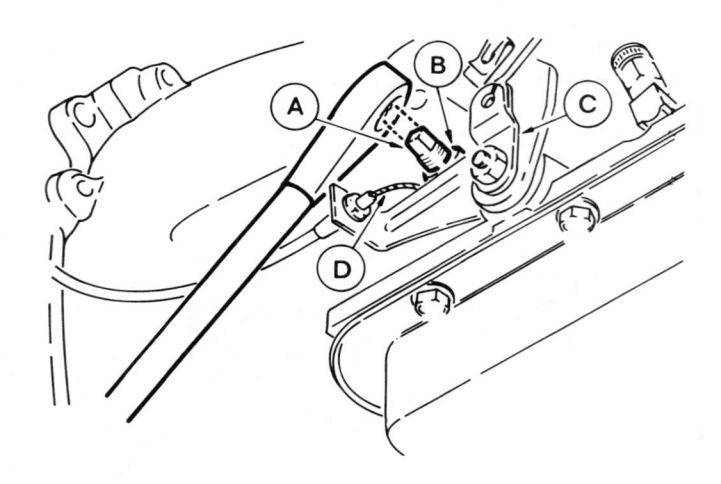

Fig. 6.26 Front brake band adjustment (Sec 19)

A Adjuster screw   C Downshift lever
B Locknut   D Downshift cable

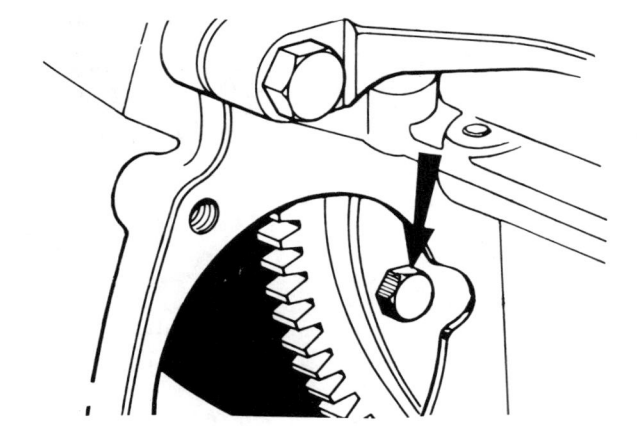

Fig. 6.27 Torque converter to driveplate bolt – arrowed (Sec 20)

Chapter 6 Manual gearbox and automatic transmission

5 Unscrew and remove the torque converter-to-driveplate bolts through the starter motor opening, turning the engine as necessary to position the bolts in the opening.
6 Remove the propeller shaft, referring to Chapter 7 as necessary. Place a polythene bag over the end of the transmission to prevent dirt from entering.
7 Detach the speedometer cable from the extension housing.
8 Disconnect the shift rod at the transmission manual lever, and the downshift cable at the transmission downshift lever. If necessary, the shift rod may be completely removed by disconnecting it from the selector lever.
9 Disconnect the downshift outer cable from the support bracket.
10 Disconnect the starter inhibitor (neutral start) switch leads.
11 Disconnect the vacuum lines from the vacuum unit.
12 Position a trolley jack beneath the transmission and raise it to *just* take the transmission weight.
13 Remove the engine rear support to crossmember nut and the transmission extension housing crossmember.
14 Disconnect both exhaust pipes at the manifolds (Chapter 3) and support them away from the transmission.
15 Lower the transmission and also support the front end of the engine.
16 Disconnect the oil cooler lines at the transmission and plug them to prevent dirt from entering.
17 Remove all the converter housing-to-engine bolts, and the transmission filler tube.

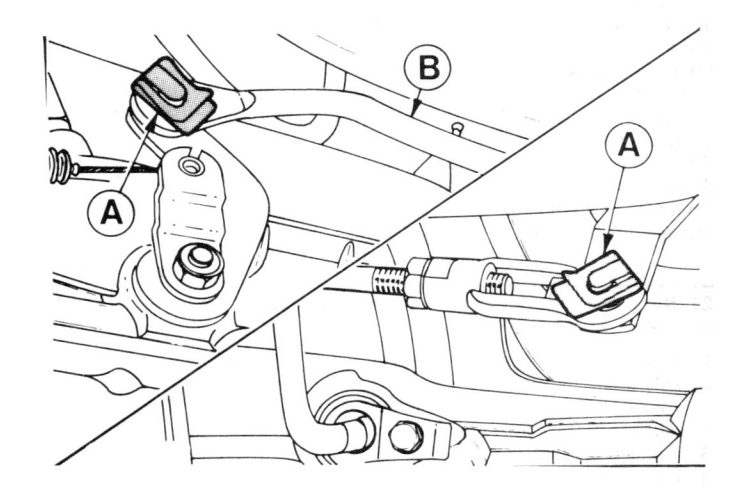

Fig. 6.28 Connecting clips A and shift rod B (Sec 20)

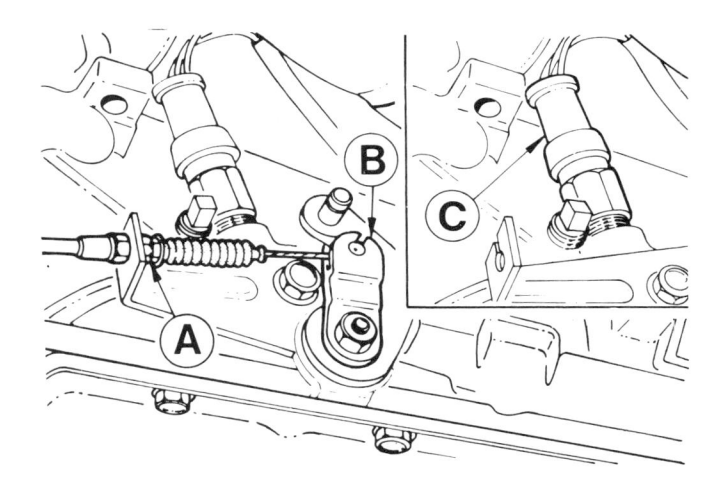

Fig. 6.29 Downshift cable locknut A, transmission lever B and starter inhibitor switch C (Sec 20)

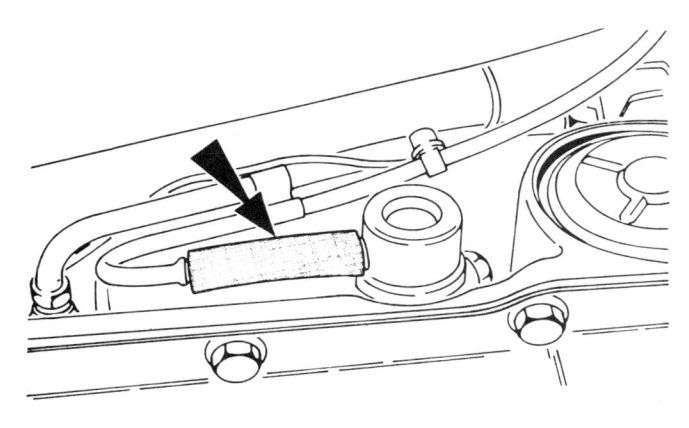

Fig. 6.30 Vacuum line connection to the vacuum unit – arrowed (Sec 20)

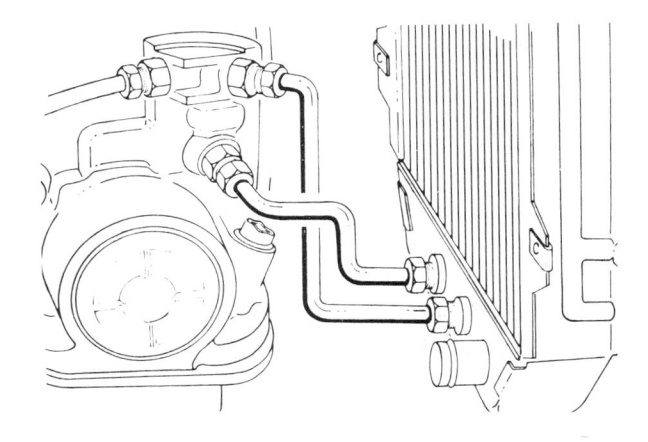

Fig. 6.31 Oil cooler pipes (Sec 20)

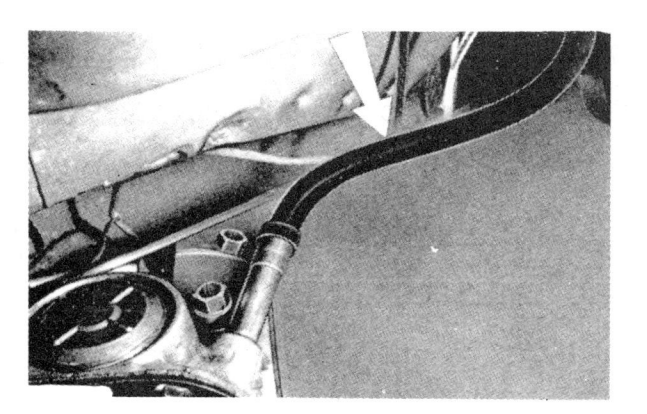

Fig. 6.32 Oil filler tube – arrowed (Sec 20)

Chapter 6 Manual gearbox and automatic transmission

18 Carefully move the transmission rearwards and downwards, and away from the car. Make sure that the torque converter remains fully in the transmission.
19 Refitting is the reverse of removal but ensure that the torque converter drain plug is in line with the hole in the driveplate. To check that the torque converter is positively engaged, measure the distance A between the converter housing to engine mating face and the end of the stub shaft. This should be at least 10 mm (0.4 in) (Fig. 6.33). Adjust the selector cable and inhibitor switch as described later in this Chapter. Refill the transmission unit with the specified fluid before starting the engine and check the level as described in Section 19.

21 Selector mechanism – removal, overhaul and refitting

1 Apply the handbrake then jack up the front of the car and support on axle stands.
2 Inside the car, prise off the selector lever escutcheon, and withdraw the illumination mounting from the selector lever.
3 Beneath the car, remove the spring clip and detach the selector rod from the selector lever.

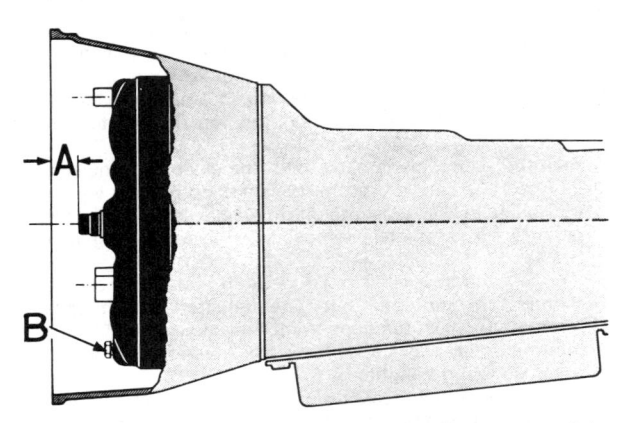

Fig. 6.33 Checking torque converter engagement with transmission oil pump (Sec 20)

A = At least 10.0 mm (0.4 in)   B   Drain plug

Fig. 6.34 Selector mechanism (Secs 21 and 22)

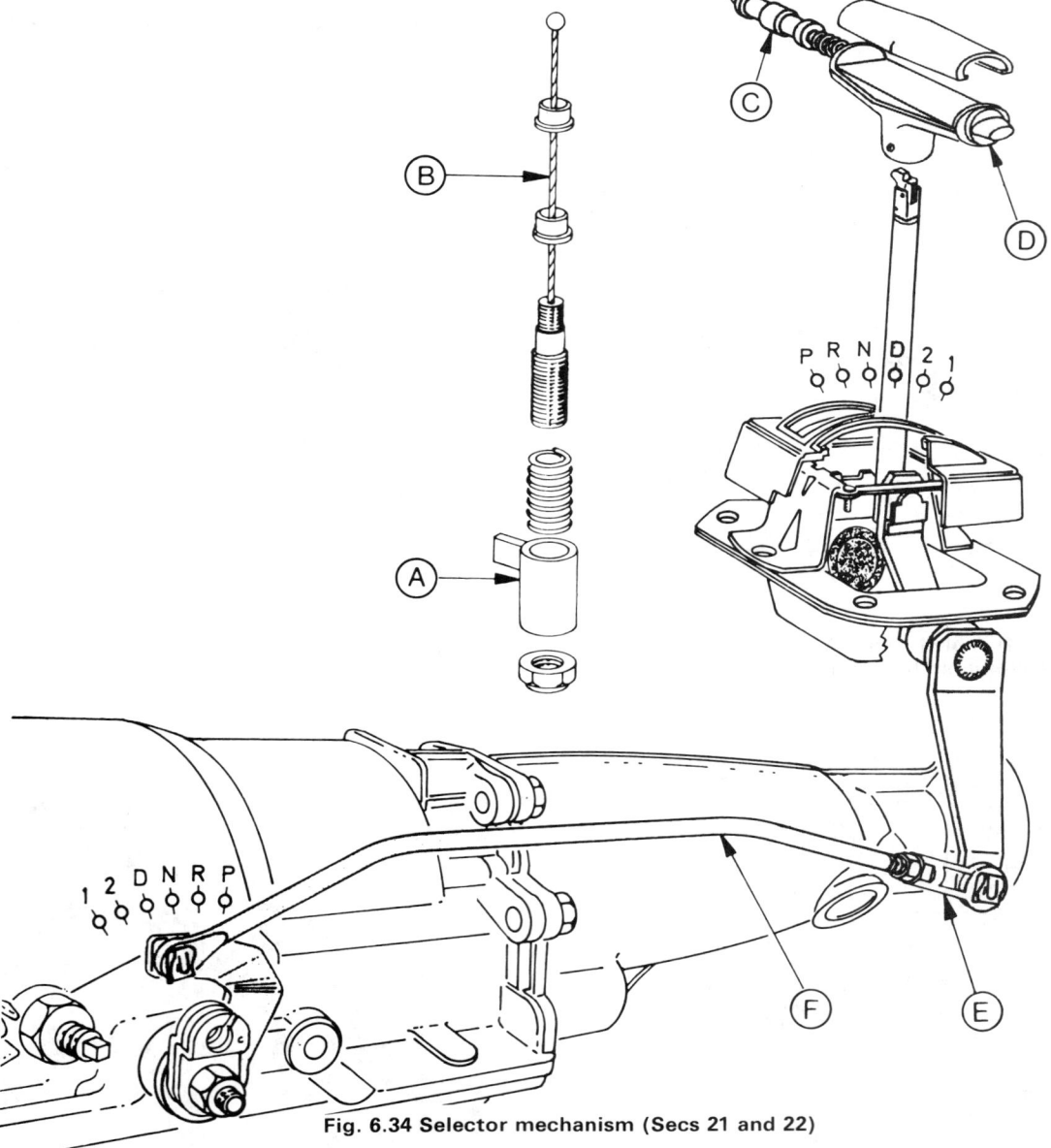

A   Selector pawl   C   Pushbutton   E   Connecting clip
B   Selector cable   D   Handle   F   Shift rod

Chapter 6 Manual gearbox and automatic transmission

4 If required, remove the spring clip and detach the selector rod from the transmission selector lever.
5 Remove the four bolts retaining the selector lever housing to the transmission tunnel and remove the housing.
6 Remove the rubber plug from the side of the selector lever housing, unscrew the nut and press the lower lever out of the housing.
7 Unscrew the locknut on the inhibitor cable and withdraw the pawl, spring and guide bush.
8 Remove the Allen screw from the T-handle and remove the handle. Remove the push button and spring.
9 Using a pin punch, drive out the retaining pin and remove the inhibitor mechanism and inhibitor cable.
10 Refitting is a reversal of the above procedure, noting that the T-handle Allen screw is inserted from the front, so that the pushbutton is nearest the driver.
11 Adjust the mechanism as described in Section 22.

22 Selector mechanism – adjustment

1 Apply the handbrake then jack up the front of the car and support on axle stands.

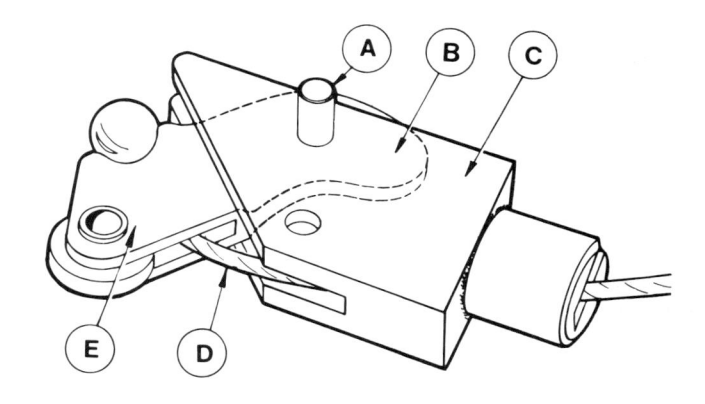

Fig. 6.35 Removing the inhibitor cable (Sec 21)

A Straight pin   D Operating cable
B Segment   E Roller assembly
C Guide piece

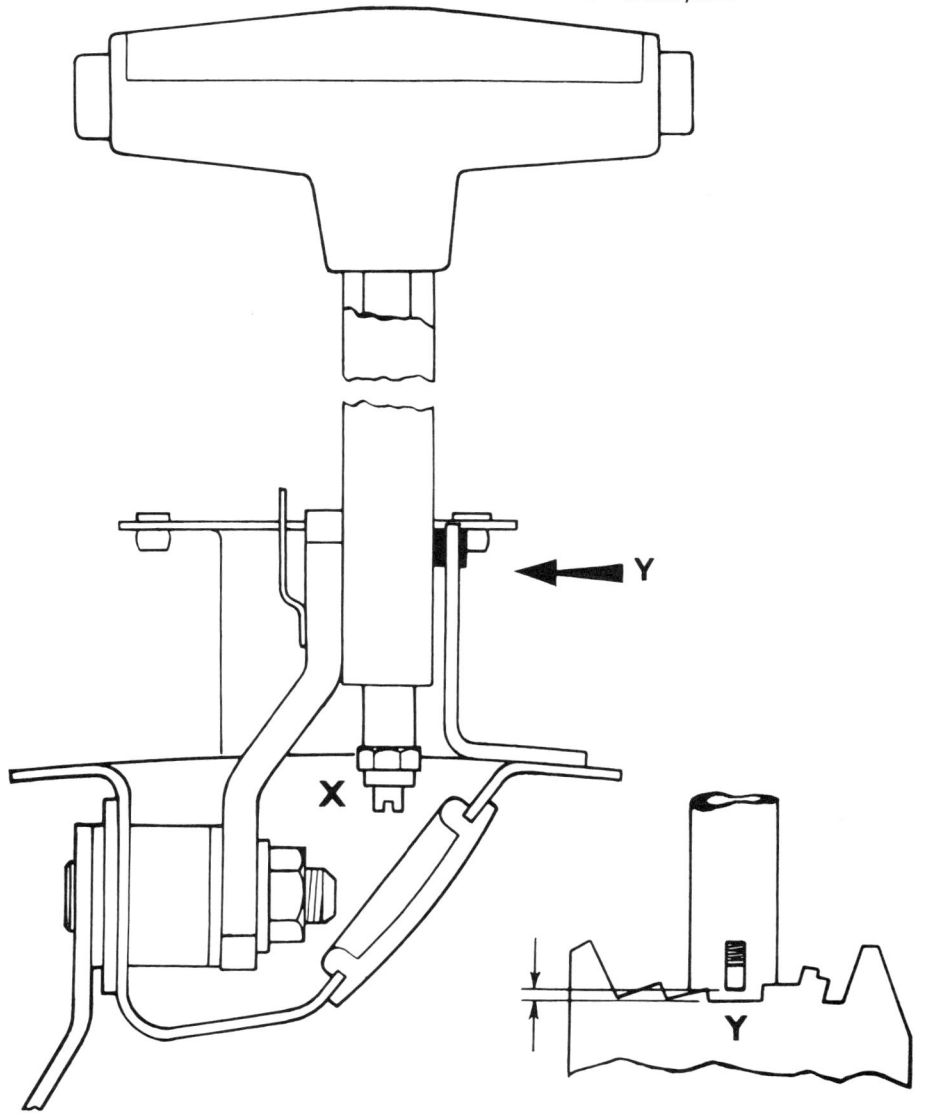

Fig. 6.36 Selector mechanism adjustment (Sec 22)

X Locknut
Y = 0.1 to 0.2 mm (0.004 to 0.008 in)

2 Select position 'D' then prise off the selector lever escutcheon and, from under the car, remove the plug in the side of the selector housing.
3 Adjust the inhibitor cable locknut X to give a dimension of 0.1 to 0.2 mm (0.004 to 0.008 in) (Fig. 6.36).
4 Refit the plug and escutcheon.
5 With the transmission shift lever and the manual selector lever in 'D', adjust the selector rod link until it can be reconnected without strain.

23 Downshift cable – removal, refitting and adjustment

1 Support the front of the car on axle stands. Remove the split pin and disconnect the cable from the carburettor linkage.
2 Slacken the adjusting nut from the upper mounting bracket, pull back the outer sheath and unhook the cable from the slot.
3 Unscrew the nut at the transmission mounting bracket and unhook the outer sheath.
4 Unhook the inner cable from the transmission lever.
5 Refitting is a reversal of the above, but the inner nut at the upper end should be screwed on completely, and the outer nut only a few turns.
6 Depress the accelerator pedal fully, and check that the throttle plate is fully open.
7 Using a screwdriver, lever the downshift cable lever upwards, pulling the inner cable fully upwards.
8 Turn the adjusting nut to lengthen or shorten the cable to give a clearance of 0.5 to 1.3 mm (0.02 to 0.05 in) between the downshift lever and the accelerator shaft (Fig. 6.37). Tighten the locknut.

24 Starter inhibitor/reverse lamp switch – removal and refitting

1 Pull the cable connector off the switch.
2 Unscrew the starter inhibitor switch and remove the O-ring.
3 To refit the switch, fit a new O-ring and screw the switch into the housing finger-tight. Tighten the switch to the torque wrench setting given in the Specifications. It is important that this torque is not exceeded.
4 Refit the cable plug connector to the switch.
5 Check that the engine will start only with the lever in the 'P' and 'N' positions and in no other position.
6 Turn on the ignition without starting the engine. Check that the reversing light comes on in the 'R' position, but not in any other.

25 Extension housing oil seal – renewal

1 Remove the propeller shaft, as described in Chapter 7.
2 Jack the rear of the car to minimise the loss of oil from the transmission, then carefully prise out the oil seal, being careful not to damage the machined recess into which the seal fits.
3 Ensure that the recess is clean and undamaged, then press in a new seal with the lip of the seal inwards. Tap the seal home fully with a tube of suitable diameter, or with a hammer and a block of wood, being careful that the seal is inserted squarely.
4 Refit the propeller shaft, lower the car to the ground and then check the level of the transmission fluid as described in Section 19.

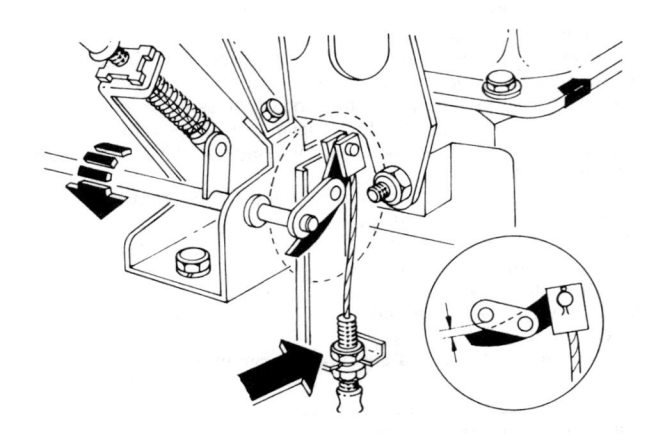

Fig. 6.37 Adjusting the downshift cable (Sec 23)

Depress throttle fully (broken arrow). Pivot the downshift lever away from the accelerator shaft and adjust the clearance at the adjusting nut (solid arrow)

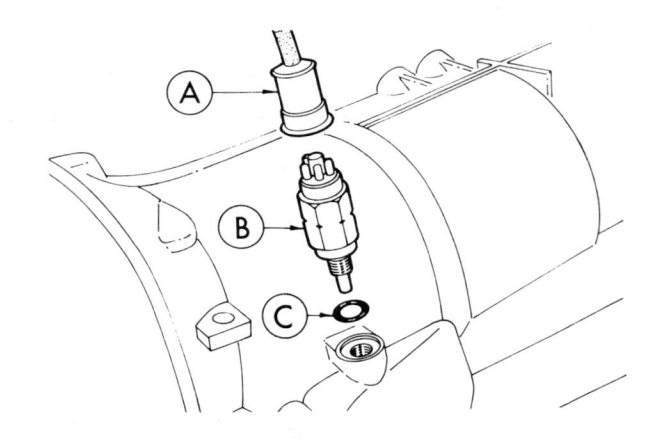

Fig. 6.38 Inhibitor switch renewal (Sec 24)

A Wiring connector   C O-ring
B Inhibitor switch

26 Fault diagnosis – automatic transmission

Fault in these units are most commonly the result of low fluid level, incorrect adjustment of the selector linkage or downshift cable, or a faulty inhibitor switch. If the problem still exists after these points have been checked, it will be necessary to take the car to a dealer with equipment to diagnose internal faults.

Chapter 7 Propeller shaft

Contents

Centre bearing – renewal 3
Fault diagnosis – propeller shaft 5
General description 1
Propeller shaft – removal and refitting 2
Universal joints – inspection 4

Specifications

Type .... Two-piece tubular, with two Hardy-Spicer type universal joints and one constant velocity (CV) centre joint, centre support bearing

Torque wrench settings

	Nm	lbf ft
Centre bearing	18 to 23	13 to 17
Propeller shaft to rear axle flange	60 to 65	44 to 48

1 General description

Drive is transmitted from the gearbox to the rear axle by means of a finely balanced two-piece tubular propeller shaft. The universal joint at the front of the shaft incorporates a splined yoke which is a sliding fit on the gearbox mainshaft. A further universal joint at the rear of the shaft incorporates a flanged yoke which is bolted to the rear axle drive flange.

The front and rear shafts are connected via a constant velocity joint, and a centre bearing located on the rear of the front shaft supports the assembly on the underbody.

The propeller shaft universal joints cannot be renewed without special equipment because the joint spiders are staked into the yokes in a position determined during electronic balancing. When joint wear is detected (see Section 4), either a new propeller shaft should be obtained, or the complete propeller shaft should be passed to a suitably equipped engineering workshop for repair.

2 Propeller shaft – removal and refitting

1 Chock both front wheels then jack up the rear of the car and

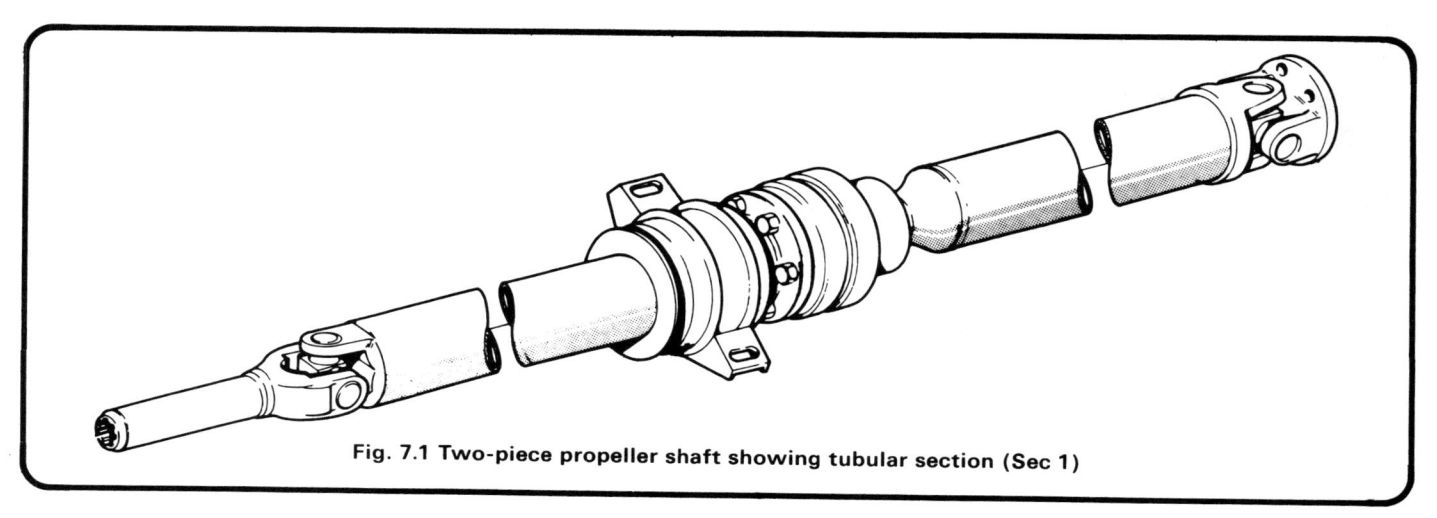

Fig. 7.1 Two-piece propeller shaft showing tubular section (Sec 1)

support on axle stands positioned on the underbody main channels.
2 Mark the propeller shaft and rear axle drive flanges in relation to each other (photo).
3 Unscrew the four flange bolts while holding the shaft with a metal bar inserted through the universal joint or by applying the handbrake.
4 Unscrew the centre bearing mounting bolts (photo), but note the location of any shims fitted between the bearing bracket and underbody. The shims must be refitted in their original locations to maintain the correct shaft alignment.
5 Place a container beneath the rear of the gearbox to catch any oil which may leak as the propeller shaft is removed.
6 Push the shaft forward slightly to separate the two flanges at the rear, then lower the end of the shaft and pull it rearwards to disengage it from the gearbox mainshaft splines (photo).
7 Refitting the propeller shaft is a reversal of the above procedure. Ensure that the mating marks scratched on the propeller shaft and rear axle flanges line up, and that any shims at the centre bearing are refitted. To ensure the correct positioning of the centre bearing and constant velocity joint, first fit the centre bearing bolts finger tight then tighten the rear flange bolts to the specified torque. With the rear axle hanging freely, pull the front shaft and centre bearing forward until the constant velocity joint is felt to bear against the rear of the centre bearing. With the front shaft held in this position tighten the centre bearing bolts to the specified torque; making sure that the bearing housing is at 90° to the shaft and the bearing support rubber is not distorted. Finally check and if necessry top up the gearbox oil level.

3 Centre bearing – renewal

1 Remove the propeller shaft as decribed in Section 2.
2 Mark the constant velocity joint housing and the flange on the front shaft in relation to each other. Also mark the front shaft in relation to the flange.
3 Unscrew the constant velocity joint housing bolts and separate the front and rear shafts (photo).
4 Unscrew the centre bolt, remove the washer, and withdraw the flange from the splines on the front shaft.
5 Remove the dust excluder, followed by the centre bearing housing with its rubber insulator. Note that the recess on the housing faces the front of the shaft.
6 Using a two-legged puller, or suitable levers, withdraw the bearing followed by the inner dust excluder.
7 Prise open the metal tabs and remove the rubber insulator from the housing.
8 Clean the components then fit the new rubber insulator to the housing so that the boss is uppermost. Bend the metal tabs down to secure the insulator.
9 Refit the inner dust excluder.

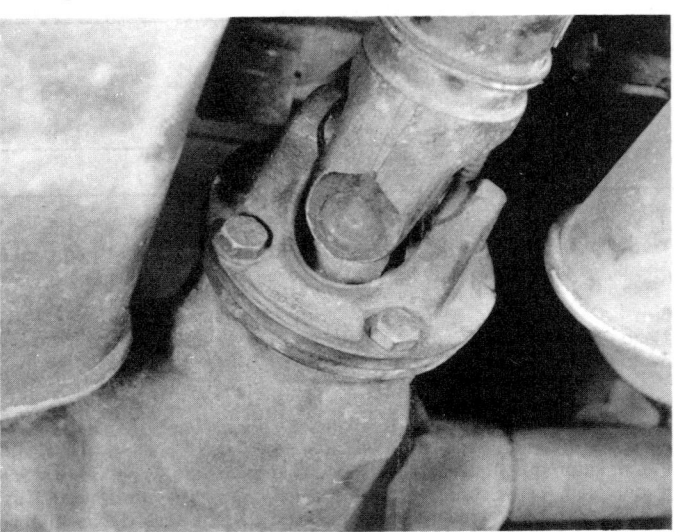

2.2 Propeller shaft rear universal joint and drive flanges

2.4 Centre bearing and mounting bolts

2.6 Withdrawing the propeller shaft from the gearbox mainshaft splines

3.3 The constant velocity joint housing bolts

Chapter 7 Propeller shaft

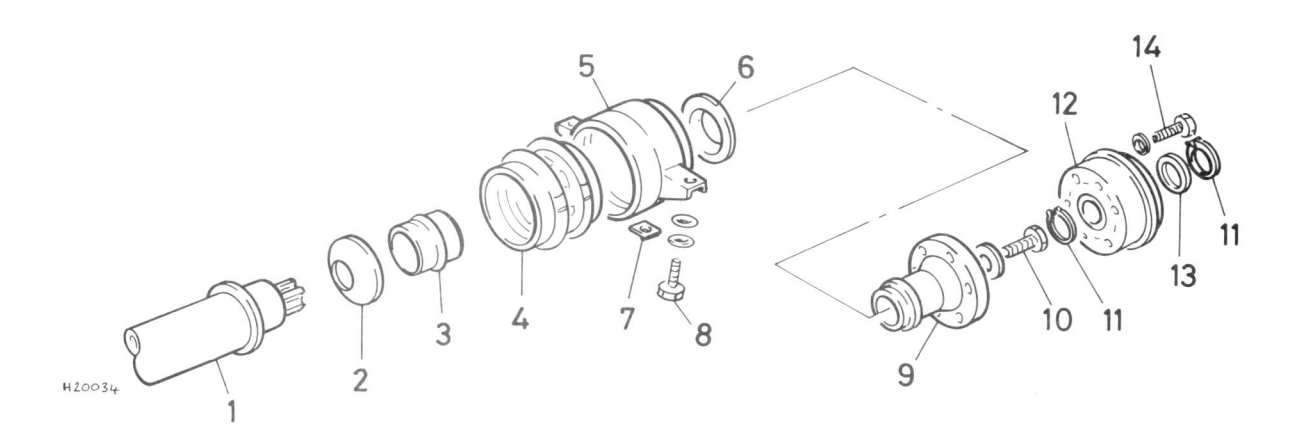

Fig. 7.2 Centre bearing and CV joint components (Sec 3)

1 Front shaft
2 Inner dust excluder
3 Centre bearing
4 Rubber insulator
5 Housing
6 Outer dust excluder
7 Alignment shim
8 Bolt
9 Flange
10 Centre bolt
11 Circlips
12 CV joint assembly
13 Spring cap
14 Bolt

Fig. 7.3 The recess (arrowed) in the centre bearing housing must face forward (Sec 3)

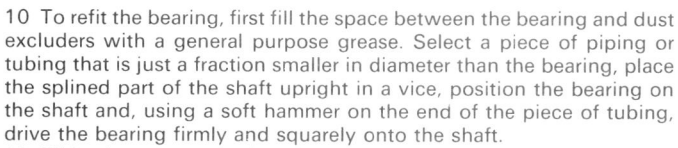

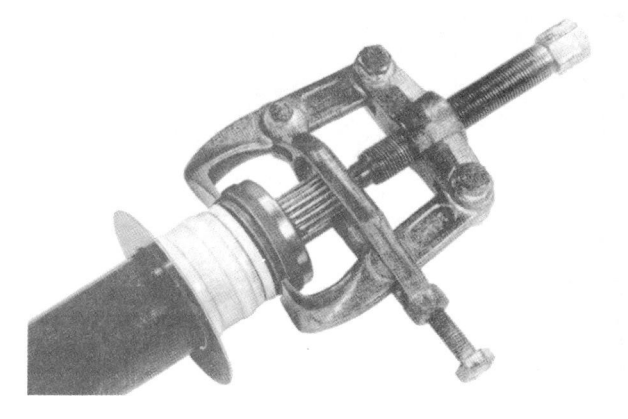

Fig. 7.4 Using a puller to withdraw the centre bearing (Sec 3)

10 To refit the bearing, first fill the space between the bearing and dust excluders with a general purpose grease. Select a piece of piping or tubing that is just a fraction smaller in diameter than the bearing, place the splined part of the shaft upright in a vice, position the bearing on the shaft and, using a soft hammer on the end of the piece of tubing, drive the bearing firmly and squarely onto the shaft.
11 Slide the centre bearing housing and rubber insulator over the bearing, making sure that the previously noted recess faces the front of the shaft. Refit the outer dust excluder.
12 Locate the flange on the splines of the front shaft, making sure that the previously made marks are aligned. Refit the centre bolt and washer and tighten the bolt.
13 Assemble the two shafts together with the previously made marks aligned, then refit and tighten the constant velocity joint housing bolts.
14 Refit the propeller shaft with reference to Section 2.

4 Universal joints – inspection

1 Wear in the needle roller bearings is characterised by vibration in the transmission, 'clonks' on taking up the drive, and in extreme cases of lack of lubrication, metallic squeaking, and ultimately grating and shrieking sounds as the bearings break up.

Fig. 7.5 Bending the metal tabs to secure the insulator (Sec 3)

2 It is easy to check if the needle roller bearings are worn with the propeller shaft in position, by trying to turn the shaft with one hand, the other holding the rear axle flange when the rear universal is being checked, and the front yoke when the front universal is being checked. Any movement between the propeller shaft and the front and the rear yokes is indicative of considerable wear. If worn, a new propeller shaft must be obtained, or the existing shaft overhauled by a suitably equipped engineering workshop.

5 Fault diagnosis – propeller shaft

Symptom	Reason(s)
Vibration	Worn universal joints or centre bearing Propeller shaft out of balance Deteriorated rubber insulator on centre bearing
Knock or 'clunk' when taking up drive	Worn universal joints Loose constant velocity or rear drive flange bolts
Excessive 'rumble' increasing with roadspeed	Worn centre bearing

Chapter 8 Rear axle

Contents

Axle rear cover – removal and refitting	6
Axleshaft bearing – renewal	5
Axleshaft – removal and refitting	4
Fault diagnosis – rear axle	11
General description	1
Limited-slip differential – general	10
Pinion oil seal – renewal	7
Rear axle – overhaul	8
Rear axle – removal and refitting	3
Routine maintenance	2
Wheel stud – renewal	9

Specifications

General

Axle designation	D (Salisbury)
Type	Hypoid, semi-floating, integral differential, limited-slip differential on 2.8 Special
Ratio	3.09:1
Lubricant capacity	1.1 litre (1.9 pint)
Lubricant type/specification	Gear oil, viscosity SAE 90 to Ford spec SQM-2C 9007-AA (Duckhams Hypoid 90S – Hypoid 90DL for limited slip differential)
Differential side gear endfloat	0.15 mm (0.006 in)
Pinion turning torque (setting)	2.5 to 2.7 Nm (1.8 to 2.0 lbf ft)
Crownwheel and pinion backlash	0.10 to 0.20 mm (0.004 to 0.008 in)

Torque wrench settings

	Nm	lbf ft
Bearing cap to axle casing	60 to 68	44 to 50
Cover to axle casing	35 to 45	26 to 33
Crownwheel	80 to 87	59 to 64
Axleshaft bearing retainer/backplate	27 to 31	20 to 23
Drive pinion nut	100 to 120	74 to 89
Rear spring U-bolt	25 to 36	18 to 27
Stabiliser bar clamp bolt	40 to 50	30 to 37
Propeller shaft to rear axle drive flange	60 to 65	44 to 48

1 General description

The rear axle is of the hypoid semi-floating type and is located by semi-elliptic road springs, in conjunction with a stabiliser bar.

The differential is integral with the axle housing and the rear axle must be removed to carry out any repairs to the differential which may become necessary.

The drive pinion runs in two tapered roller bearings, which are preloaded by a spacer. Early axles have a fixed length spacer and later ones a collapsible spacer (Figs. 8.1 and 8.2). Both types of spacer are interchangeable, but care must be taken to ensure that the appropriate method of determining drive pinion bearing preload is used.

The correct depth of mesh of the pinion in the crownwheel is obtained by the use of selected shims inserted between the pinion and rear tapered roller bearing.

The crownwheel is bolted to the differential case, which is mounted on two tapered roller bearings. These bearings are preloaded by selective-fit shims, positioned on either side of the differential housing. The differential assembly is of the two pinion type.

2.8 Special models are fitted with a limited-slip differential.

The axleshafts are splined to the differential side gears and supported at the outer ends by ball-bearings. Oil seals are fitted at each end of the axle housing.

Overhauling the rear axle requires a variety of special tools and it is not recommended that any attempt should be made to do this unless these tools are at hand.

Chapter 8 Rear axle

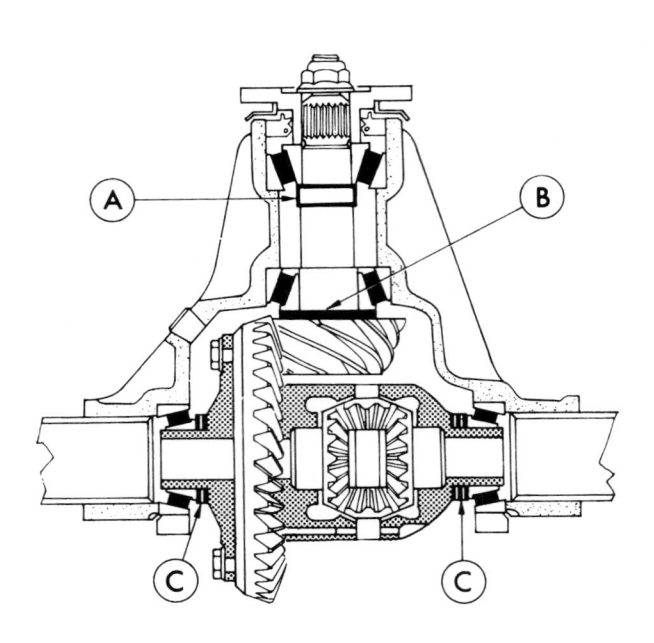

Fig. 8.1 Cross-section of early rear axle (Sec 1)

A Fixed length spacer
B Pinion shim
C Differental shims

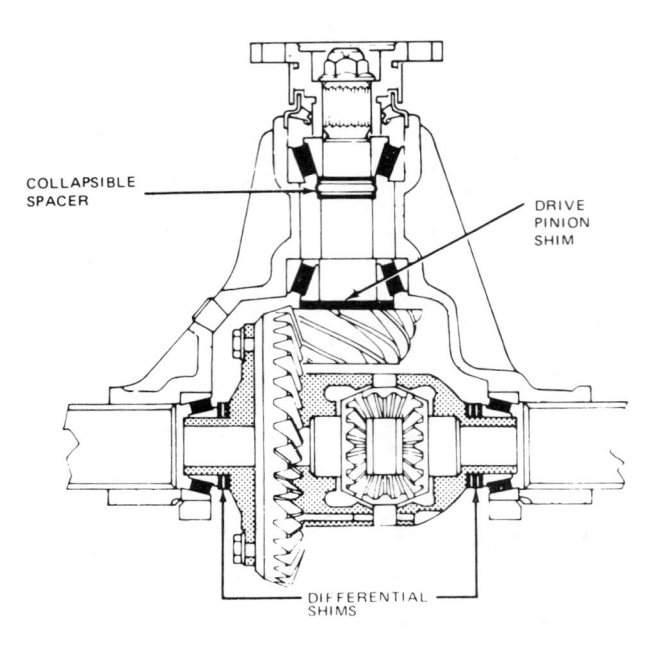

Fig. 8.2 Cross-section of later rear axle (Sec 1)

2 Routine maintenance

At the intervals specified in the Routine Maintenance section in the front of the manual carry out the following procedure.

Check rear axle oil level
1 Jack up the rear of the car and support on axle stands. To ensure that the car is level also jack up the front of the car. Alternatively, position the car over an inspection pit or on ramps.
2 Using a suitable square key, unscrew the filler plug from the left-hand side of the differential housing.
3 Using a finger or a piece of bent metal rod, check that the oil is level with the bottom of the filler plug hole or a maximum of 10.0 mm (0.39 in) below the hole.
4 If necessary top up with the correct grade of oil as given in 'Recommended lubricants and fluids'. Do not overfill the rear axle, otherwise the oil will overheat and be lost through the breather vent.
5 Refit and tighten the filler plug then lower the car to the ground.

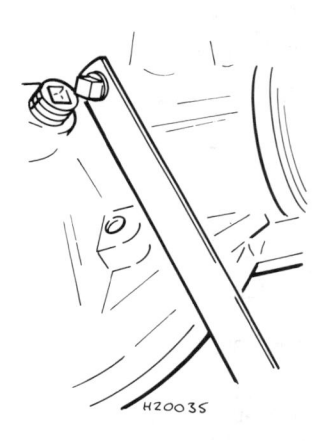

Fig. 8.3 Removing the rear axle filler plug (Sec 2)

3 Rear axle – removal and refitting

1 Jack up the rear of the car and support with axle stands or blocks positioned beneath the side-members. Chock the front wheels and remove the rear wheels.
2 Mark the propeller shaft and rear axle drive flanges in relation to each other.
3 Unscrew the four flange bolts while holding the shaft with a metal bar inserted through the universal joint, or by applying the handbrake. Lower the rear of the propeller shaft.
4 Disconnect the handbrake cable and, where applicable, the transverse rod from the rear axle with reference to Chapter 9.
5 Clamp the flexible brake hose to the rear axle then unscrew the nuts and disconnect it from the rigid brake pipe and axle bracket. If a hose clamp is not available, tighten a piece of polythene sheet beneath the master cylinder cap to reduce any fluid loss and also plug the disconnected hose.
6 Jack up the rear axle a few inches then unscrew and remove the shock absorber lower mounting bolts. Prise the shock absorbers from the brackets on the rear axle.

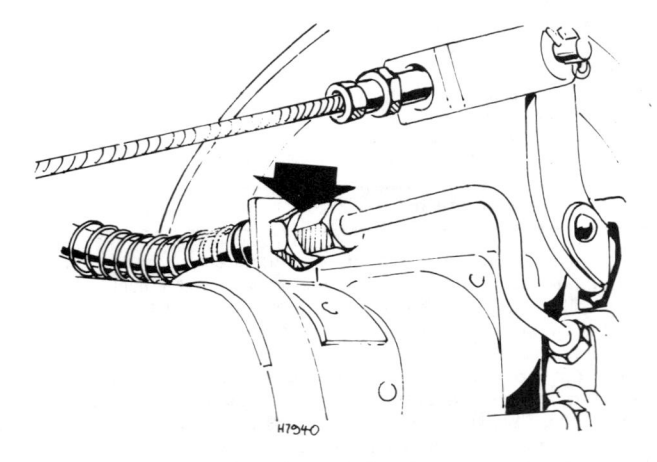

Fig. 8.4 Flexible brake hose connection to the rear axle – arrowed (Sec 3)

Chapter 8 Rear axle

Fig. 8.5 Disconnecting the rear stabiliser bar (Sec 3)

4.3 Axleshaft bearing retainer and backplate bolt location (arrowed)

7 Using grips (Fig. 8.5) pull the stabiliser bar to the rear while unscrewing the mounting bolts from the rear axle. Remove the clamps.
8 With the rear axle suitably supported, unscrew the nuts from the bottom of the U-bolts and remove the lower spring plates and insulators.
9 With the help of an assistant, lift the rear axle over the rear springs and withdraw it from under the car.
10 To refit the rear axle, first check that the upper insulators and brackets are correctly located on the springs.
11 With the help of an assistant, lift the rear axle into position on the rear springs.
12 Connect the propeller shaft to the drive flange with the previously made marks aligned, and loosely fit the four bolts, together with new lockwashers.
13 Locate the insulators and lower spring plates on the springs then fit the U-bolts, together with new self-locking nuts. Tighten the nuts progressively to the specified torque.
14 Locate the clamps on the stabiliser bar mounting rubbers then pull the bar into position with the grips and fit the mounting bolts. Tighten the bolts to the specified torque.
15 Jack up the rear axle as necessary and push the shock absorbers into their brackets. Insert the mounting bolts and tighten the nuts.
16 Refit the flexible brake hose and tighten the nuts.
17 Remove the hose clamp or polythene sheet then bleed the brakes as described in Chapter 9.
18 Refit the handbrake cable and, where applicable, the transverse rod to the rear axle with reference to Chapter 9.
19 Tighten the drive flange-to-propeller shaft bolts to the specified torque.
20 Refit the rear wheels and lower the car to the ground.
21 Check the rear axle oil level and top up if necessary.

4 Axleshaft – removal and refitting

1 Chock the front wheels then jack up the rear of the car and support with axle stands. Remove the rear wheel.
2 Release the handbrake and remove the brake drum.
3 Using a socket through the special holes in the drive flange, unscrew the bolts securing the backplate and bearing retainer to the axle casing (photo).
4 Withdraw the axleshaft, together with the bearing. If it is tight due to the outer track seizing in the casing temporarily refit the wheel and pull on the wheel. If this is unsuccessful, use a slide hammer attached to the drive flange, or alternatively screw two long bolts through the casing at diametrically opposite points (Fig. 8.7) and tighten them progressively until the bearing is free.
5 To refit the axleshaft, first wipe clean the bearing recess in the casing and apply a little lithium based grease both to the casing and bearing outer track.

Fig. 8.6 Removing the axleshaft retaining bolts (Sec 4)

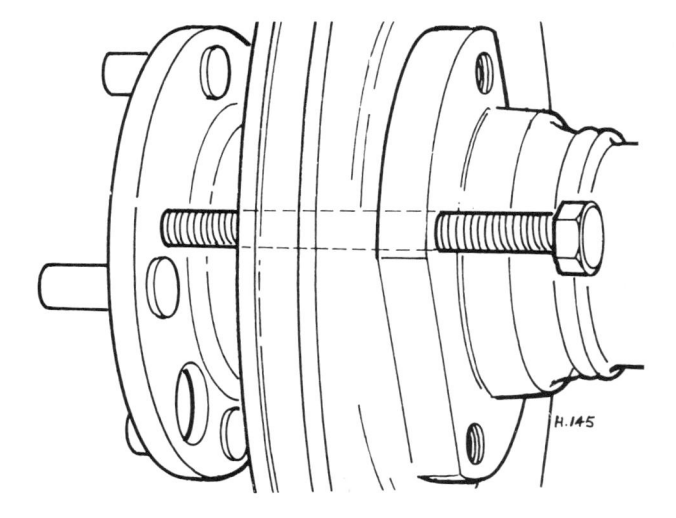

Fig 8.7 Using long bolts to withdraw the axleshaft (Sec 4)

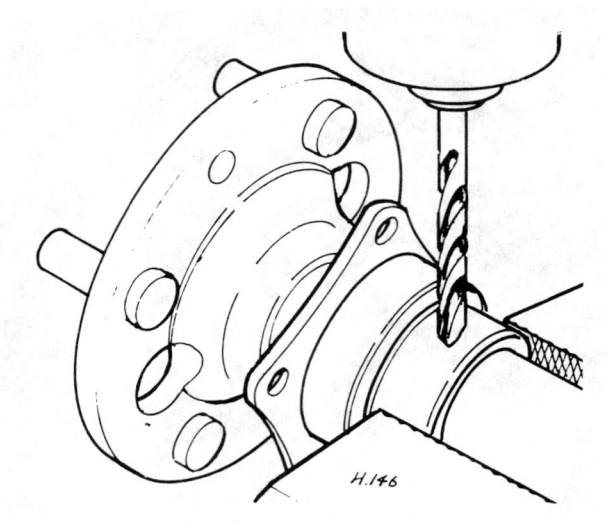

Fig. 8.8 Drilling the bearing retaining ring prior to splitting with a cold chisel (Sec 5)

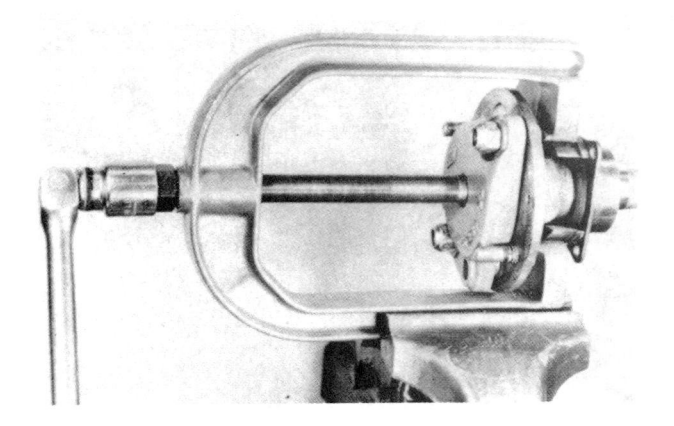

Fig. 8.9 Using the special Ford tool to remove the axleshaft bearing (Sec 5)

6 Insert the axleshaft and engage the splined end of the shaft with the differential side gear. Tap the shaft with a mallet until the bearing is fully entered in the casing.
7 Align the backplate and bearing retainer holes with the casing holes, insert the bolts and tighten them evenly to the specified torque.
8 Refit the brake drum and rear wheel and lower the car to the ground.
9 Check the rear axle oil level and top up if necessary.

5 Axleshaft bearing – renewal

Note: *A special tool is required to remove the bearing from the axleshaft (see Fig. 8.9).*
1 Remove the axleshaft as described in Section 4.
2 Using an 8 mm (0.31 in) drill, make a hole in the bearing retaining ring – taking care not to penetrate into the axleshaft.
3 Support the ring on a vice then split it across the hole with a sharp cold chisel. Remove the ring from the shaft.
4 Press the bearing off the shaft using a suitable puller or press. The special Ford tool is shown in Fig. 8.9.
5 Remove the bearing retainer.
6 Clean the axleshaft, bearing retainer and axle casing.
7 Locate the bearing retainer on the axleshaft, followed by the new bearing with the oil seal facing away from the axleshaft flange.
8 Slide on the new bearing retaining ring and press it fully against the bearing inner track using a suitable puller or a length of tube.
9 Refit the axleshaft with reference to Section 4.

6 Axle rear cover – removal and refitting

1 Chock the front wheels then jack up the rear of the car and support on axle stands.
2 Clean the rear cover and surrounding area (photo).
3 Remove the handbrake relay lever and secondary cable, or transverse rod (as applicable) with reference to Chapter 9.
4 Place a suitable container beneath the rear axle to catch the oil as the rear cover is removed.
5 Undo and remove the ten bolts and spring washers that secure the rear cover to the final drive housing. Lift away the rear cover and its gasket.
6 Before refitting the rear cover make sure that the mating faces are free of the old gasket or jointing compound.
7 Fit a new gasket and then the rear cover and secure with the ten bolts and spring washers. The cover bolts protrude into the final drive housing so it is important that a suitable oil-resistant sealing compound is smeared onto the threads of each bolt before it is fitted.
8 Tighten the cover securing bolts to the specified torque wrench setting.
9 Reconnect the handbrake components with reference to Chapter 9.
10 Refill the rear axle with the correct grade of oil then lower the car to the ground.

6.2 Axle rear cover

7 Pinion oil seal – renewal

1 This operation may be performed with the rear axle in position or on the bench.
2 Undo and remove the two bolts, spring and plain washers that secure the centre bearing support to the underside of the body (refer to Chapter 7).
3 With a scriber or file, mark a line across the propeller shaft and rear axle drive flanges so that they may be refitted together in their original positions.
4 Undo and remove the four bolts and spring washers securing the propeller shaft and rear axle driving flanges and carefully lower the propeller shaft to the floor.

5 Carefully clean the front of the final drive housing as there will probably be a considerable amount of dirt and oil if the seal has been leaking for a while (photo).
6 Remove the wheels and brake drums then using either a spring balance and a length of cord wrapped round the drive pinion, or a torque wrench, check and record the turning torque of the pinion. On early models fitted with a fixed length spacer (see Section 1) this procedure is not necessary. However, unless the final drive has been completely dismantled previously, there is no way of determining which spacer is fitted.
7 Using a suitable long-handled tool or large wrench, grip the rear axle pinion flange and with a socket undo and remove the pinion flange retaining self-locking nut. This nut must be discarded and a new one obtained ready for reassembly.
8 Place a container under the front of the final drive housing to catch any oil that may come out, then using a universal puller and suitable thrust pad pull off the rear axle pinion flange from the drive pinion.
9 Using a screwdriver or small chisel carefully remove the old oil seal. It will probably be necessary to partially destroy it. Note the correct way round is with the lip facing inwards.
10 Before fitting a new seal apply some grease to the inner face between the two lips of the seal.
11 Apply a little jointing compound to the outer face of the seal.
12 Using a tubular drive of suitable diameter, carefully drive the oil seal into the final drive housing until flush with the casing.
13 Refit the rear axle pinion flange and once again hold squarely with the tool or wrench.
14 Screw on a new self-locking nut until the endfloat is just eliminated, then tighten the nut a fraction at a time while checking the pinion turning torque with the spring balance or torque wrench. Continue tightening until the torque equals that recorded in paragraph 6 plus a further 0.2 to 0.4 Nm (0.15 to 0.30 lbf ft) to allow for the additional friction of the new oil seal. Do not tighten the nut further, otherwise, if a collapsible spacer is fitted, it will be over-compressed and the assembly will then have to be dismantled to fit a new spacer. However, if the minimum specified torque wrench setting for the pinion nut has not been reached, the spacer fitted is probably of the fixed length type and the nut should be tightened further. While doing this check that the pinion turning torque does not increase.
15 Refit the brake drum and wheels.
16 Reconnect the propeller shaft aligning the previously made marks on the flanges, and refit the bolts with new spring washers. Tighten to the specified torque wrench setting.
17 Refit the centre bearing support securing bolts, spring and plain washers and tighten.
18 Finally check the oil level in the rear axle and top up if necessary.

8 Rear axle – overhaul

Note: *Most garages will prefer to renew the complete rear axle as a unit if it is worn rather than to dismantle the unit to renew any damaged or worn parts. To do the job correctly requires the use of special and expensive tools which the majority of garages do not have.*

The primary object of these special tools is to enable the mesh of the crownwheel to the pinion to be very accurately set and thus ensure that noise is kept to a minimum. If any increase in noise cannot be tolerated (provided that the rear axle is not already noisy due to a defective part) then it is best to purchase an exchange built up rear axle or a good secondhand unit.
1 Remove the rear axle from the car (Section 3) and thoroughly clean it.
2 Remove the axleshafts (Section 4) and the axle rear cover (Section 6). If necessary remove the brake backplates.
3 Working inside the axle casing, undo and remove the four bolts that hold the two U-shaped differential bearing caps in the casing.
4 With a scriber, mark the relative positions of the two bearing caps so that they can be refitted in their original positions. Lift away the two end caps: there may already be mating numbers shown in Fig. 8.11.
5 Obtain two pieces of 50 mm (2 in) square wood at least 300 mm (12 in) long and, with a sharp knife, taper the ends along a length of 150 mm (6 in).
6 Place the tapered ends of the wooden levers in the two cut-aways of the differential casing and, using the rear cover face of the final drive housing as a fulcrum, carefully lever the differential assembly from the final drive housing.

7.5 Front of final drive housing showing pinion flange and nut

7 If it is necessary to remove the two differential case bearings these may be removed next using a universal two-legged puller and suitable thrust pad. Carfully ease each bearing from its location. Recover the shim packs from behind each bearing noting from which side they came.
8 Using a scriber, mark the relative positions of the crownwheel and differential housing so that the crownwheel may be fitted in its original position, unless it is to be renewed.
9 Undo and remove the eight bolts that secure the crownwheel to the differential housing. Using a soft-faced hammer, tap out the crownwheel from its location on the differential housing.
10 Using a suitable diameter parallel pin pinch, tap out the pin that locks the differential pinion gear shaft to the differential housing. **Note:** The hole into which the peg fits is slightly tapered, and the opposite end may be lightly peened over and should be cleaned with a suitable diameter drill.
11 Using a soft metal drift, tap out the differential pinion gear shaft. Lift away the differential pinion gears, side gears and thrust washers; taking care to ensure that the thrust washers are left with their relative gears.
12 Professional fitters at the dealers use a special tool for holding the pinion drive flange stationary whilst the nut in the centre of the flange is unscrewed. Since it is very tight it will require some force to undo it. The average owner will not normally have the use of this special tool so, as an alternative method, clamp the pinion flange in a vice and then undo the nut. Any damage caused to the edge of the flange by the vice should be carefully filed smooth. This nut must not be used again so a new one will be required during reassembly.
13 Using a universal two-legged puller and suitable thrust pad draw the pinion drive flange from the end of the pinion shaft.
14 The pinion shaft may now be removed from the final drive housing. Carefully inspect the large taper roller bearing behind the pinion gear and if it shows signs of wear or pitting on the rollers or cage the bearing must be renewed.
15 Using a universal two-legged puller and suitable thrust pad, draw the bearing from the pinion shaft.
16 The smaller taper roller bearing and oil seal may next be removed from the final drive housing; pinion drive flange end. To do this use a soft metal drift with a tapered end or suitable diameter tube and, working inside the housing, tap the bearing circumference outwards so releasing first the oil seal and then the bearing.
17 Again using the soft metal drift and working inside the housing, drift out the bearing cups. These must not be used with new bearings.
18 The final drive assembly is now dismantled and should be washed and dried with a clean lint-free rag ready for inspection. Check all bearings for signs of wear or pitting and if evident a new set of bearings should be obtained.
19 Examine the teeth of the crownwheel and pinion or pitting, score marks, chipping and general wear. If a crownwheel and pinion is required, a mated crownwheel and pinion must be fitted and under no

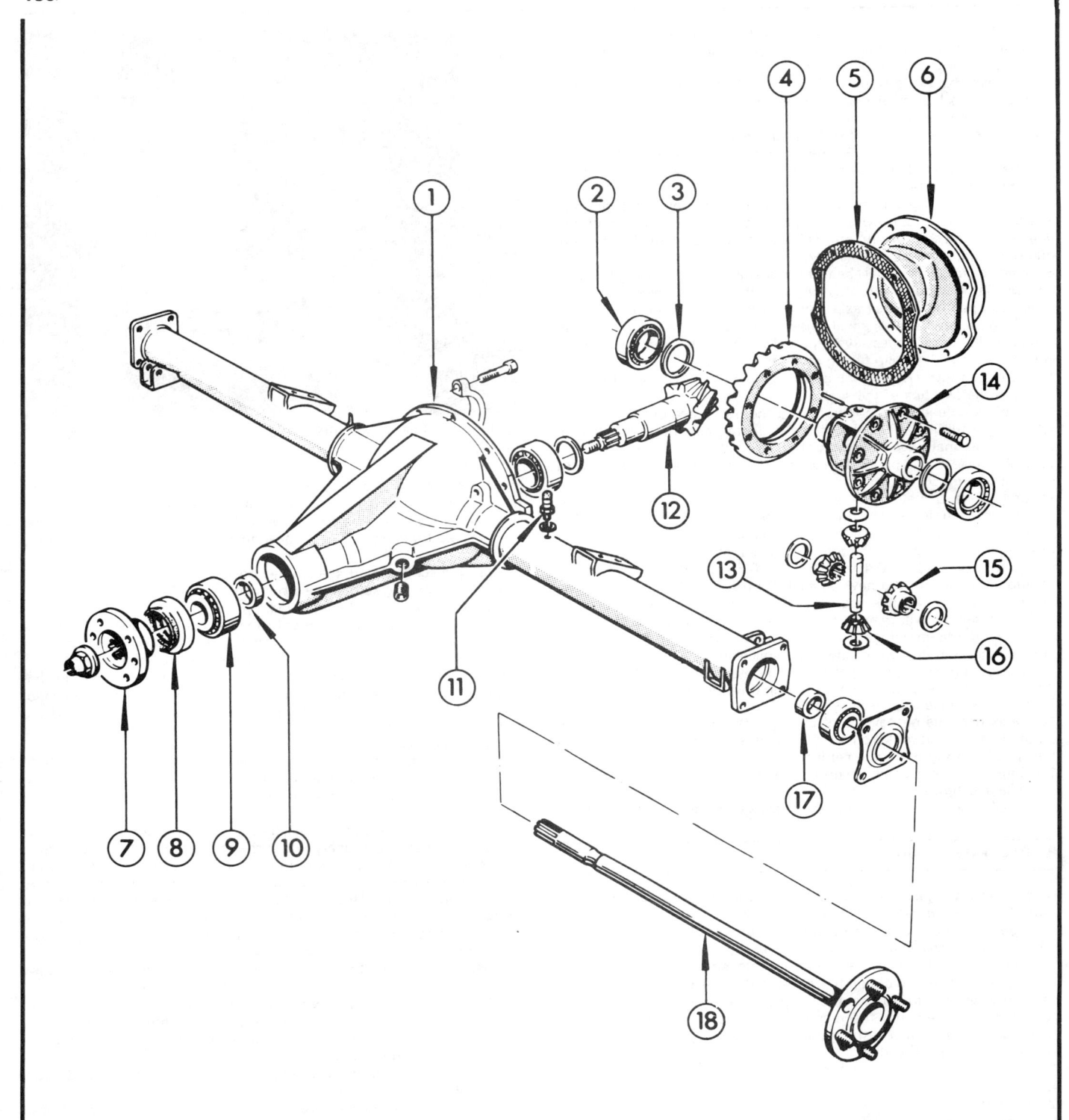

Fig. 8.10 Exploded view of the rear axle (Sec 8)

1 Axle housing
2 Differential case taper roller bearing
3 Differential case shim
4 Crownwheel
5 Gasket
6 Cover
7 Drive pinion flange
8 Oil seal
9 Drive pinion taper roller bearing
10 Drive pinion spacer – fixed length or collapsible
11 Vent valve
12 Drive pinion
13 Pinion gear shaft
14 Differential case
15 Side gear
16 Differential pinion
17 Retaining ring
18 Axleshaft

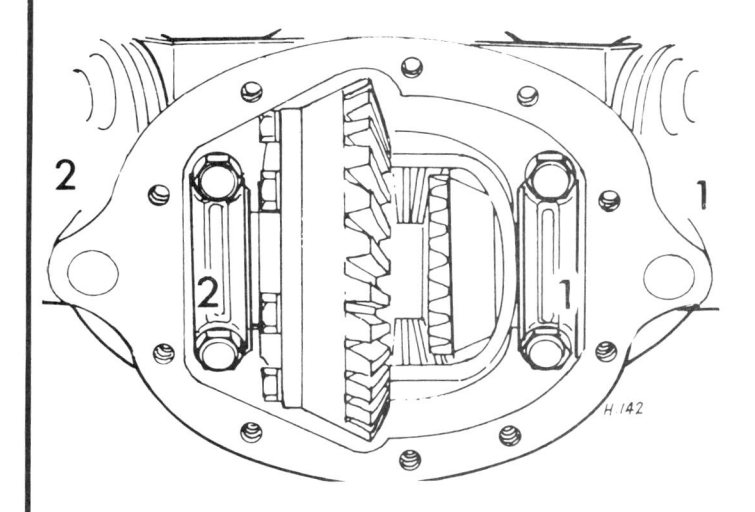

Fig. 8.11 Differential casing and bearing cap identification marks (Sec 8)

Fig. 8.12 Removing the differential assembly (Sec 8)

Fig. 8.13 Removing the locking pin from the pinion gear shaft (Sec 8)

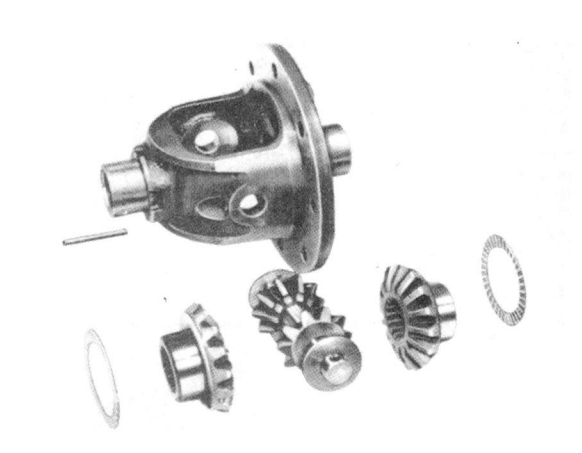

Fig. 8.14 Differential components (Sec 8)

Fig. 8.15 Checking the differential side gear endfloat (Sec 8)

Fig. 8.16 Checking the crownwheel and pinion backlash (Sec 8)

circumstances may only one part of the two be renewed. Inspect the differential pinions and side gears for signs of pitting, score marks, chipping and general wear. Obtain new gears as necessary. Inspect the thrust washers for signs of wear or deep scoring. Obtain new thrust washers as necessary. Once the pinion oil seal has been disturbed it must be discarded and a new one obtained.

20 When new parts have been obtained as required, reassembly can begin. First fit the thrust washers to the side gears and place them in position in the differential housing. The ground side of the thrust washers must face the side gears.

21 Place the thrust washers behind the differential pinion gears and mesh these two gears with the side gears through the two apertures in the differential housing. Make sure they are diametrically opposite to each other. Rotate the differential pinion gears through 90° so bringing them into line with the pinion gear shaft bore in the housing.

22 Insert the pinion gear shaft with the locking pin hole in line with the pin hole.

23 Using feeler gauges, measure the endfloat of each side gear. If the specified figure is exceeded, new thrust washers must be obtained. Dismantle the assembly again and fit new thrust washers.

24 Lock the pinion gear shaft using the pin which should be tapped fully home using a suitable diameter parallel pin punch. Peen over the end of the pin hole to stop the pin working its way out.

25 The crownwheel may next be refitted. First heat it for approximately 10 minutes in boiling water. Wipe the mating faces of the crownwheel and differential housing and, if original parts are being used, place the crownwheel into position with the previously made marks aligned. Refit the eight bolts that secure the crownwheel and tighten these in a progressive and diagonal manner to the specified torque.

26 Place the shim packs back in their original fitted position on the differential housing bearing location. Using a piece of suitable diameter tube very carefully fit the differential housing bearings with the smaller diameter of the taper outwards. The bearing cage must not in any way be damaged.

27 Place the shims behind the head of the pinion gear and, using a suitable diameter tube, carefully fit the larger taper roller bearing onto the pinion shaft. The larger diameter of the bearing must be next to the pinion head.

28 Using suitable diameter tubes, fit the two taper roller bearing cones into the final drive housing, making sure that they are fitted the correct way round.

29 Slide the shim and spacer onto the pinion shaft and insert into the final drive housing.

30 Refit the second and smaller diameter taper roller bearing onto the end of the pinion shaft and follow this with a new oil seal. Before the seal is actually fitted, apply some grease to the inner face between the two lips of the seal.

31 Apply a little jointing compound to the outer face of the seal.

32 Using a tubular drift of suitable diameter, carefully drive the oil seal into the final drive housing, Make quite sure that it is fitted squarely and flush with the housing.

33 Refit the drive pinion flange and hold securely in a bench vice. If a fixed length spacer is fitted, fit and tighten the pinion nut to the specified torque. If a collapsible spacer is fitted, follow the procedure given in Section 7 paragraph 14, setting the pinion turning torque to between 2.5 and 2.7 Nm (1.8 to 2.0 lbf ft) which includes 0.2 to 0.4 Nm (0.15 to 0.30 lbf ft) for the additional friction of the new oil seal. If new bearings have been fitted, together with a fixed length spacer, check that the turning torque is between the limits just stated. If not, a longer or shorter spacer must be fitted until the setting is correct.

34 Fit the bearing cones to the differential housing bearings and carefully ease the housing into position in the final drive housing.

35 Refit the bearing caps in their original positions. Smear a little jointing compound on the threads of each cap securing bolt and fit into position. When all four bolts have been refitted tighten these up in a diagonal and progressive manner to the specified torque.

36 If possible, mount a dial indicator gauge so that the probe is resting on one of the teeth of the crownwheel and determine the backlash between the crownwheel and pinion. The backlash may be varied by decreasing the thickness of the shims behind one bearing and increasing the thickness of shims behind the other, thus moving the crownwheel into or out of mesh as required. The total thickness of the shims must not be changed.

37 The best check the do-it-yourself owner can make to ascertain the correct meshing of the crownwheel and pinion is to smear a little engineer's blue onto the crownwheel and pinion and then rotate the pinion. The contact mark should appear right in the middle of the

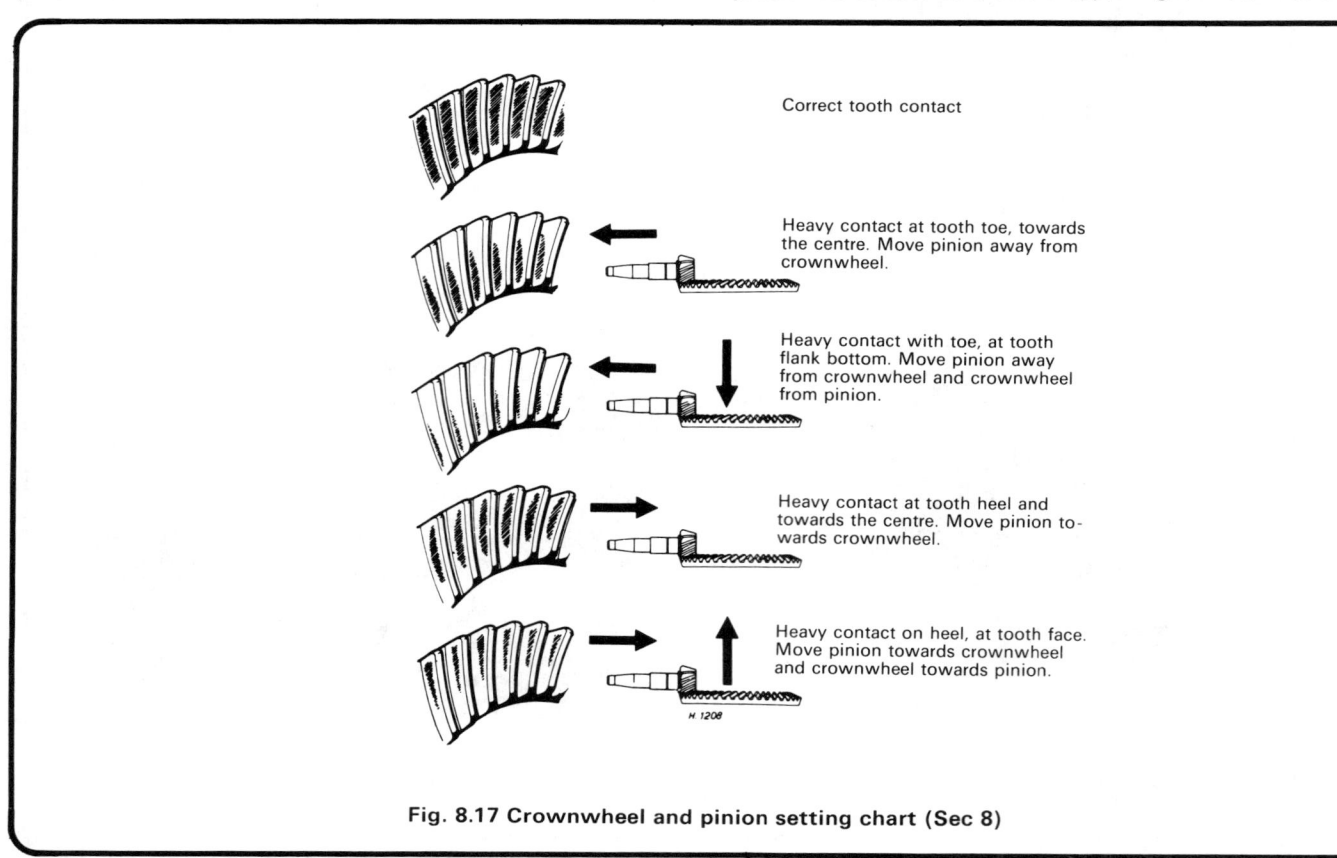

Fig. 8.17 Crownwheel and pinion setting chart (Sec 8)

crownwheel teeth. Refer to Fig. 8.17 where the correct tooth pattern is shown. Also shown are incorrect tooth patterns and the method of obtaining the correct pattern. Obviously this will take time and further dismantling but will be worth it.
38 Refit the brake backplates followed by the rear cover (Section 6) and axleshafts (Section 4).
39 Refit the rear axle (Section 3).

9 Wheel stud – renewal

1 Chock the front wheels then jack up the rear of the car and support with axle stands. Remove the rear wheel.
2 Release the handbrake and remove the brake drum.
3 Apply penetrating oil to the stud to be removed and allow it to soak for several minutes.
4 Press or drive the wheel stud through the axleshaft flange.
5 Clean the hole in the flange then insert the new stud from behind and align the splines with those made by the original stud.
6 Pull the stud fully into the flange using a wheel nut and washers as shown in Fig. 8.18.
7 Refit the brake drum and rear wheel then lower the car to the ground.

10 Limited-slip differential – general

The limited-slip differential is fitted to 2.8 Special models and its main purpose is to prevent wheel spin of the inside rear wheel when driving round bends or out of bends with full engine power applied.

Fig. 8.18 Using a wheel nut to pull a stud into the axle flange (Sec 8)

Under these conditions there is a weight transfer to the outside wheel, and the design of the standard differential is such that an adverse torque can be applied to the inside wheel causing wheel spin. The limited-slip differential uses multi-plate clutches to counteract wheel spin and the unit also has a further benefit in that the car can pull away more positively on slippery surfaces.
The multi-plate clutches are connected between the axleshafts and two thrust rings which support the pinion gear shafts in special square cut-outs (Fig. 8.19). When power is applied, the shafts force the thrust

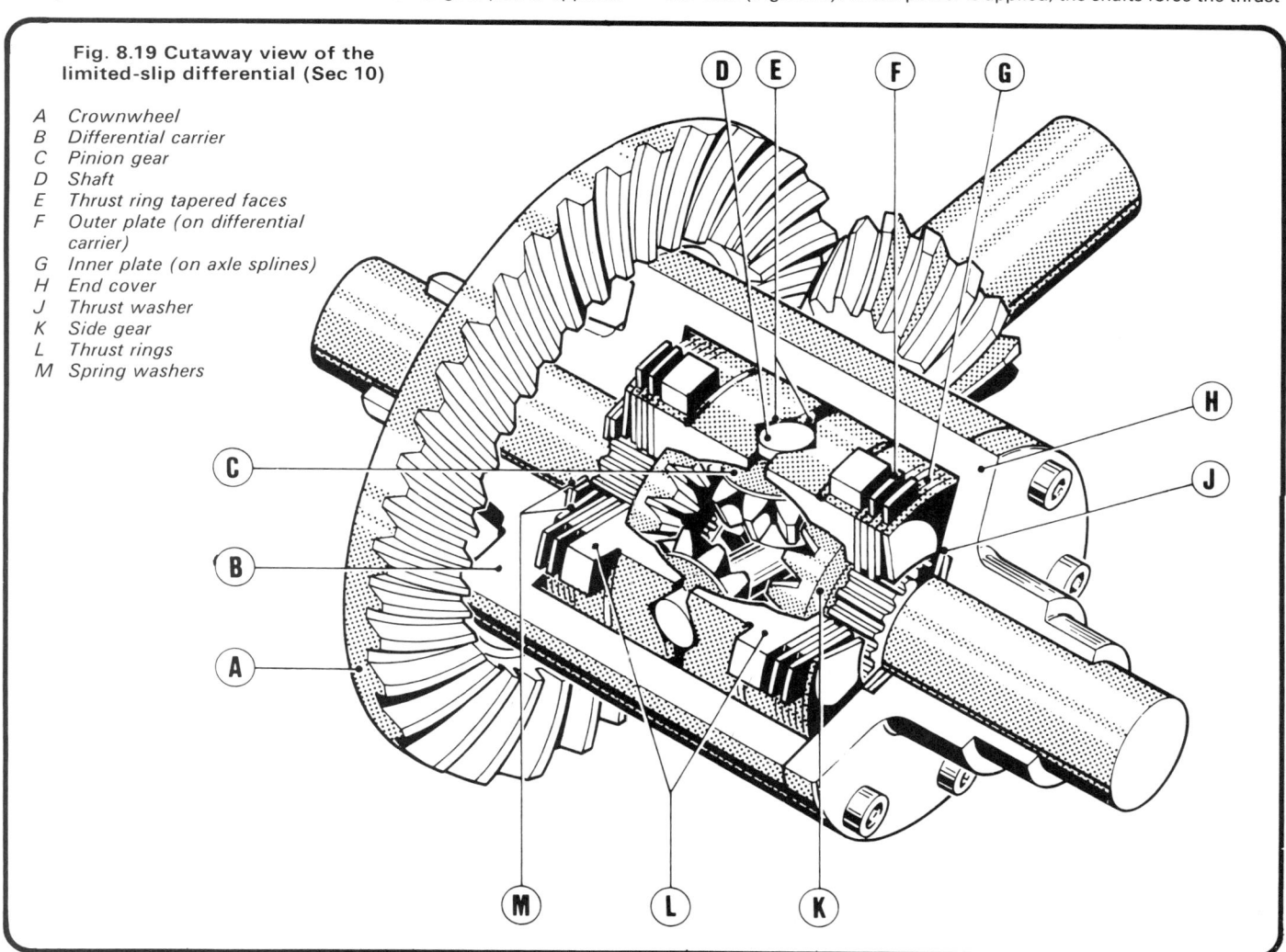

Fig. 8.19 Cutaway view of the limited-slip differential (Sec 10)

A Crownwheel
B Differential carrier
C Pinion gear
D Shaft
E Thrust ring tapered faces
F Outer plate (on differential carrier)
G Inner plate (on axle splines)
H End cover
J Thrust washer
K Side gear
L Thrust rings
M Spring washers

rings apart and compress the clutch plates. The action is variable, depending on the power applied, so that under full power the differential will lock completely and full torque will be applied to both rear wheels. Under coasting or overrun conditions the differential will operate as a normal standard unit. The unit incorporates four differential pinions as against the two pinions fitted to standard units.

11 Fault diagnosis – rear axle

Symptom	Reason(s)
Noise	Worn axleshaft bearing
	Insufficient lubricant
	Worn differential housing bearings
	Worn crown wheel and pinion
	Incorrect crown wheel and pinion mesh
'Clonk' on acceleration and deceleration	Worn axleshaft splines
	Excessive crownwheel and pinion backlash
Oil leakage	Pinion oil seal leaking
	Rear cover gasket leaking
	Oil seal on axleshaft bearing leaking

Chapter 9 Braking system

Contents

Fault diagnosis – braking system	23
Footbrake pedal – removal and refitting	16
Front brake caliper – overhaul	5
Front brake caliper – removal and refitting	4
Front brake discs – inspection, removal and refitting	6
Front disc pads – renewal	3
General description	1
Handbrake – adjustment	18
Handbrake cable(s) and rod – removal and refitting	20
Handbrake lever – removal and refitting	19
Hydraulic pipes and hoses – removal and refitting	14
Hydraulic system – bleeding	15
Master cylinder – overhaul	11
Master cylinder – removal and refitting	10
Pressure differential switch – description, removal and refitting	12
Pressure reducing valve – description, removal and refitting	13
Rear backplate – removal and refitting	9
Rear brake shoes – renewal	7
Routine maintenance	2
Stop-lamp switch – removal, refitting and adjustment	17
Vacuum servo unit – description	21
Vacuum servo unit – removal and refitting	22
Wheel cylinder – removal, overhaul and refitting	8

Specifications

System type .... Dual hydraulic circuit, split front and rear with pressure reducing valve in rear circuit on later models. Disc front brakes and self-adjusting rear drum brakes, all with servo assistance. Cable or cable and rod operated handbrake on rear brakes

Brake fluid type/specification .... Hydraulic fluid to Ford spec SAM-6C 9103-A (Duckhams Universal Brake and Clutch Fluid)

Front brakes
Disc thickness:
- New .... 12.7 mm (0.5 in)
- Minimum .... 11.4 mm (0.45 in)

Disc run-out (maximum including hub) .... 0.09 mm (0.0035 in)
Minimum disc pad lining thickness .... 3.0 mm (0.118 in)

Rear brakes
Minimum brake shoe lining thickness .... 1.5 mm (0.059 in)

Torque wrench settings

	Nm	lbf ft
Caliper	47 to 68	35 to 50
Brake disc	41 to 46	30 to 34
Backplate	27 to 31	20 to 23
Hydraulic unions	7 to 10	5 to 7
Bleed screw (max)	10	7
Recuperating valve retainer (Girling)	49 to 62	36 to 46

1 General description

Disc brakes are fitted to the front wheels and drum brakes to the rear wheels. The cable or cable and rod operated handbrake operates independently on the rear wheels. A vacuum servo unit is fitted between the master cylinder and footbrake pedal to provide assistance when the pedal is depressed.

The hyralic circuit is split front and rear. Under normal conditions both circuits operate in unison, however, in the event of failure of one circuit, the remaining circuit will provide adequate braking to stop the car in an emergency. Later models incorporate a pressure reducing valve in the rear hydraulic circuit to prevent rear wheel locking during heavy braking.

When working on the braking system, remember that brake fluid is toxic and will also damage paintwork. If spilt it should be washed away immediately with cold water.

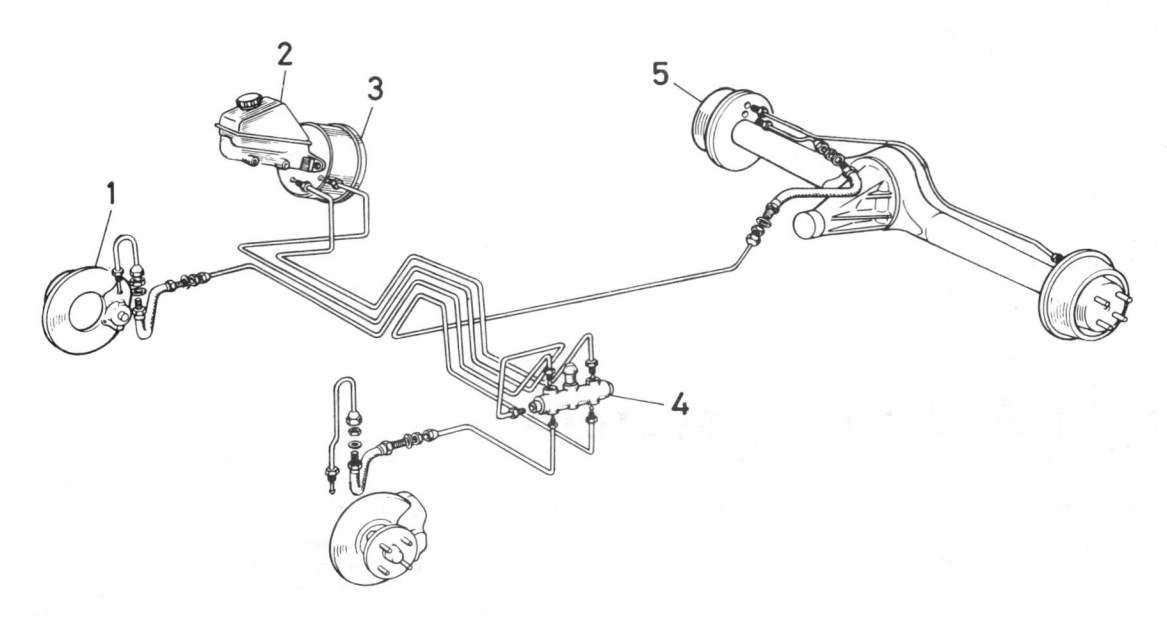

142 Chapter 9 Braking system

Fig. 9.1 Dual line hydraulic brake circuit (Sec 1)

1   Front brake disc and caliper
2   Master cylinder and fluid reservoir
3   Vacuum servo unit
4   Pressure differential warning switch
5   Rear drum brake

2 Routine maintenance

At the intervals specified in the Routine Maintenance section in the front of the manual carry out the following procedures.

Check brake fluid level

1 Check that the brake fluid level in the reservoir is at or near the maximum mark. The reservoir is located on the master cylinder on the right-hand side of the bulkhead in the engine compartment. Slight variations of level will occur according to the wear of the brake linings. If the level drops towards the minimum mark the complete hydraulic system should be checked for leaks.
2 With the car raised, carefully inspect all the hydraulic pipes, hoses and unions for chafing, cracks, leaks and corrosion.

3 To top up the fluid unscrew the filler cap while holding the cable connector stationary, and add brake fluid as necessary. Refit the cap on completion.
4 When checking the brake fluid level it is a good practice to also check the low level warning light. With the handbrake released and the ignition switched on, depress the rubber membrane on the reservoir filler cap (photo). If the warning light fails to come on check the bulb and associated circuit.

Check disc pad and rear brake shoe linings

5 Jack up the front and rear of the car and support on axle stands. Release the handbrake.
6 Although it is possible to check the front disc pad linings with a mirror and torch, it is recommended that the front wheels are first removed.
7 Measure the thickness of lining on each disc pad and if any is less than 3.0 mm (0.118 in) renew the complete set of front disc pads.
8 If the rear brake backplates incorporate inspection holes, remove the rubber plugs and check that the thickness of the linings is not less than 1.5 mm (0.059 in).

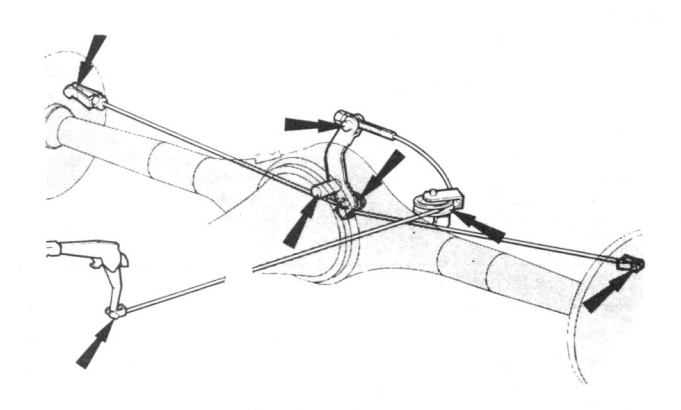

2.4 Depress the rubber membrane to check the fluid low level warning light

Fig. 9.2 Handbrake linkage lubrication points (arrowed) on early models (Sec 2)

Chapter 9 Braking system

9 If there are no inspection holes, remove the wheels and drums to inspect the linings.
10 Where any lining is less than the minimum thickness, renew the complete set of rear brake shoes.
11 Refit the drums and wheels as necessary then lower the car to the ground.

Check operation of servo unit
12 Visually inspect the vacuum hose leading from the inlet manifold or air box to the servo unit for deterioration and security.
13 With the engine stopped, depress the brake pedal several times to dissipate the vacuum from the servo unit.
14 Depress the brake pedal with moderate pressure then start the engine. As the vacuum builds up, the pedal should move slightly towards the floor – indicating that the servo is providing resistance.

Lubricate handbrake linkage
15 Jack up the rear of the car and support on axle stands.
16 Oil the primary cable connection to the bottom of the handbrake lever.
17 On 1974/1975 models, grease the primary cable where it passes over the pulley wheel, and oil the pulley and lever pivot points.
18 On later models grease the transverse rod support bearing.
19 Oil the clevis pins at the rear backplates.
20 Lower the car to the ground.

Renew the brake fluid
21 Bleed the hydraulic system as described in Section 15 allowing for the new brake fluid to completely fill the circuit being bled.
22 At the same time remove the rear brake drums and check the wheel cylinders for leakage.

3 Front disc pads – renewal

1 Apply the handbrake then jack up the front of the car and support on axle stands.
2 Remove the front wheels.
3 If the fluid level in the master cylinder reservoir is high, when the pistons are moved into their respective bores to accommodate new pads, the level could rise sufficiently for the fluid to overflow. Place absorbent cloth around the reservoir, or syphon a little fluid out, so preventing paintwork damage being caused by the hydraulic fluid.
4 Using a pair of long-nosed pliers, extract the two small clips that hold the main retaining pins in place (photo).
5 Remove the main retaining pins which run through the caliper and the metal backing of the pads and the shims.
6 The friction pads can now be removed from the caliper. If they prove difficult to remove by hand, a pair of long-nosed pliers can be used. Lift away the shims and tension springs (where fitted) (photos).
7 Carefully clean the recesses in the caliper in which the friction pads and shims lie, and the exposed faces of each piston, removing all traces of dirt or rust.
8 Using a piece of wood, carefully retract the pistons.
9 Place the brake pad tension springs on the brake pads and shims and locate in the caliper. Note that, where applicable, the arrows on the shims must face the forward rotation of the front wheels (photo).
10 Insert the main pad retaining pins, making sure that the tangs of the tension springs are under the retaining pins. Secure the pins with the small wire clips.
11 Repeat the procedure on the remaining front brake then refit the wheels and lower the car to the ground.
12 To correctly seat the pistons, pump the brake pedal several times and finally top up the hydraulic fluid level in the master cylinder reservoir as necessary.

4 Front brake caliper – removal and refitting

1 Remove the front disc pads as described in Section 3.
2 If it is intended to fit new caliper pistons and/or seals, depress the brake pedal to bring the pistons into contact with the disc and assist subsequent removal of the pistons.
3 Fit a hose clamp to the flexible hose leading to the caliper. If a hose clamp is not available, tighten a piece of polythene sheet beneath the

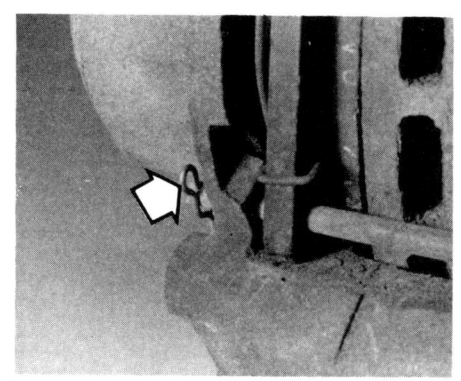

3.4 A pad retaining pin and clip (arrowed)

3.6A Removing a disc pad

3.6B The two disc pads removed

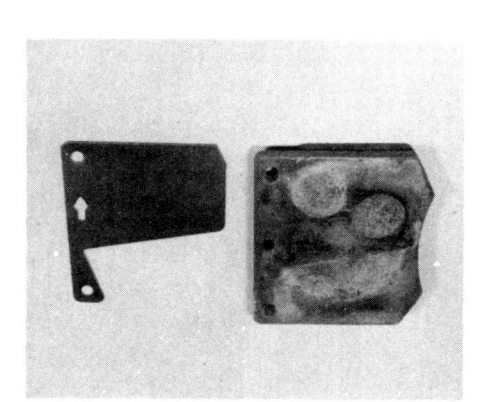

3.6C Disc pad and shim

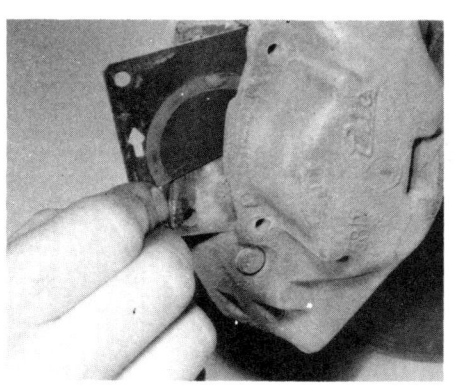

3.9A The shim arrow must face the forward rotation of the front wheel

3.9B Fitting the tension springs

fluid reservoir filler cap to help reduce any fluid loss.
4 Where a rigid brake line is connected to the caliper, completely unscrew the union nut. Where the flexible hose is connected directly to the caliper, loosen only the union and unscrew the caliper from the hose later.
5 Prise back the tabs on the locking plate and unscrew the mounting bolts (photo). The caliper may now be withdrawn over the disc and where necessary unscrewed from the flexible hose.
6 Refitting is a reversal of removal, but fit a new locking plate and tighten the mounting bolts to the specified torque. Bend the tabs to lock the bolts. On completion bleed the hydraulic system with reference to Section 15.

5 Front brake caliper – overhaul

1 With the caliper removed, clean away all external dirt.
2 Where applicable, prise out the dust cover circlips and release the dust covers from the external annular groove.
3 Extract the pistons; keeping them identified side for side. If necessary, use low air pressure from a foot pump to force the pistons from the bores.
4 Remove the dust covers from the pistons or bores as applicable.
5 Prise the piston sealing rings from the bores taking care not to scratch the bore surface.
6 It is important that the two halves of the caliper are not separated under any circumstances. If hydraulic fluid leaks are evident from the joint, the caliper must be renewed complete. Thoroughly wash all parts in methylated spirits, or clean hydraulic fluid. During reassembly, new rubber seals must be fitted. These should be well lubricated, with clean hydraulic fluid. Inspect the pistons and bores for signs of wear, score marks, or damage and if evident, new parts should be obtained ready for fitting, or a new caliper obtained.
7 Reassemble the components using a reversal of the dismantling procedure, but before inserting the pistons dip them in clean hydraulic fluid to assist entry through the seals. On completion press the pistons fully into the bores.

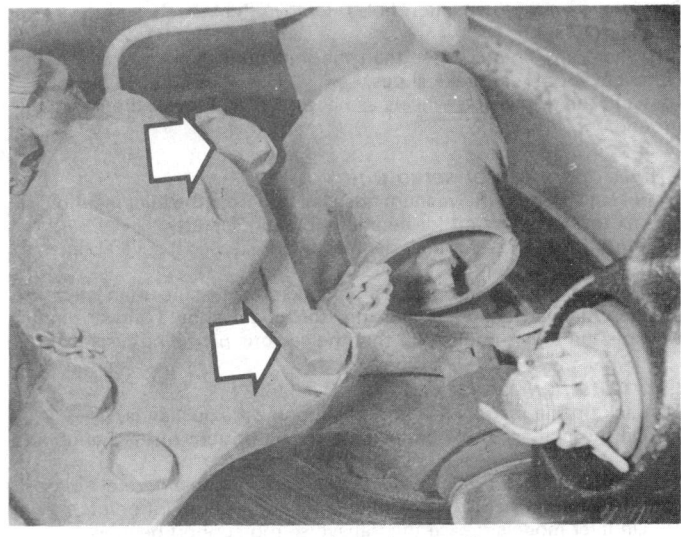

4.5 Front brake caliper mounting bolts (arrowed)

6 Front brake discs – inspection, removal and refitting

1 Apply the handbrake then jack up the front of the car and support on axle stands. Remove the front roadwheels.
2 Rotate the disc by hand and examine it for deep scoring, grooving or cracks. Light scoring is normal, but if excessive the disc must be renewed.
3 Scrape away any loose rust from the outer edge of the disc, then measure the disc thickness using a micrometer or calipers. If less than the minimum specified thickness the disc must be renewed.

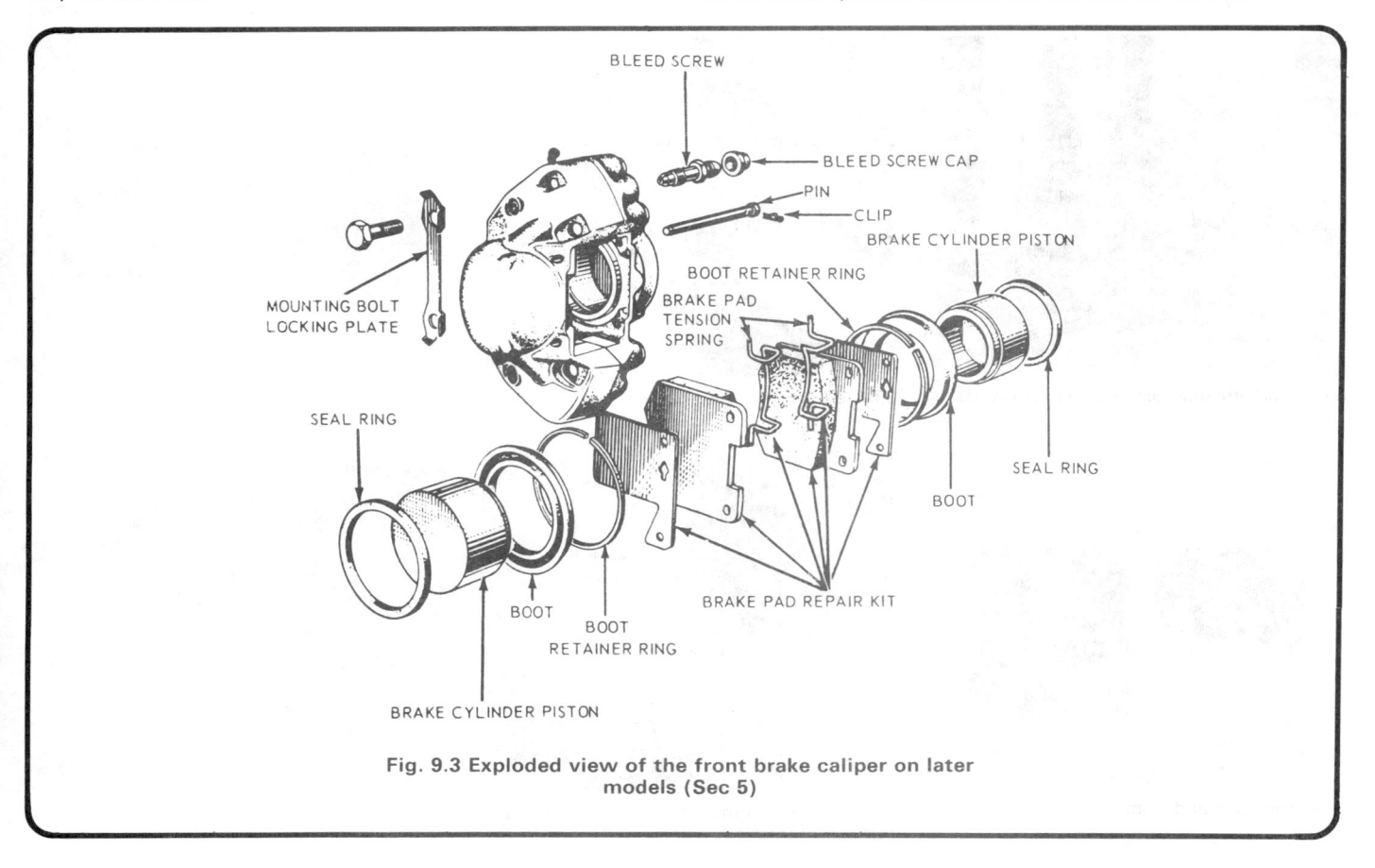

Fig. 9.3 Exploded view of the front brake caliper on later models (Sec 5)

Chapter 9 Braking system

4 Note that in order to maintain even braking both discs should be renewed at the same time.
5 To remove a disc, first remove the hub as described in Chapter 10.
6 Prise up the locking tabs, unscrew the four bolts, and remove the disc from the hub.
7 Clean the mating faces of the disc and hub then locate the disc on the hub and align the bolt holes.
8 Insert new bolts, together with new locking plates, and tighten them to the specified torque. Lock by bending the tabs.
9 The hub may now be refitted with reference to Chapter 10.

7 Rear brake shoes – renewal

1 Check the front wheels then jack up the rear of the car and support on axle stands. Release the handbrake and remove the rear roadwheels.
2 Unscrew the brake drum retaining screw then tap off the drum using a soft-faced mallet (photos).
3 Brush away all dust from the brake shoes, backplate and drum, *taking care not to inhale it as it is injurious to health.*
4 Depress each shoe hold-down spring (photo) and rotate the spring retaining washer through 90°, to disengage it from the pin secured to the backplate. Lift away the washer and spring.

Early models (approximately pre December 1974)
5 Ease each shoe from its location slot in the fixed pivot and then detach the other end of each shoe from the wheel cylinder.
6 Note which way round and into which holes in the shoes the two retracting springs fit.
7 Lift away the two brake shoes and retracting springs.
8 Turn the self-adjusting ratchet wheel until it abuts the shoulder on the slot head bolt.

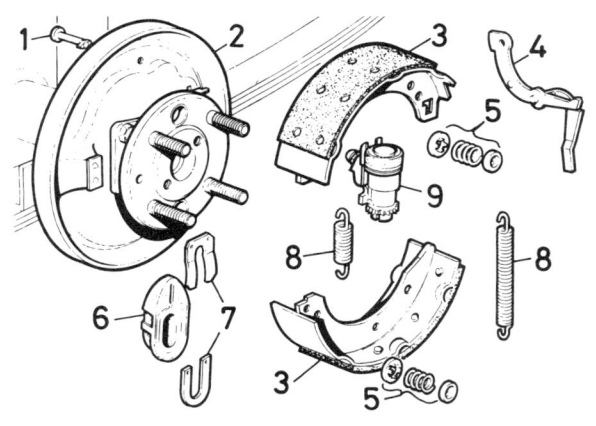

Fig. 9.4 Rear brake shoe components on early models (Sec 7)

1 Pin for hold-down spring
2 Backplate
3 Brake shoes
4 Handbrake operating lever
5 Hold-down springs
6 Rubber dust cover
7 Retaining plates
8 Retracting springs
9 Wheel cylinder

Later models
9 Note the location of the shoe retracting springs and then unhook them using a screwdriver or suitable pliers. Remove the forward shoe. Unhook the small return spring from the rear shoe and then remove the shoe leaving the adjuster in position (photos).
10 To remove the adjuster unit, detach the handbrake cable or rod at the clevis (inner side of backplate) and withdraw the adjuster mechanism (photos).

7.2A Rear brake shoes with drum removed (early models)

7.2B Rear brake shoes with drum removed (later models)

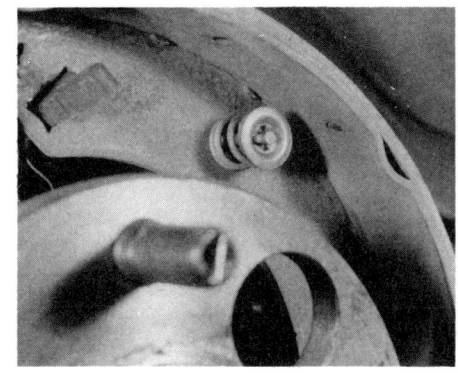

7.4 Brake shoe hold-down spring

7.9A Brake shoe attachment to wheel cylinder (later models)

7.9B Brake shoe attachment to anchor (later models)

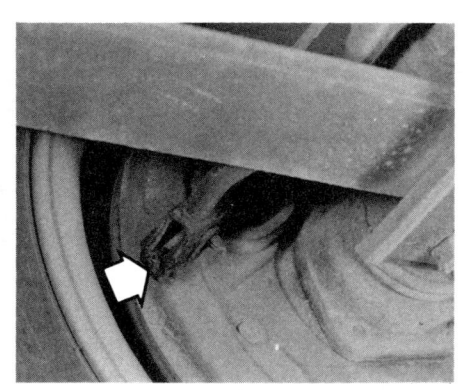

7.10 Handbrake transverse rod connection to operating lever on backplate (arrowed)

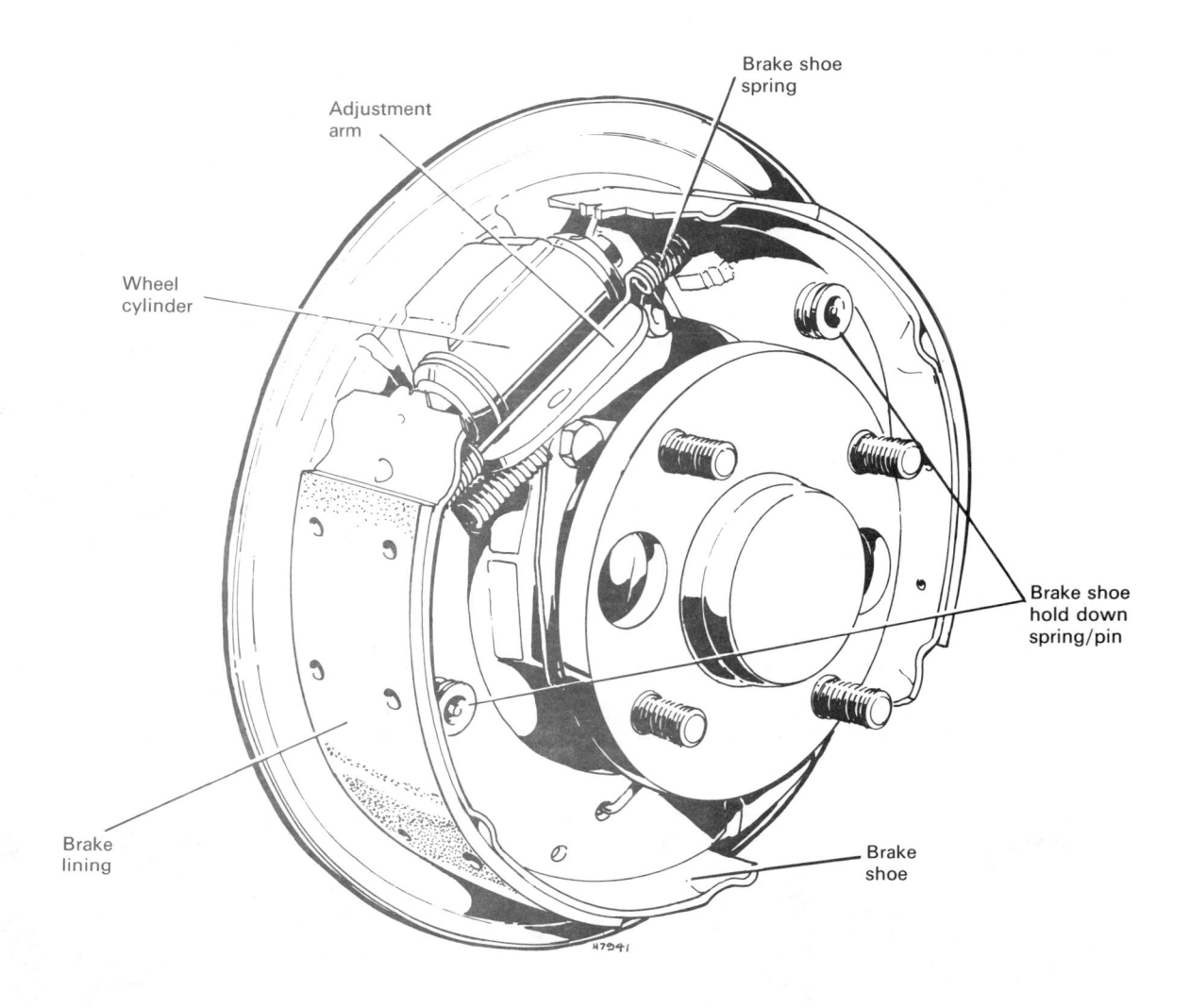

Fig. 9.5 Rear brake shoe components on later models (Sec 7)

All models
11 Check the wheel cylinder for leaks and for free movement of the piston(s). Apply a little brake grease to the shoe contact points on the backplate, wheel cylinder, anchor and self-adjusting mechanism. On early models check that the wheel cylinder is free to move up and down in the backplate and apply a little brake grease to the sliding surfaces.
12 Fit the new brake shoes using a reversal of the removal procedure, but check that the self-adjusting mechanism is set to the minimum position before refitting the drum.
13 Finally apply the footbrake firmly several times. On later models this will actuate the self-adjusting mechanism; however, on early models, release the footbrake then apply the handbrake several times until the ratchet is heard to stop turning.

8 Wheel cylinder – removal, overhaul and refitting

1 Remove the rear brake shoes as described in Section 7.
2 Fit a hose clamp to the flexible hose leading to the rear brakes. If a hose clamp is not available, tighten a piece of polythene sheet beneath the fluid reservoir filler cap to help reduce any fluid loss.
3 Unscrew the union nut(s) connecting the rigid brake line(s) to the rear of the wheel cylinder.
4 On pre December 1974 models remove the split pin and clevis pin from the handbrake lever and pull off the rubber dust cover. Pull out the two U-shaped plates and withdraw the wheel cylinder. Note that the spring plate is inserted from the piston end of the wheel cylinder.
5 On later models unbolt the wheel cylinder from the backplate.
6 To dismantle the wheel cylinder, remove the clip where fitted then prise the rubber boot(s) from the cylinder body.
7 Extract the piston(s) and separate the rubber boot(s).
8 Remove the piston return spring.
9 Prise the seal(s) from the piston(s).
10 Clean all the components in methylated spirits or clean hydraulic fluid then inspect them for wear and damage. Check the cylinder bore and piston(s) for scoring and corrosion, and if evident renew the complete wheel cylinder. If the components are in good condition obtain a repair kit of seals.

Chapter 9 Braking system

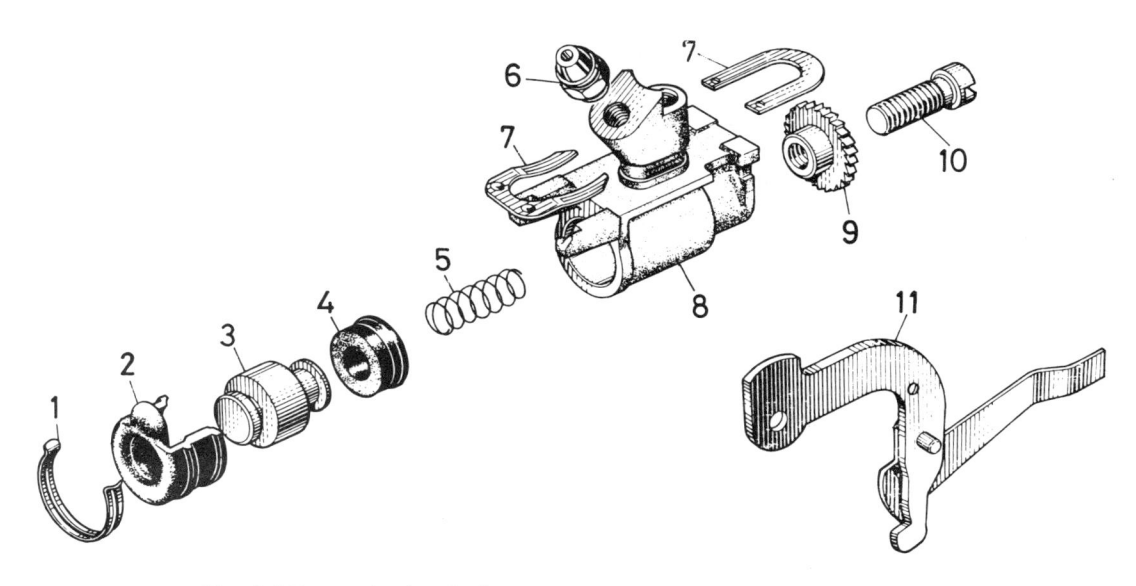

Fig. 9.6 Rear wheel cylinder components on early models (Sec 8)

1 Clip
2 Rubber boot
3 Piston
4 Seal
5 Return spring
6 Bleed screw
7 Retaining plates
8 Wheel cylinder body
9 Self-adjusting ratchet
10 Adjusting boot
11 Handbrake operating lever

11 Dip the piston(s) and piston seal(s) in hydraulic fluid then reassemble the wheel cylinder in reverse order to dismantling. Note that the seal lip must face into the cylinder.
12 Refit the wheel cylinder using a reversal of the removal procedure. Note that on early models the self-adjusting ratchet is handed – RH thread on the RH side, and LH thread on the LH side.
13 On completion bleed the hydraulic system as described in Section 15.

9 Rear backplate – removal and refitting

1 Remove the wheel cylinder as described in Section 8.
2 Remove the axleshaft (halfshaft), with reference to Chapter 8, then lift off the backplate.
3 Thoroughly clean the backplate and axle casing then refit the backplate with reference to Chapter 8 and Section 8 of this Chapter.

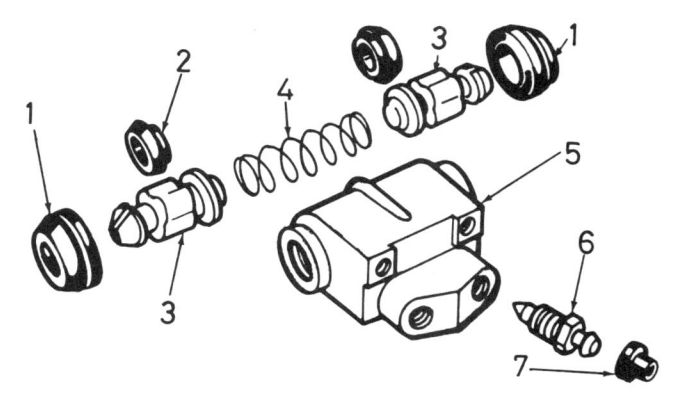

Fig. 9.7 Rear wheel cylinder components on later models (Sec 8)

1 Rubber boots
2 Seals
3 Pistons
4 Return spring
5 Wheel cylinder body
6 Bleed screw
7 Dust cover

10 Master cylinder – removal and refitting

1 Place a switchable container beneath the master cylinder to catch any spilled fluid, and also place some cloth on the surrounding bodywork and components.
2 Disconnect the wiring from the low fluid warning switch on the revervoir filler cap (photo).
3 Unscrew the reservoir filler cap and draw out the fluid using a syphon or suction tube.
4 Unscrew the union nuts and disconnect the brake lines.
5 Unscrew and remove the mounting nuts and washers and lift the master cylinder from the studs on the servo unit.
6 Empty the fluid into the container and remove the master cylinder from the engine compartment; taking care not to spill fluid on the bodywork.
7 Refitting is a reversal of removal, but on completion bleed the hydraulic system as described in Section 15.

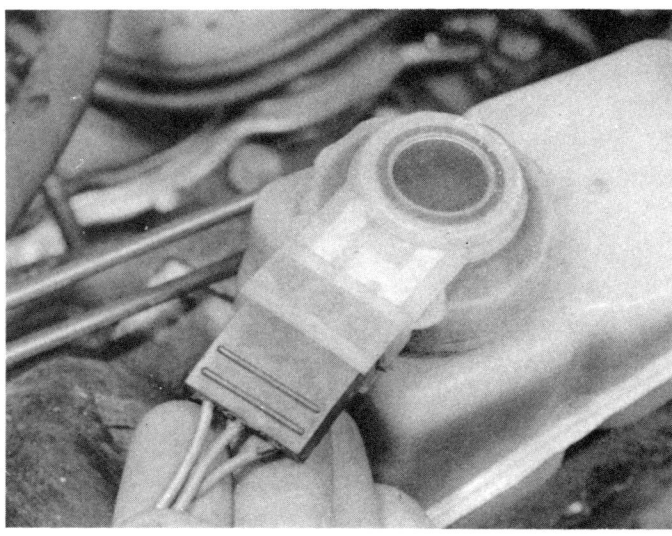

10.2 Disconnecting the low fluid warning switch wiring

11 Master cylinder – overhaul

1 With the master cylinder removed (Section 10) clean the exterior surfaces.

Girling type
2 Remove the screws and withdraw the fluid reservoir, twisting it as necessary from the end seal.
3 Remove the two seals after extracting the circlip over the end seal.
4 Using a hexagon key, unscrew the retainer and lift out the primary recuperating valve. Depress the primary piston during this procedure.
5 Extract the primary and secondary pistons, together with their associated components, noting the order of removal. If necessary top the master cylinder on a block of wood or use air pressure from a foot pump through the inlet ports.
6 Separate the primary and secondary pistons from the intermediate spring. Use the fingers to remove the gland seal from the primary piston.
7 The secondary piston assembly should be separated by lifting the thimble leaf over the shouldered end of the piston. Using the fingers, remove the seal from the secondary piston.
8 Depress the secondary spring, allowing the valve stem to slide through the keyhole in the thimble, thus releasing the tension on the spring.
9 Detach the valve spacer, taking care of the spring washer which will be found located under the valve head.

Teves (ATE) type
10 Depress the primary piston, then extract the circlip from the mouth of the cylinder and also unscrew the stop screw.
11 Extract the primary piston components noting the order of removal, followed by the secondary piston components. If necessary top the master cylinder on a block of wood or use air pressure from a foot pump through the inlet ports.
12 Prise the seals from the pistons, noting which way round they are fitted, then if necessary remove the springs and seats. The primary piston spring is retained by a screw.

Both types
13 Clean all the components in methylated spirits or brake fluid then inspect them for wear and damage. Check the cylinder bore and pistons for scoring and corrosion, and if evident renew the complete master cylinder. If the components are in good condition obtain a repair kit of seals.
14 Dip the piston and seals in hydraulic fluid then reassemble the master cylinder in reverse order to dismantling. Make sure that the seals are assembled the correct way round as previously noted. Note that the secondary seals may be identified with a silver band. On the Girling type make sure that the shim on the end valve is fitted with its convex side against the shoulder flange. Before refitting the fluid reservoir on the Girling type check that the recuperating valve is open with the primary piston released, and closed when the piston is depressed.

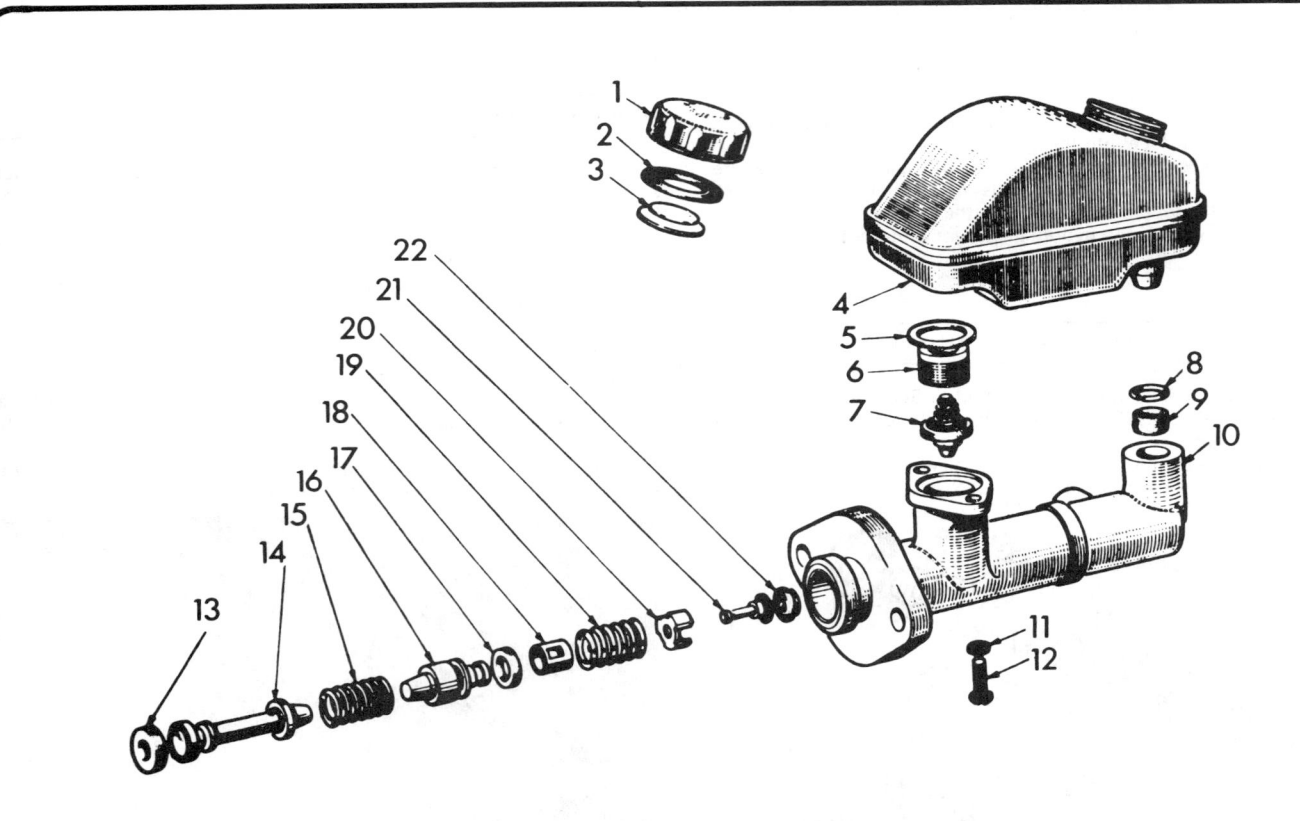

Fig. 9.8 Exploded view of the Girling master cylinder (Sec 11)

1 Filler cap
2 Seal
3 Seal retainer
4 Fluid reservoir
5 Seal
6 Retainer
7 Primary recuperating valve
8 Circlip
9 Seal
10 Master cylinder body
11 Washer
12 Reservoir retaining screw
13 Seal
14 Primary piston
15 Spring
16 Secondary piston
17 Seal
18 Retainer
19 Spring
20 Spring retainer
21 Valve
22 Seal

Chapter 9 Braking system

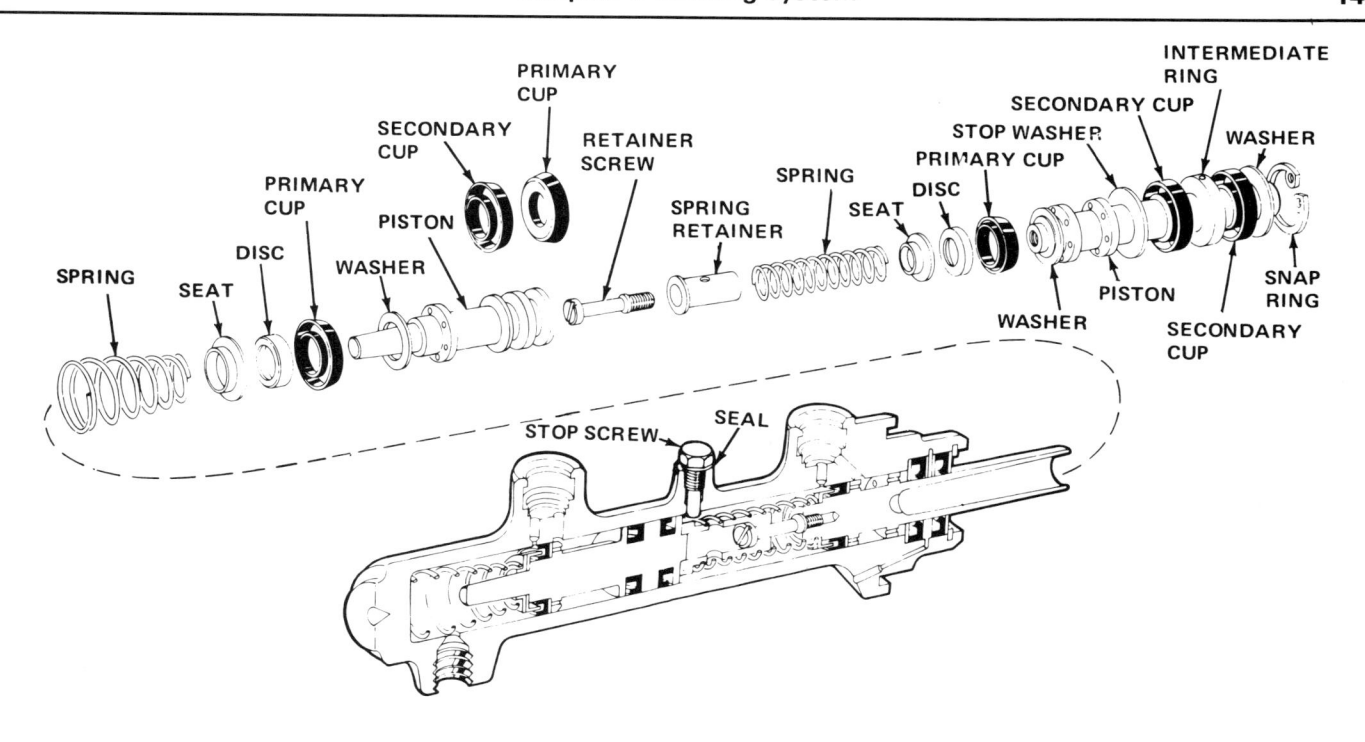

Fig. 9.9 Exploded view of the Teves (ATE) master cylinder (Sec 11)

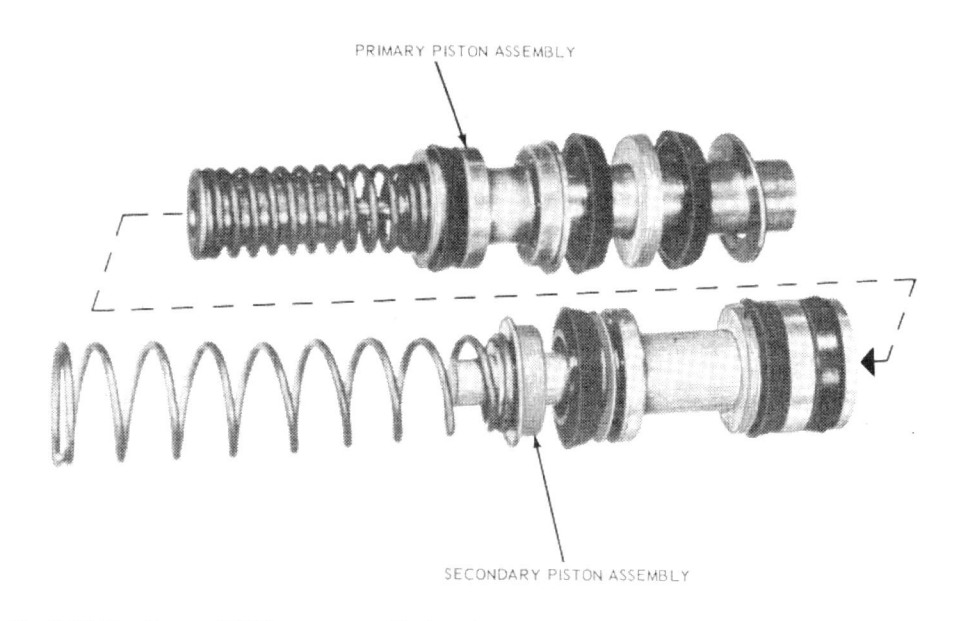

Fig 9.10 The Teves (ATE) master cylinder pistons assembled ready for refitting (Sec 11)

12 Pressure differential switch – description, removal and refitting

1 Early models were fitted with a pressure differential switch to illuminate a warning light on the instrument panel in the event of a drop in pressure in either the front or rear hydraulic circuits. The switch incorporates a piston which is kept 'in balance' if the pressures in the front and rear circuits are equal. However, if there is a pressure drop in one circuit, the piston is displaced and the switch contacts close.

2 The switch is located on the lower left-hand side of the engine compartment and although it was once possible to obtain new internal seals these are no longer available. The switch may be renewed alone (without the hydraulic body) by disconnecting the wiring and unscrewing it.

3 To remove the hydraulic body, first tighten a piece of polythene sheeting beneath the brake fluid reservoir filler cap to reduce the loss of fluid.

4 Disconnect the switch wiring.

5 Unscrew the five union nuts then disconnect and plug the pipes.

6 Unscrew the mounting bolt and remove the unit.

7 Refitting is a reversal of removal, but on completion bleed the hydraulic system as described in Section 15.

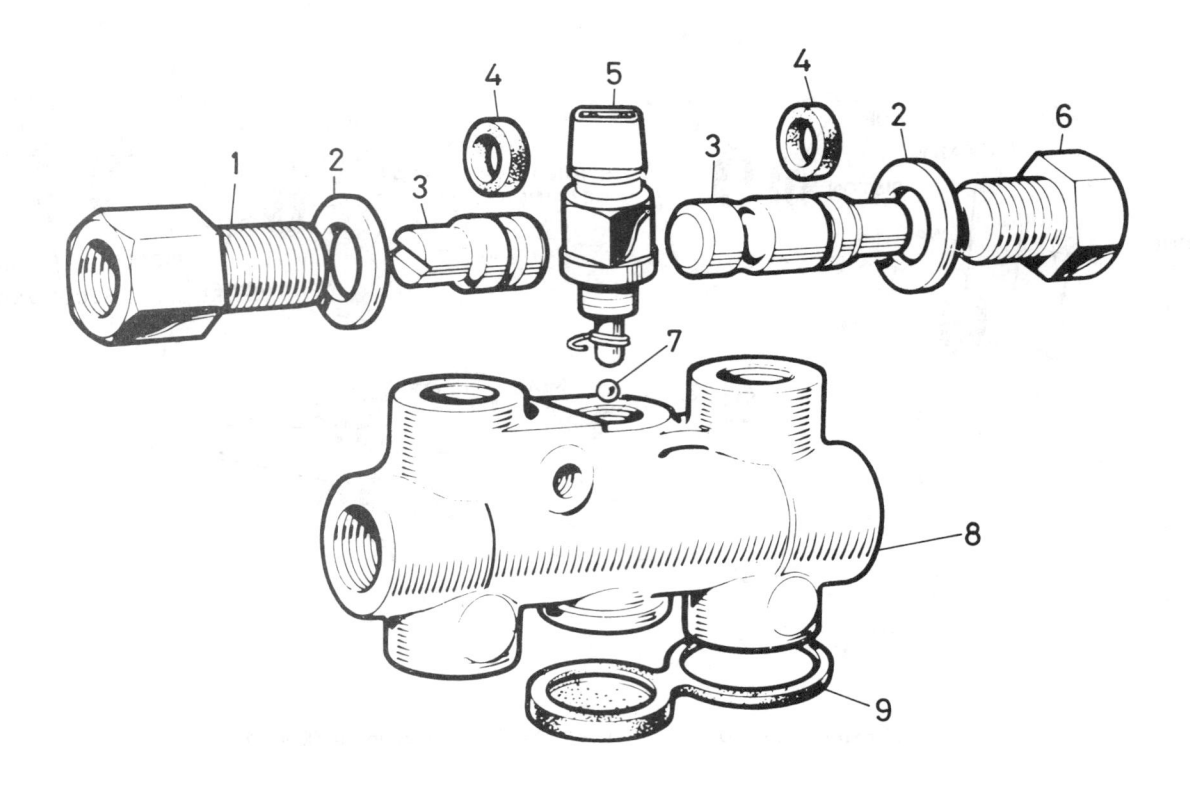

Fig. 9.11 Exploded view of the pressure differential switch (Sec 12)

1 Front brake inlet union
2 Washers
3 Pistons
4 Seals
5 Warning switch
6 End plug
7 Ball
8 Body
9 Rubber cover

13 Pressure reducing valve – description, removal and refitting

1 All models from approximately 1978 are fitted with a pressure reducing valve in the rear hydraulic circuit. The valve is located on the lower left-hand side of the engine compartment (photo) and it effectively limits the braking force available to the rear wheels during heavy braking to prevent rear wheel locking.
2 Checking of the valve requires the use of a pressure gauge and, therefore, this work should be carried out by a Ford garage.
3 To remove the valve, first tighten a piece of polythene sheeting beneath the brake fluid reservoir filler cap to reduce the loss of fluid.
4 Unscrew the two union nuts then disconnect and plug the pipes.
5 Unscrew the mounting bolts and remove the unit.
6 Refitting is a reversal of removal, but on completion bleed the hydraulic system as described in Section 15.

14 Hydraulic pipes and hoses – removal and refitting

1 First take measures to reduce the loss of hydraulic fluid. If a rigid pipe downstream of a flexible hose is being removed fit a hose clamp to the hose. If this is not possible tighten a piece of polythene sheeting beneath the fluid reservoir filler cap.
2 To remove a rigid pipe, unscrew the union nuts at each end and release the pipe from the retaining clips, where fitted.
3 To remove a flexible hose, first disconnect the rigid pipes by unscrewing the union nuts then unscrew the nuts securing the hose to the mounting brackets.
4 Rigid brake pipes are normally made up from straight pipe using special bending and flare-forming tools. The old pipe should therefore

13.1 Rear brake pressure reducing valve

be retained as a guide.
5 Refitting is a reversal of removal, but make sure that flexible hoses are positioned clear of suspension or steering components which could chafe them. Finally bleed the hydraulic system as described in Section 15.

Chapter 9 Braking system

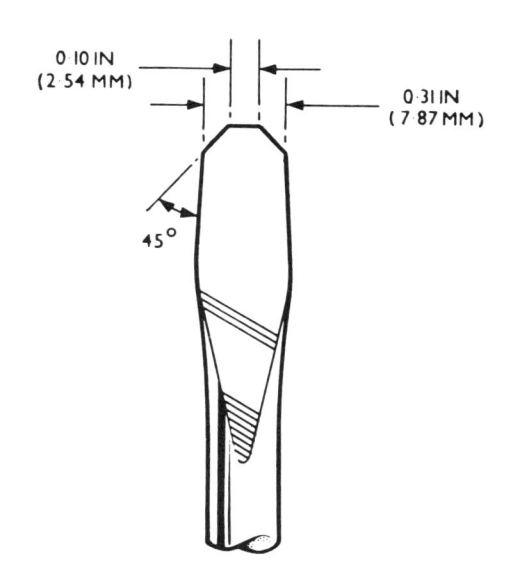

Fig. 9.12 Modified screwdriver for centalising the pressure differential switch piston (Sec 15)

15.12 Rubber dust cap on front brake caliper

15 Hydraulic system – bleeding

1 The correct functioning of the brake hydraulic system is only possible after removal of all air from the components and circuit; this is achieved by bleeding the system. Note that only clean unused brake fluid, which has remained unshaken for at least 24 hours, must be used.
2 If there is any possibility of incorrect fluid having been used in the system, the brake lines and components must be completely flushed with uncontaminated fluid and new seals fitted to the components.
3 **Never** reuse brake fluid which has been bled from the system.
4 During the procedure, do not allow the level of brake fluid to drop more than halfway down the reservoir.
5 Before starting work, check that all pipes and hoses are secure, unions tight and bleed screws closed. Take great care not to allow brake fluid to come into contact with the car paintwork, otherwise the finish will be seriously damaged. Wash off any spilled fluid immediately with cold water.
6 If the hydraulic system has only been partially disconnected and suitable precautions were taken to prevent further loss of fluid, then it should only be necessary to bleed the brake concerned, or that part of the circuit 'downstream' from where the work took place.
7 Where a pressure differential switch is fitted (Section 12) it will be necessary for an assistant to hold the internal piston in the central position during the bleeding operation, otherwise it will be difficult to get the warning light to go out and stay out. A screwdriver modified as shown in Fig. 9.12 should be inserted through the base of the hydraulic body after removing the rubber cover.
8 If possible, one of the DIY brake bleeding kits available from accessory shops should be used, particularly the type which pressurises the system, as they greatly simplify the task and reduce the risk of expelled air and fluid being drawn back into the system. Where one of these kits is being used, follow the manufacturer's instructions concerning its operation.
9 If a brake bleeding kit is not being used, the system may be bled in the conventional way as follows:
10 Obtain a clean glass jar, a suitable length of plastic or rubber tubing which is a tight fit over the bleed screws, a tin of the specified brake fluid and the help of an assistant.
11 Depress the brake pedal several times to dissipate any vacuum from the servo unit.
12 Clean the area around the bleed screw and remove the rubber dust cap (photo). If the complete system is to be bled, start at the right-hand front brake.
13 Fit the tube to the bleed screw, insert the free end of the tube into the jar and pour in sufficient brake fluid to keep the end of the tube submerged. Open the bleed screw half a turn and have your assistant depress the brake pedal to the floor and then release it. Tighten the bleed screw at the end of each downstoke to prevent the expelled air and fluid from being drawn back into the system. Repeat the procedure until clean brake fluid, free from air bubbles, can be seen emerging from the tube, then finally tighten the bleed screw, remove the tube and refit the dust cap. Remember to check the brake fluid level in the reservoir periodically and keep it topped up.
14 Repeat the procedure on the left-hand front brake followed by the left-hand rear brake. Note that there is no bleed screw on the right-hand rear brake as the fluid is supplied direct through the wheel cylinder to the left-hand rear brake. When completed, recheck the fluid level in the reservoir, top up if necessary and refit the cap.
15 Depress the brake pedal several times; it should feel firm and free from 'sponginess' which would indicate air is still present in the system.

16 Footbrake pedal – removal and refitting

1 Working inside the car, remove the lower facia panel below the steering column.
2 Extract the clip and pull out the clevis pin attaching the servo pushrod to the pedal.
3 Unhook the return spring.
4 Remove the spring clip and slide the pedal from the shaft.
5 Clean the components then examine them for wear and damage. The bushes may be renewed separately if necessary by driving them from the pedal.
6 Refitting is a reversal of removal, but lubricate the bushes, shaft and clevis pin with a little grease.

17 Stop lamp switch – removal, refitting and adjustment

1 Working inside the car, remove the lower facia panel below the steering column.
2 Disconnect the wiring plug, unscrew the locking nut and detach the switch (photo).
3 Refitting is a reversal of removal, but adjust the position of the switch so that when the footbrake pedal is at the rest position the switch is depressed half its total travel.
4 On completion, check that the switch operates correctly.

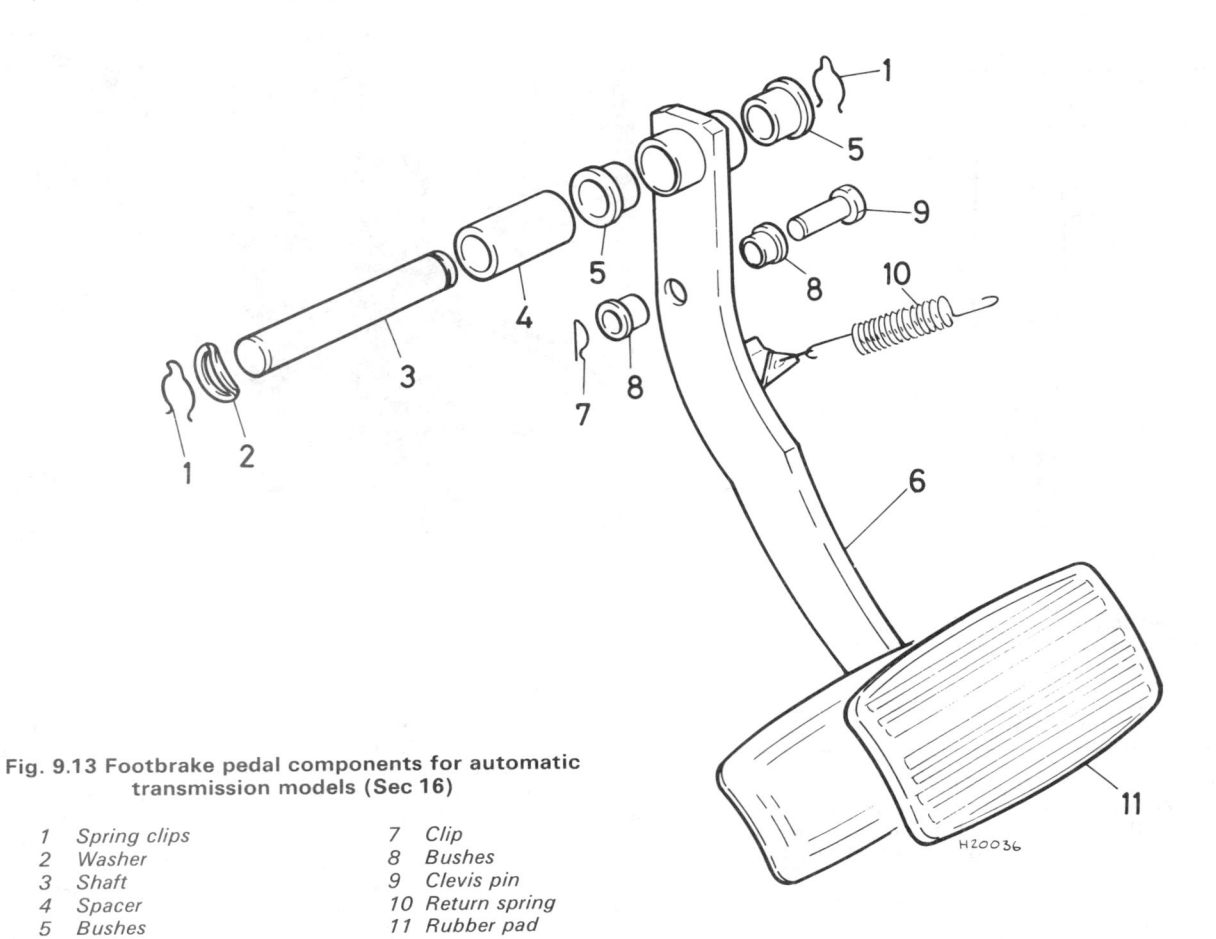

Fig. 9.13 Footbrake pedal components for automatic transmission models (Sec 16)

1 Spring clips
2 Washer
3 Shaft
4 Spacer
5 Bushes
6 Pedal
7 Clip
8 Bushes
9 Clevis pin
10 Return spring
11 Rubber pad

17.2 Stop-lamp switch and wiring plug

Early (1974/1975) models

2 Loosen the locknut on the end of the primary cable then turn the adjustment nut until the lever is just clear of the stop on the rear axle bracket. Tighten the locknut.

3 Loosen the locknut on the right-hand end of the secondary cable then turn the adjustment ferrule until all cable slack is just eliminated. Make sure that the operating levers on the backplates remain on their stops (ie fully released).

18 Handbrake – adjustment

1 Jack up the rear of the car and support on axle stands. Chock the front wheels. Fully release the handbrake lever.

Fig. 9.14 Handbrake primary cable adjustment nut and locknut (arrowed) on early models (Sec 18)

Chapter 9 Braking system

Fig. 9.15 Handbrake secondary cable adjustment ferrule and locknut (arrowed) on early models (Sec 18)

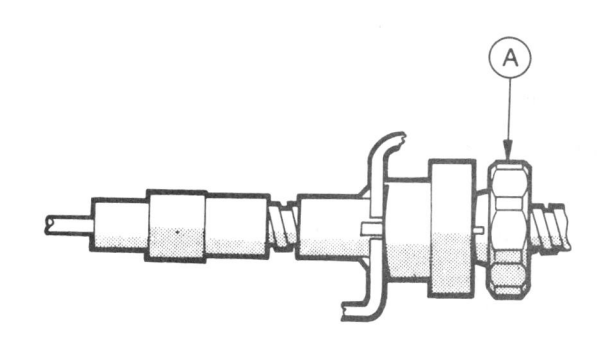

Fig 9.16 Handbrake cable adjusting nut (A) on later models (Sec 18)

Fig. 9.17 Handbrake operating lever and abutment adjustment gap 'A' on later models (Sec 18)

Later models
4 Locate the cable adjuster just forward of the rear axle differential casing (photo), then separate the adjusting nut and sleeve using a screwdriver. Engage the keyed sleeve in the bracket.
5 Turn the adjusting nut until all cable slack is eliminated and a clearance of 1.0 to 1.5 mm (0.039 to 0.059 in) exists at the handbrake operating lever abutment points (Fig. 9.17). Temporarily apply the handbrake to engage the adjusting nut with the sleeve then recheck the adjustment.

All models
6 Lower the car to the ground.

19 Handbrake lever – removal and refitting

1 Chock the front wheels then jack up the rear of the car and support on axle stands. Release the handbrake.
2 Where applicable remove the centre console with reference to Chapter 11.
3 Disconnect the primary cable from the lower end of the handbrake lever beneath the car. On early models back off the cable adjustment and unhook the clevis (photo). On later models remove the split pin and withdraw the clevis pin (photo).
4 Remove the screws and withdraw the gaiter over the top of the lever.
5 Unbolt the handbrake lever from the floor.
6 Refitting is a reversal of removal, but check the handbrake adjustment as described in Section 18.

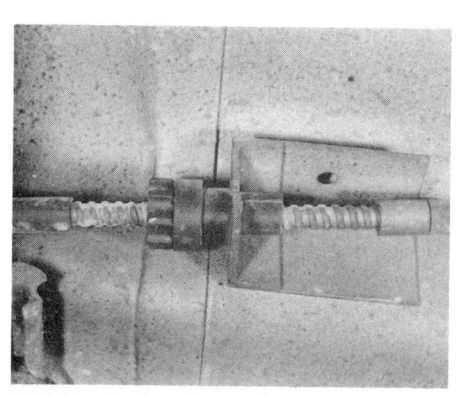

18.4 Handbrake adjuster (later models)

19.3A Primary cable connection to handbrake lever (early models)

19.3B Primary cable connection to handbrake lever (later models)

Chapter 9 Braking system

20 Handbrake cable(s) and rod – removal and refitting

1 Chock the front wheels then jack up the rear of the car and support on axle stands. Release the handbrake.

Primary cable

2 Disconnect the cable from the lower end of the handbrake lever. On early models back off the cable adjustment and unhook the clevis. On later models remove the split pin and withdraw the clevis pin.
3 On early models, unscrew the locknut and adjustment nut from the rear end of the cable, disconnect the cable from the lever, and withdraw it from the underbody guides and pulley.
4 On later models remove the clevis pin securing the cable to the right-hand rear operating lever, unclip the cable from the transverse rod, then withdraw the cable from the underbody guides and bracket.

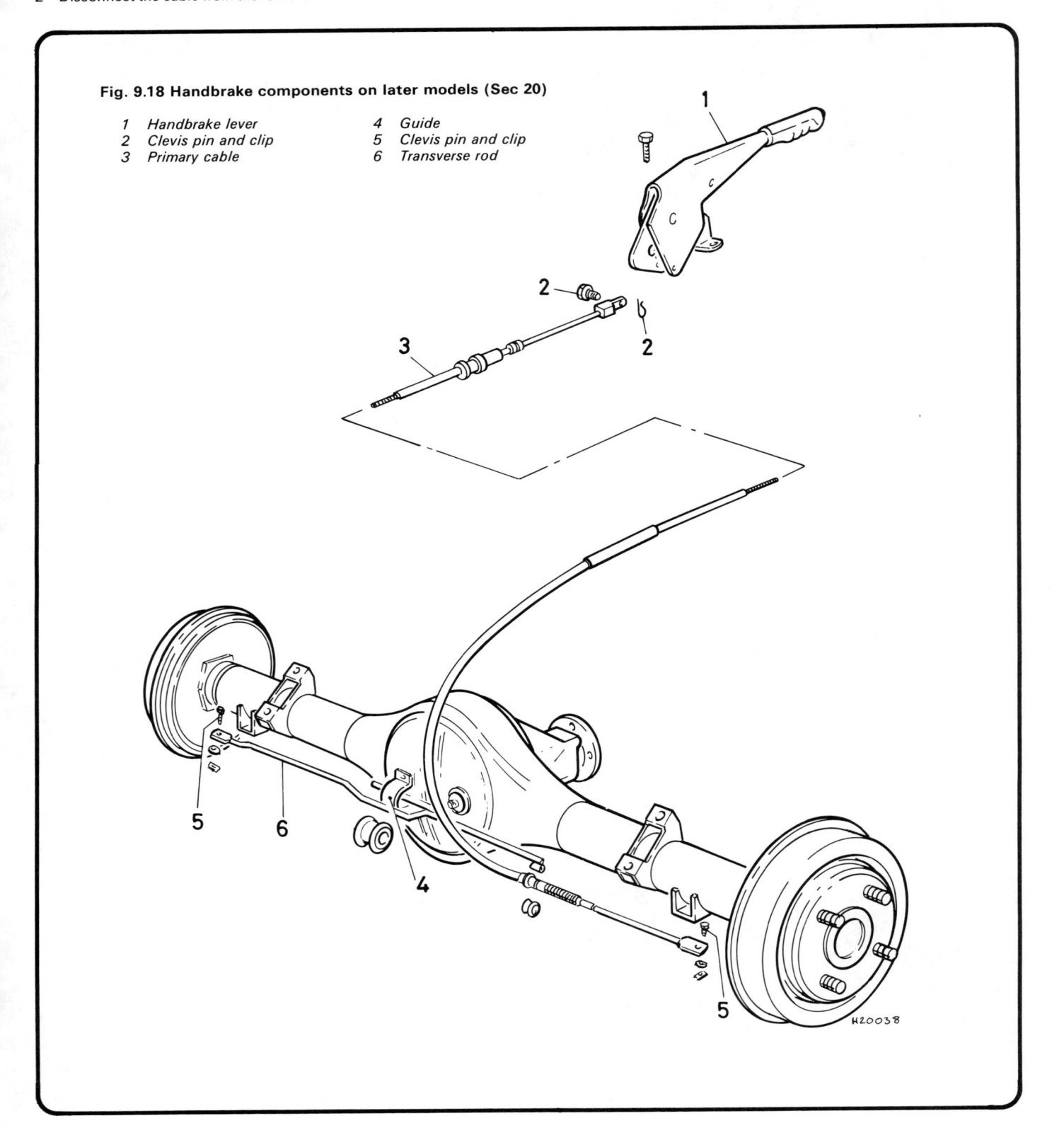

Fig. 9.18 Handbrake components on later models (Sec 20)

1 Handbrake lever
2 Clevis pin and clip
3 Primary cable
4 Guide
5 Clevis pin and clip
6 Transverse rod

Chapter 9 Braking system

Secondary cable (early models)
5 Loosen the locknut and unscrew the adjustment ferrule from the right-hand rear clevis.
6 Remove the split pin and clevis pin securing the secondary cable to the left-hand rear operating lever.
7 Extract the clip, remove the clevis pin, and withdraw the pulley wheel from the lever on the rear axle.
8 Remove the cable from the lever.

Transverse rod (later models)
9 Disconnect the cable from the right-hand rear operating lever by removing the split pin and clevis pin.
10 Unclip the cable from the transverse rod.
11 Remove the split pin and clevis pin securing the rod to the left-hand rear operating lever.
12 Slide the transverse rod from the guide on the rear axle.
13 Check the guide bush for wear and damage and renew it if necessary by prising it out and pressing a new bush into position.

All models
14 Refitting is a reversal of removal, but lubricate the linkage as described in Section 2, and adjust the cable as described in Section 18.

21 Vacuum servo unit – description

A vacuum servo unit is fitted into the brake hydraulic circuit, in series with the master cylinder, to provide assistance to the driver when the brake pedal is depressed. This reduces the effort required by the driver to operate the brakes, under all braking conditions.

This unit operates by vacuum, obtained from the induction manifold and comprises basically a booster diaphragm and check valve. The servo unit and hydraulic master cylinder are connected together so that the servo unit piston rod acts as the master cylinder pushrod. The driver's braking effort is transmitted through another pushrod to the servo unit piston and its built-in control system. The servo unit piston does not fit tightly into the cylinder, but has a strong diaphragm to keep its edges in constant contact with the cylinder wall, so ensuring an airtight seal between the two parts. The forward chamber is held under vacuum conditions created in the inlet manifold of the engine and during periods when the brake pedal is not in use, the controls open a passage to the rear chamber, so placing it under vacuum conditions as well. When the brake pedal is depressed, the vacuum passage to the rear chamber is cut off and the chamber exposed to atmospheric pressure. The consequent rush of air pushes the servo piston forward in the vacuum chamber and operates the main pushrod to the master cylinder.

The controls are designed so that assistance is given under all conditions. When the brakes are not required, vacuum in the rear chamber is re-established when the brake pedal is released. All air from the atmosphere entering the rear chamber is passed through a small air filter.

Under normal operating conditions the vacuum servo unit is very reliable and does not require overhaul except at very high mileages. In this case it is far better to obtain a service exhange unit, rather than repair the original unit. Although internal components were once available for overhauling purposes this service has now been discontinued.

22 Vacuum servo unit – removal and refitting

1 Remove the master cylinder as described in Section 10.
2 Disconnect the vacuum hose from the non-return valve on the servo unit.
3 Working inside the car, remove the lower facia panel below the steering column.
4 Extract the clip and pull out the clevis pin attaching the pushrod to the footbrake pedal.
5 Unscrew the nuts securing the mounting bracket to the bulkhead, then withdraw the servo unit and bracket from the engine compartment.
6 Unscrew the nuts and separate the braket from the servo unit.
7 Refitting is a reversal of removal with reference also to Section 10 when refitting the master cylinder.

23 Fault diagnosis – braking system

Symptom	Reason(s)
Excessive pedal travel	Rear brake self-adjusting mechanism faulty Air in hydraulic system
Brakes pull to one side	Worn or contaminated linings Seized caliper or wheel cylinder
Brake judder	Excessive run-out or distortion of discs or drums Worn or contaminated linings Brake backplate or caliper loose Worn suspension balljoints
Excessive effort to stop car	Servo unit faulty Worn or contaminated linings

Chapter 10 Suspension and steering

Contents

Fault diagnosis – suspension and steering	22
Front coil spring – removal and refitting	6
Front hub – removal, refitting and bearing adjustment	3
Front hub bearings – renewal	4
Front stabilisr bar – removal and refitting	7
Front suspension strut – removal and refitting	5
Front track control arm – removal and refitting	8
Front wheel alignment and steering angles	20
General description	1
Power steering – bleeding	15
Power steering gear – removal and refitting	16
Power steering pump – removal and refiting	17
Rear leaf spring – removal, bush renewal and refitting	10
Rear shock absorber – removal and refitting	9
Rear stabiliser bar – removal and refitting	11
Routine maintenance	2
Steering column – removal, overhaul and refitting	19
Steering gear (manual) – overhaul and adjustment	13
Steering gear (manual) – removal and refitting	12
Steering wheel – removal and refitting	18
Track rod and balljoint – renewal	14
Wheels and tyres – general care and maintenance	21

Specifications

Front suspension .... Independent, MacPherson struts with integral telescopic shock absorbers and stabiliser bar

Rear suspension .... Semi-elliptic leaf springs with rigid axle, telescopic shock absorbers and stabiliser bar

Steering gear (manual)
Type .... Rack and pinion
Steering wheel turns (lock to lock) 3.36
Lubricant type .... Gear oil, viscosity SAE 90EP (Duckhams Hypoid 90S)
Lubricant capacity .... 0.15 litre (0.26 pint)

Chapter 10 Suspension and steering

Power steering gear
Type	Hydraulic assisted rack and pinion
Steering wheel turns (lock to lock)	3.25
Lubricant type	SAE 40 oil
Lubricant capacity	0.20 litre (0.35 pint)
Fluid type	ATF to Ford spec SQM-2C 9007-AA (Duckhams Q-Matic)
Fluid capacity:	
Separate reservoir	2.2 litre (3.9 pint)
Integral reservoir	0.5 litre (0.9 pint)

Front wheel alignment
Camber	+0° 30'
Maximum difference side to side	1° 0'
Castor	+1° 0'
Maximum difference side to side	0° 45'
Toe-in	0 to 7.0 mm (0 to 0.28 in)

Tyres
Size .... 185/70HR13, 185/70VR13, 205/60VR13 or 195/50VR15

Pressures lbf/in² (kgf/cm²):

	Front	Rear
185/70HR13:		
Up to 3 persons	23 (1.6)	23 (1.6)
More than 3 persons	28 (2.0)	28 (2.0)
185/70VR13:		
Up to 3 persons	24 (1.7)	28 (2.0)
More than 3 persons	28 (2.0)	32 (2.2)
205/60VR13:		
Up to 3 persons	28 (2.0)	28 (2.0)
More than 3 persons	32 (2.2)	32 (2.2)
195/50VR15:		
Up to 3 persons	26 (1.8)	26 (1.8)
More than 3 persons	32 (2.2)	32 (2.2)

Torque wrench settings

	Nm	lbf ft
Front suspension		
Strut upper mounting	20 to 24	15 to 18
Piston rod nut	40 to 45	30 to 33
Suspension balljoint	41 to 48	30 to 35
Stabiliser bar mounting	28 to 33	21 to 24
Stabiliser bar to track control arm	21 to 61	15 to 45
Track control arm pivot	25 to 30	18 to 22
Rear suspension		
Shock absorber lower mounting	53 to 63	39 to 46
Shock absorber upper mounting	27 to 33	20 to 24
Stabiliser bar to axle	40 to 50	30 to 37
Stabiliser bar to underbody	35 to 41	26 to 30
U-bolt	25 to 36	18 to 27
Spring front mounting	35 to 41	26 to 30
Spring rear shackle	11 to 14	8 to 10
Steering		
Steering gear (manual)	20 to 24	15 to 18
Track rod end balljoint	24 to 30	18 to 22
Coupling shaft to pinion	16 to 20	12 to 15
Coupling shaft to column	23 to 30	17 to 22
Steering wheel	27 to 34	20 to 25
Steering column	17 to 20	13 to 15
Roadwheels		
Wheel nuts/bolts:		
Conical face	85 to 115	63 to 85
Flat face	115 to 140	85 to 103

1 General description

The front suspension is of independent MacPherson strut type incorporating integral telescopic shock absorbers and a stabiliser bar.

The rear suspension is of semi elliptic leaf spring-with-'live' axle type, incorporating telescopic shock absorbers and a stabiliser bar.

The steering is of rack and pinion type, but may be either manual or power-assisted. The steering wheel incorporates a convoluted cam which is designed to collapse in the event of an accident to prevent injury to the driver. It is not recommended that the power steering gear is dismantled on this work is best entrusted to a specialist.

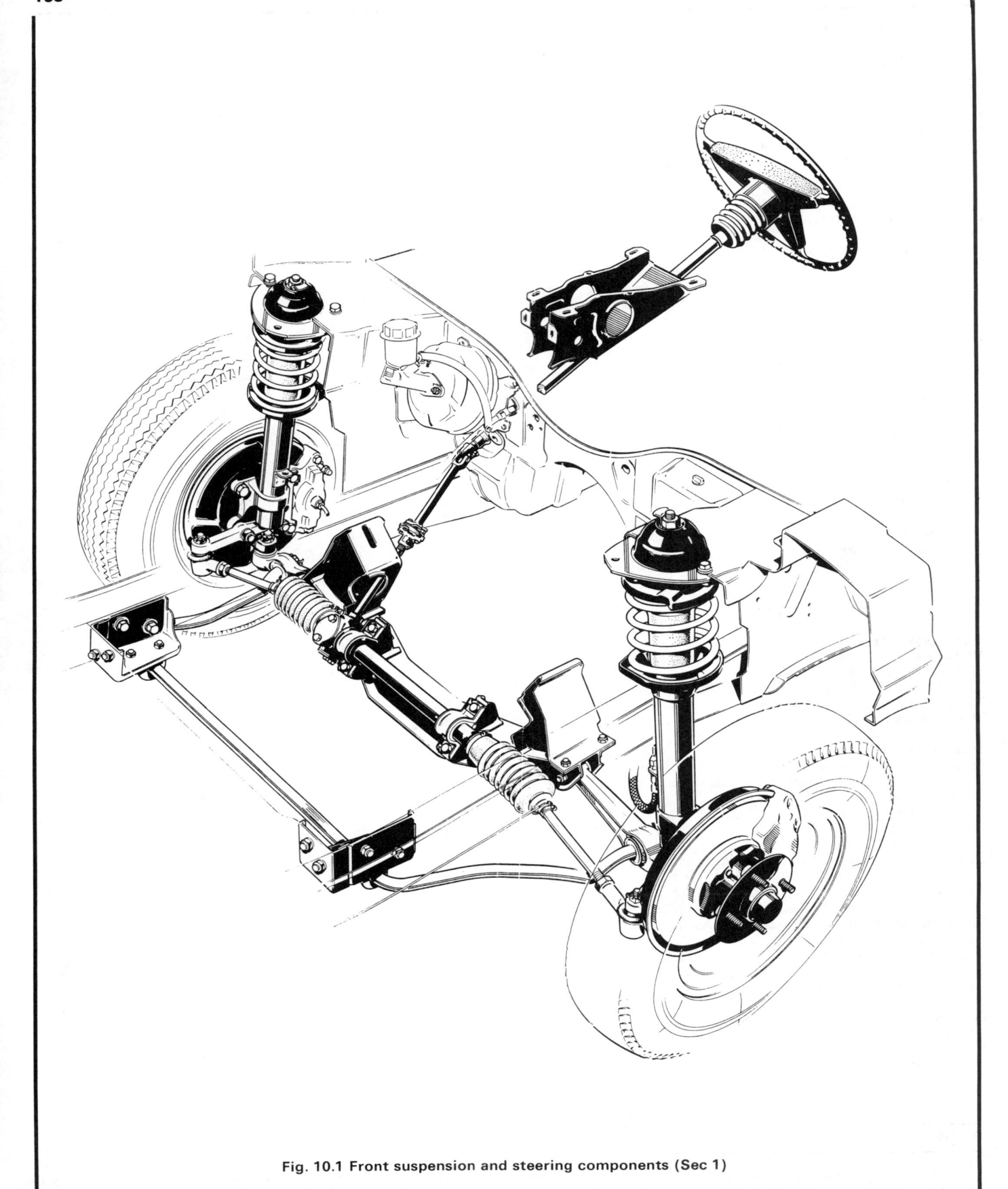

Fig. 10.1 Front suspension and steering components (Sec 1)

159

Fig. 10.2 Rear suspension components (Sec 1)

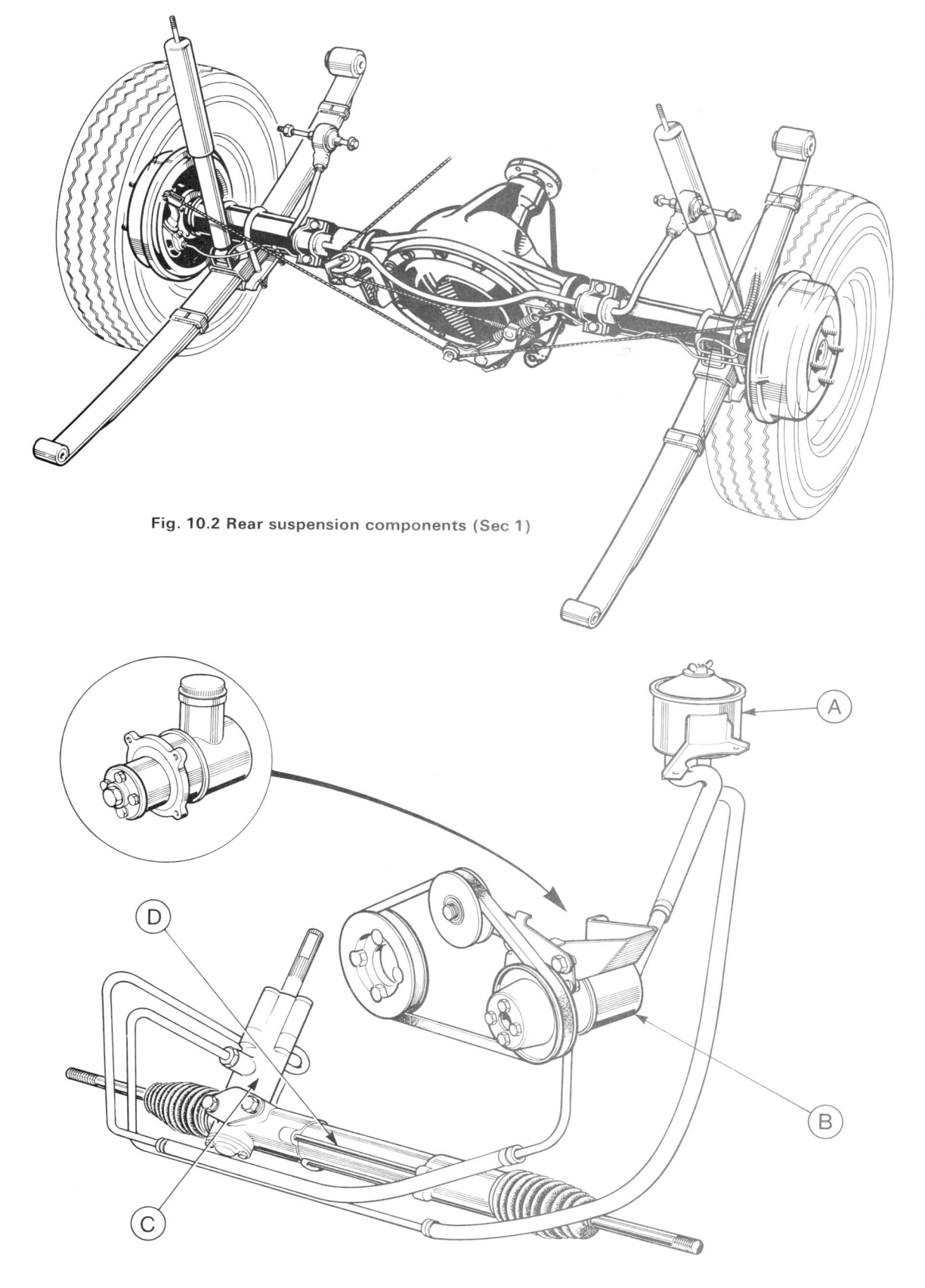

Fig. 10.3 Power steering components (Sec 1)

A   Reservoir *B   Pump* *C   Control valve* *D   Steering gear*

Chapter 10 Suspension and steering

Fig. 10.4 Drivebelt adjustment pulley bolts on 3.0 litre models – arrowed (Sec 2)

Fig. 10.5 Topping-up the power steering reservoir on early models (Sec 2)

2 Routine maintenance

At the intervals specified in the Routine Maintenance section in the front of the manual carry out the following procedures.

Check tyres for inflation, wear and condition
1 Check and if necessary adjust the tyre pressures.
2 Check all tyres for wear and condition with reference to Section 21.

Check power steering pump drivebelt tension
3 Check that the deflection of the drivebelt midway along its longest run between upper pulleys is approximately 13.0 mm (0.5 in) under moderate thumb pressure.
4 If adjustment is necessary, loosen the idler pulley bracket bolts, re-position the bracket then re-tighten the bolts. On 3.0 litre models the bracket must be pivoted around the end bolt, whereas on 2.8 litre models the bracket must be moved within the elongated bolt holes (photo).

Check power steering fluid level and hydraulic hoses
5 If possible the level should be checked with the fluid at normal operating temperature.
6 On early models unscrew the wing nut and remove the reservoir cover and gasket. The fluid should be up to the MAX level mark on the reservoir.
7 On later models, turn the filler cap/dipstick anti-clockwise and withdraw it from the reservoir. Wipe clean the dipstick, fully refit it then remove it again. The fluid should be up to the FULL hot or cold level mark on the dipstick as applicable (photo).

8 Where necessary add the specified fluid to bring the level up to the correct mark.
9 Check the hydraulic hoses for condition and security.

Check steering and suspension components
10 Raise the front and rear of the car or position over an inspection pit.
11 Check the steering rack and tie-rods for damage and wear. Also check the rack rubber bellows and track rod end rubber boots for splits and leakage.
12 Check the front and rear suspension components for damage and wear. In particular check the front suspension track control arm and stabilizer rubber mounting bushes. Use a lever to check that there is no excessive wear in the front suspension lower balljoints which are integral with the track control arms.
13 Check the operation of the front and rear shock absorbers. This is best achieved with the car on the ground by depressing each corner of the car in turn. On releasing the corner being checked the body should rise then settle immediately on its way down. If there are several oscillations the shock absorber is weak and should be renewed.

3 Front hub – removal, refitting and bearing adjustment

1 Apply the handbrake then jack up the front of the car and support on axle stands. Remove the roadwheel.
2 Remove the front brake caliper with reference to Chapter 9. Where the flexible brake hose is connected directly to the caliper there is no need to disconnect the hose as the caliper can be suspended by wire away from the disc.
3 By careful tapping and levering, remove the dust cap from the centre of the hub (photo).

2.4 Adjusting the power steering pump drivebelt tension on the 2.8 litre engine

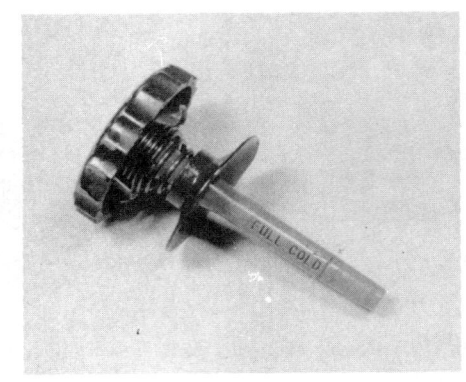

2.7 Power steering filler cap/dipstick for the 2.8 litre engine

3.3 Removing the dust cap from the front hub

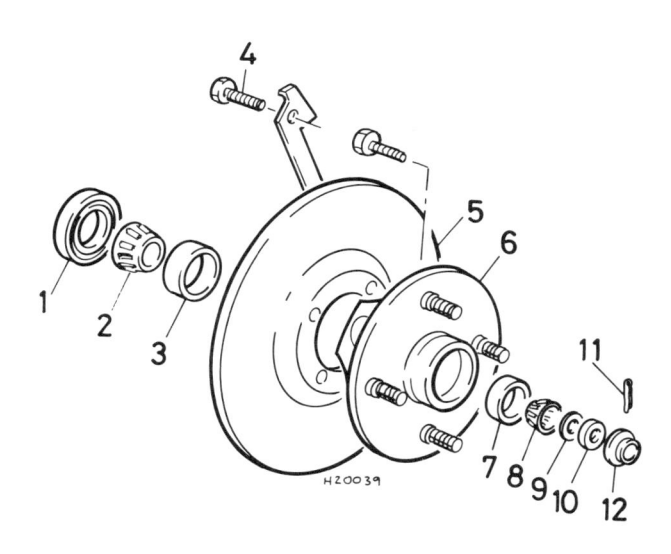

Chapter 10 Suspension and steering

Fig. 10.6 Front hub components (Sec 3)

1 Oil seal
2 Inner bearing race
3 Inner bearing track
4 Screw
5 Disc
6 Hub
7 Outer bearing track
8 Outer bearing race
9 Thrust washer
10 Nut
11 Split pin
12 Dust cap

5.6 Front suspension strut upper mounting

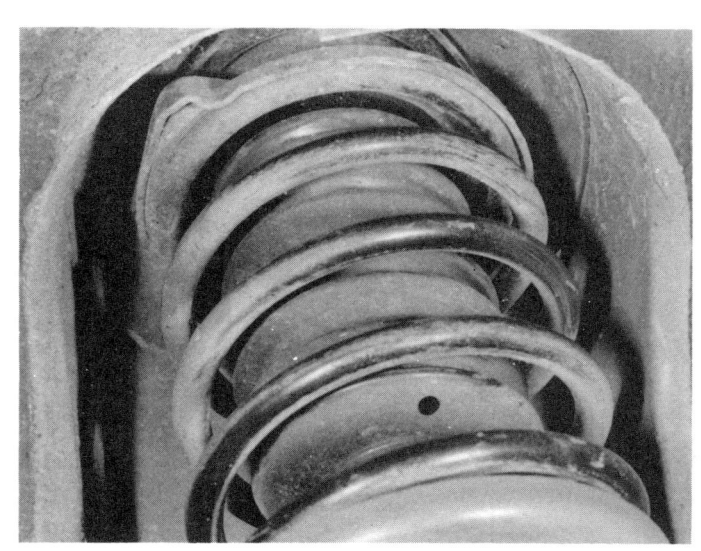

5.7 Lowering the front suspension strut from under the wheel arch

4 Remove the split pin from the nut retainer and undo the adjusting nut from the stub axle.
5 Withdraw the thrust washer and the outer tapered bearing.
6 Pull off the complete hub and disc assembly from the stub axle.
7 To refit, locate the hub assembly on the stub axle and slide on the outer bearing and thrust washer.
8 Fit the nut then tighten it to a torque of 20 to 25 Nm (15 to 18 lbf ft) while at the same time turning the hub anti-clockwise.
9 Loosen the nut by 180° (half a turn) then retighten using only the fingers and thumb.
10 Refit the nut retainer so that the split pin holes are aligned. Fit a new split pin and refit the dust cap.
11 Refit the front brake caliper with reference to Chapter 9.
12 Refit the roadwheel and lower the car to the ground.

4 Front hub bearings – renewal

1 Remove the front hub as described in Section 3.
2 Carefully prise out the grease seal from the back of the hub assembly and remove the inner tapered bearing.
3 Using a brass drift, or bearing puller, remove the bearing outer tracks from the ends of the hub.
4 Scoop out the grease from inside the hub and wash clean.
5 Drive the new bearing tracks into the hub with their tapers facing outwards using suitable metal tubing.
6 Working the grease well into the bearings, fully pack the bearing cages and rollers with wheel bearing grease. **Note:** *Leave the hub cavity half empty, to allow for subsequent expansion of the grease.*
7 Fit the inner tapered bearing in its track then press a new grease seal squarely into the hub using a piece of wood.
8 Refit the front hub with reference to Section 3.

5 Front suspension strut – removal and refitting

1 Apply the handbrake then jack up the front of the car and support on axle stands. Remove the roadwheel.
2 Remove the front hub with reference to Section 3.

3 On early models, remove the rigid brake pipe and detach the flexible hose from the bracket on the strut (if necessary refer to Chapter 9).
4 Disconnect the track rod end from the steering arm (Section 14).
5 Extract the split pin and unscrew the nut securing the track control arm balljoint to the strut. Separate the joint using a balljoint separator tool.
6 Working under the bonnet, unscrew the three mounting nuts securing the strut to the wheel arch (photo). Where applicable note the location of the accelerator cable support bracket beneath one of the nuts.
7 Lower the strut assembly from under the wheel arch (photo).
8 Refititng is a reversal of removal, but tighten all nuts to the specified torque. Refer to Sections 14 and 3 as necessary.

6 Front coil spring – removal and refitting

Note: *In order to remove the coil spring from the strut it will be necessary to obtain a suitable spring compressor tool.*

1 Remove the front suspension strut as described in Section 5.

Chapter 10 Suspension and steering

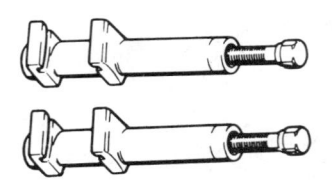

Fig. 10.7 Coil spring compressor tools (Sec 6)

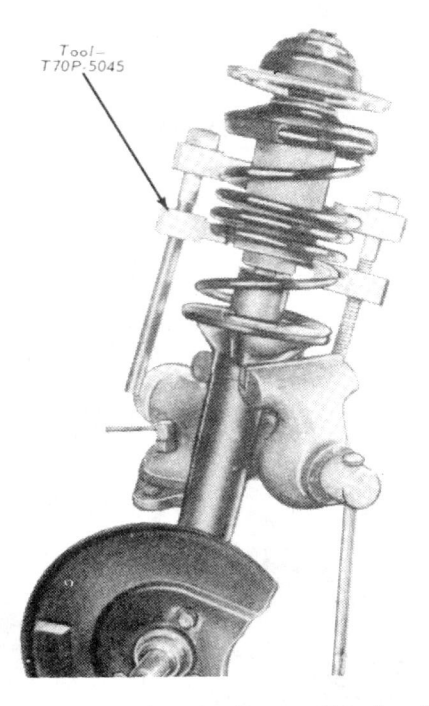

Fig. 10.8 Compressing the front coil spring (Sec 6)

7.2 Front stabiliser bar clamp mounting

2 Mount the strut firmly in a vice.
3 Fit the spring compressor tool(s) to the coil spring and compress the spring until it is just clear of the upper mounting. Make sure that the tool is fully engaged with the spring.
4 Prise off the plastic cap then unscrew the piston nut and remove the cranked retainer.

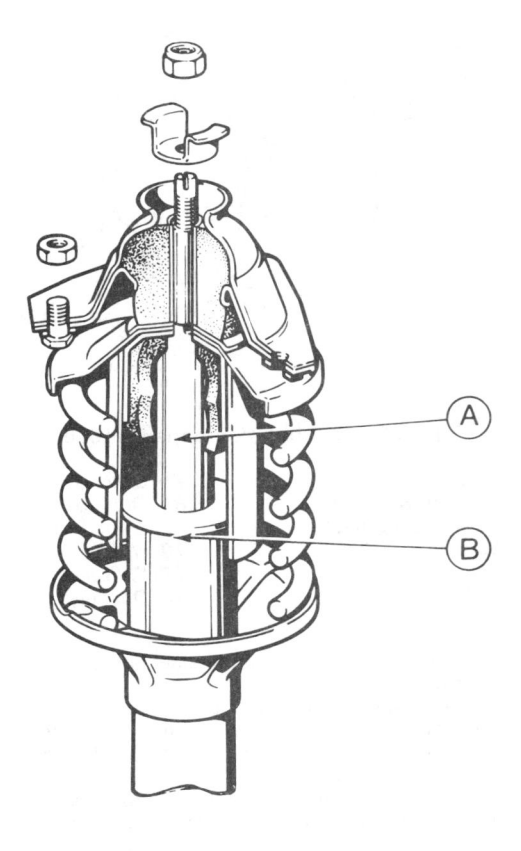

Fig. 10.9 Cut-away view of the front coil spring and strut mounting components (Sec 6)

A Shock absorber piston B Strut

5 Remove the upper mounting followed by the coil spring, rebound rubber and lower spring seat.
6 The coil springs are colour-coded and if a new spring is being fitted it must have the same colour code.
7 Refitting is a reversal of removal, but it will be necessary to first fit the compressor tool to the coil spring. Where a dished washer is fitted to the piston rod, the convex side must face upwards. Fit a new piston nut – using locking fluid on its threads. Do not fully tighten the piston nut until the weight of the car is on the suspension and, when tightening it, position the cranked retainer with its two ears at 90° to the centre-line of the car. Where the piston rod incorporates a slot, crimp the nut flange into the slot.

7 Front stabiliser bar – removal and refitting

1 Apply the handbrake then jack up the front of the car and support on axle stands. Remove the front roadwheels.
2 Working under the car at the front, knock back the locking tabs on the four bolts securing the two front clamps that hold the stabiliser bar to the frame and then undo the four bolts and remove the clamps and rubber insulators (photo).
3 Remove the split pins from the castellated nuts retaining the stabiliser bar to the track control arms, then undo the nuts and pull off

Chapter 10 Suspension and steering

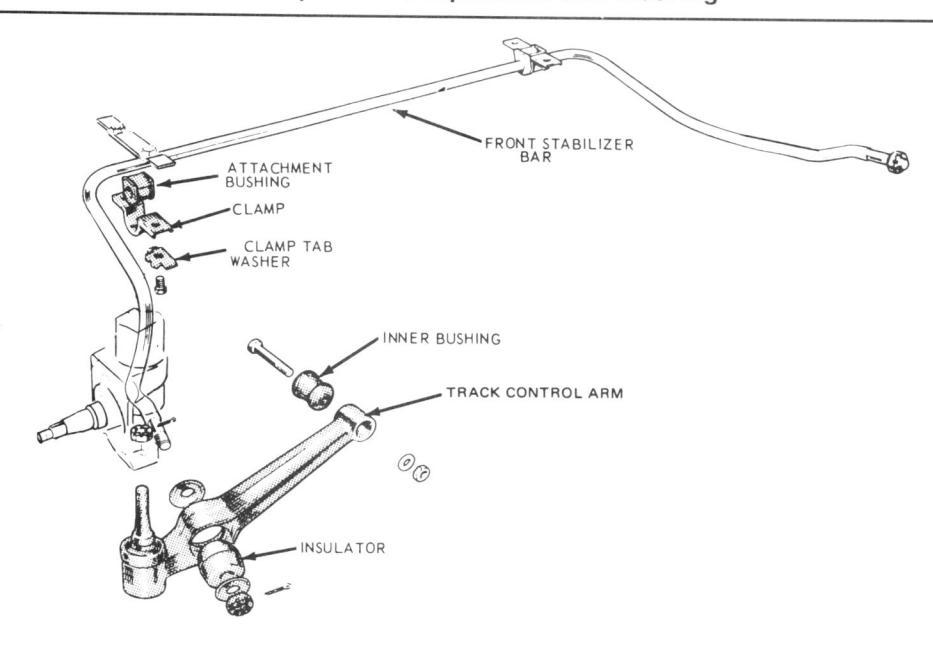

Fig. 10.10 Front stabiliser bar and track control arm components (Secs 7 and 8)

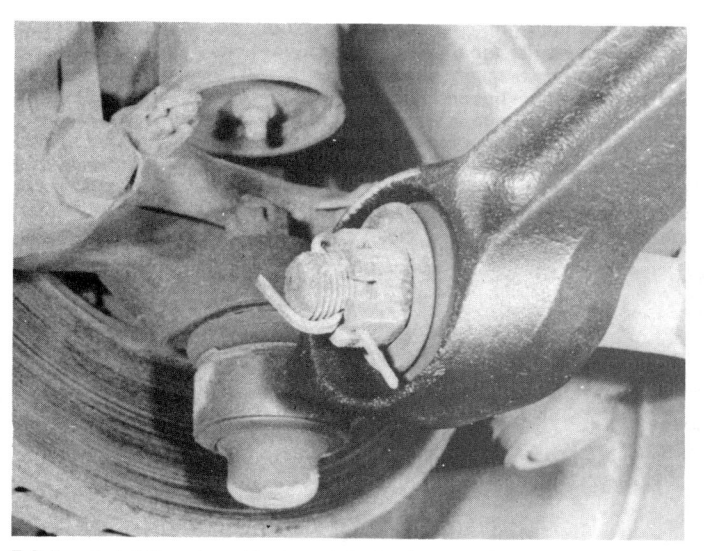

7.3 Front stabiliser bar attachment to track control arm

8.4 Track control arm inner pivot bolt

the large washers, carefully noting the way in which they are fitted (photo).
4 Pull the stabiliser bar forward out of the two track control arms and remove from the car.
5 With the stabiliser bar out of the car, remove the sleeve and large washer from each end of the bar, again noting the correct fitting positions. If necessary renew the mounting rubbers.
6 Reassembly is a reversal of the above procedure, but new locking tabs must be used on the front clamp bolts and new split pins on the castellated nuts. The bolts on the clamps and the castellated nuts on each end of the stabiliser bar must not be fully tightened down until the car is resting on its wheels.
7 Once the car is on its wheels, the castellated nuts on the ends of the stabiliser bar should be tightened down to the specified torque and new split pins fitted. The four clamp bolts on the front mounting points must be tightened to the specified torque and the locking tabs knocked up.

8 Front track control arm – removal and refitting

1 Apply the handbrake then jack up the front of the car and support on axle stands. Remove the roadwheel.
2 Working under the car, remove the split pin and unscrew the castellated nut that secures the track control arm to the stabiliser bar.
3 Lift away the large dished washer, noting which way round it is fitted.
4 Remove the self-lock nut and flat washer from the back of the track control arm pivot bolt (photo). Release the inner end of the track control arm.
5 Withdraw the split pin and unscrew the nut securing the track control arm balljoint to the base of the suspension unit. Separate the joint, using a balljoint separator, or wedges.
6 To refit the track control arm, first assemble the track control arm ball stud to the base of the suspension unit.
7 Refit the nut and tighten to the specified torque. Secure with a new split pin.
8 Place the track control arm so that it correctly locates over the stabiliser bar and then secure the inner end.
9 Slide the pivot bolt into position from the front and secure with the flat washer and a new self-locking nut. The nut must be to the rear. Tighten the nut to the specified torque, when the car is on the ground.
10 Fit the dished washer to the end of the stabiliser bar, making sure it is the correct way round and secure with the castellated nut. This must

Chapter 10 Suspension and steering

be tightened when the car is on the ground, to the specified torque. Lock the castellated nut with a new split pin.

9 Rear shock absorber – removal and refitting

Early models
1 Remove the back seat, after having removed the two screws from the floor assembly crossmember.
2 Remove the screws securing the seat belt to the top of the B-pillar.
3 Detach the B-pillar cover (2 screws).
4 Remove the top trim from the side window (4 screws)
5 Remove the two screws from the rocker panel at the rear end and pull off the door weatherstrip, in the region of the side trim.
6 Take out the boot side trim (2 screws) and the carpet.
7 Remove the lining of the rear panel (5 screws) and of the side panel (10 screws), folding the rear seat forward for access.

Later models
8 Remove the cover from the access hole in the rear side trim (photo).

All models
9 Check the front wheels then jack up the rear of the car and support on axle stands. remove the roadwheel.
10 Unscrew the nuts from the top of the shock absorber and remove the washer and insulator rubber.
11 Unscrew and remove the bottom mounting bolt and withdraw the shock absorber from the car. Note the location of the spacer in the mounting rubber.
12 If necessary the mounting rubbers may be renewed separately.
13 Refitting is a reversal of removal but tighten the mounting nuts to the specified torque.

10 Rear leaf spring – removal, bush renewal and refitting

1 Chock the front wheels then jack up the rear of the car and support on axle stands positioned on the body jacking points. Remove the roadwheel.
2 Place a trolley jack underneath the differential housing to support the rear axle assembly when the spring is removed. Do not raise the jack under the differential housing so that the springs are flattened, but raise it just enough to take the full weight of the axle with the springs fully extended.
3 Undo the rear shackle nuts and remove the combined shackle bolt and plate assemblies, then remove the rubber bushes.
4 Undo the nut from the front mounting and take out the bolt running through the mounting (photo).
5 Undo the nuts on the ends of the four U-bolts and remove the U-bolts, together with the attachment plate and rubber spring insulators (photo).
6 The rubber bushes can be pressed or driven out, and new bushes fitted as described for the bushes in the stabiliser bar and end pieces in Section 11. A little glycerine, or brake fluid, will allow the bushes to be pressed in more easily. Note that the front bushes are 11 mm (0.43 in) diameter and the rear bushes are 8 mm (0.31 in) diameter.
7 Refitting the spring is the reverse of the removal procedure. The nuts on the U-bolts, spring front mounting and rear shackles must be torqued to the figures given in the Specifications at the beginning of this Chapter only after the car has been lowered onto its wheels.

11 Rear stabiliser bar – removal and refitting

1 Chock the front wheels then jack up the rear of the car and support on axle stands. Remove the rear wheels for better access.
2 Using a multi-grip wrench, or similar tool, to hold the stabiliser bar towards the axle tube, remove the two bolts at each stabiliser bar-to-axle tube bracket (photo).
3 Disconnect the nut and bolt at each end of the stabiliser bar, where it is attached to the floor assembly, and withdraw the bar (photo).
4 To renew a stabiliser bar mounting bush, remove the locknut at one end and unscrew the end piece. Remove the nut and withdraw both rubber bushes from the stabiliser bar.
5 Dip the new rubber bushes in glycerine, or brake fluid; ensure that the stabiliser bar surface is clean and not scored, then slide on the

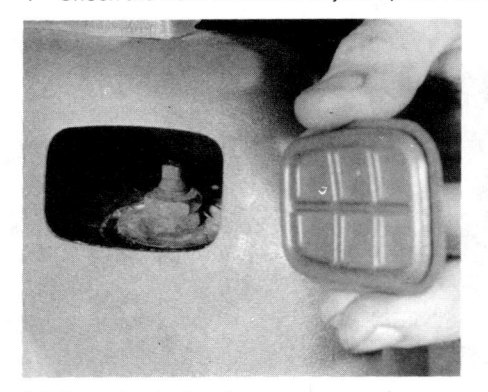

9.8 Rear shock absorber upper mounting access hole

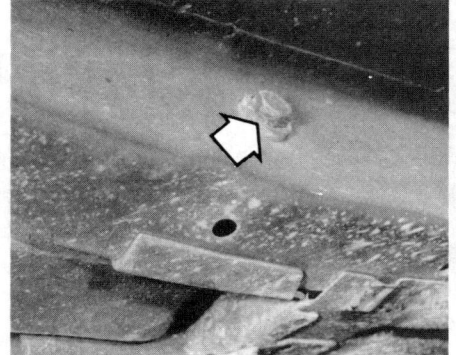

10.4 Rear spring front mounting bolt (arrowed)

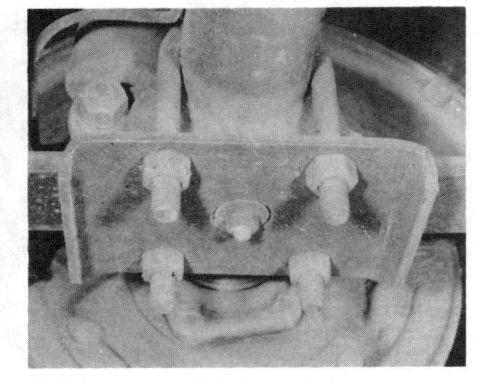

10.5 Rear spring U-bolts and plate

11.2 Rear stabiliser bar mounting on the rear axle

11.3 Rear stabiliser bar mounting on the underbody

Chapter 10 Suspension and steering

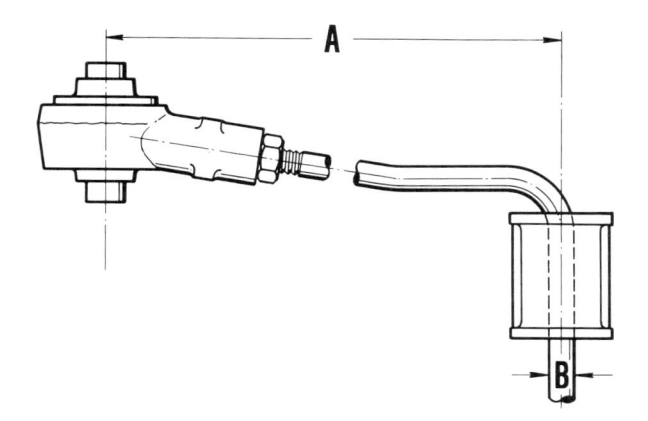

Fig. 10.11 Rear stabiliser bar end piece positioning (Sec 11)

A = 262.0 ± 2.5 mm (10.3 ± 0.1 in)
B = 10.0 mm (0.4 in)

bushes and refit the end piece. When fitted, the endpiece should be positioned as shown in Fig. 10.11. Dimension A should be between 25.59 and 264.5 mm (10.2 and 10.4 in) and the difference between the two sides must not be greater than 2.5 mm (0.1 in).
6 If the bushes in the end pieces require renewal, it may be found more convenient to remove the end pieces from the stabiliser bar, although this is not essential. The bushes can be pressed out using a suitable drift whilst the end piece is supported on a suitable diameter tube. Fitting is straightforward, the new bushes being pressed in until the steel case on the outside of the bush is flush with the inside of the end piece. Note the position of the semi-circular recess in the bush as shown in Fig. 10.12.
7 When refitting the stabiliser bar, it should be fitted at the floor end first, with the washers and self-locking nuts loosely installed.
8 The bar is then fitted to the axle tube, using a suitable tool to pull it towards the axle. The brackets, clamps and rubber insulators should now be fitted and the bolts tightened to the specified torque.
9 Lower the vehicle to the ground, then load the vehicle so that the centre of the axle tube and the spring rear eye are on the same horizonal level (the weight required is approximately that of two adults). The nuts and bolts securing the stabiliser bar to the floor can now be tightened to the specified torque.

12 Steering gear (manual) – removal and refitting

1 Before starting set the front wheels in the straight-ahead position. Jack up the front of the car and place stands under the track control arms. Lower the car slightly on the jack so that the track rods are in a near-horizontal position. Apply the handbrake and remove the front roadwheels.
2 Remove the nut and bolt from the clamp at the front of the flexible coupling on the steering column. This clamp holds the coupling to the pinion splines.
3 Working on the front crossmember, knock back the locking tabs on the two bolts on each rack housing U-clamp, undo the bolt and remove the locking tabs and clamps.
4 Remove the split pins and castellated nuts from the ends of each trackrod, where they join the steering arms. Separate the track rods from the steering arms, using a balljoint separator or wedges, and lower the steering gear downwards out of the car.
5 Before refitting the steering gear, make sure that the wheels have remained in the straight-ahead position. Also check the condition of the mounting rubbers round the housing and if they appear worn, or damaged, renew them.
6 Check that the steering gear is also in the straight-ahead position. This can be done by ensuring that the distances between the ends of both track rods and the steering gear housing on both sides are the same.
7 Place the steering gear in its location on the crossmember and at the same time mate up the splines on the pinion, with the splines in the clamp on the steering column flexible coupling.
8 Refit the two U-clamps, using new locking tabs under the bolts.

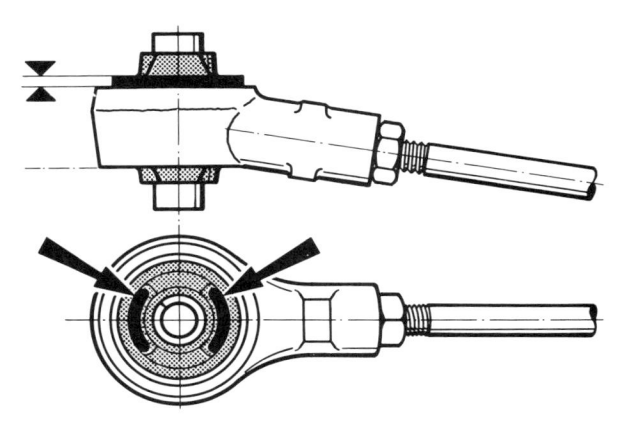

Fig. 10.12 Correct position of stabiliser bar rubber bush (Sec 11)

Tighten the bolts to the specified torque.
9 Refit the track rod ends into the steering arms, refit the castellated nuts and tighten them to the specified torque. Use new split pins to retain the nuts.
10 Tighten the clamp bolt on the steering column flexible coupling to the specified torque, having first made sure that the pinion is correctly located in the splines.
11 Lower the car to the ground, then check and if necessary adjust the front wheel alignment as described in Section 20.

13 Steering gear (manual) – overhaul and adjustment

1 Remove the track rod ends (Section 14) then remove and discard the wire retaining clips, remove the bellows and drain the lubricant.
2 Mount the steering gear in a soft-jawed vice and drill out the pins securing the track rod housings to the locknuts. Centre-punch the pins before drilling them. Use a 4 mm (0.16 inch or No. 22) drill, but do not drill too deeply.
3 It is now necessary to unscrew the housings from the balljoints, so that the track rods, housings, locknuts, ball seats, washers and springs can be removed. Ideally this requires the use of special tools which should be available from a Ford Dealer, but if improvised grips or wrenches are used, take care that no parts are damaged (if parts are damaged, new items must be obtained).
4 Remove the rack slipper cover plate, shim pack, gasket and slipper.
5 Remove the pinion bearing preload cover plate, shim pack, gasket and lower bearing.
6 Using a screwdriver, or similar tool, prise out the pinion oil seal.
7 Clean all dirt and paint from the pinion shaft, then push the pinion out of the housing. Withdraw the rack from the pinion end of the steering gear housing.
8 Take out the pinion upper bearing and washer.
9 Clean and inspect all the parts for damage and wear. Examine the bush in the end of the rack tube furthest from the pinion. If worn, it can be pressed out and a new bush fitted.
10 Commence reassembly by fitting the pinion upper bearing and washer into the housing.
11 Position the rack into the housing and leave it in the central position.
12 Fit the pinion, ensuring that, after fitting, the flat is towards the right-hand side of the vehicle (irrespective of right or left-hand drive vehicles).
13 Fit the pinion lower bearing cover plate.
14 Assemble the rack slipper, spring, gasket, shim pack and cover plate.
15 Lubricate the ball seats, balls and housings, with the specified gear oil. Screw the locknuts onto the ends of the steering rack.
16 Assemble the springs, washers, ball seats, trackrod ends and housing. Tighten the housings, to obtain a rotational torque of 7 Nm (5 lbf ft), then lock them with the locknuts. Recheck the torque after tightening the locknut.
17 Drill new holes (even if the old holes are in alignment), 4 mm (0.16

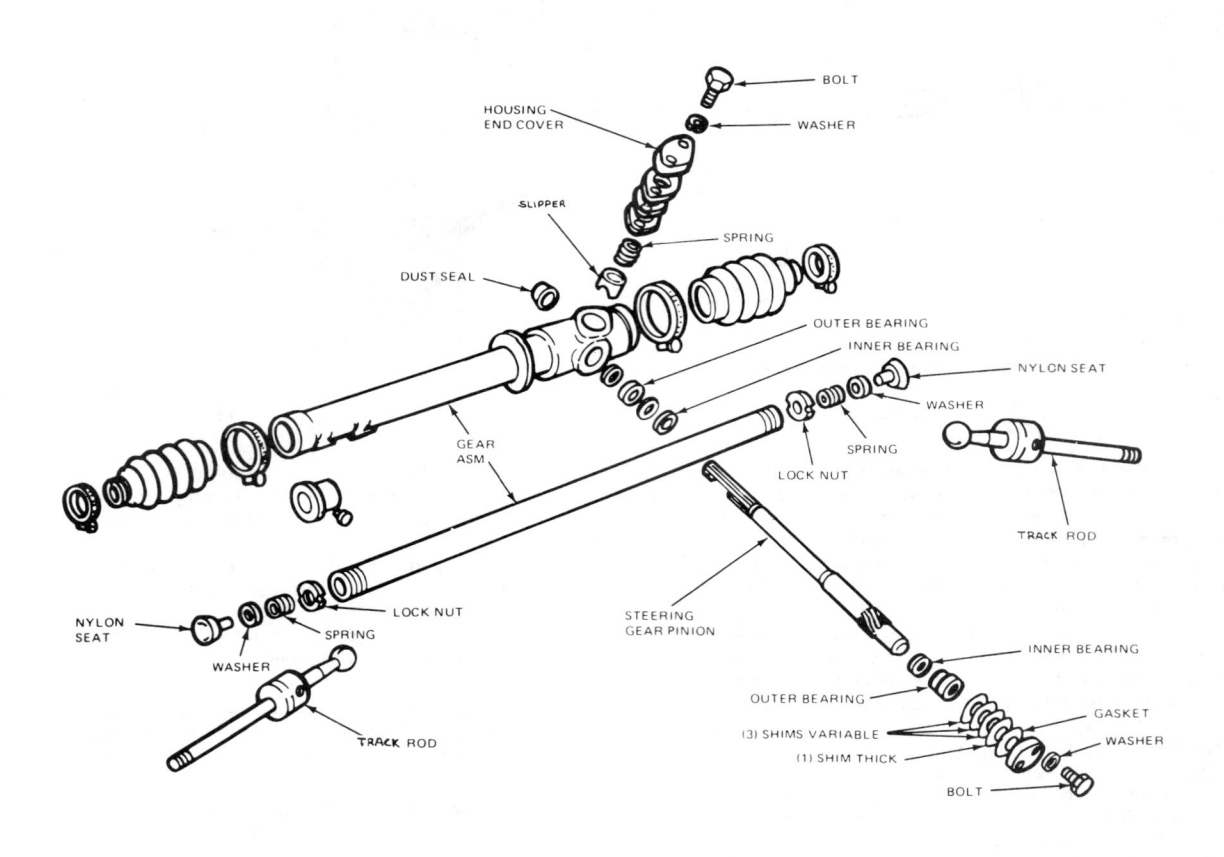

Fig. 10.13 Exploded view of the manual steering gear (Sec 13)

in or No. 22 drill) diameter x 9 mm (0.35 inch) deep along the break lines between the housing and the locknut, approximately opposite the spanner locating hole in the housing.
18 Fit new retaining pins and peen over the surrounding metal, to retain them.
19 Lightly grease the inside of the bellows, where they will contact the track rods. Install one bellows, ensuring that it locates in the track rod groove, then fit a new retaining clip.
20 Add the specified quantity of steering gear oil, operating the rack over its range of travel to assist the lubricant in flowing. Do not overfill.
21 Fit the other bellows, then carry out the following adjustment procedures. Ideally this will require the use of a dial gauge and mounting block, a surface table, a torque gauge and a splined adaptor. It is felt that most people will be able to suitably improvise, using other equipment, but if this cannot be done and the equipment listed is not available, the job should be entrusted to your local Ford Dealer.
22 With the steering gear mounted in a soft-jawed vice, remove the rack slipper cover plate, shim pack, gasket, and spring. Also remove the pinion bearing cover plate, shim pack and gasket. If any oil is lost during the adjustment procedures the steering gear must be replenished before fitting to the car.

Pinion bearing preload
23 Place the shim pack and cover plate on the bearing, tighten the bolts, then slacken them so that the cover plate touches the shim. The shim pack must comprise at least three shims, one of which must be 2.35 mm (0.093 in), this being immediately against the cover plate.
24 Measure the cover plate-to-housing gap, and if outside the range 0.28 to 0.33 mm (0.011 to 0.013 in) reduce the shim pack thickness (if the gap is too large), or increase it (if the gap is too small) until this gap is obtained. Remember that the 2.35 mm (0.093 in) shim must remain immediately against the cover plate.
25 When the correct gap is obtained, remove the cover plate, install the gasket and refit the cover plate. Apply locking fluid to the cover bolt threads, fit them and tighten to 8 to 11 Nm (6 to 8 lbf ft).

Rack slipper adjustment
26 Having set the pinion bearing preload, measure the height of the slipper above the main body of the rack, as the rack is moved from lock to lock by turning the pinion. Note the height reading obtained.
27 Prepare a shim pack which, including the thickness of the rack slipper bearing gasket, is 0.05 to 0.15 mm (0.002 to 0.006 in) thicker than the dimension noted in paragraph 26.
28 Fit the spring, gasket, shim pack and cover plate to the rack housing (gasket nearest housing). Apply locking fluid to the cover bolt threads, fit them and torque tighten to 8 to 11 Nm (6 to 8 lbf ft).
29 Measure the torque required to turn the pinion throughout its range of travel. This should be 1.13 to 2.03 Nm (0.83 to 1.50 lbf ft); if outside this range, faulty components, lack of lubricant etc, should be suspected.

14 Track rod end balljoint – renewal

1 Jack up the front of the car and support on axle stands. Apply the handbrake and remove the appropriate front roadwheel.
2 Slacken the balljoint locknut on the track rod by a quarter of a turn.
3 Extract the split pin and unscrew the nut securing the track rod end

Chapter 10 Suspension and steering

to the steering arm (photo).
4 Using a balljoint separator tool (photo), release the track rod end from the steering arm.
5 Unscrew the balljoint from the track rod, noting the exact number of turns required.
6 Screw the new balljoint onto the track rod the same number of turns.
7 Fit the track rod end in the steering arm, screw on the nut, and tighten it to the specified torque. Align the holes and insert a new split pin.
8 Tighten the balljoint locknut while holding the track rod end horizontal.
9 Refit the roadwheel and lower the car to the ground.
10 Check and adjust the front wheel alignment as described in Section 20.

15 Power steering – bleeding

1 The power steering system will only need bleeding in the event of air being introduced into the system, eg where pipes have been disconnected or where a leakage has occurred. To bleed the system proceed as described in the following paragraphs.
2 Open the bonnet and check the fluid level in the reservoir. Top-up if necessary using the specified type of fluid.
3 If fluid is added, allow two minutes, then run the engine at approximately 1500 rpm. Slowly turn the steering wheel from lock to lock, whilst checking and topping-up the fluid until the level remains steady and no more bubbles appear in the reservoir.
4 Clean and refit the reservoir cap and close the bonnet.

16 Power steering gear – removal and refitting

1 The procedure is similar to that described in Section 12 for the manual steering gear. Note, however, that the mounting bolts go completely through the front crossmember, and the rear nuts are locked with split pins. Additionally the fluid lines must be disconnected, and after refitting the system should be topped up and bled with reference to Section 15 (photos).

17 Power steering pump – removal and refitting

1 Loosen the idler pulley bracket bolts, release the drivebelt tension, then remove the drivebelt from the pump pulley.
2 On 3.0 litre models, jack up the front of the car and support on axle stands, then remove the engine splash shield. Remove the fuel pump from the engine (Chapter 3) but leaving the fuel lines connected.
3 Disconnect the fluid lines from the pump and drain the fluid into a suitable container. The pressure line has a screwed union, but the return line is a rubber hose connection (photos).
4 Unbolt the power steering pump, together with the mounting bracket and stay, where applicable, and remove from the engine (photos). The bracket can then be unbolted from the pump if necessary.
5 Refitting is a reversal of removal, but tension the drivebelt and bleed the system with reference to Sections 2 and 15.

14.3 Track rod end and steering arm

14.4 Typical balljoint separator tool

18 Steering wheel – removal and refitting

1 Disconnect the battery negative lead.
2 Prise the motif from the centre of the steering wheel (photos).

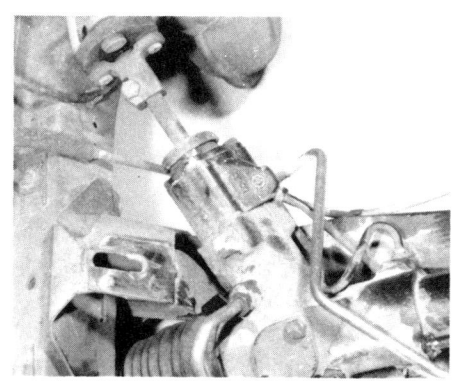

16.1A Power steering gear showing pinion and lower coupling shaft universal joint

16.1B Power steering gear mounting bolt (arrowed) ...

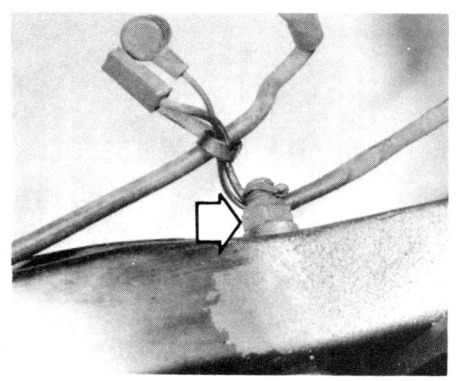

16.1C .. and nut (arrowed)

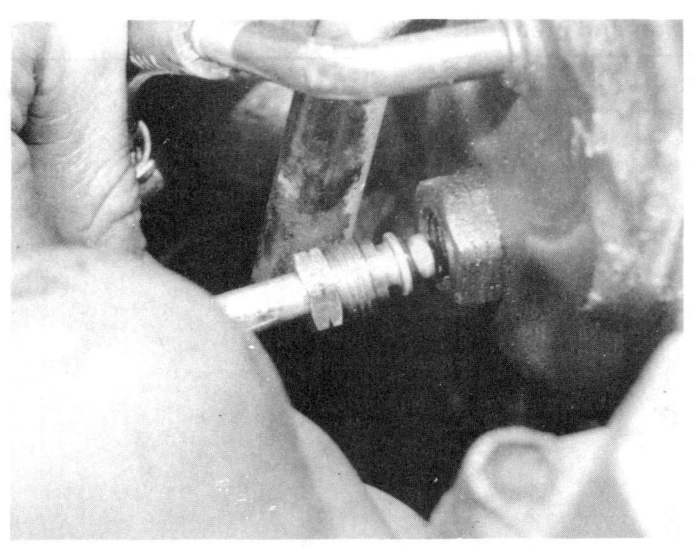

17.3A Removing the pressure line from the power steering pump

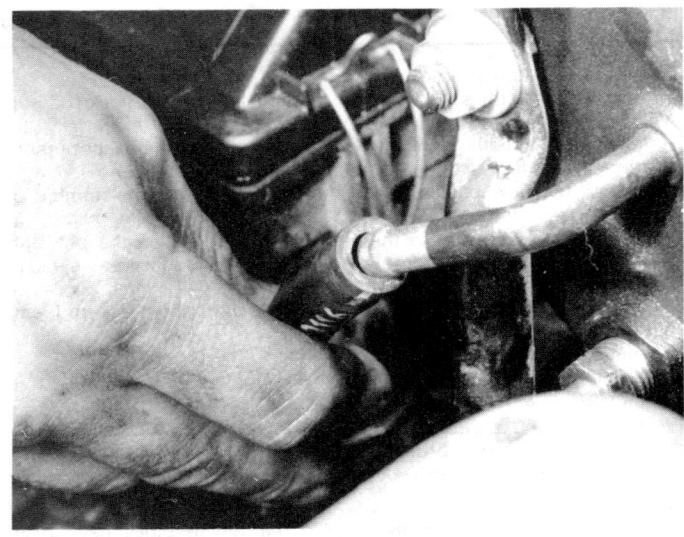

17.3B Removing the return line from the power steering pump

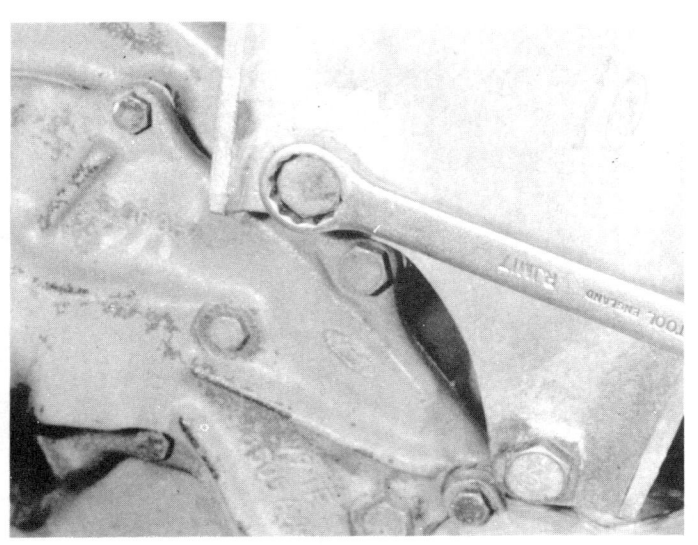

17.4A Unbolting the power steering pump mounting bracket ...

17.4B ... and stay

18.2A Removing the steering wheel motif (early models)

18.2B Removing the steering wheel motif (later models)

Chapter 10 Suspension and steering

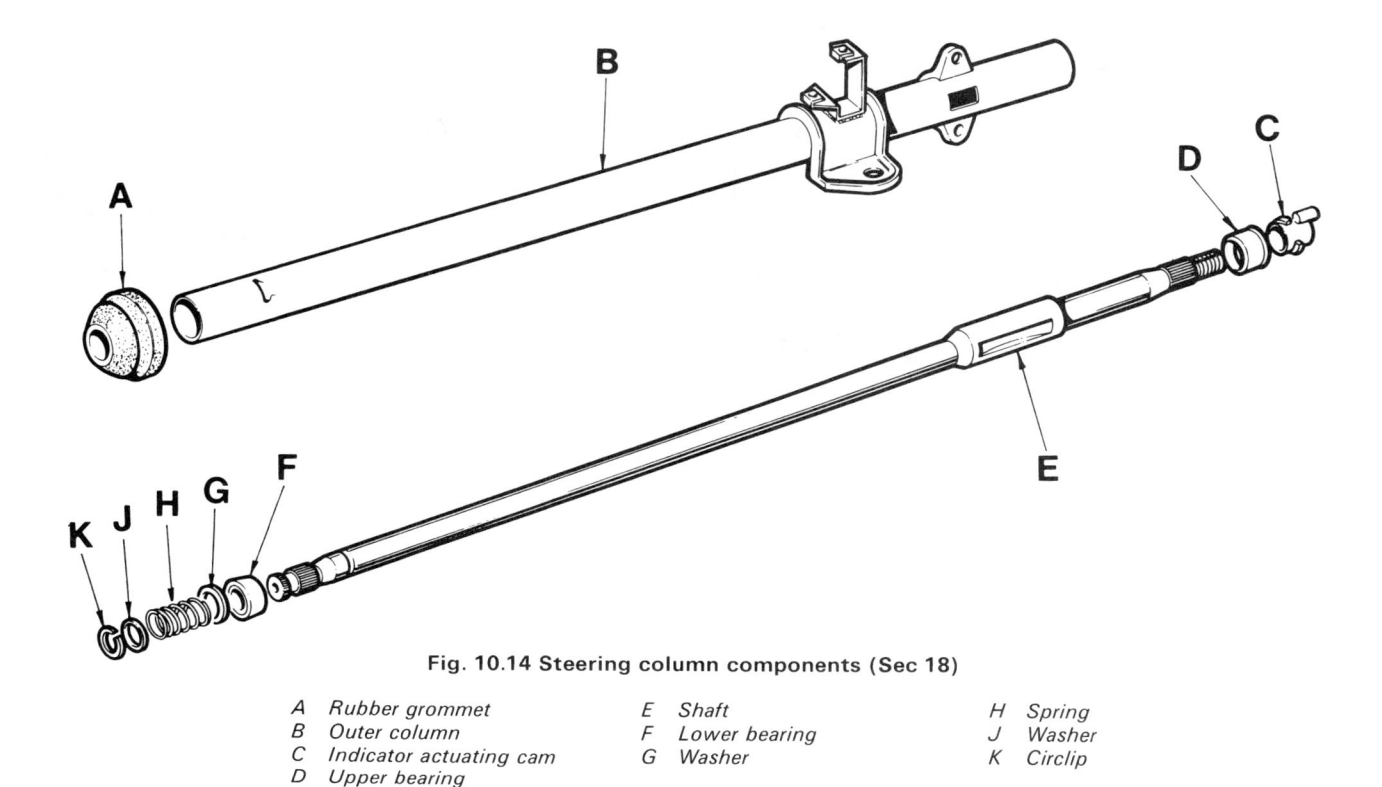

Fig. 10.14 Steering column components (Sec 18)

- A Rubber grommet
- B Outer column
- C Indicator actuating cam
- D Upper bearing
- E Shaft
- F Lower bearing
- G Washer
- H Spring
- J Washer
- K Circlip

3 With the front wheels straightahead, unscrew the nut from the top of the column.
4 Mark the steering wheel and column in relation to each other then pull the steering wheel from the splines, taking care not to extend the special convoluted section which acts as a safety device to minimise driver injury in the event of an accident.
5 Refitting is a reversal of removal, making sure that the previously made marks are aligned. Tighten the nut to the specified torque.

19 Steering column – removal, overhaul and refitting

1 Remove the steering wheel as described in Section 18.
2 Remove the upper and lower steering coupling clamp bolts and tap the coupling shaft down the pinion shaft to disconnect the coupling shaft from the column (photo).
3 Remove the direction indicator actuator cam.
4 Remove the steering column shroud (2 screws at the bottom, then pull out at the top) and lower the dash panel trim (photos).
5 Disconnect the direction indicator switch from the column (two bolts).
6 Disconnect the loom wiring from the ignition switch (photo).
7 Remove the two steering column retaining bolts and pull the column assembly from the vehicle (photo). Push the grommet out of the floorpan.
8 Drill off the steering column lockbolt heads, or tap them round with a pin punch, then use suitable grips to pull out the bolt shanks. Remove the steering lock (refer to Chapter 12, if necessary).
9 Remove the circlip, washer and spring from the lower end of the column.
10 Tap the lower end of the shaft with a soft-faced hammer to remove the shaft and bearing from the top of the column.
11 Using the shaft as a drift, tap the lower bearing out of the column.
12 Inspect all the parts for wear and damage, renewing if necessary.
13 Commence reassembly by positioning the shaft in the column, then assemble the lower bearing (small diameter towards the column), spring, washer and circlip to the shaft. Push the assembly into the column to locate the bearing against the stops.
14 Press the upper bearing onto the column.

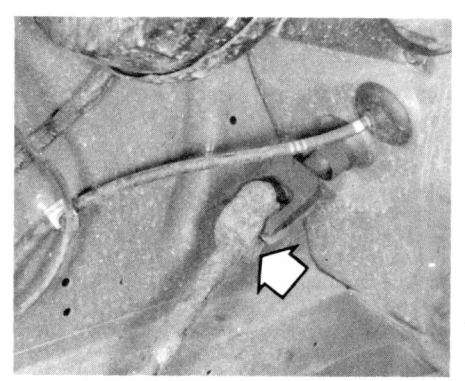

19.2 Steering coupling shaft (arrowed)

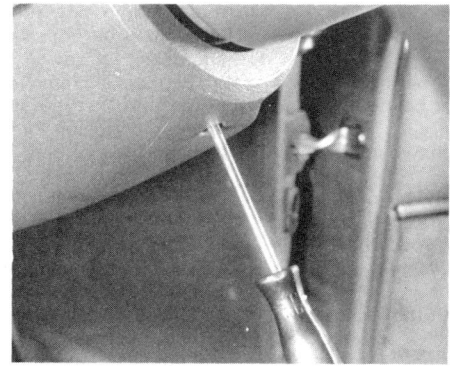

19.4A Remove the screw ...

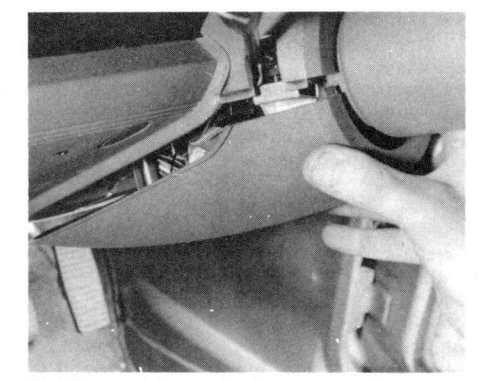

19.4B ... and lower the bottom steering column shroud

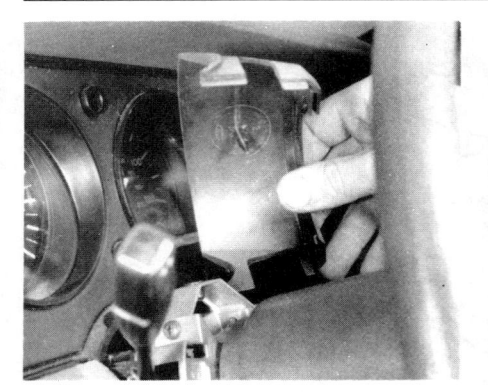

19.4C Removing the top steering column shroud

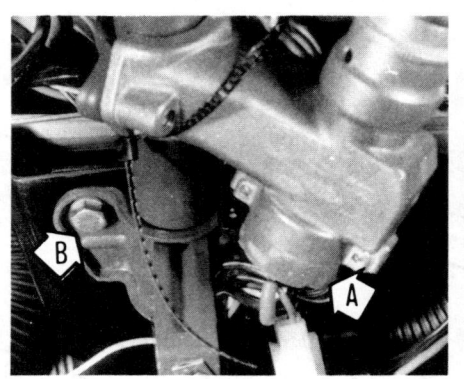

19.6 Ignition switch and wiring (A), also column upper mounting bolt (B)

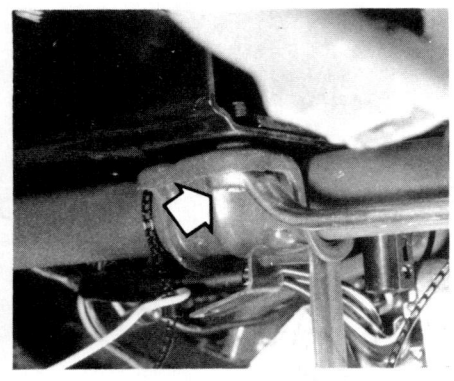

19.7 Removing the steering column upper mounting bolt (arrowed)

15 Secure the steering lock to the column and shear the heads from the new bolts.
16 Use the steering lock to locate the shaft in the column, then fit the direction indicator actuating cam and steering wheel. Check that the roadwheels are still in the straight-ahead position.
17 Install the steering column grommet at the lower end.
18 Locate the column assembly and secure it with the two mounting bolts.
19 The remainder of the refitting procedure is the reverse of the removal procedure.

20 Front wheel alignment and steering angles

1 Accurate front wheel alignment is essential for good steering and tyre wear. Before considering the steering angle, check that the tyres are correctly inflated, that the front wheels are not buckled, that the hub bearings are not worn or incorrectly adjusted and that the steering linkage is in good order, without slackness or wear at the joints.
2 Wheel alignment consists of four factors:
Camber, is the angle at which the front wheels are set from the vertical when viewed from the front of the car. Positive camber is the amount (in degrees) that the wheels are tilted outwards at the top from the vertical.
Castor, is the angle between the steering axis and a vertical line when viewed from each side of the car. Positive castor is when the steering axis is inclined rearwards at the top.
Steering axis inclination is the angle, when viewed from the front of the car, between the vertical and an imaginary line drawn between the upper and lower suspension strut pivots.
Toe-in is the amount by which the distance between the front inside edges of the roadwheels (measured at hub height) is less than the distance measured between the rear inside edges.
3 The angles of camber, castor and steering axis are set in production and are not adjustable.
4 Front wheel alignment (toe-in) checks are best carried out with modern setting equipment, but a reasonably accurate alternative is by means of the following procedure.
5 Place the car on level ground with the wheels in the straight-ahead position.
6 Obtain or make a toe-in gauge. One may easily be made from a length of rod or tubing, cranked to clear the sump or bellhousing and having a setscrew and locknut at one end.
7 With the gauge, measure the distance between the two inner wheel rims at hub height at the front of the wheel.
8 Rotate the roadwheel through 180° (half a turn), by pushing or pulling the car and then measure the distance again at hub height between the inner wheel rims at the rear of the roadwheel. This measurement should either be the same as the one just taken, or greater by not more than 7 mm (0.28 in).
9 Where the toe-in is found to be incorrect, slacken the locknuts on each track rod, also the flexible bellows clips and rotate each track rod by an equal amount until the correct toe-in is obtained. Tighten the track rod and locknuts, while the balljoints are held in the centre of their arcs of travel. It is imperative that the lengths of the track rods are always equal otherwise the wheel angles on turns will be incorrect. If new components have been fitted, set the roadwheels in the straight-ahead position and also centralise the steering wheel. Adjust the lengths of the track rods by turning them so that the track rod end balljoint studs will drop easily into the eyes of the steering arms. Measure the distances between the centres of the balljoints and the grooves on the inner ends of the track rods and adjust, if necessary, so that they are equal. This is an initial setting only and precise adjustment must be carried out as described earlier.

21 Wheels and tyres – general care and maintenance

Wheels and tyres should give no real problems in use provided that a close eye is kept on them with regard to excessive wear or damage. To this end, the following points should be noted.
Ensure that tyre pressures are checked regularly and maintained correctly. Checking should be carried out with the tyres cold and not immediately after the vehicle has been in use. If the pressures are checked with the tyres hot, an apparently high reading will be obtained owing to heat expansion. Under no circumstances should an attempt be made to reduce the pressures to the quoted cold reading in this instance, or effective underinflation will result.
Underinflation will cause overheating of the tyre owing to excessive flexing of the casing, and the tread will not sit correctly on the road surface. This will cause a consequent loss of adhesion and excessive wear, not to mention the danger of sudden tyre failure due to heat build-up.
Overinflation will cause rapid wear of the centre part of the tyre tread coupled with reduced adhesion, harsher ride, and the danger of shock damage occurring in the tyre casing.
Regularly check the tyres for damage in the form of cuts or bulges, especially in the sidewalls. Remove any nails or stones embedded in the tread before they penetrate the tyre to cause deflation. If removal of a nail *does* reveal that the tyre has been punctured, refit the nail so that its point of penetration is marked. Then immediately change the wheel and have the tyre repaired by a tyre dealer. Do *not* drive on a tyre in such a condition. In many cases a puncture can be simply repaired by the use of an inner tube of the correct size and type. If in any doubt as to the possible consequences of any damage found, consult your local tyre dealer for advice.
Periodically remove the wheels and clean any dirt or mud from the inside and outside surfaces. Examine the wheel rims for signs of rusting, corrosion or other damage. Light alloy wheels are easily damaged by 'kerbing' whilst parking, and similarly steel wheels may become dented or buckled. Renewal of the wheel is very often the only course of remedial action possible.
The balance of each wheel and tyre assembly should be maintained to avoid excessive wear, not only to the tyres but also to the steering and suspension components. Wheel imbalance is normally signified by vibration through the vehicle's bodyshell, although in many cases it is particularly noticeable through the steering wheel. Conversely, it should be noted that wear or damage in suspension or steering components may cause excessive tyre wear. Out-of-round or out-of-true tyres, damaged wheels and wheel bearing wear/maladjustment also fall into this category. Balancing will not usually cure vibration caused by such wear.

Chapter 10 Suspension and steering

Wheel balancing may be carried out with the wheel either on or off the vehicle. If balanced on the vehicle, ensure that the wheel-to-hub relationship is marked in some way prior to subsequent wheel removal so that it may be refitted in its original position.

General tyre wear is influenced to a large degree by driving style – harsh braking and acceleration or fast cornering will all produce more rapid tyre wear. Interchanging of tyres may result in more even wear, but this should only be carried out where there is no mix of tyre types on the vehicle. However, it is worth bearing in mind that if this is completely effective, the added expense of replacing a complete set of tyres simultaneously is incurred, which may prove financially restrictive for many owners.

Front tyres may wear unevenly as a result of wheel misalignment. The front wheels should always be correctly aligned according to the settings specified by the vehicle manufacturer.

Legal restrictions apply to the mixing of tyre types on a vehicle. Basically this means that a vehicle must not have tyres of differing construction on the same axle. Although it is not recommended to mix tyre types between front axle and rear axle, the only legally permissible combination is crossply at the front and radial at the rear. When mixing radial ply tyres, textile braced radials must always go on the front axle, with steel braced radials at the rear. An obvious disadvantage of such mixing is the necessity to carry two spare tyres to avoid contravening the law in the event of a puncture.

In the UK, the Motor Vehicles Construction and Use Regulations apply to many aspects of tyre fitting and usage. It is suggested that a copy of these regulations is obtained from your local police if in doubt as to the current legal requirements with regard to tyre condition, minimum tread depth, etc.

22 Fault diagnosis – suspension and steering

Symptom	Reason(s)
Vehicle pulls to one side	Incorrect wheel alignment Wear in suspension or steering components Accident damage to suspension or steering components Incorrect tyre pressures
Steering stiff or heavy	Lack of steering gear lubricant Seized balljoint Incorrect wheel alignment Power steering faulty
Excessive play in steering	Worn steering or suspension joints Wear in coupling shaft universal jonts Worn rack and pinion Incorrectly adjusted rack slipper
Wheel wobble and vibration	Roadwheels out of balance Roadwheels buckled or distorted Faulty or damaged tyre Worn steering or suspension joints Weak shock absorbers
Tyre wear uneven	Incorrect wheel alignment Wear in suspension or steering components Roadwheels out of balance Weak shock absorbers

Chapter 11 Bodywork and fittings

Contents

Adjustable exterior mirror – removal and refitting	37
Bonnet – removal, refitting and adjustment	26
Bonnet release cable – removal and refitting	25
Bumpers – removal and refitting	8
Centre console – removal and refitting	32
Demister nozzles – removal and refitting	46
Door – removal and refitting	20
Door lock assembly – removal and refitting	16
Door private lock – removal and refitting	21
Door rattles – tracing and rectification	7
Door remote control handle – removal and refitting	12
Door trim panel – removal and refitting	13
Door window frame – removal and refitting	18
Door window glass – removal and refitting	17
Exterior door handle – removal and refitting	15
Face level vents – removal and refitting	47
Facia panel – removal and refitting	31
Fuel filler flap – removal and refitting	30
General description	1
Heater assembly – removal and refitting	43
Heater assembly (Behr) – dismantling and reassembly	44
Heater assembly (Smiths standard and heavy duty) – dismantling and reassembly	45
Heater controls – adjustment	40
Heater controls – removal and refitting	41
Heater water valve (heavy duty heater) – removal and refitting	42
Interior mirror – removal and refitting	38
Load space trim panel – removal and refitting	24
Maintenance – bodywork and underframe	3
Maintenance – upholstery and carpets	4
Major body damage – repair	6
Minor body damage – repair	5
Opening rear quarter glass assembly – removal and refitting	23
Radiator grille – removal and refitting	9
Rear quarter trim panel – removal and refitting	22
Routine maintenance	2
Seat belts – general	36
Seats – removal and refitting	39
Sunroof – removal and refitting	34
Sunroof bracket and drive assembly – removal and refitting	35
Sunroof panel – adjustments	33
Tailgate – removal and refitting	27
Tailgate private lock – removal and refitting	28
Tailgate striker plate – removal and refitting	29
Tailgate window glass – removal and refitting	11
Window frame moulding and door weatherstrips – removal and refitting	19
Windscreen – removal and refitting	10

Specifications

Torque wrench settings	**Nm**	**lbf ft**
Seatbelts:	
At pillar	20 to 30	15 to 22
At floor	35 to 42	26 to 31
Opening rear quarter glass toggle to C-pillar	4.5	3.3

Chapter 11 Bodywork and fitting

1 General description

The body is of all-steel welded construction with impact absorbing front and rear sections. Anti-burst locks are fitted to the doors, and the tailgate is supported by gas-filled dampers. All window glass is of toughened safety type.

The front seats are of reclining bucket type, while the rear seats may be of individual bucket type or bench type.

The level of trim both interior and exterior varies according to the particular model.

2 Routine maintenance

At the intervals specified in the Routine Maintenance section in the front of the manual carry out the following procedures.

Check seat belts
1 Thoroughly check all seat belts for fraying, cuts, damage or deterioration and if necessary renew them. Check for security.

Check and lubricate bonnet and door locks
2 Oil the bonnet lock and safety catch and check that they operate correctly.
3 Oil all the door check straps and lock barrels.
4 Oil the fuel filler cap hinge and lock.

Check underbody protective coating
5 With the car supported on axle stands, check the condition of the underbody protective coating, and where necessary clean any damaged areas and apply a new coating. If the car is still within the corrosion warranty period check the terms and conditions of the warranty first.

3 Maintenance – bodywork and underframe

The general condition of a vehicle's bodywork is the one thing that significantly affects its value. Maintenance is easy but needs to be regular. Neglect, particularly after minor damage, can lead quickly to further deterioration and costly repair bills. It is important also to keep watch on those parts of the vehicle not immediately visible, for instance the underside, inside all the wheel arches and the lower part of the engine compartment.

The basic maintenance routine for the bodywork is washing – preferably with a lot of water, from a hose. This will remove all the loose solids which may have stuck to the vehicle. It is important to flush these off in such a way as to prevent grit from scratching the finish. The wheel arches and underframe need washing in the same way to remove any accumulated mud which will retain moisture and tend to encourage rust. Paradoxically enough, the best time to clean the underframe and wheel arches is in wet weather when the mud is thoroughly wet and soft. In very wet weather the underframe is usually cleaned of large accumulations automatically and this is a good time for inspection.

Periodically, except on vehicles with a wax-based underbody protective coating, it is a good idea to have the whole of the underframe of the vehicle steam cleaned, engine compartment included, so that a thorough inspection can be carried out to see what minor repairs and renovations are necessary. Steam cleaning is available at many garages and is necessary for removal of the accumulation of oily grime which sometimes is allowed to become thick in certain areas. If steam cleaning facilities are not available, there are one or two excellent grease solvents available which can be brush applied. The dirt can then be simply hosed off. Note that these methods should not be used on vehicles with wax-based underbody protective coating or the coating will be removed. Such vehicles should be inspected annually, preferably just prior to winter, when the underbody should be washed down and any damage to the wax coating repaired. Ideally, a completely fresh coat should be applied. It would also be worth considering the use of such wax-based protection for injection into door panels, sills, box sections, etc, as an additional safeguard against rust damage where such protection is not provided by the vehicle manufacturer.

After washing paintwork, wipe off with a chamois leather to give an unspotted clear finish. A coat of clear protective wax polish will give added protection against chemical pollutants in the air. If the paintwork sheen has dulled or oxidised, use a cleaner/polisher combination to restore the brilliance of the shine. This requires a little effort, but such dulling is usually caused because regular washing has been neglected. Care needs to be taken with metallic paintwork, as special non-abrasive cleaner/polisher is required to avoid damage to the finish. Always check that the door and ventilator opening drain holes and pipes are completely clear so that water can be drained out. Bright work should be treated in the same way as paint work. Windscreens and windows can be kept clear of the smeary film which often appears by the use of a proprietary glass cleaner. Never use any form of wax or other body or chromium polish on glass.

4 Maintenance – upholstery and carpets

Mats and carpets should be brushed or vacuum cleaned regularly to keep them free of grit. If they are badly stained remove them from the vehicle for scrubbing or sponging and make quite sure they are dry before refitting. Seats and interior trim panels can be kept clean by wiping with a damp cloth. If they do become stained (which can be more apparent on light coloured upholstery) use a little liquid detergent and a soft nail brush to scour the grime out of the grain of the material. Do not forget to keep the headlining clean in the same way as the upholstery. When using liquid cleaners inside the vehicle do not over-wet the surfaces being cleaned. Excessive damp could get into the seams and padded interior causing stains, offensive odours or even rot. If the inside of the vehicle gets wet accidentally it is worthwhile taking some trouble to dry it out properly, particularly where carpets are involved. *Do not leave oil or electric heaters inside the vehicle for this purpose.*

5 Minor body damage – repair

The photographic sequences on pages 174 and 175 illustrate the operations detailed in the following sub-sections.
Note: *For more detailed information about bodywork repair, the Haynes Publishing Group publish a book by Lindsay Porter called The Car Bodywork Repair Manual. This incorporates information on such aspects as rust treatment, painting and glass fibre repairs, as well as details on more ambitious repairs involving welding and panel beating.*

Repair of minor scratches in bodywork
If the scratch is very superficial, and does not penetrate to the metal of the bodywork, repair is very simple. Lightly rub the area of the scratch with a paintwork renovator, or a very fine cutting paste, to remove loose paint from the scratch and to clear the surrounding bodywork of wax polish. Rinse the area with clean water.

Apply touch-up paint to the scratch using a fine paint brush; continue to apply fine layers of paint until the surface of the paint in the scratch is level with the surrounding paintwork. Allow the new paint at least two weeks to harden; then blend it into the surrounding paintwork by rubbing the scratch area with a paintwork renovator or a very fine cutting paste. Finally, apply wax polish.

Where the scratch has penetrated right through to the metal of the bodywork, causing the metal to rust, a different repair technique is required. Remove any loose rust from the bottom of the scratch with a penknife, then apply rust inhibiting paint to prevent the formation of rust in the future. Using a rubber or nylon applicator fill the scratch with bodystopper paste. If required, this paste can be mixed with cellulose thinners to provide a very thin paste which is ideal for filling narrow scratches. Before the stopper-paste in the scratch hardens, wrap a piece of smooth cotton rag around the top of a finger. Dip the finger in cellulose thinners and then quickly sweep it across the surface of the stopper-paste in the scratch; this will ensure that the surface of the stopper-paste is slightly hollowed. The scratch can now be painted over as described earlier in this Section.

Repair of dents in bodywork
When deep denting of the vehicle's bodywork has taken place, the first task is to pull the dent out, until the affected bodywork almost

attains its original shape. There is little point in trying to restore the original shape completely, as the metal in the damaged area will have stretched on impact and cannot be reshaped fully to its original contour. It is better to bring the level of the dent up to a point which is about 1/8 in (3 mm) below the level of the surrounding bodywork. In cases where the dent is very shallow anyway, it is not worth trying to pull it out at all. If the underside of the dent is accessible, it can be hammered out gently from behind, using a mallet with a wooden or plastic head. Whilst doing this, hold a suitable block of wood firmly against the outside of the panel to absorb the impact from the hammer blows and thus prevent a large area of the bodywork from being 'belled-out'.

Should the dent be in a section of the bodywork which has a double skin or some other factor making it inaccessible from behind, a different technique is called for. Drill several small holes through the metal inside the area – particularly in the deeper section. Then screw long self-tapping screws into the holes just sufficiently for them to gain a good purchase in the metal. Now the dent can be pulled out by pulling on the protruding heads of the screws with a pair of pliers.

The next stage of the repair is the removal of the paint from the damaged area, and from an inch or so of the surrounding 'sound' bodywork. This is accomplished most easily by using a wire brush or abrasive pad on a power drill, although it can be done just as effectively by hand using sheets of abrasive paper. To complete the preparation for filling, score the surface of the bare metal with a screwdriver or the tang of a file, or alternatively, drill small holes in the affected area. This will provide a really good 'key' for the filler paste.

To complete the repair see the Section on filling and re-spraying.

Repair of rust holes or gashes in bodywork

Remove all paint from the affected area and from an inch or so of the surrounding 'sound' bodywork, using an abrasive pad or a wire brush on a power drill. If these are not available a few sheets of abrasive paper will do the job just as effectively. With the paint removed you will be able to gauge the severity of the corrosion and therefore decide whether to renew the whole panel (if this is possible) or to repair the affected area. New body panels are not as expensive as most people think and it is often quicker and more satisfactory to fit a new panel than to attempt to repair large areas of corrosion.

Remove all fittings from the affected area except those which will act as a guide to the original shape of the damaged bodywork (eg headlamp shells etc). Then, using tin snips or a hacksaw blade, remove all loose metal and any other metal badly affected by corrosion. Hammer the edges of the hole inwards in order to create a slight depression for the filler paste.

Wire brush the affected area to remove the powdery rust from the surface of the remaining metal. Paint the affected area with rust inhibiting paint; if the back of the rusted area is accessible treat this also.

Before filling can take place it will be necessary to block the hole in some way. This can be achieved by the use of aluminium or plastic mesh, or aluminium tape.

Aluminium or plastic mesh is probably the best material to use for a large hole. Cut a piece to the approximate size and shape of the hole to be filled, then position it in the hole so that its edges are below the level of the surrounding bodywork. It can be retained in position by several blobs of filler paste around its periphery.

Aluminium tape should be used for small or very narrow holes. Pull a piece off the roll and trim it to the approximate size and shape required, then pull off the backing paper (if used) and stick the tape over the hole; it can be overlapped if the thickness of one piece is insufficient. Burnish down the edges of the tape with the handle of a screwdriver or similar, to ensure that the tape is securely attached to the metal underneath.

Bodywork repairs – filling and re-spraying

Before using this Section, see the Sections on dent, deep scratch, rust holes and gash repairs.

Many types of bodyfiller are available, but generally speaking those proprietary kits which contain a tin of filler paste and a tube of resin hardener are best for this type of repair. A wide, flexible plastic or nylon applicator will be found invaluable for imparting a smooth and well contoured finish to the surface of the filler.

Mix up a little filler on a clean piece of card or board – measure the hardener carefully (follow the maker's instructions on the pack) otherwise the filler will set too rapidly or too slowly. Using the applicator apply the filler paste to the prepared area; draw the applicator across the surface of the filler to achieve the correct contour and to level the filler surface. As soon as a contour that approximates to the correct one is achieved, stop working the paste – if you carry on too long the paste will become sticky and begin to 'pick up' on the applicator. Continue to add thin layers of filler paste at twenty-minute intervals until the level of the filler is just proud of the surrounding bodywork.

Once the filler has hardened, excess can be removed using a metal plane or file. From then on, progressively finer grades of abrasive paper should be used, starting with a 40 grade production paper and finishing with 400 grade wet-and-dry paper. Always wrap the abrasive paper around a flat rubber, cork, or wooden block – otherwise the surface of the filler will not be completely flat. During the smoothing of the filler surface the wet-and-dry paper should be periodically rinsed in water. This will ensure that a very smooth finish is imparted to the filler at the final stage.

At this stage the 'dent' should be surrounded by a ring of bare metal, which in turn should be encircled by the finely 'feathered' edge of the good paintwork. Rinse the repair area with clean water, until all of the dust produced by the rubbing-down operation has gone.

Spray the whole repair area with a light coat of primer – this will show up any imperfections in the surface of the filler. Repair these imperfections with fresh filler paste or bodystopper, and once more smooth the surface with abrasive paper. If bodystopper is used, it can be mixed with cellulose thinners to form a really thin paste which is ideal for filling small holes. Repeat this spray and repair procedure until you are satisfied that the surface of the filler, and the feathered edge of the paintwork are perfect. Clean the repair area with clean water and allow to dry fully.

The repair area is now ready for final spraying. Paint spraying must be carried out in a warm, dry, windless and dust free atmosphere. This condition can be created artificially if you have access to a large indoor working area, but if you are forced to work in the open, you will have to pick your day very carefully. If you are working indoors, dousing the floor in the work area with water will help to settle the dust which would otherwise be in the atmosphere. If the repair area is confined to one body panel, mask off the surrounding panels; this will help to minimise the effects of a slight mis-match in paint colours. Bodywork fittings (eg chrome strips, door handles etc) will also need to be masked off. Use genuine masking tape and several thicknesses of newspaper for the masking operations.

Before commencing to spray, agitate the aerosol can thoroughly, then spray a test area (an old tin, or similar) until the technique is mastered. Cover the repair area with a thick coat of primer; the thickness should be built up using several thin layers of paint rather than one thick one. Using 400 grade wet-and-dry paper, rub down the surface of the primer until it is really smooth. While doing this, the work area should be thoroughly doused with water, and the wet-and-dry paper periodically rinsed in water. Allow to dry before spraying on more paint.

Spray on the top coat, again building up the thickness by using several thin layers of paint. Start spraying in the centre of the repair area and then, using a circular motion, work outwards until the whole repair area and about 2 inches of the surrounding original paintwork is covered. Remove all masking material 10 to 15 minutes after spraying on the final coat of paint.

Allow the new paint at least two weeks to harden, then, using a paintwork renovator or a very fine cutting paste, blend the edges of the paint into the existing paintwork. Finally, apply wax polish.

Plastic components

With the use of more and more plastic body components by the vehicle manufacturers (eg bumpers, spoilers, and in some cases major body panels), rectification of damage to such items has become a matter of either entrusting repair work to a specialist in this field, or renewing complete components. Repair by the DIY owner is not really feasible owing to the cost of the equipment and materials required for effecting such repairs. The basic technique involves making a groove along the line of the crack in the plastic using a rotary burr in a power drill. The damaged part is then welded back together by using a hot air gun to heat up and fuse a plastic filler rod into the groove. Any excess plastic is then removed and the area rubbed down to a smooth finish. It is important that a filler rod of the correct plastic is used, as body components can be made of a variety of different types (eg polycarbonate, ABS, polypropylene).

Chapter 11 Bodywork and fitting

If the owner is renewing a complete component himself, he will be left with the problem of finding a suitable paint for finishing which is compatible with the type of plastic used. At one time the use of a universal paint was not possible owing to the complex range of plastics encountered in body component applications. Standard paints, generally speaking, will not bond to plastic or rubber satisfactorily. However, it is now possible to obtain a plastic body parts finishing kit which consists of a pre-primer treatment, a primer and coloured top coat. Full instructions are normally supplied with a kit, but basically the method of use is to first apply the pre-primer to the component concerned and allow it to dry for up to 30 minutes. Then the primer is applied and left to dry for about an hour before finally applying the special coloured top coat. The result is a correctly coloured component where the paint will flex with the plastic or rubber, a property that standard paint does not normally possess.

6 Major body damage – repair

Where serious damage has occurred or large areas need renewal due to neglect, it means certainly that completely new sections or panels will need welding in and this is best left to professionals. If the damage is due to impact it will also be necessary to completely check the alignment of the bodyshell structure. Due to the principle of construction the strength and shape of the whole can be affected by damage to a part. In such instances the services of a Ford agent with specialist checking jigs are essential. If a frame is left misaligned it is, first of all, dangerous as the vehicle will not handle properly and, secondly, uneven stresses will be imposed on the steering, engine and transmission, causing abnormal wear or complete failure. Tyre wear may also be excessive.

7 Door rattles – tracing and rectification

1 The most common cause of door rattles is a misaligned, loose or worn striker plate. Other causes may be:

(a) Loose door or window winder handles
(b) Loose or misaligned door lock components
(c) Loose or worn remote control mechanism

2 It is quite possible for rattles to be the result of a combination of the above faults, so a careful examination should be made to determine the exact cause.
3 If it is found necessary to adjust the striker plate (photo), close the door to the first of the two locking positions. Visually check the relative attitude of the striker outside edge to the lock support plate edge. The edges A and B (Fig. 11.1) should be parallel and can be checked by shining a torch through the door gap from above and below the striker.
4 Check the amount by which the door stands proud of the adjacent panel. Adjust the striker plate as necessary to obtain a dimension of 6.0 mm (0.24 in).

7.3 Door striker plate

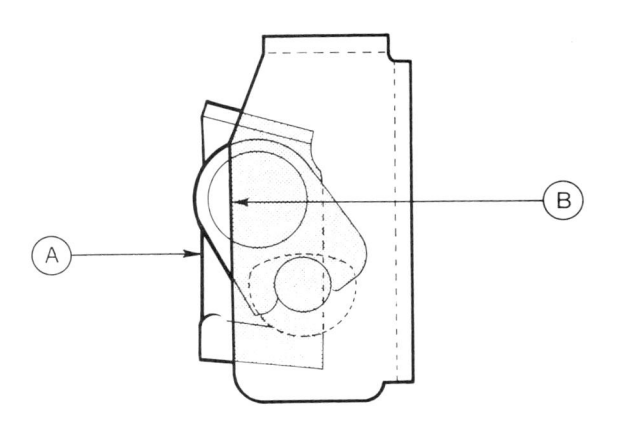

Fig. 11.1 Aligning the striker to lock support plate (Sec 7)

Edges A and B should be parallel

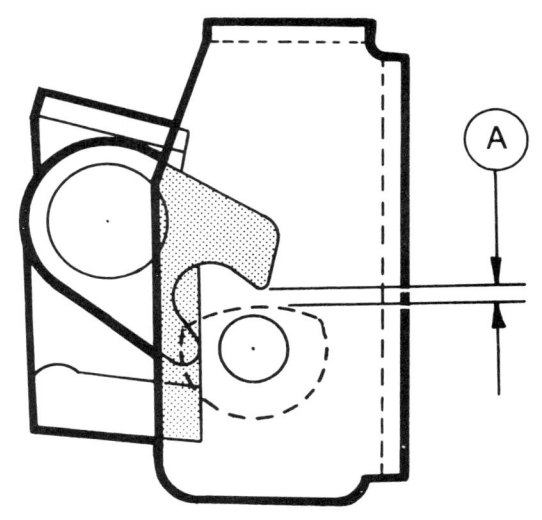

Fig. 11.2 Lock claw-to-striker clearance (Sec 7)

A = 7.0 mm (0.28 in)

5 With the lock in the open position, check the lock claw striker clearance (dimension A in Fig. 11.2). This should be 7.0 mm (0.28 in) and can be checked by placing a small ball of plasticine, or similar material on the striker post and checking its height after gently closing the door. The striker plate can be repositioned vertically to obtain this dimension, but take care to not to disturb any previous initial settings of the plate.

8 Bumpers – removal and refitting

Front bumper

1 Disconnect the battery earth lead. Also where necessary disconnect the indicator wiring.
2 Remove the radiator grille (Section 9).
3 Remove the four bumper retaining screws (photo) and lift away the bumper. On later models disengage the bumper corner sections from the body side clips as the bumper is withdrawn.
4 Refitting is the reverse of the removal procedure, but do not fully tighten the bolts until the bumper is correctly aligned.

Rear bumper

5 Open the tailgate, then remove the mat and the sub-floor.
6 Remove the jack and washer water reservoir (where applicable).
7 Remove the two nuts, spring washer and flat washers, at each end and lift away the bumper (photo).

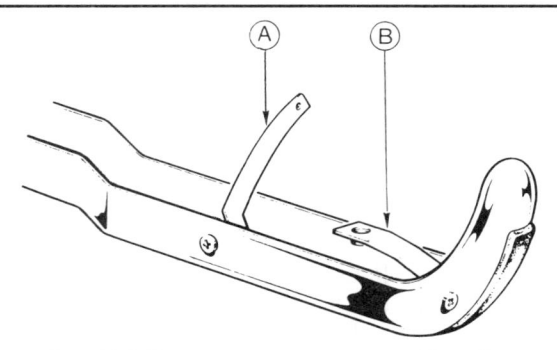

176   Chapter 11 Bodywork and fitting

Fig. 11.3 Front bumper mounting (Sec 8)

A   Inner mounting bracket   B   Outer mounting bracket

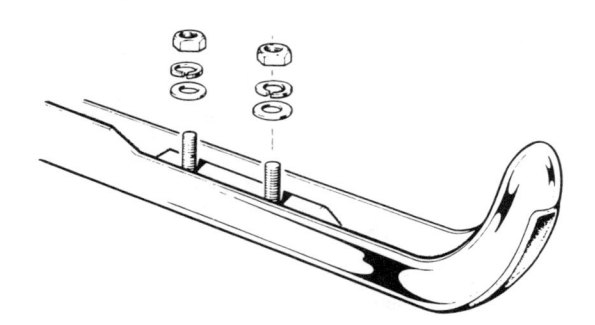

Fig. 11.4 Rear bumper mounting (Sec 8)

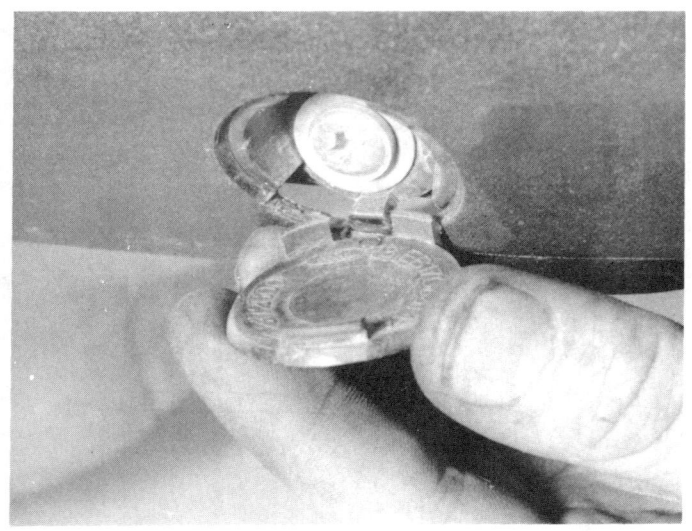

8.3 On later models the bumper corner section screws are located beneath plastic caps

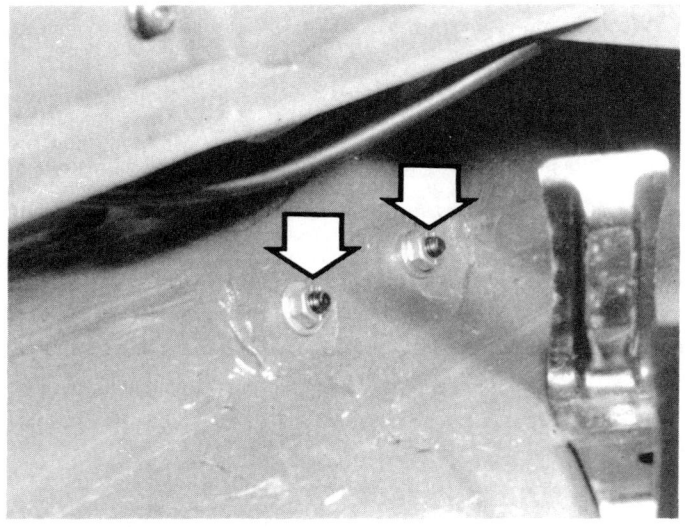

8.7 Rear bumper retaining nuts (arrowed)

9.2A Removing the upper ...

9.2B ... and lower radiator grille mounting screws

8 Refitting is the reverse of the removal procedure but do not fully tighten the nuts until the bumper is correctly aligned.

Bumper trim strips
9 The bumper trim strips can be removed, and replacements fitted, by prising them in and out of the retaining grooves. The job is made a little easier if a soap and water solution is applied to the T-shaped retaining groove.

9   Radiator grille – removal and refitting

1   Disconnect the battery earth lead. Open the bonnet.
2   Remove the mounting screws and washers and lift away the grille (photo). On early models it is necessary to disconnect the indicator wiring before the grille can be removed fully.
3   Refitting is a reversal of removal.

10 Windscreen – removal and refitting

1 If you are unfortunate enough to have a windscreen shatter, or should you wish to renew your present windscreen, fitting a replacement is one of the few jobs which the average owner is advised to leave to a professional, but for the owner who wishes to attempt the job himself, the following instructions are given.
2 Cover the bonnet with a blanket, or cloth to prevent accidental damage and remove the windscreen wiper blades and arms as detailed in Chapter 12.
3 Put on a pair of lightweight shoes and get into one of the front seats. With a piece of soft cloth between the sole of your shoes and the windscreen glass, place both feet in one top corner of the windscreen and push firmly (See Fig. 11.5).
4 When the weatherstrip has freed itself from the body flange in that area, repeat the process at frequent intervals along the top edge of the windscreen until, from outside the car, the glass and weatherstrip can be removed together.
5 If you are having to renew your windscreen due to a shattered screen, remove all traces of sealing compound and broken glass from the weatherstrip and body flange.
6 Gently prise out the clip which covers the joint of the chromium finisher strip and pull the finisher strip out of the weatherstrip. Then remove the weatherstrip from the glass or, if it is on the car (as in the case of a shattered screen) remove it from the body flange.
7 To fit a new windscreen start by fitting the weatherstrip around the new windscreen glass.
8 Apply a suitable sealant to the weatherstrip-to-body groove. Fit a fine, but strong, piece of cord right the way round the groove, allowing an overlap of about 150 mm (6 in) at the joint.
9 From outside the car, place the windscreen in its correct position, making sure that the loose end of the cord is inside the car.
10 With an assistant pressing firmly on the outside of the windscreen, get into the car and slowly pull the cord thus drawing the weatherstrip over the body flange (See Fig. 11.6).
11 Apply a further layer of sealant to the underside of the rubber-to-glass groove, from outside the car.
12 Refit the chromium finisher strip into its groove in the weatherstrip and fit the clip which covers the joint.
13 Carefully clean off any surplus sealant from the windscreen glass, before it has a chance to harden and then refit the windscreen wiper arms and blades.

11 Tailgate window glass – removal and refitting

1 Where applicable, remove the window glass wiper arm and blade, and carefully disconnect the heater element connections.
2 Carefully prise out the mylar insert from the rubber moulding.
3 If possible, obtain help from an assistant and carefully use a blunt bladed screwdriver to push the weatherstrip lip along the upper transverse section under the tailgate aperture flange. When approximately two thirds of the weatherstrip lip has been treated in this manner, pressure should be applied to the glass from inside the car. The glass and weatherstrip can then be removed from the outside.
4 Clean the lip of the window aperture and the glass and weatherstrip if they are to be used again. Do not use solvents such as petrol or white spirit on the weatherstrip, as these may cause deterioration of the rubber.
5 When refitting, fit the weatherstrip to the glass, then insert a drawcord in the rubber-to-body groove, so that the cord ends emerge at the bottom centre with approximately 150 mm (6 in) of overlap. During this operation it may help to retain the weatherstrip to the glass by using short lengths of masking tape.
6 Apply a suitable sealant to the body flange. Position the glass and weatherstrip assembly to the body aperture and push up until the weatherstrip groove engages the top tranverse flange of the body aperture. Ensure that the ends of the drawcord are inside the car, then get the assistant to push the window firmly at the base whilst one end of the drawcord is pulled from the weatherstrip groove. Ensure that the cord is pulled at right-angles to the flange (ie towards the centre of the glass) and that pressure is always being applied on the outside of the glass in the vicinity of the point where the drawcord is being pulled.
7 When the glass is in position, remove the masking tape which may have been used, then seal the weatherstrip to the glass.

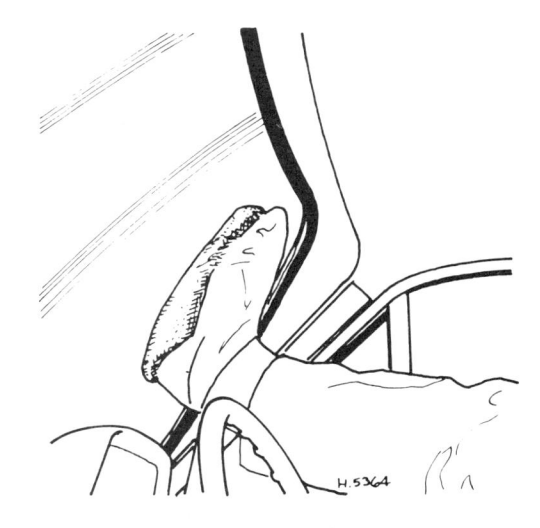

Fig. 11.5 Windscreen removal (Sec 10)

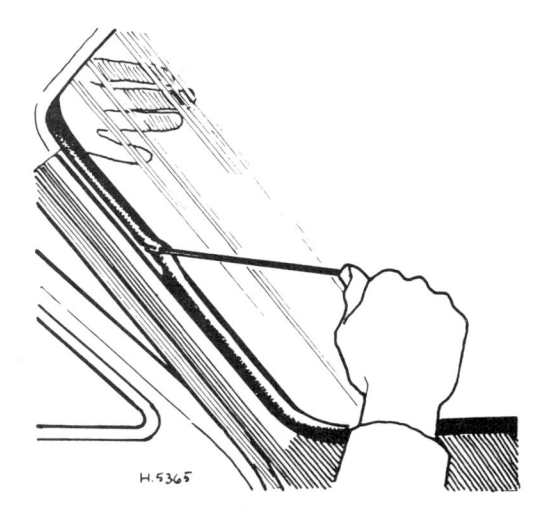

Fig. 11.6 Windscreen fitting using a drawcord (Sec 10)

8 Lubricate the mylar insert with a rubber lubricant and refit it.
9 Refit the wiper arm and blade (where applicable) and reconnect the heater element connections.

12 Door remote control handle – removal and refitting

1 Remove the door trim panel, as described in Section 13.
2 Push the remote control handle assembly towards the front of the car and pull it out of the opening in the door inner panel (photo).
3 Lift the protective cap off the rear and twist the assembly to disengage it from the operating rod (photo).
4 Remove the protective cap from the operating rod.
5 Refitting is the reverse of the removal procedure.

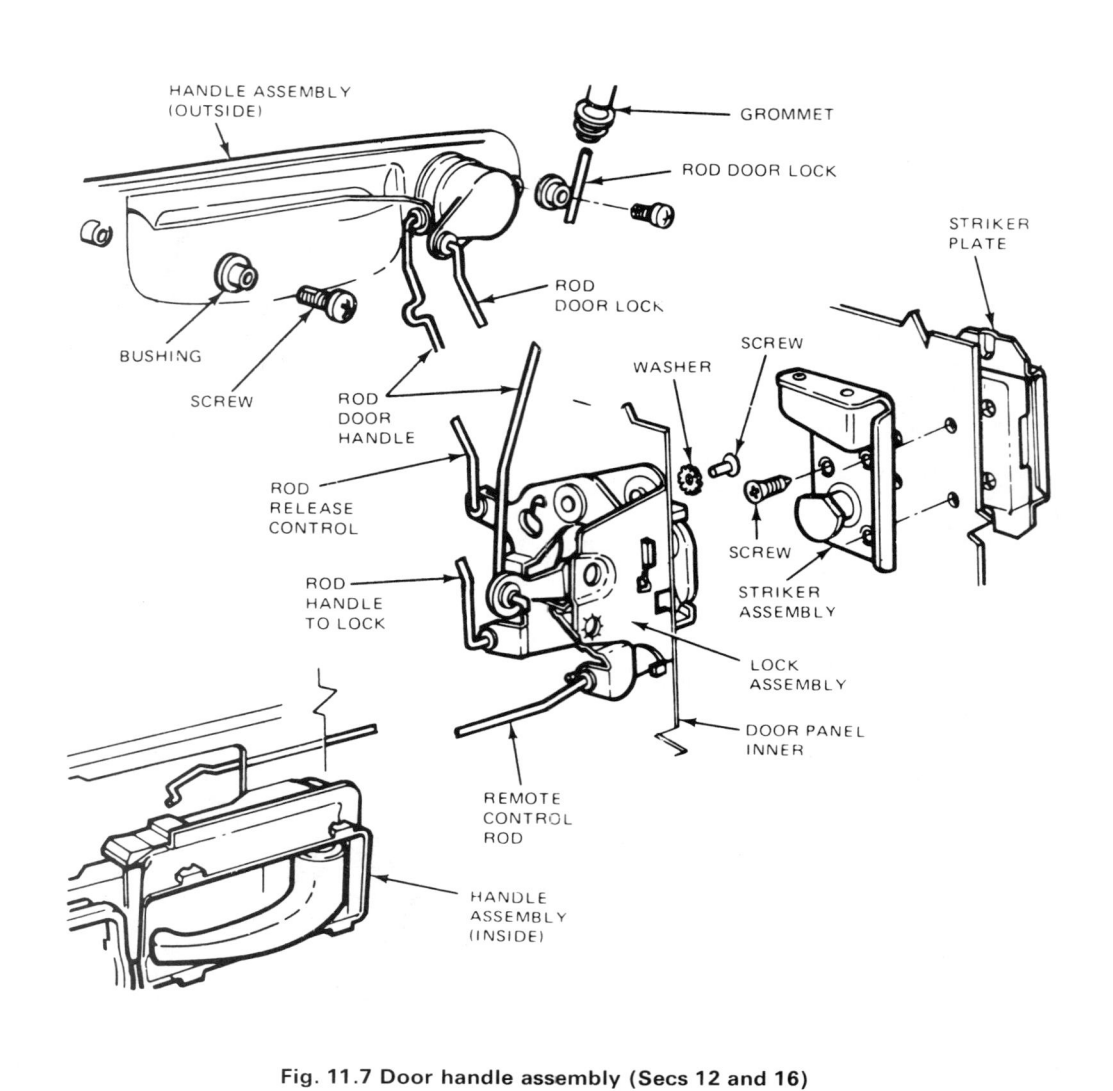

Fig. 11.7 Door handle assembly (Secs 12 and 16)

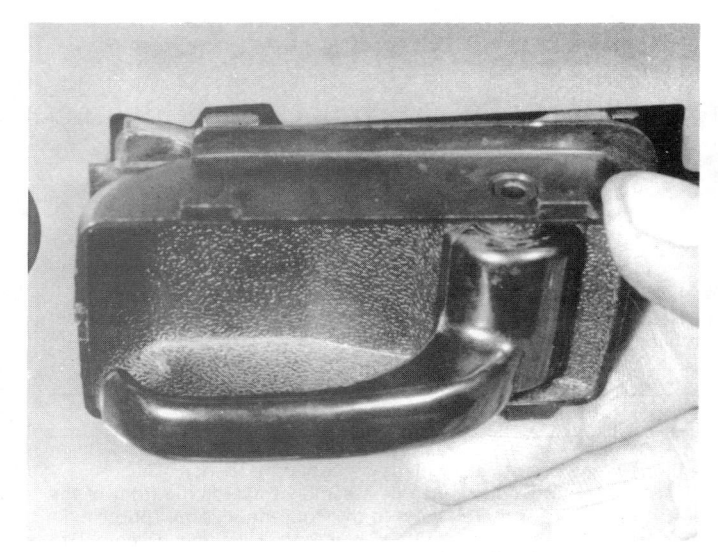

12.2 Removing the remote control handle from the door

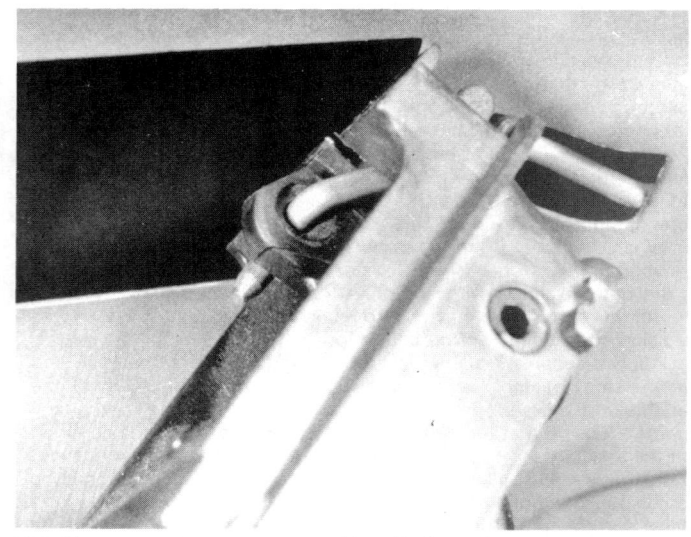

12.3 Disengage the remote control handle from the operating rod

Chapter 11 Bodywork and fitting

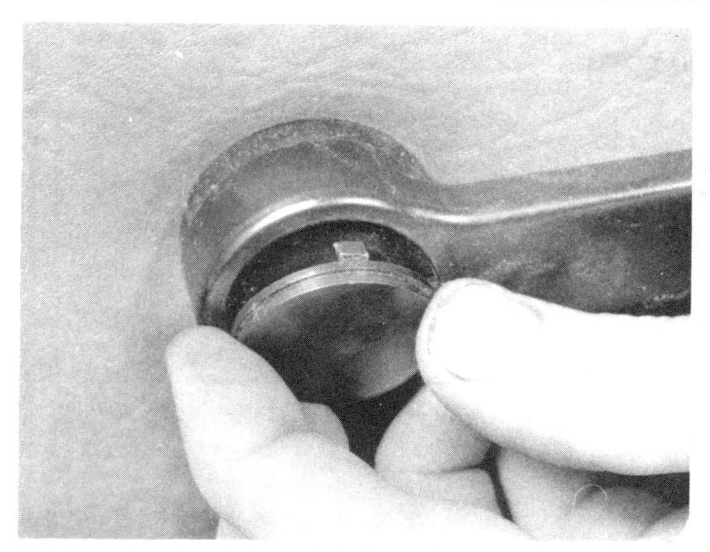

13.1 Removing the window winder insert (later models)

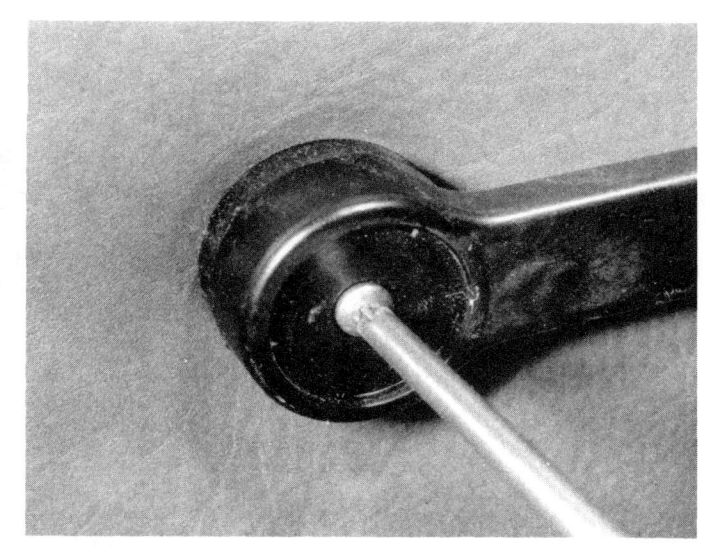

13.2A Remove the screw ...

13 Door trim panel – removal and refitting

1 Carefully lift up and remove the window winder handle insert (photo).
2 Remove the winder handle retaining screw and pull off the handle and escutcheon (photos).
3 Remove the two armrest retaining screws, turn the armrest through 90° and pull out the top fixing (photos).
4 Carefully remove the remote control bezel by sliding it forwards and unscrew the private lock button (photos).
5 Where applicable, unscrew the nut from the exterior mirror control knob (photo).
6 Taking care that no damage to the panel or paintwork occurs, carefully prise the trim panel from the door panel. Where applicable disconnect the speaker wiring (photos).
7 Refitting is a reversal of removal, but fit the window winder handle so that, when the window is closed, the winder is in the lower vertical position.

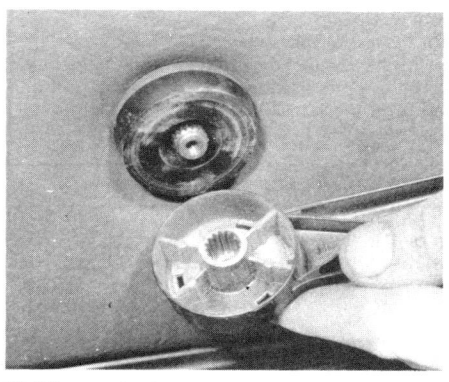

13.2B ... and window winder handle

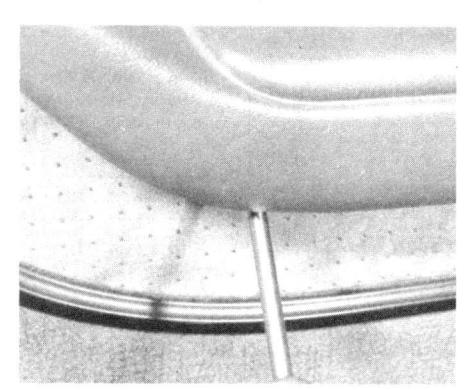

13.3A Remove the screws ...

13.3B ... then turn the armrest 90° to remove it

13.4A Slide the remote control bezel forwards to remove it

13.4B Door lock button

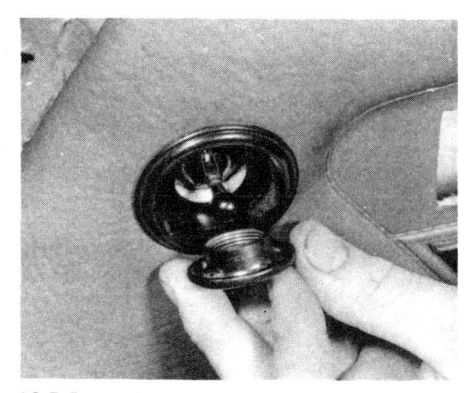

13.5 Removing the nut from the exterior mirror control knob

13.6A Door trim panel retaining clip

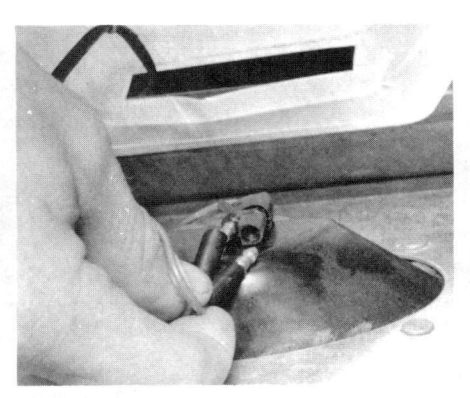

13.6B Disconnecting speaker wiring when removing the door trim panel

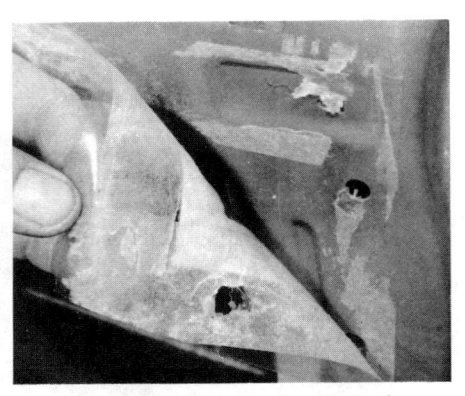

14.2 Removing the plastic sheet from the inner door

14.4A On later models the window regulator is riveted to the door

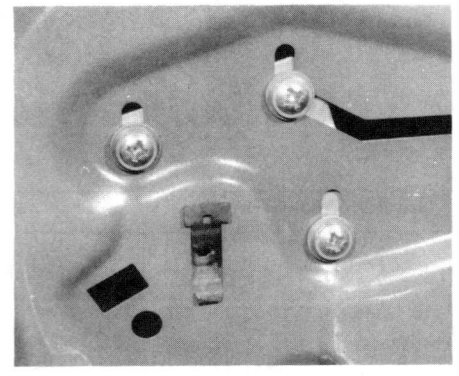

14.4B Window regulator relay bracket screws

14 Door window regulator assembly – removal and refitting

1 Remove the door trim panel, as described in the previous Section.
2 Peel off the plastic sheet (photo).
3 Temporarily refit the winder handle and lower the window.
4 On early models remove the screws securing the regulator assembly to the inner door panel. On later models the main assembly is riveted in position and the rivets must be drilled out. The relay bracket is secured by screws (photos).
5 Draw the regulator assembly towards the rear of the door, to disengage if from the runner at the base of the window.
6 Push the window glass up and use adhesive tape on each side of the glass and over the window frame to retain it. If it is to be left for any length of time, use a wooden support as well.
7 Withdraw the regulator from the door.
8 Refitting is the reverse of the removal procedure, alignment being obtained by adjusting the pivot plate as necessary. A riveter tool and new rivets will be required to fit the later assembly.

15 Exterior door handle – removal and refitting

1 Remove the door trim panel, as previously described.
2 Pull away from the plastic sheet the exterior handle then disconnect the two connecting links from the door lock to the exterior handle.

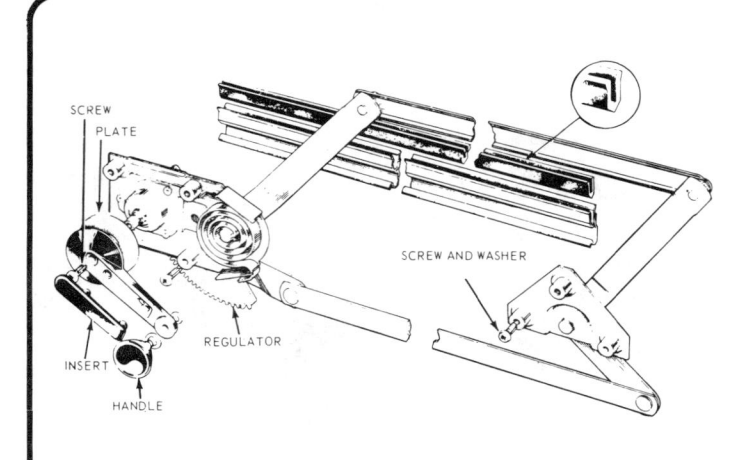

Fig. 11.8 Door window regulator assembly (Sec 14)

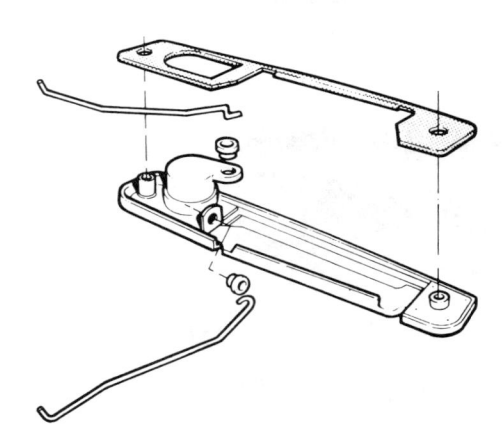

Fig. 11.9 Exterior door handle (Sec 15)

Chapter 11 Bodywork and fitting

3 Remove the two handle retaining bolts and withdraw the handle.
4 Refitting is the reverse of the removal procedure, but do not forget to install the bushes for the link rods. A little petroleum jelly on the rod ends will assist with their installation.

16 Door lock assembly – removal and refitting

1 Remove the door trim panel as previously described and remove the plastic sheeting.
2 Remove the remote control handle and two window frame bolts.
3 Using a screwdriver, prise the clips from the exterior handle rod and detach the rods from the lock.
4 Remove the cross-head screws securing the lock to the shell and the plastic clips securing the remote control rod to the inner panel (photo).
5 Remove the lock from the door through the lower rear access aperture.
6 When refitting, insert the remote control rod through the door aperture, ensuring that the rod lies against the door inner panel. Locate the lock on the door shell, pushing the frame towards the outer panel, to enable the lock to be correctly positioned on the rear shell.
7 Secure the lock with the three screws and secure the remote control rod to the inner panel, with the two plastic clips.
8 Refit the exterior handle rods in their respective lock locations. Position the black bush. A and white bushes B as shown in Fig. 11.10.
9 The remainder of the refitting procedure is the reverse of the removal procedure.

17 Door window glass – removal and refitting

1 Remove the door trim panel, as previously described, then peel the plastic sheeting away from the door panel apertures.
2 Remove the door belt moulding/weatherstrip assembly (see Fig. 11.11).
3 Wind up the window glass, then remove the pivot plate screws. Remove the four regulator gear plate securing screws or rivets. Disengage the studs and rollers of the regulator arms from the door glass channel and carefully lift out the glass. Allow the regulator to fall away, pivoting on the regulator handle shaft.
4 When refitting, initially insert a small block of wood in the bottom of the door assembly. Locate the glass in the door panel, so that it is resting on the wooden block.
5 Locate the studs and rollers of the regulator arm into the door glass channel, then temporarily install the window handle and turn it to align the gear plate with the panel fixings. Secure the plate to the inner panel.
6 Loosely assemble the pivot plate, then wind up the glass and align it in the frame. Tighten the pivot plate screws.
7 The remainder of the refitting procedure is the reverse of the removal procedure.

16.4 Door lock and retaining screws

Fig. 11.10 The black (A) and white (B) door lock bushes (Sec 16)

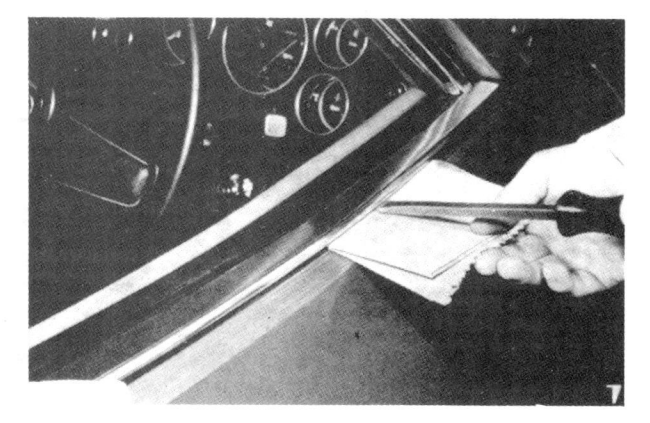

Fig. 11.11 Removing the door belt moulding (Sec 17)

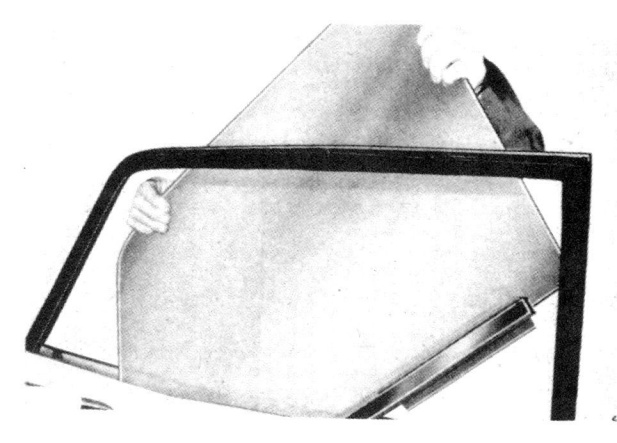

Fig. 11.12 Removing the door window glass (Sec 17)

18 Door window frame – removal and refitting

1 Remove the door trim panel as previously described, then peel the plastic sheeting away from the lower door panel apertures.
2 Remove the door belt moulding/weatherstrip assembly.
3 Lower the window glass, then peel back the lower front corner of the plastic sheeting and remove the reflector (where applicable) to gain access to the front and rear lower fixing bolts.
4 Remove the five bolts and frame seals, to free the frame from the shell. Push the glass out of the frame at the rear of the door, so that the frame lies between the glass and the outer panel. Repeat for the front of the door.
5 Pull the rear of the frame from the shell, whilst guiding the front of the frame rearwards past the first door bolt moulding retaining clip. This enables the frame to be lifted clear.
6 When refitting, insert the front of the frame so that the vertical leg lies to the rear of the first moulding clip.
7 Spring the rear of the frame into the shell, so that the frame lies between the glass and the outer panel, whilst springing the frame front vertical leg past the moulding clip, so that this also lies between the glass and the inner panel.
8 Spring the frame around the glass and secure it with the five bolts.
9 Pull the weatherstrip from the door aperture flange, then shut the door and adjust the frame to obtain a gap between the frame and flange (in and out) of 10 to 14 mm (0.4 to 0.6 in), and between the frame and the A-pillar (fore and aft) of 8 to 12 mm (0.3 to 0.5 in). Tighten the bolts.
10 The remainder of the refitting procedure is the reverse of the removal procedure.

19 Window frame moulding and door weatherstrips – removal and refitting

1 Where applicable, wind the window down to its fullest extent. Carefully prise the weatherstrip out of the groove in the door outer bright metal finish moulding.
2 When refitting, correctly position the weatherstrip over its groove. With the thumbs, carefully prise the strip fully into the groove.
3 Wind the window up (where applicable) and check that the weatherstrip is fitted correctly.

20 Door – removal and refitting

1 With the door open, remove the clip from the bottom of the check arm pivot pin then remove the pin (photo).
2 Support the door on blocks of wood.
3 Where door speakers are fitted remove the trim panel (Section 13) then remove the wiring from the front edge of the door.
4 Where roll pin type hinges are fitted (early models) use a narrow drift to drive out the roll pins, then remove the door from the car. Where solid hinge pins are fitted (photo) the pins cannot be removed and the hinges must therefore be detached from the body pillar as follows.

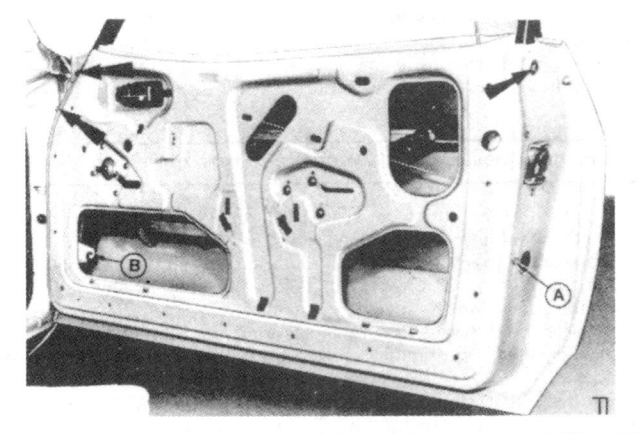

Fig. 11.13 Window frame retaining bolts – arrowed (Sec 18)

Prise out reflector for access to A and pull back plastic sheet over B

5 Remove the scuff plate front screw and locally lift the weatherstrip.
6 Remove the lower facia panel.
7 On the right-hand side, remove the instrument panel as described in Chapter 12. On the left-hand side remove the glove compartment.
8 Remove the face level vent.
9 Unscrew the nuts and remove the plates from the inner sides of the hinges, then lift the door from the car.
10 Refitting is a reversal of removal, but check the door striker adjustment as described in Section 7.

21 Door private lock – removal and refitting

1 Remove the door trim panel as described in Section 13.
2 Prise the clip from the control rod and disconnect it from the main door lock (photo).
3 Slide the retaining clip down then withdraw the private lock from the outside of the door.
4 Refitting is a reversal of removal.

22 Rear quarter trim panel – removal and refitting

1 Remove the screws retaining the window quarter trim and lift away the trim.
2 Remove the B-pillar vertical trim and the seatbelt screw (where applicable).
3 Remove the two screws and remove the rear seat cushion. Where applicable, feed the seatbelt and buckle assemblies through the opening in the cushion.
4 Remove the three trim panel screws and the step plate. Where

20.1 Door check arm (arrowed)

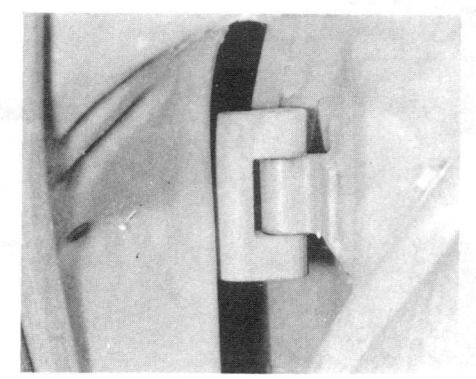

20.4 Door hinge on a later model

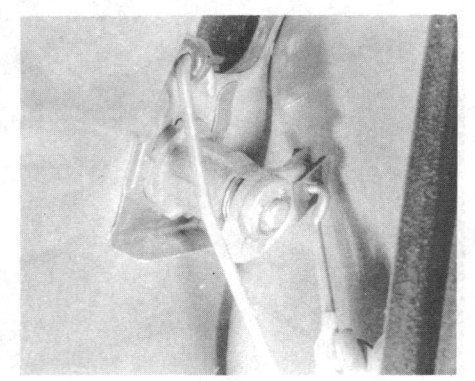

21.2 Inner view of the private door lock

Chapter 11 Bodywork and fitting

applicable, remove the luggage compartment hook. Carefully prise away the trim panel.
5 Refitting is the reverse of the removal procedure, but on completion, tighten the seatbelt bolt to the specified torque.

23 Opening rear quarter glass assembly – removal and refitting

1 Remove the trim covers from the B-panel and the quarter window surround (trim). Where applicable first remove the rear seatbelt inertia reel (Section 36).
2 Remove the two toggle retaining screws and remove the toggle from the rear C-pillar.
3 Remove the window frame weatherstrip then drive out the toggle-to-catch retaining pin and remove the toggle.
4 Refitting is the reverse of the removal procedure, but lubricate the B-pillar hinge pivots with a soap solution prior to fitting the glass assembly. Adjust the toggle, or weatherstrip flange, to achieve 8 to 10 mm (0.3 to 0.4 in) gap between the glass and the weatherstrip flange. Tighten the toggle screws to the specified torque.

24 Load space trim panel – removal and refitting

1 Remove the rear quarter trim panel, as previously described.
2 Pull the seat forward and remove the ten securing screws. It may also be necessary to detach the back trim panel (five screws).
3 Remove the interior light connection and the seatback lock knob and then remove the panel.
4 Refitting is the reverse of the removal procedure. Ensure that the sound deadening material is correctly positioned and that the trim panel does not foul the seat release hinge mechanism.

25 Bonnet release cable – removal and refitting

1 In the event of the release cable breaking, it is possible to remove the radiator grille to operate the lock spring by hand. Grille removal is dealt with in Section 9, but since it is not possible to open the bonnet, it will be found a little difficult (though not impossible) to gain access to the upper retaining screws.
2 To remove the release cable in normal circumstances, remove the radiator upper cowl plate.
3 From inside the car, remove both the clevis pins and the spring and disconnect the release cable from the control lever (photo).
4 Slacken the cable adjuster clamp and release the cable from the hook lock spring.
5 Remove the cable retaining clips, then pull the cable through the dash panel to remove it.
6 Refitting of the cable is essentially the reverse of the removal procedure, adjusting as necessary, to remove any cable slack. For further information on this, refer to paragraph 6 of the following Section.

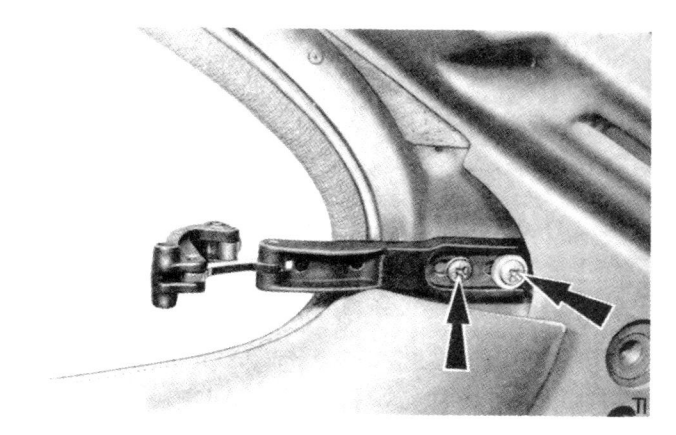

Fig. 11.14 Quarter window fixing screws – arrowed (Sec 23)

26 Bonnet – removal, refitting and adjustment

1 Open the bonnet to its fullest extent. Using a suitable implement, scribe a line around the hinges. Remove the screw securing the earth lead where applicable (photo).
2 Remove the two bolts and washers securing each side of the bonnet to its hinges (photo). With assistance, it can now be lifted off.
3 Refitting is a reversal of removal procedure. Before fully tightening the securing bolts, ensure that the hinges are aligned with the scribed marks. This will ensure correct alignment.
4 If it is found that the bonnet requires adjustment, this can be effected in the vertical plane by slackening the catch post locknut and screwing the catch post in or out. Fore-and-aft adjustment can be effected by slackening the hinge bolts.
5 Adjustable bump rubbers are also provided and these should be positioned as necessary to stop vibration, but at the same time must allow the bonnet to be closed easily.
6 Adjustment of the bonnet locking spring can be made by slackening the cable clamp on the upper crossmember and sliding the outer cable through the clip as necessary. When correctly positioned, the hood lock spring/cable setting dimension should be as shown in Fig. 11.15. Tighten the clamp screw when the adjustment is satisfactory.

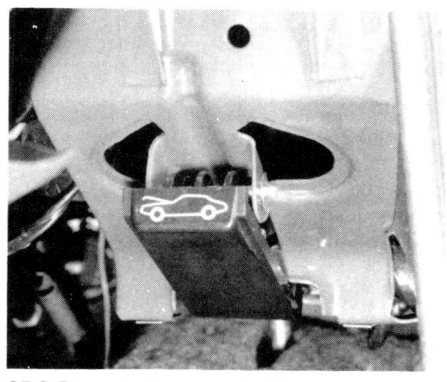

25.3 Bonnet release control lever

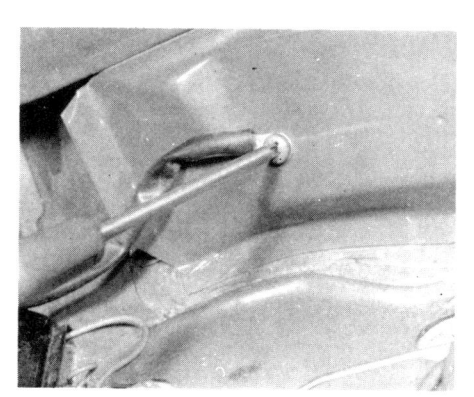

26.1 Removing bonnet earth lead

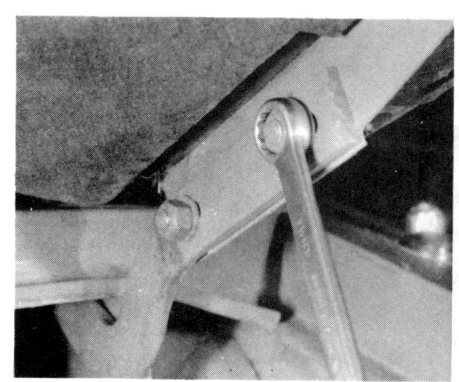

26.2 Removing bonnet hinge bolts

Chapter 11 Bodywork and fitting

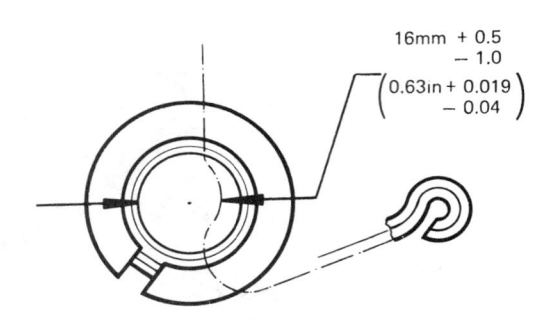

Fig. 11.15 Bonnet locking spring setting dimension (Sec 26)

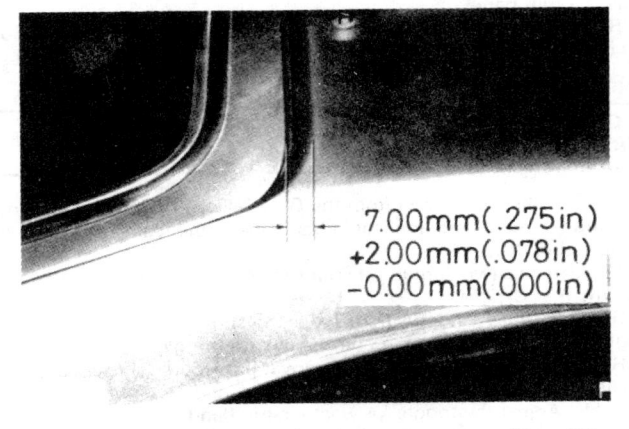

Fig. 11.16 Tailgate-to-roof edge alignment (Sec 27)

27 Tailgate – removal and refitting

1 Open the tailgate and detach the in-line connectors from the damper(s)/strut(s) (photos).
2 Support the tailgate, then disconnect the dampers by prising out the plastic wedges with a small screwdriver (photo).
3 With help from an assistant, support the tailgate and remove it by removing the hinge bolts (photo).
4 When installing, align the tailgate so that the edges are flush with the rear of the roof and the C-pillar sides, and the gap between the tailgate and the roof edge is 7 to 9 mm (0.3 to 0.4 in) – see Fig. 11.16.
5 Align the lower edges of the tailgate, so that it is flush with the rear corners of the body, with the striker plate in the upper central position.
6 Pads can be used beneath the C-pillar bumpers for alignment of the tailgate sides with the C-pillar slope (photo). Note that the thick end of the bumper faces towards the front of the vehicle on early models.
7 On completion, refit the dampers and wiring connectors.

28 Tailgate private lock – removal and refitting

1 Open the tailgate and prise off the trim, then remove the latch (three bolts and washers) (photo).
2 Remove the lock cylinder retaining nut, then turn the lock spider, to

27.1A Tailgate damper connection to body

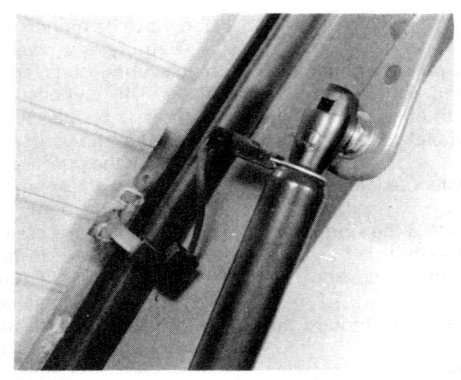

27.1B Tailgate damper connection to tailgate

27.2 Method of disconnecting the tailgate dampers

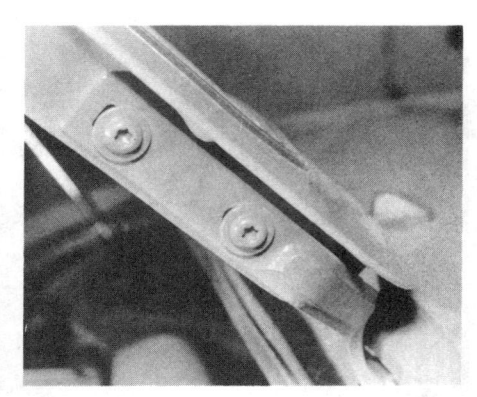

27.3 Tailgate hinge bolts

27.6 Tailgate bumper pad

28.1 Tailgate latch

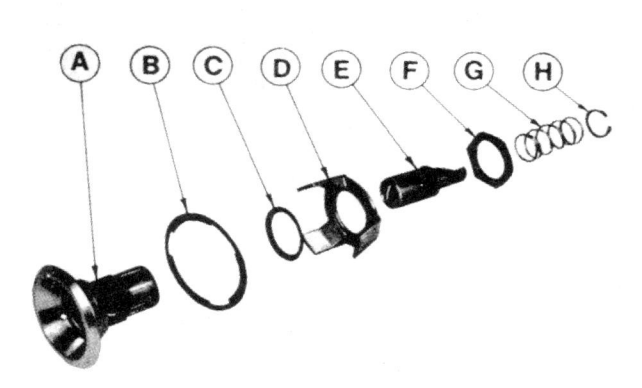

Chapter 11 Bodywork and fitting

Fig. 11.17 Tailgate private lock components (Sec 28)

- A Outer body
- B Seal
- C Washer
- D Spider
- E Lock barrel
- F Nut
- G Spring
- H Circlip

29.1 Tailgate striker

detach it from the lock cylinder and outer body.
3 Refitting is the reverse of the removal procedure.
4 If it is found necessary to remove the lock barrel, this can be done after it has been removed, by removing the circlip from the barrel housing. The spring, barrel and spider can then be detached and a new barrel fitted by reversing this procedure.

29 Tailgate striker plate – removal and refitting

1 Open the tailgate then carefully scribe a mark around the striker to facilitate refitting (photo).
2 Remove the single bolt and washer and take off the striker plate.
3 Refitting is the reverse of the removal procedure, following which adjustment can be made, if found necessary, to obtain satisfactory opening and closing of the tailgate.

30 Fuel filler flap – removal and refitting

1 Remove the load space trim panel, as previously described.
2 Remove the two screws from inside the car and lift away the filler flap.
3 Refitting is the reverse of the removal procedure.

31.5A Prise out the centre vent ...

31 Facia panel – removal and refitting

1 Disconnect the battery earth lead.
2 Remove the steering column shroud retaining screws. Remove the lower half and release the upper half retaining lug from its spring clip by pulling sharply upwards.
3 Remove the instrument cluster, as described in Chapter 12.
4 Detach the flexible pipes from the dash panel vents and defrosters.
5 Where applicable remove the ashtray and centre vent and remove the concealed screws (photos).
6 Remove the left and right-hand A-pillar trims (2 screws each) – also the grab handle, if fitted.
7 Remove the instrument panel pad retaining screws and lift the pad away. If appropriate, remove the dash panel vents and transfer them to the new instrument panel.
8 Installation is essentially the reverse of the removal procedure, but connect the flexible pipes to the vents and defrosters before the crash pad is fitted.

31.5B ... and remove the screws

186 Chapter 11 Bodywork and fitting

32.8A Prise out the cap ...

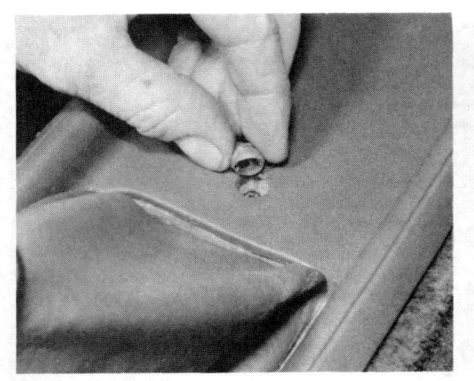

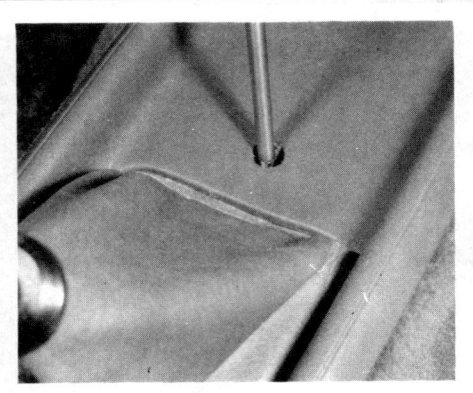

32.8B ... and remove the console screw

32.9 Removing the handbrake lever boot and bezel

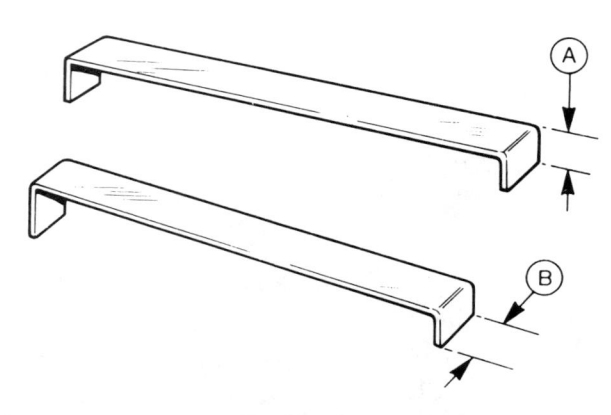

Fig. 11.18 Clock removal tools (Sec 32)

A = 4.0 mm (0.16 in)   B = 10.0 mm (0.40 in)

32 Centre console – removal and refitting

Basic type

1 Lift the carpet from around the front of the console, then push the clock and bezel out of the housing. Disconnect the clock leads.
2 Remove the two screws at the rear end and two more from the clock end, to release the console.
3 Remove the gear lever knob, or T-handle.
4 Lift off the console. As applicable, remove the clock mounting plate screws and/or gearshift lever boot.
5 Refitting is the reverse of the removal procedure.

Ghia type (also fitted to later non-Ghia models)

6 Lift the carpet from around the front of the console.
7 Fabricate two small brackets from 1.5 mm (0.06 in) sheet steel (see Fig. 11.18) and insert these behind the black bezel to remove the clock. Disconnect the wiring.
8 Remove the two console retaining screws, then prise out the cap in front of the gearshift lever and remove the screws beneath (photos).
9 Carefully prise the handbrake lever boot and bezel assembly from the console and remove it (photo).
10 Carefully prise out the centre panel cover from behind the gear lever.
11 Lift up the centre armrest and remove the screw from the bottom of the compartment.
12 Remove the armrest hinge screws and place the armrest in the compartment.
13 Remove the two screws at the front end of the armrest.
14 Prise out the cap at the rear of the console and remove the screw.
15 Carefully lift out the rear section, front section and console compartment.
16 Remove the heat insulating pad from beneath the console compartment area.

17 As applicable, remove the electrical connector from the base of each seatbelt stalk and remove the stalk fixing bolt through the top of the console. Remove the stalk assembly through the inside console support.
18 Pull the carpet away from the sides of the support assembly, then detach the support by removing the eight securing screws.
19 Refitting is basically the reverse of the removal procedure, but first ensure that all the spire nuts are correctly located on the brackets. Tighten the seatbelt stalks to the specified torque.

33 Sunroof panel – adjustments

Gap adjustment

1 When closed, a gap of 5.8 to 6.9 mm (0.23 to 0.27 in) should exist between the sunroof and the car roof. If a vinyl roof cover is fitted, this gap should be reduced by 1.78 mm (0.07 in). If adjustment is required, proceed as detailed below.
2 With the sunroof half open, remove the sunroof headlining frame.
3 Close the sunroof, then pull the headlining from the rear and remove it.
4 To adjust the front of the sunroof, loosen the front guide retaining screws, then whilst pressing the guides inwards, tighten the screws.
5 To adjust the rear, loosen the two fixing screws at each side of the adjusting base, then press the rear slides inwards and tighten.
6 Fit the headlining and headlining frame.
7 If the adjustment is still incorrect, remove the weatherstrip from the roof panel and bend its mounting flange as necessary.

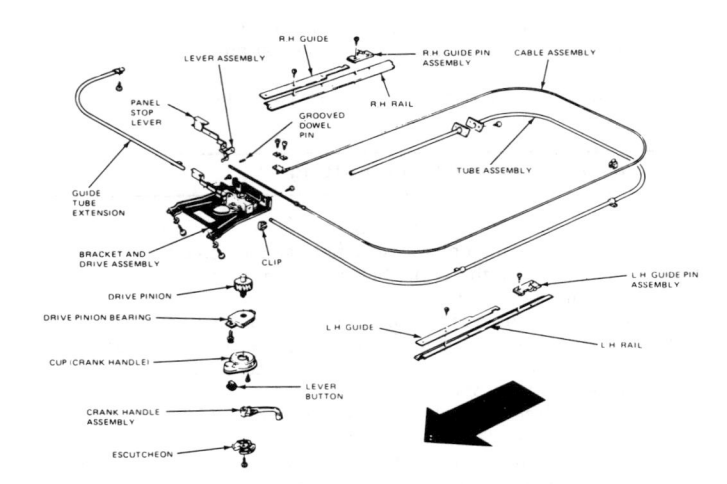

Fig. 11.19 Sunroof control mechanism (Secs 33, 34 and 35)

Chapter 11 Bodywork and fitting   187

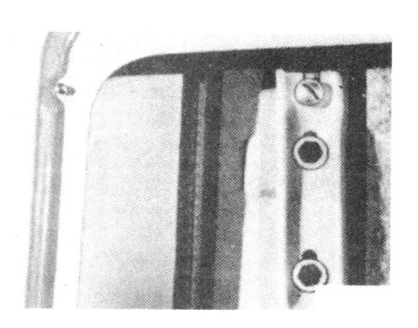

Fig. 11.20 Sunroof front gap adjustment (Sec 33)

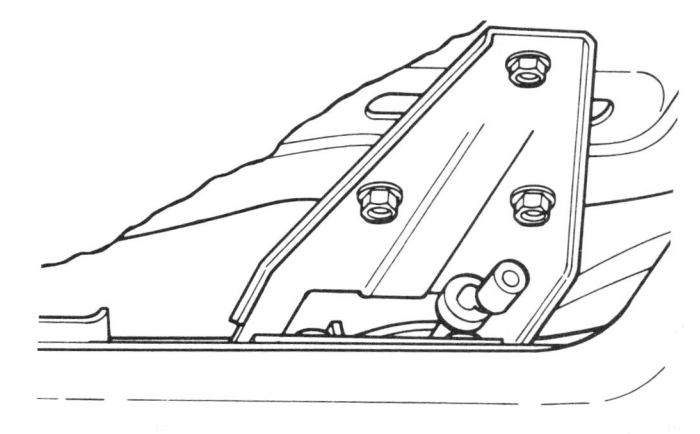

Fig. 11.21 Sunroof rear gap adjustment (Sec 33)

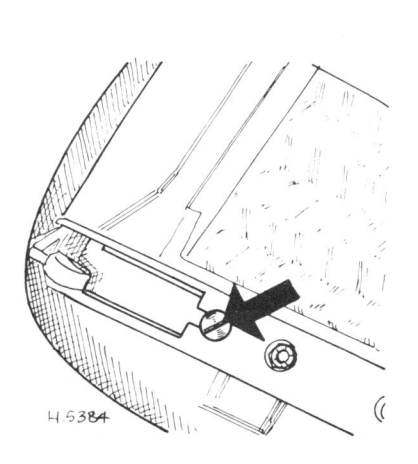

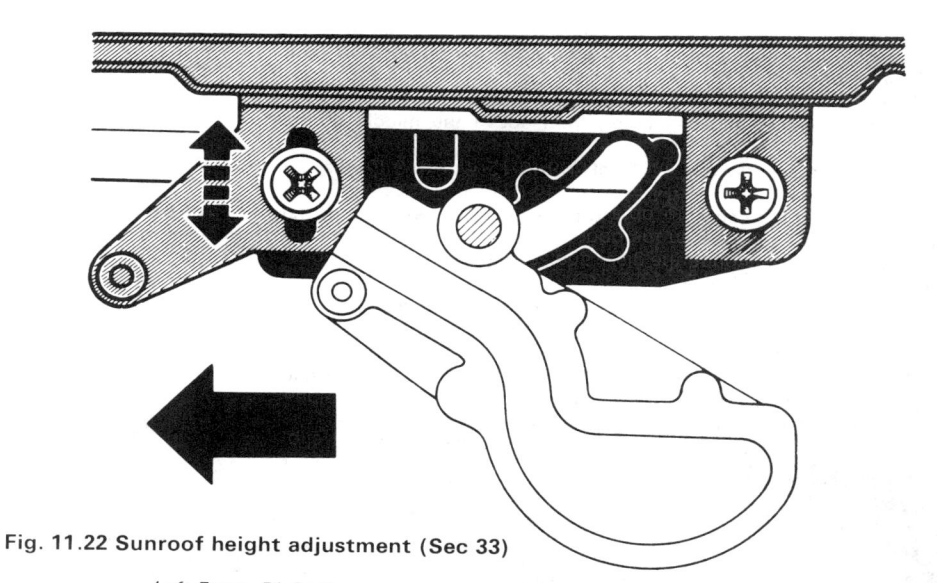

Fig. 11.22 Sunroof height adjustment (Sec 33)

Left Front, Right Rear

Height adjustment
8 The front edge of the sunroof should be flush with, or 1 mm (0.04 in) below, the edge of the car roof. The rear edge should be flush with, or 1 mm (0.04 in) above, the car roof. If adjustment is required, proceed as described below.
9 Proceed as described in paragraphs 1 to 4 inclusive, but before tightening the screws, adjust the sunroof height by turning the adjusting screw (Fig. 11.22) as necessary. Tighten the guide retaining screws.
10 To adjust the rear end, loosen the adjusting screws on the link assemblies (on the inner side of the panel). Adjust the height and tighten the screws.
11 Check for leaks and noises, adjusting again, if necessary.
12 Fit the headlining and headlining frame.

34 Sunroof – removal and refitting

1 Open the sunroof, then mark the position of both guide pin assemblies. Remove the guide pins and guides.
2 Close the sunroof, then push the button upwards to the tilt position.
3 Lift the front of the sunroof out of its opening, whilst turning the handle until the screws fastening the cable to the base assembly are accessible.
4 Unscrew the cable and lift out the complete sunroof, including the base assembly.
5 Unscrew the handle assembly, pull out the handle and escutcheon then remove the button control. Remove the cup.
6 Unscrew the gear bearing, then remove it, together with the pinion.
7 Move the cable so that the grooved dowel pin appears in the opening of the pinion. Pull out the pin and discard it. The operating cable can now be pulled out of the tube assembly.
8 To refit the sunroof, install the operating cable into the tube assembly. Turn the cable at its T-formed end and fit a new grooved dowel pin. Smear a little general purpose grease on the cable and where it will contact the pinion, then assemble the pinion and bearing.
9 Fit the handle cup and fit the button to the lever.
10 Fit the handle and escutcheon, then check the handle position as follows:

(a) *Pull the button down and turn the handle fully clockwise, which should now be approximately 30 degrees ahead of its original position opposite the cup of the handle (Fig. 11.23).*

(b) *Push the button up and turn the handle fully anti-clockwise, which should now be approximately 30 degrees ahead of its original position. The handle can be reset on the pinion splines as necessary.*

11 Insert the sunroof into the opening and attach the cable to the base assembly.
12 Carefully move the sunroof to the rear, by turning the handle, then screw on the guide pin assemblies with the pin outwards, to coincide

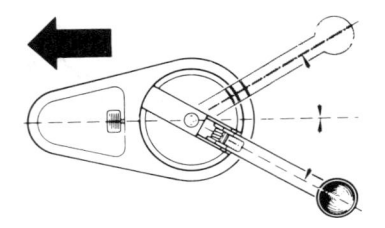

188   Chapter 11 Bodywork and fitting

Fig. 11.23 Checking the sunroof handle alignment (Sec 34)

with the marks made when removed. Note that the left and right-hand guides are not interchangeable.
13 Adjust the roof, as described in the previous Section.

35 Sunroof bracket and drive assembly – removal and refitting

1 Remove the crank handle, the pinion drive and the bearing.
2 Remove the covering strip, then remove the mirror, courtesy light and sun visors.
3 Remove the windscreen, as described in Section 10.
4 Carefully remove the headlining from above the middle of the windscreen.
5 Pull the grooved dowel pin out of the actuating cable and discard it.
6 Pull back on the cable slightly, to clear the drive assembly of the cable.
7 Remove the tube guide screw (right-hand side) and clip (left-hand side), and remove the tubes.
8 Remove the screws, to release the bracket and drive assembly.
9 Refitting the bracket and drive assembly is the reverse of the removal procedure, using a new grooved dowel pin. Check the handle position, as described in paragraph 10 of the previous Section and adjust the roof, if necessary, as described in Section 33.

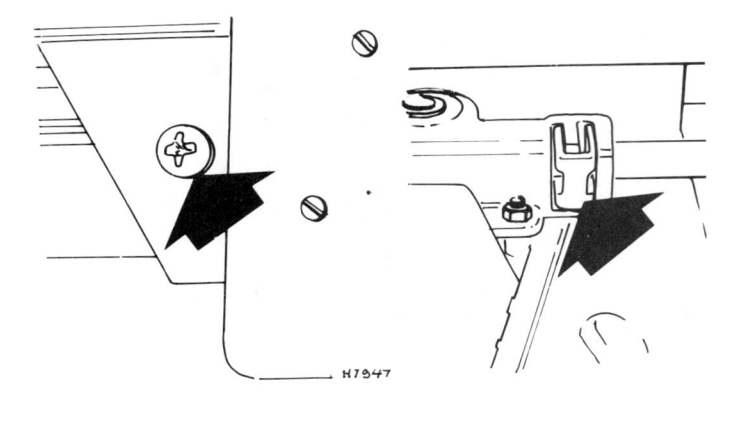

Fig. 11.24 Tube guide screw and clip – arrowed (Sec 35)

36 Seat belts – general

1 All models are fitted with inertia reel seat belts for the front seats. Rear seat belts of a similar type are also available.
2 Fig. 11.25 shows the floor-mounted front seat belt stalk. Removal of the basic type fixing is straightforward, but for Ghia and later models the centre console must be partly removed, as described in Section 32.
3 To remove the front seat belt inertia reel, remove the rear quarter trim panel, as described in Section 22, then remove the three nuts from the anchor plate and lower the assembly through the aperture, whilst feeding the belt through the slot in the inner panel (photos).
4 To remove the rear seat belts remove the cushion and unscrew the anchor bolts. The inertia reels can be unbolted from the rear side panels (photos).

36.3A Front seat belt anchor plate and nuts

36.3B Feeding the front seat belt through the slot in the inner panel

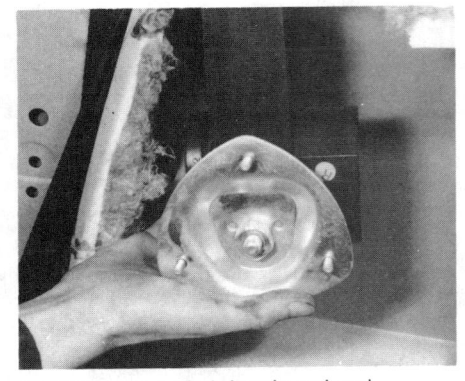

36.3C Front seat belt inertia reel and anchor plate

36.4A Rear seat belt outer anchor

36.4B Rear seat belt middle anchor

36.4C Rear seat belt inertia reel

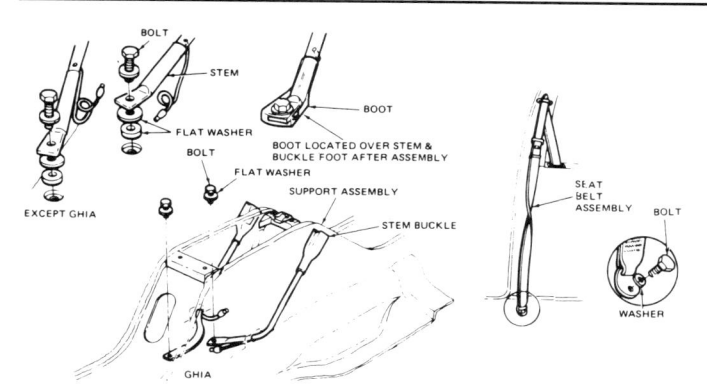

Chapter 11 Bodywork and fitting

Fig. 11.25 Front seat belt anchorage points (Sec 36)

5   When refitting the seat belts, make sure that the anchor plates are positioned at the correct angles and tighten the bolts to the specified torque.

37 Adjustable exterior mirror – removal and refitting

1   Remove the door trim panel as described in Section 13.
2   Unscrew the mounting nuts and withdraw the mirror from the outside of the door, at the same time feeding the cables through the hole (photo).
3   Refitting is a reversal of removal.

38 Interior mirror – removal and refitting

1   Prise the plastic cover from the mirror base (photo).

2   Unscrew the cross-head screws and remove the mirror and base from the roof (photo).
3   Refitting is a reversal of removal.

39 Seats – removal and refitting

Front seats
1   Refit the seat fully to the rear then unscrew the mounting bolts from the front of the slide rails (photo).
2   Adjust the seat fully to the front, unscrew the rear mounting bolts, then remove the seat from the car.
3   Refitting is a reversal of removal.

Rear seats
4   Remove the screws from the front of the seat cushion, and lift the cushion from the car (photo).
5   Release the seat from the upper latch(es), then unbolt the hinges and remove the seat backrest.
6   Refitting is a reversal of removal.

40 Heater controls – adjustment

1   Disconnect the battery earth lead, then remove the glove compartment by unscrewing 7 screws at the top and 2 nuts at the bottom. Also disconnect the glove compartment lighting leads.
2   Move the heater controls to a point 2 mm (0.08 in) from the end position, then remove both outer cable clips (see Figs. 11.26, 11.27 and 11.28 as appropriate).

Standard heater
3   Check that the distributor and regulator flap levers are at the end of their travel and clamp the outer cables in this position.

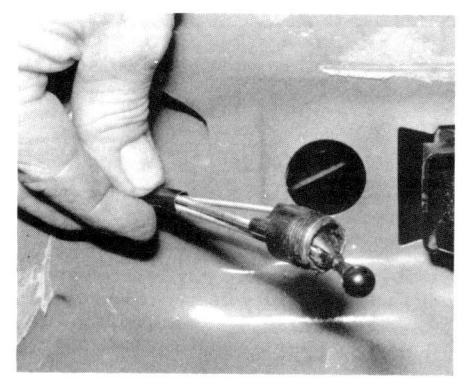

37.2 Adjustable exterior mirror cable control

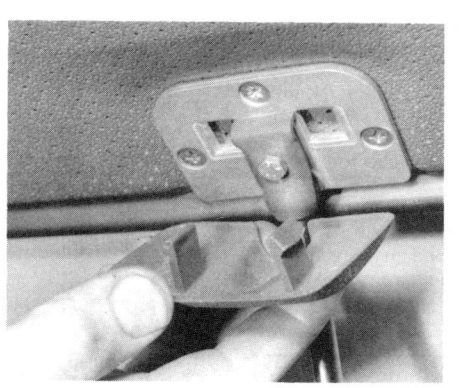

38.1 Remove the plastic cover ...

38.2 ... and unscrew the mirror base

39.1 Front seat front mounting bolt

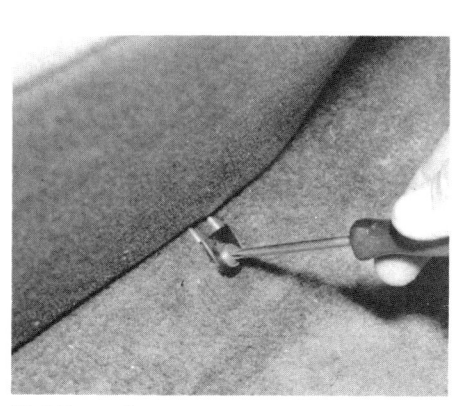

39.4 Removing rear seat cushion mounting screws

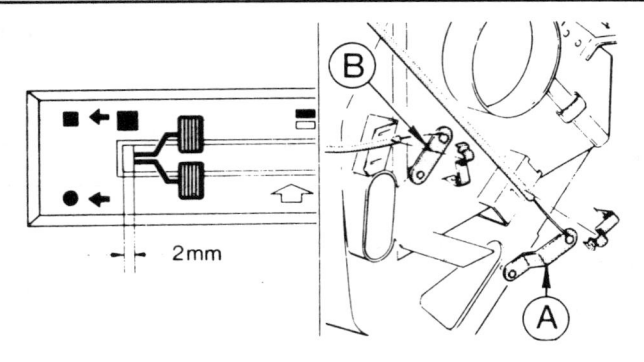

Fig. 11.26 Heater control cable adjustment on standard heater (Sec 40)

A Distributor flap control lever   B Regulator flap control lever

Heavy duty heater
4 Check that the distributor flap lever and water control lever (Fig. 11.28) are at the end of their travel and clamp the outer cables in this position.

All heaters
5 On completion, reconnect the glove compartment lighting leads, then refit the glove compartment and reconnect the battery earth lead.

41 Heater controls – removal and refitting

1 Disconnect the battery earth lead.
2 Remove the steering column shroud (2 screws at the bottom, then pull out at the top). Lower the steering column (leaving the two bolts in position) sufficiently to allow the instrument cluster trim to be removed.
3 Disconnect the switch leads and remove the instruments cluster trim, complete with cowl trim (11 screws). Remove the instrument cluster bezel (3 screws).
4 Pull off the heater control knobs. If this is not possible break them with a large pair of pliers then remove the controls (4 screws). Do not disconnect the control cables at the heater.
5 Remove the heater control panel (2 screws). Remove the blower switch and lighting leads.
6 Disconnect the cables from the heater controls.
7 Refitting is the reverse of the removal procedure, during which it will be necessary to adjust the cables, as described in the previous Section. Also it will be necessary to obtain new heater control knobs where applicable.

42 Heater water valve (heavy duty heater) – removal and refitting

1 Drain the engine coolant and disconnect the lower hose from the radiator (refer to Chapter 2, if necessary).
2 Disconnect the three water hoses from the water valve.
3 Remove the outer cable from the clip on the water valve bracket, then remove the assembly from the bulkhead (2 screws).
4 Twist the water valve, to disconnect the cable from the operating lever.
5 Refitting is the reverse of the removal procedure, during which adjustment should be made, as described in Section 40 for the heavy duty heater. On completion, refit the radiator hose and fill the cooling system, as described in Chapter 2.

43 Heater assembly – removal and refitting

1 Disconnect the battery earth lead.
2 Drain the coolant, referring to Chapter 2, if necessary.
3 Disconnect the water hoses from the heater heat exchanger (photo). If practicable, blow through the heat exchanger with

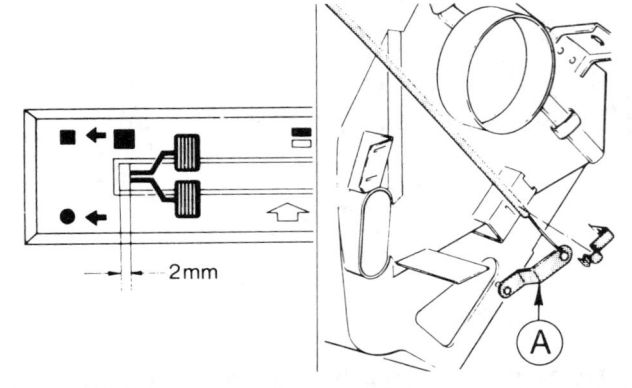

Fig. 11.27 Heater control cable adjustment on heavy duty heater (Sec 40)

A Distributor flap control lever

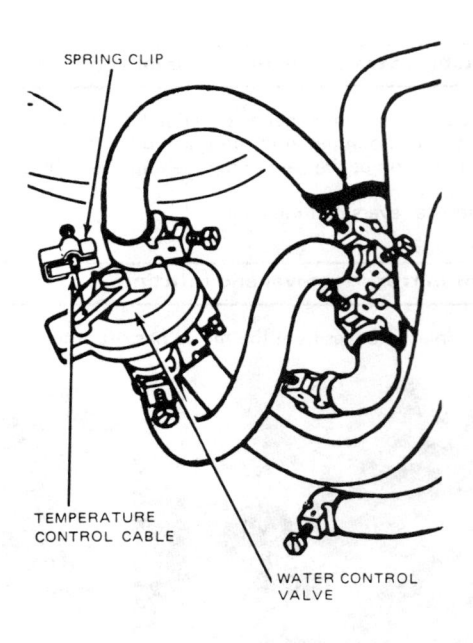

Fig. 11.28 Water valve on heavy duty heater (Sec 40)

43.3 Heater water hoses

Chapter 11 Bodywork and fitting

compressed air to remove any coolant remaining. Alternatively place cloths and/or newspapers beneath to absorb any spillage.
4 Remove the cover panel, together with the heat exchanger-to-water connection gasket, from the bulkhead (2 screws).
5 Slacken the gearlever locknut, then remove the gearlever. The locknut requires a special peg spanner, available from Ford, but it is not difficult to fabricate a tool which which will do the job.
6 Remove the parcel tray (4 screws). Where there is a centre console, this must be removed also (refer to Section 32).
7 Remove the steering column shroud (2 screws at the bottom, then pull out at the top). Lower the steering column (leaving the two bolts in position), sufficiently to allow the instrument cluster to be removed.
8 Remove the lower facia panel, complete with cover panel. Remove the ashtray and cigarette lighter and disconnect the switches (photos).
9 Remove the glove compartment by unscrewing the 7 screws at the top and 2 nuts at the bottom. Also disconnect the glove compartment lighting leads.
10 Disconnect the demister nozzle hoses, together with their connections (1 screw each).
11 Disconnect the facia vent hoses from the heater. There are 2 on the standard heater and 4 on the heavy duty heater.
12 Remove the lower dash panel support stay (1 screw).
13 Disconnect the control cables from the heater and the heater blower leads.
14 Remove the demister nozzles (refer to Section 46, if necessary).
15 Remove the windscreen wiper motor bracket from its mounting.
16 Remove the 4 heater securing screws, (photo) then pull the heater far enough rearward for the water connection pipes to clear the bulkhead. Tilt the top of the heater upward and forward, and withdraw it sideways, then remove the foam gasket also.
17 Refitting is the reverse of the removal procedure, during which it will be necessary to adjust the heater controls as described in Section 40. Do not forget to tighten the gearlever locknut. On completion, fill the cooling system, as described in Chapter 2.

44 Heater assembly (Behr) – dismantling and reassembly

1 Remove the distributor flapshaft (1 clip). Note that the flap remains in the lower section of the housing.
2 Remove the clamps securing the two halves of the housing, using circlip pliers. Remove the upper section, complete with the motor, from the lower section.
3 Remove the heat exchanger and from the frame the lower section

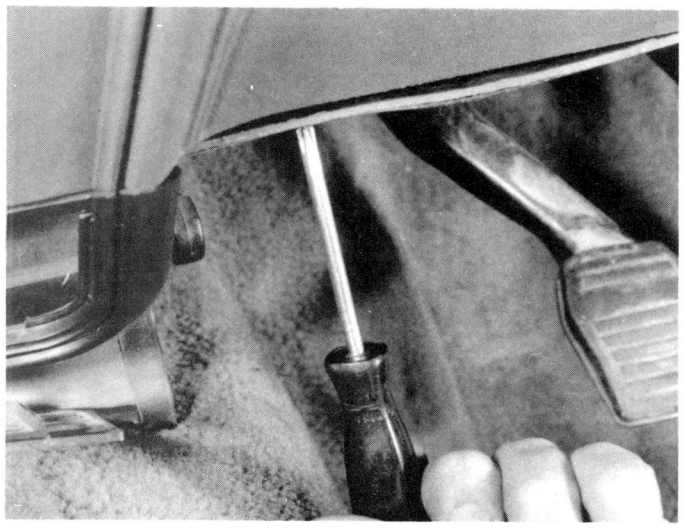

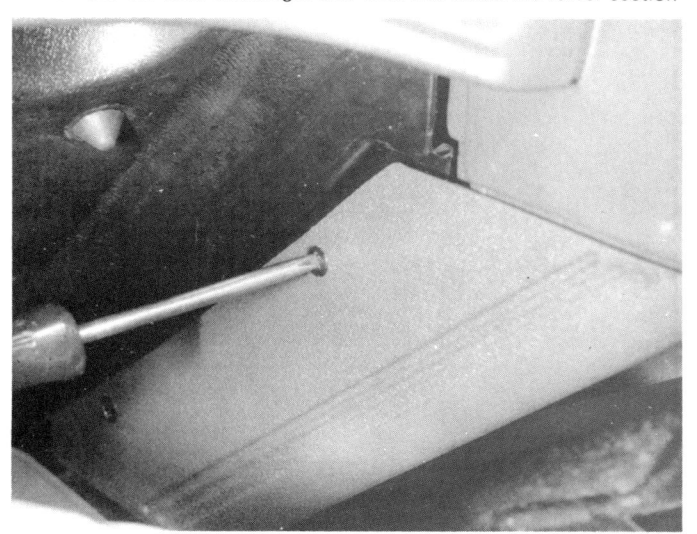

43.8A Removing the lower facia panel ...

43.8B .. and centre cover panel

43.8C Removing the ashtray

43.16 Heater mounting bolt (arrowed)

192

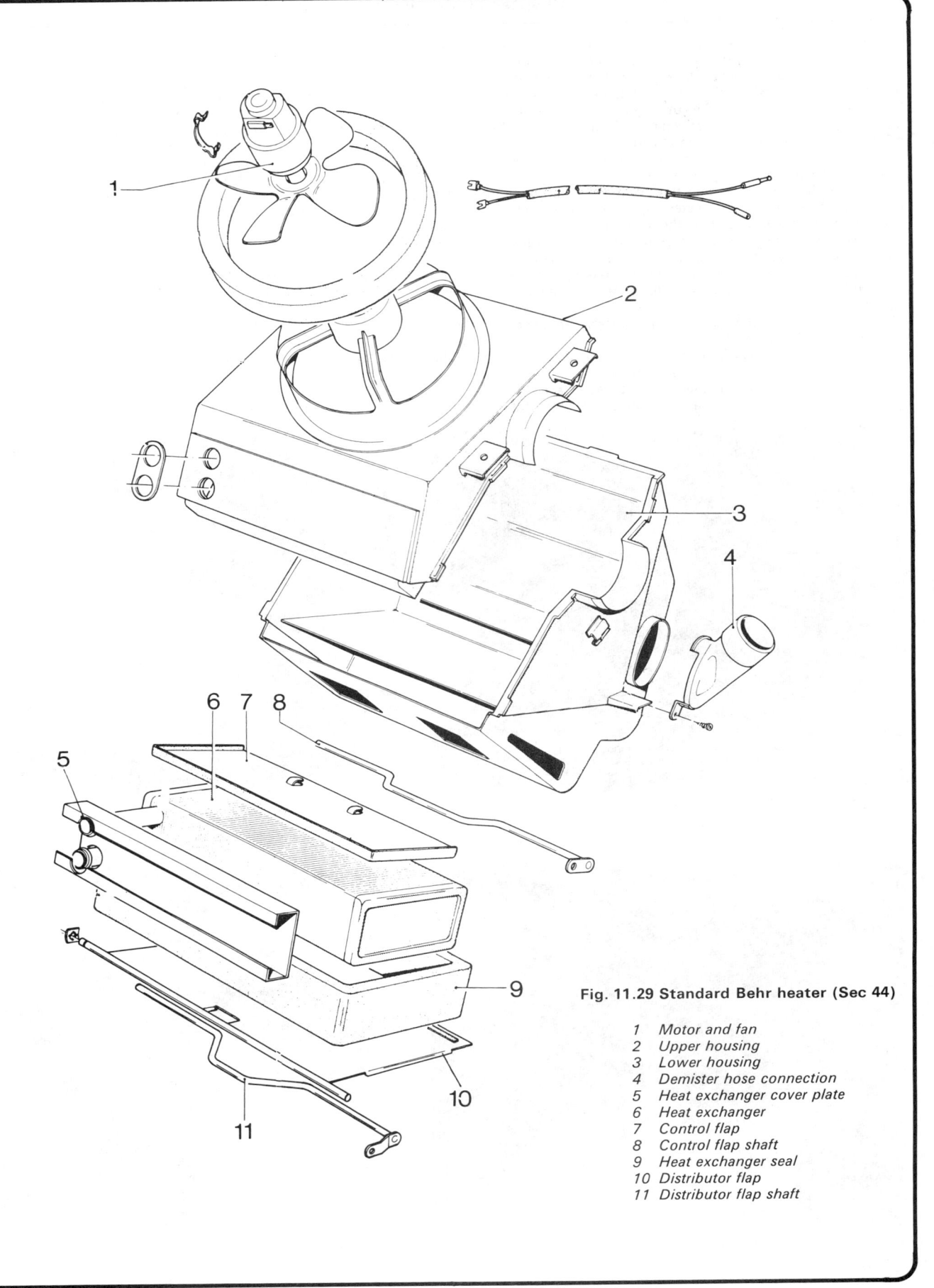

Fig. 11.29 Standard Behr heater (Sec 44)

1 Motor and fan
2 Upper housing
3 Lower housing
4 Demister hose connection
5 Heat exchanger cover plate
6 Heat exchanger
7 Control flap
8 Control flap shaft
9 Heat exchanger seal
10 Distributor flap
11 Distributor flap shaft

Chapter 11 Bodywork and fittings

of the housing, then remove the heat exchanger from the frame and take off the foam packing.

4 To remove the distributor flap from the housing, remove the clip and withdraw the control lever sideways.

5 Remove the regulator flap from the lower section of the housing. Bend back the 2 clamping straps sufficiently to enable the control lever to be withdrawn after it has been turned towards the side, then remove the regulating flap.

6 Remove the retaining straps for the blower motor cap, by pressing outwards from the inside using a screwdriver.

7 Detach the motor from the upper section. Disconnect the motor leads, remove the 4 retaining clamps and remove the motor and fan inwards.

8 When reassembling, position the blower motor so that the electrical connections face towards the cable fastening at the upper section. Secure the motor and connect the leads, then fit the motor cap.

9 Position the regulating flap in the lower section and insert the control lever by turning it from the side as necessary and swing it round into the straps. Close the straps using pliers.

10 Position the distributor flap in the lower section and insert the control lever from the side.

11 The remainder of the reassembly procedure is the reverse of the removal procedure.

45 Heater assembly (Smiths standard and heavy duty) – dismantling and reassembly

Standard heater

1 Remove clips A and B (Fig. 11.32) and remove the heater housing side cover complete with flaps (15 screws).

Heavy duty heater

2 Remove clip A (Fig. 11.32) and remove the heater housing side cover complete with flaps (15 screws).

All heaters

3 Remove the heat-exchanger and foam seal.
4 Prise off the circlip and remove the fan from the blower motor shaft.
5 Detach the blower motor from the support (3 nuts and bolts).
6 Reassembly is the reverse of the dismantling procedure.

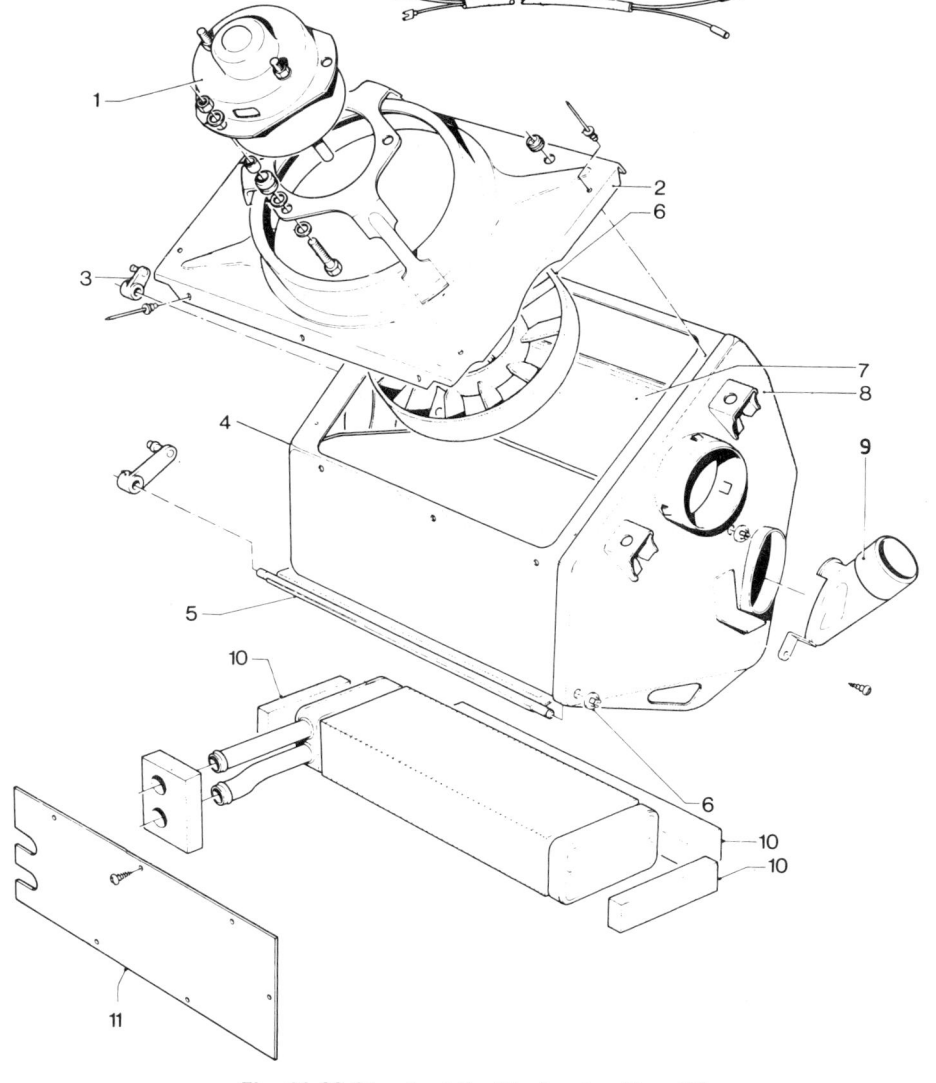

Fig. 11.30 Standard Smiths heater (Sec 45)

1 Motor assembly	4 Right-hand housing cover	7 Control valve	10 Heat exchanger seal
2 Cover and bracket assembly	5 Distributor flap	8 Housing	11 Plenum chamber cover
3 Control valve operating lever	6 Fan	9 Demister hose connection	

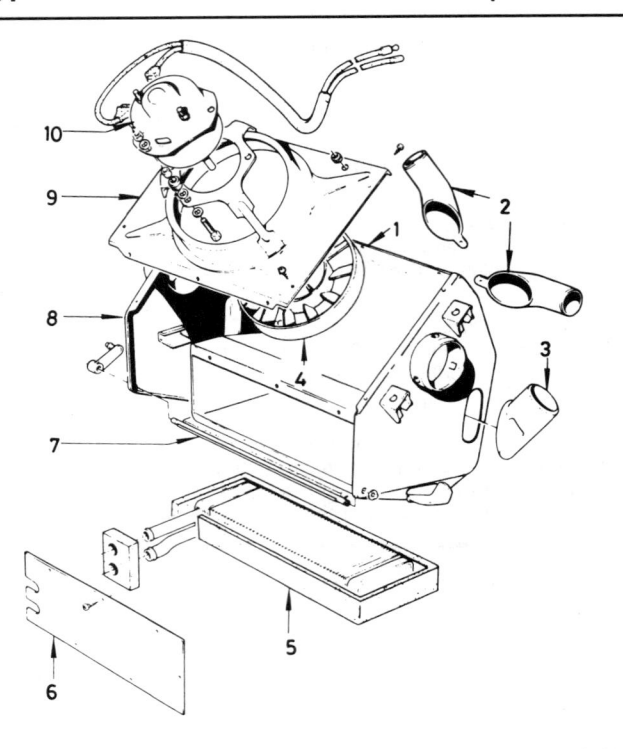

Fig. 11.31 Heavy duty Smiths heater (Sec 45)

1 Housing
2 Hot air supply to facia connection
3 Demister hose connection
4 Fan
5 Heat exchanger seal
6 Plenum chamber cover
7 Distributor flap
8 Right-hand housing cover
9 Cover and bracket assembly
10 Motor assembly

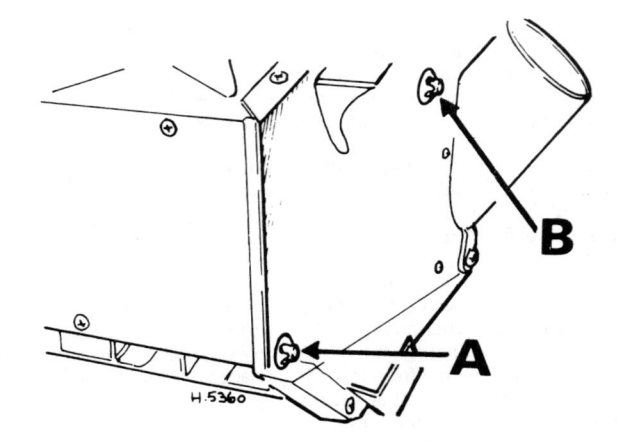

Fig. 11.32 Heater distributor flap (A) and regulator flap (B) clips (Sec 45)

5 Remove the instrument cluster (4 screws), disconnecting the speedometer drive cable and electrical connections. If there is any doubt about the position of any of the electrical connections, make a note of them first of all.
6 Withdraw the demister nozzle hose and remove the demister nozzle (1 screw), turning it upward and outwards so that the inlet side of the nozzle can come out first from the instrument cluster opening.
7 Refitting is the reverse of the removal procedure.

46 Demister nozzles – removal and refitting

Passenger's side
1 Remove the glove compartment by unscrewing the 7 screws at the top and 2 nuts at the bottom. Also disconnect the glove compartment lighting leads.
2 Withdraw the hose from the demister nozzle and remove the nozzle (1 screw).
3 Refitting is the reverse of the removal procedure.

Driver's side
4 Initially proceed as described in paragraphs 1, 2 and 3 of Section 41. Additionally remove the ashtray and cigar lighter.

47 Face level vents – removal and refitting

Passenger's side
1 Remove the glove compartment by unscrewing 7 screws at the top and 2 nuts at the bottom. Also disconnect the glove compartment lighting leads.
2 Withdraw the hose(s) from the vent.
3 Remove the vent by unscrewing the 2 nuts which are accessible from the rear of the panel.
4 Refitting is the reverse of the removal procedure.

Driver's side
5 Initially proceed as described in paragraphs 1, 2 and 3 of Section 41. Additionally remove the 9 screws at the top of the instrument panel trim.
6 With the hose(s) from the vent.
7 Remove the vent by usncrewing the 2 nuts which are accessible from the rear of the panel.
8 Refitting is the reverse of the removal procedure.

Chapter 12 Electrical system

Contents

Alternator – dismantling and reassembly	11
Alternator – removal and refitting	7
Alternator brushes (Bosch) – inspection, removal and refitting	9
Alternator brushes (Femsa) – inspection, removal and refitting	10
Alternator brushes (Lucas) – inspection, removal and refitting	8
Battery – charging	5
Battery – description and testing	4
Battery – removal and refitting	6
Central facia switches – removal and refiting	43
Cigar lighter – removal and refitting	48
Clock (console mounted) – removal and refitting	37
Dim-dip lighting system – general	51
Door pillar switch – removal and refitting	41
Electrical system – precautions	2
Fault diagnosis – electrical system	54
Flasher unit and relays – general	47
Front direction indicator lamp and bulb – removal and refitting	19
Front driving lamp and bulb – removal and refitting	21
Front foglamp and bulb – removal and refitting	20
Fuses	35
General description	1
Handbrake warning light switch – removal and refitting	45
Hazard warning switch – removal and refitting	38
Headlight and auxiliary (driving) light alignment	17
Headlight and sidelight bulbs – renewal	18
Headlight assembly – removal and refitting	16
Horn – fault tracing, removal and refitting	34
Ignition switch and lock – removal and refitting	40
Instrument cluster – removal and refitting	36
Instrument cluster illumination switch – removal and refitting	42
Instrument voltage regulator – removal and refitting	50
Interior lamp and bulb – removal and refitting	24
Light and windscreen wiper switches (Series II models) – removal and refitting	46
Load space lamp switch – removal and refitting	44
Mobile radio equipment – interference-free installation	53
Radio/cassette player – removal and refitting	52
Rear lamp and bulb – removal and refitting	22
Rear number plate lamp and bulb – removal and refitting	23
Routine maintenance	3
Speedometer cable – renewal	49
Starter motor – dismantling and reassembly	15
Starter motor – general description	12
Starter motor – removal and refitting	14
Starter motor – testing on engine	13
Steering column multi-function switch – removal and refitting	39
Tailgate washer pump – removal and refitting	26
Tailgate washer nozzle – removal and refitting	27
Tailgate wiper motor and linkage – removal and refitting	25
Windscreen washer nozzle – removal and refitting	30
Windscreen washer pump – removal and refitting	29
Windscreen wiper arms and blades – removal and refitting	33
Windscreen wiper motor – dismantling and reassembly	32
Windscreen wiper motor and linkage – removal and refitting	28
Windscreen wipers – fault tracing	31

Specifications

System type .... 12 volt, negative earth

Battery
Capacity (amp hour) .... 44, 55 or 66

Alternator (Bosch)

Type	G1-28A	K1-35A	K1-55A
Output at 13.5V and 6000 rpm (Nominal)	28 amp	35 amp	55 amp
Stator winding resistance per phase	0.20 to 0.210 ohm	0.130 to 0.137 ohm	0.010 to 0.017 ohm
Rotor winding resistance at 20°C (68°F)	4.0 to 4.4 ohm	4.0 to 4.4 ohm	4.0 to 4.4 ohm
Minimum brush protrusion in free position	5.0 mm (0.197 in)	5.0 mm (0.197 in)	5.0 mm (0.197 in)
Regulating voltage at 4000 rpm, 3 to 7 amp load	13.7 to 14.5 volt	13.7 to 14.5 volt	13.7 to 14.5 volt

Alternator (Femsa)

Type	ALD 12-32 or ALD 12-33
Output at 13.5V and 6000 rpm (nominal)	32 amp
Stator winding resistance per phase	0.163 to 0.183 ohm
Rotor winding resistance at 20°C (68°F)	4.85 to 5.15 ohm
Minimum brush protrusion in free position	7.0 mm (0.276 in)
Regulating voltage at 4000 rpm, 3 to 7 amp load	13.7 to 14.5 volt

Alternator (Lucas)

Type	15 ACR	17 ACR
Output at 13.5V and 6000 rpm (nominal)	28 amp	35 amp
Stator winding resistance per phase	0.188 to 0.208 ohm	0.126 to 0.140 ohm
Rotor winding resistance at 20°C (68°F)	3.11 to 3.43 ohm	3.04 to 3.36 ohm
Minimum brush protrusion in free position	5.0 mm (0.197 in)	5.0 mm (0.197 in)
Regulating voltage at 4000 rpm, 3 to 7 amp load	14.2 to 14.6 volt	14.2 to 14.6 volt

Starter motor (Bosch)
Type .... EF 0.7, GF 1.0 or GF 1.2
Minimum brush length .... 10.0 mm (0.39 in)
Brush spring pressure .... 900 to 1300 gf (32 to 46 ozf)
Commutator:
    Minimum diameter .... 32.8 mm (1.29 in)
    Maximum out-of-round .... 0.3 mm (0.012 in)
Armature endfloat .... 0.1 to 0.3 mm (0.004 to 0.012 in)

Starter motor (Lucas)

	M35J	5M90	2M100
Type			
Minimum brush length	8.0 mm (0.31 in)	8.0 mm (0.31 in)	9.5 mm (0.37 in)
Brush spring pressure	480 gf (17 ozf)	850 gf (30 ozf)	480 gf (17 ozf)
Commutator:			
Minimum diameter	Not specified	34.0 mm (1.34 in)	Not specified
Maximum out-of-round	Not specified	0.075 mm (0.003 in)	Not specified
Armature endfloat	0.1 to 0.3 mm (0.004 to 0.012 in)	0.1 to 0.3 mm (0.004 to 0.012 in)	0.1 to 0.3 mm (0.004 to 0.012 in)

Fuses (Series II)

Main fusebox	Rating	Circuits protected
1	16 amp	Cigar lighter, clock, interior light
2	8 amp	Number plate light, instrument panel illumination
3	8 amp	RH side and tail lights
4	8 amp	LH side and tail lights
5	16 amp	Horn, heater fan motor
6	16 amp	Wiper motor, reversing lights
7	8 amp	Direction indicators, stop-lights, instrument cluster

Fuses in dipper relay housing

8	16 amp	LH dipped headlamp
9	16 amp	RH dipped headlamp
10	16 amp	RH main beam
11	16 amp	LH main beam

Fuses under facia

12	8 amp	Heated rear window (within relay)
13	2 amp	Radio (medium-slow blow)
14	8 amp	Driving lamps (within relay)
15	8 amp	Foglamps (within relay)

Fuses (Series III)

Main fusebox	Rating	Circuits protected
1	16 amp	Interior light(s), hazard flashers, cigar lighter, clock, horn, windscreen washers, headlamp washers
2	8 amp	Number plate lights, glovebox light, instrument illumination
3	8 amp	RH side and tail lights
4	8 amp	LH side and tail lights
5	8 amp	Heater fan motor
6	16 amp	Reversing lights, front and rear wiper motors, tailgate washer
7	8 amp	Stop-lights, instrument cluster

Fuses beneath headlamp relay

8	16 amp	RH main beam
9	16 amp	LH main beam
10	16 amp	LH dipped beam
11	16 amp	RH dipped beam

Bulbs (typical)

	Wattage
Headlamp:	
Standard	45/40
Halogen	60/55
Foglamp	55
Driving lamp	55
Direction indicators	21
Stop-lamp	21
Tail lamp	5
Reversing lamp	21
Sidelight and number plate light	4
Interior light	10
Instrument panel warning and illumination lights	2 or 3
Clock	1.2 or 1.4
Heated rear window switch	1.2
Auotmatic transmission selector light	1.4
Cigar lighter light	1.4
Glovebox light	2

Torque wrench settings

	Nm	lbf ft
Alternator	20 to 25	15 to 18
Starter motor	20 to 25	15 to 18

Chapter 12 Electrical system

1 General description

The electrical system is of the 12 volt negative earth type and the major components are a 12 volt battery of which the negative terminal is earthed, an alternator which is belt-driven from the crankshaft pulley, and a starter motor.

The battery supplies a steady amount of current for the ignition, lighting and other electrical circuits and provides a reserve of electricity when the current consumed by the electrical equipment exceeds that being produced by the alternator.

The alternator has its own regulator which ensures a high output if the battery is in a low state of charge or the demand from the electrical equipment is high, and a low output if the battery is fully charged and there is little demand from the electrical equipment.

The starter motor is of the pre-engaged type, incorporating an integral solenoid. On starting, the solenoid moves the drive pinion into engagement with the flywheel/driveplate ring gear before the starter motor is energised. Once the engine has started, a one-way clutch prevents the motor armature being driven by the engine.

2 Electrical system – precautions

It is necessary to take extra care when working on the electrical system to avoid damage to semiconductor devices (diodes and transistors), and to avoid the risk of personal injury. In addition to the precautions given in Safety First! at the beginning of this manual, observe the following items when working on the system.
1 *Always remove rings, watches, etc before working on the electrical system.* Even with the battery disconnected, capacitive discharge could occur if a component live terminal is earthed through a metal object. This could cause a shock or nasty burn.
2 *Do not reverse the battery connections.* Components such as the alternator or any other having semiconductor circuitry could be irreparably damaged.
3 If the engine is being started using jump leads and a slave battery, connect the batteries *positive to positive and negative to negative.* This also applies when connecting a battery charger.
4 Never disconnect the battery terminals, or alternator wiring when the engine is running.
5 The battery leads and alternator wiring must be disconnected before carrying out any electric welding on the car.

3 Routine maintenance

At the intervals specified in the Routine Maintenance section at the front of the manual carry out the following procedures.

Check and if necessary top up the washer fluid reservoirs
1 Refer to Sections 26 and 29.

Check the operation of all electrical equipment
2 Check the operation of the external lights, stop-lights, indicators, horn, wipers, washers, heater, internal lights, switches, instruments etc. Refer to the appropriate Section of the Chapter if any components are found to be inoperative.

Check drivebelt condition and tension
3 Refer to Chapter 2.

Check battery electrolyte level (except maintenance-free batteries)
4 Remove the cell cover or caps and check that the cell plates are covered by approximately 6.0 mm (0.25 in) of electrolyte.
5 If necessary top up each cell using distilled water (photo).

4 Battery – description and testing

1 The battery fitted to models manufactured before August 1982 requires regular checking of the electrolyte level, as described in Section 3. However, later batteries are generally of the maintenance-

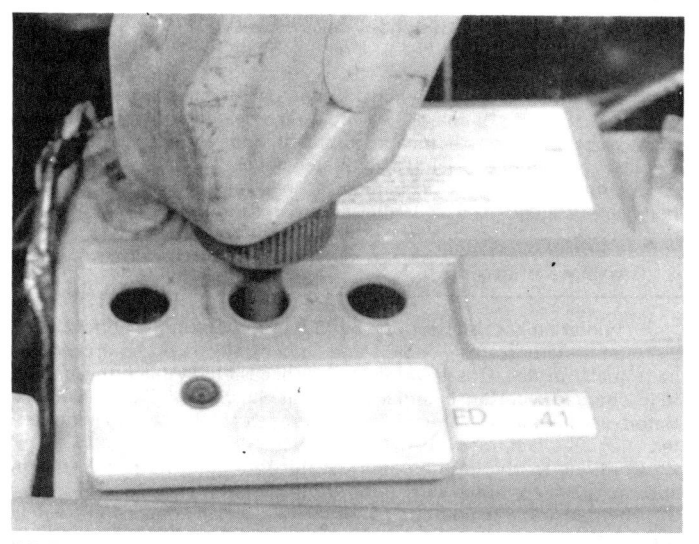

3.5 Topping-up the battery with distilled water

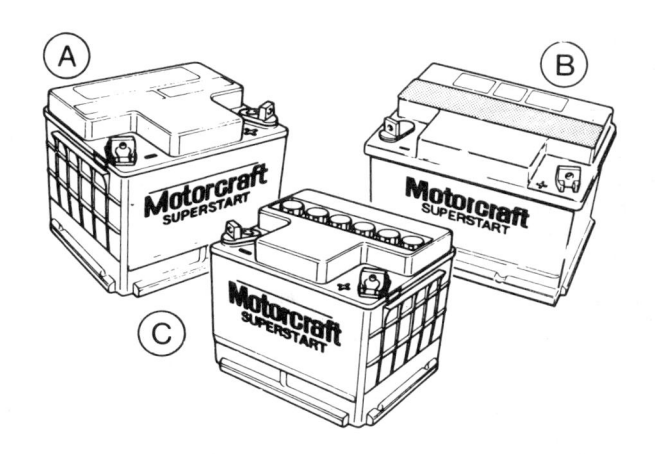

Fig. 12.1 Battery types (Sec 4)

A Maintenance-free (sealed)
B Maintenance-free (removable top)
C Conventional

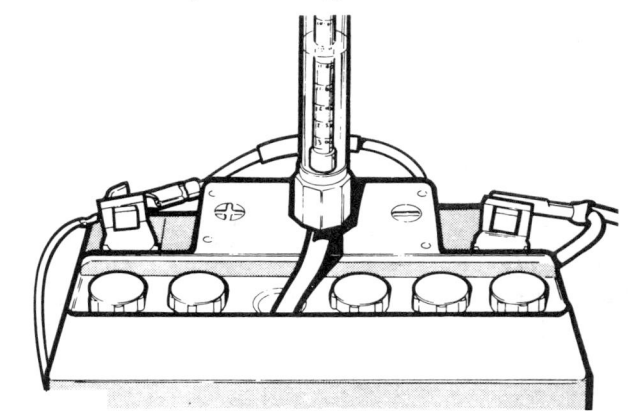

Fig. 12.2 Using a hydrometer to check the specific gravity of the electrolyte (Sec 4)

free type, where the antimony content of the cell plates is substantially less, virtually eliminating the gassing characteristics which lower the electrolyte level. Maintenance-free batteries may be of the sealed type or may incorporate a removable cell top.
2 The state of charge of the battery is best determined by checking the specific gravity of the electrolyte with a hydrometer (Fig. 12.2). For

a fully charged battery this should be between 1.270 and 1.290 at a temperature of 25°C (77°F). For each 6°C (11°F) below this temperature deduct 0.004 from the SG reading, and *vice versa*.
3 If the battery is of sealed construction use the discharge method to determine the state of charge. Switch on the headlamps for 30 seconds then switch off all electrical equipment and allow the battery to stabilise for 5 minutes. Using a voltmeter, check the battery voltage. If it is less than 12.2 volts the battery is deeply discharged, but if it is between 12.2 and 12.5 volts the battery is partially discharged.

5 Battery – charging

1 In winter time when heavy demand is placed upon the battery, such as when starting from cold, and much electrical equipment is continually in use, it is a good idea to occasionally have the battery fully charged from an external source at the rate of 3.5 to 4 amps. Before connecting the charger, both battery leads must be disconnected (negative first, then positive).
2 Continue to charge the battery at this rate, until no further rise in specific gravity is noted over a four hour period.
3 Alternatively, a trickle-charger at the rate of 1.5 amps can be safely used overnight.
4 Specially rapid 'boost' charges, which are claimed to restore the power of the battery in 1 to 2 hours, are not recommended as they can cause serious damage to the battery plates through overheating.
5 While charging the battery note that the temperature of the electrolyte should not exceed 37.8°C (100°F).

6 Battery – removal and refitting

1 The battery is located at the front of the engine compartment either on the left or right according to models.
2 First disconnect the negative (earth) lead by unscrewing the nut and pulling the lead from the terminal post.
3 Next disconnect the two positive leads in the same manner. Note that the heavy cable is nearest to the terminal post.
4 Unscrew the bolt securing the battery clamp plate to the carrier (photo). Lift away the clamp plate. Carefully lift the battery from its carrier, holding it vertically to ensure that none of the electrolyte is spilled.
5 Refitting is a direct reversal of this procedure. **Note:** *Refit the positive lead before the negative lead and smear the terminals with petroleum jelly to prevent corrosion.* **Never** *use an ordinary grease.*

7 Alternator – removal and refitting

1 Disconnect the battery.
2 Disconnect the wiring from the rear of the alternator (photo).

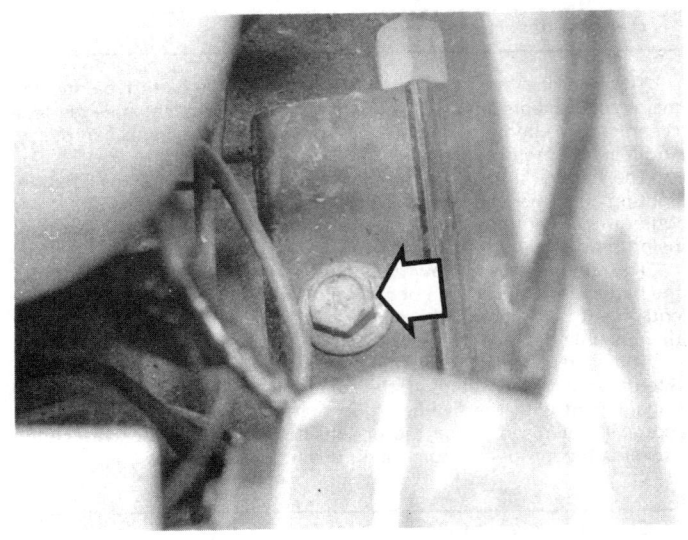

6.4 Battery clamp and bolt (arrowed)

7.2 Wiring to the rear of the alternator

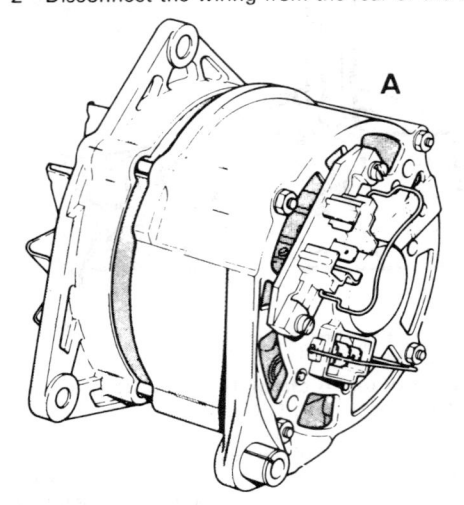

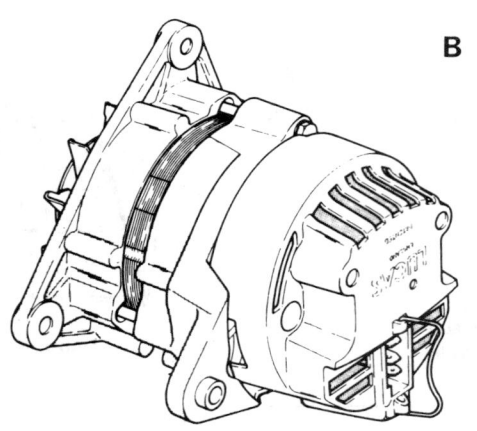

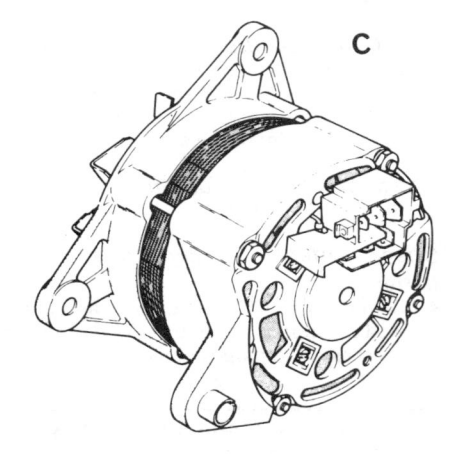

Fig. 12.3 Alternator types (Secs 7 to 11)

A Bosch   B Lucas   C Femsa

Chapter 12 Electrical system

7.3A Alternator pivot bolt (arrowed) ...

7.3B ... and adjustment bolt (arrowed) (2.8 litre engine shown)

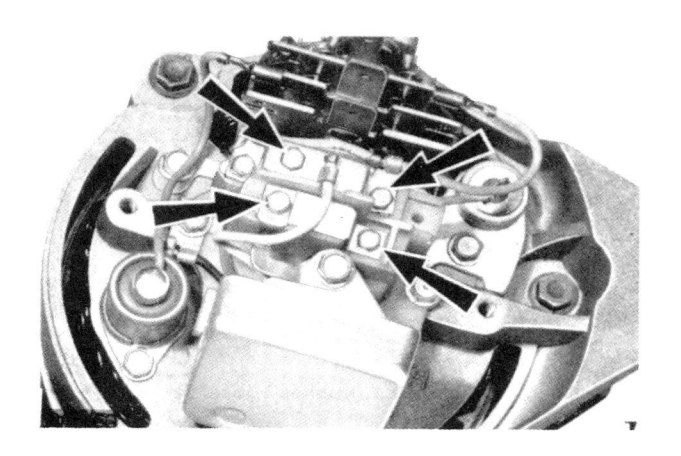

Fig. 12.4 Brush box retaining screws – arrowed (Lucas) (Sec 8)

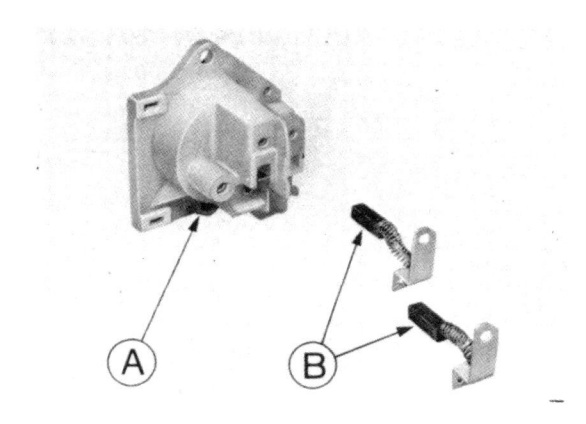

Fig. 12.5 Alternator brushes (Lucas) (Sec 8)

A Brush box   B Brushes

3 Loosen the pivot and adjustment bolts (photos), swivel the alternator towards the engine and slip the drivebelt(s) from the pulley.
4 Remove the bolts and withdraw the alternator from the engine.
5 Refitting is a reversal of removal but tension the drivebelt(s) as described in Chapter 2.

8 Alternator brushes (Lucas) – inspection, removal and refitting

1 Undo and remove the two screws and washer securing the end cover.
2 To inspect the brushes, the brush holder mountings should be removed complete by undoing the two bolts and disconnecting the Lucar connection to the diode plates (Figs. 12.4 and 12.5).
3 With the brush holder moulding removed and the brush assemblies still in position, check that they protrude from the face of the moulding by at least 5.0 mm (0.197 in). Also check that when depressed, the spring pressure is 198 to 284 gf (7 to 10 ozf) when the end of the brush is flush with the face of the brush moulding. To be done with any accuracy, this requires a push type spring gauge.
4 Should either of the foregoing requirements not be fulfilled, the brush assemblies should be renewed.

5 This can be done by simply removing the holding screws of each assembly and fitting a new assembly.
6 With the brush holder moulding removed, the slip rings on the face end of the rotor are exposed. These can be cleaned with a petrol-soaked cloth and any signs of burning may be removed with fine glass paper. On no account should any other abrasive be used, or any attempt at machining be made.
7 When the brushes are refitted, they should slide smoothly in their holders. Any sticking tendency may first be rectified by wiping with a petrol-soaked cloth or, if this fails, by carefully polishing with a very fine file where any binding marks appear.
8 Reassemble in the reverse order of dismantling.

9 Alternator brushes (Bosch) – inspection, removal and refitting

1 Undo and remove the two screws, spring and plain washers that secure the brush box to the rear of the brush end housing. Lift away the brush box (photos).
2 Check that the carbon brushes are able to slide smoothly in their guides without any sign of binding.
3 Measure the length of brushes and if they have worn down to 5.0

Chapter 12 Electrical system

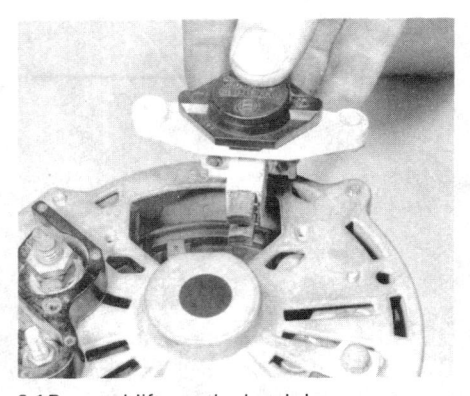

9.1A Remove the screws ...

9.1B ... and lift out the brush box

9.1C Brush box removed from the alternator

mm (0.197 in) or less, they must be renewed (photo).
4 Hold the brush wire with a pair of engineer's pliers and unsolder it from the brush box. Lift away the two brushes.
5 Insert the new brushes and check to make sure that they are free to move in their guides. If they bind, lightly polish with a very fine file.
6 Solder the brush wire ends to the brush box, taking care that solder is allowed to pass to the stranded wire.
7 Whenever new brushes are fitted, new springs should also be fitted.
8 Refitting the brush box is the reverse sequence to removal.

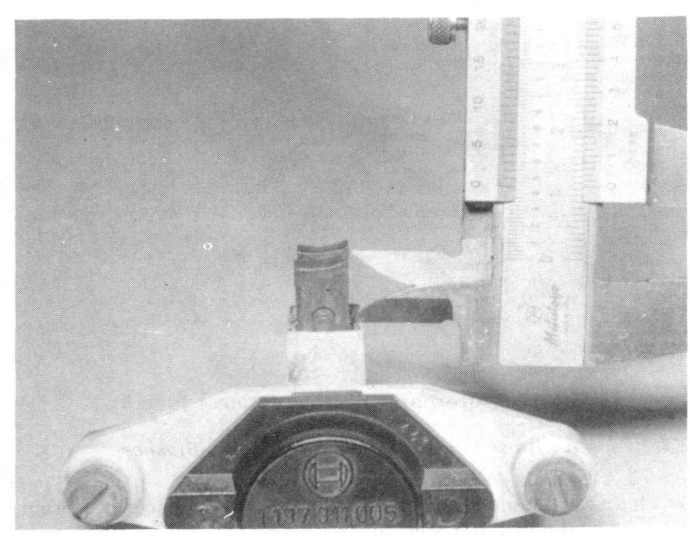

9.3 Checking the length of the brushes

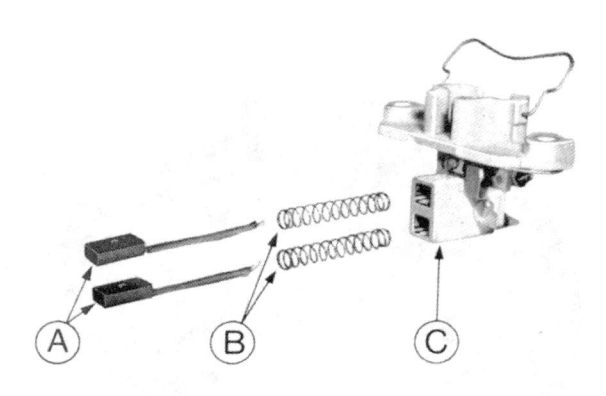

Fig. 12.6 Alternator brushes (Bosch) (Sec 9)

A Brushes
B Springs
C Brush box

10 Alternator brushes (Femsa) – inspection, removal and refitting

1 Disconnect the single wire from the brush box.
2 Remove the cross-head retaining screw, then withdraw the brush box.
3 Check that the carbon brushes are able to slide smoothly in their guides without any sign of binding.
4 Measure the amount by which the brushes protrude from the brush box. If this is less than 7.0 mm (0.276 in), obtain and fit new brushes.
5 Refitting the brush box is a straightforward reversal of the removal procedure.

11 Alternator – dismantling and reassembly

Note: *The procedure for Lucas and Femsa alternators is similar to the*

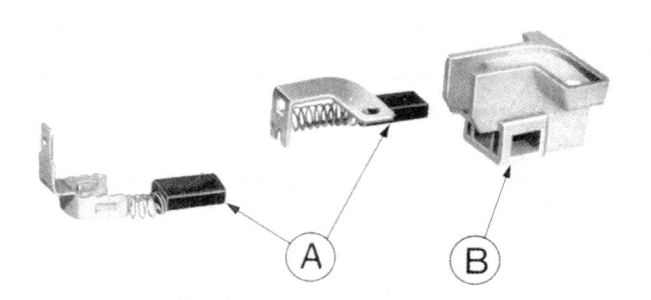

Fig. 12.7 Alternator brushes (Femsa) (Sec 10)

A Brushes
B Brush box

Chapter 12 Electrical system

procedure described in this Section for Bosch alternators but refer to Figs. 12.9 and 12.10.

1 Unscrew the pulley retaining nut while gripping the pulley in a vice (photo).

2 Remove the spring washer, pulley, Woodruff key, fan and washer(s) (photo). Note that the raised shoulder on the washer faces outwards.

3 Remove the brush box (Section 8, 9 or 10).

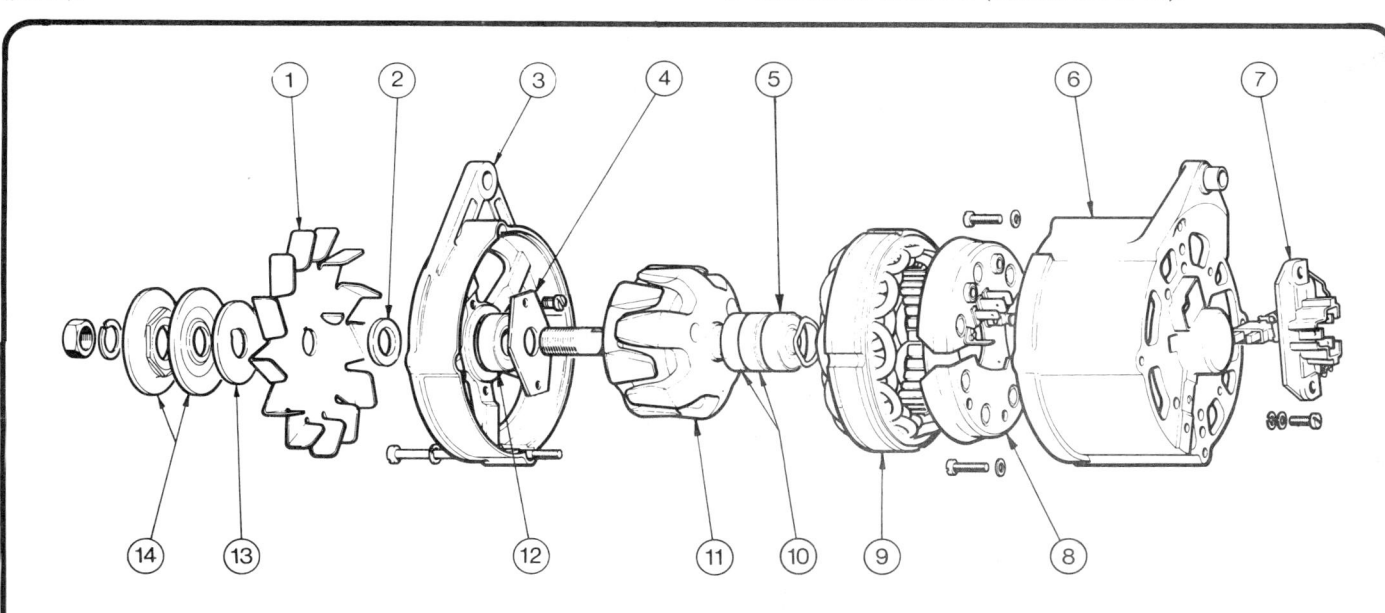

Fig. 12.8 Exploded diagram of the Bosch alternator (Sec 11)

1 Fan
2 Spacer
3 Drive end housing
4 Thrust plate
5 Slip ring end bearing
6 Slip ring end housing
7 Brush box
8 Rectifier (diode) pack
9 Stator assembly
10 Slip rings
11 Rotor
12 Drive end housing
13 Spacer
14 Pulley

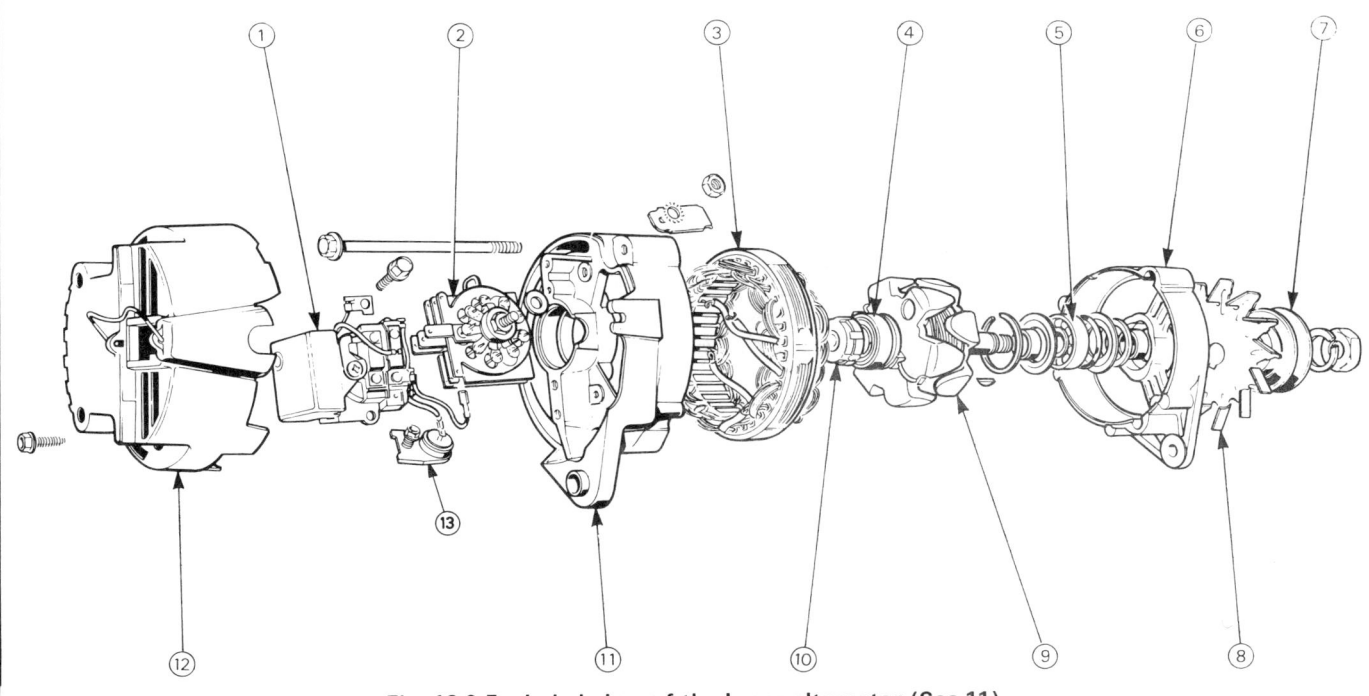

Fig. 12.9 Exploded view of the Lucas alternator (Sec 11)

1 Regulator
2 Rectifier (diode) pack
3 Stator assembly
4 Slip ring end bearing
5 Drive end bearing
6 Drive end housing
7 Pulley
8 Fan
9 Rotor
10 Slip ring
11 Slip ring end housing
12 End cover
13 Surge protection diode

11.1 Unscrew the pulley retaining nut ...

11.2A ... and remove the spring washer ...

11.2B ... pulley ...

11.2C ... Woodruff key ...

11.2D ... fan ...

11.2E ... and washer(s)

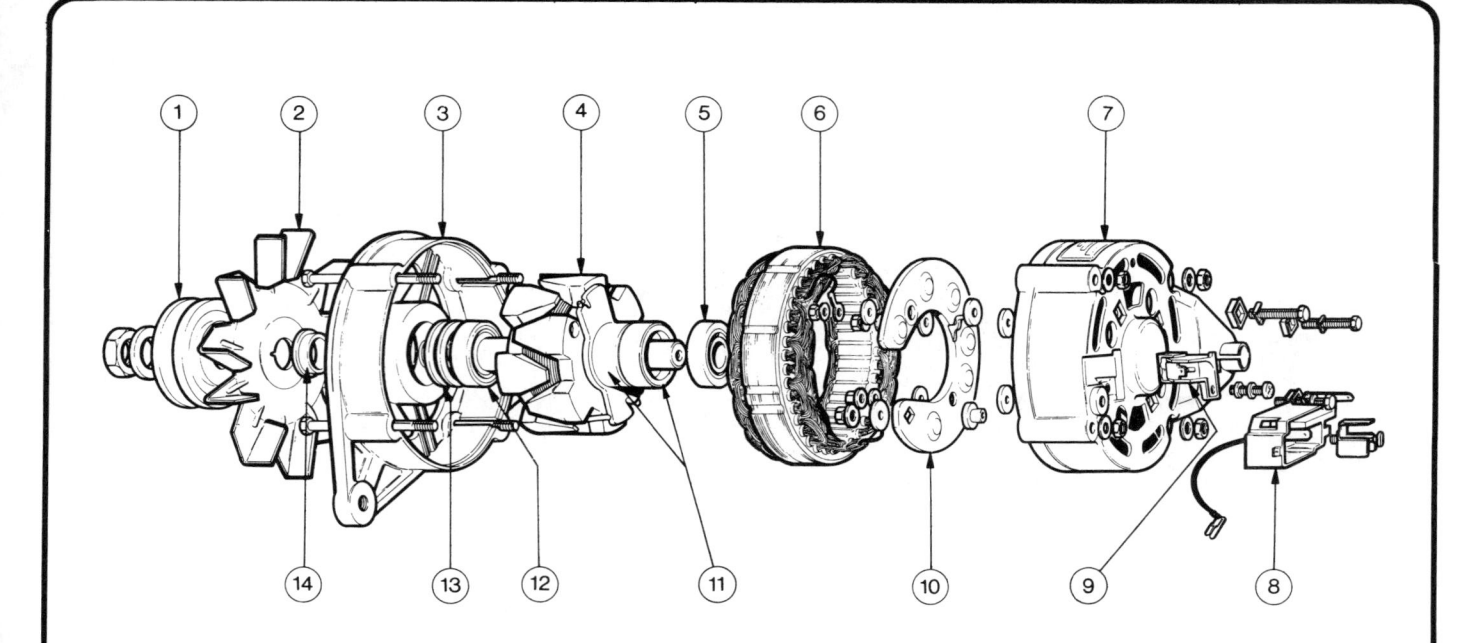

Fig. 12.10 Exploded diagram of the Femsa alternator (Sec 11)

1 Pulley
2 Fan
3 Drive end housing
4 Rotor
5 Slip ring end bearing
6 Stator assembly
7 Slip ring end housing
8 Terminal block
9 Brush box
10 Rectifier (diode) pack
11 Slip rings
12 Drive end bearing
13 Thrust washers
14 Spacer

Chapter 12 Electrical system

11.4 Removing the through-bolts

11.5 Separating the drive end housing and rotor from the slip ring end housing

4 Unscrew the through-bolts (photo).
5 Mark the end housings in relation to each other then separate the drive end housing and rotor from the slip ring end housing (photo).
6 If necessary, press the rotor from the bearing at this stage, then unscrew the end cap screws and remove the bearing (photo).
7 Using a puller, pull the bearing from the slip ring end of the rotor.
8 Unscrew the nuts from the terminals and remove the plastic cover. The stator and rectifier pack can now be removed from the slip ring end housing. Disconnect the suppressor wire where applicable, and note the location of the plastic terminal sleeves (photo).
9 The rectifier pack may be unsoldered from the stator but heat must be prevented from reaching the diodes by using pliers as a heat sink.
10 Clean all the components, with particular attention to the slip rings.
11 Reassembly is a reversal of dismantling.

11.6 Bearing end cap and screws in the drive end housing

11.8A Unscrew the nuts ...

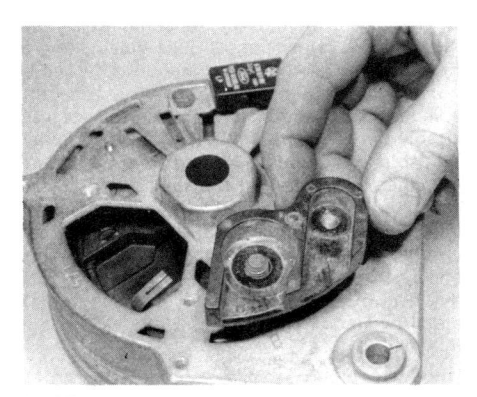

11.8B ... remove the plastic cover ...

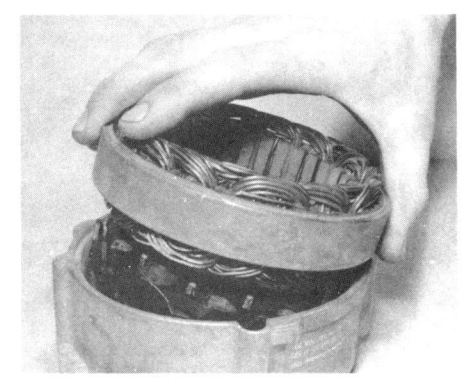

11.8C ... then separate the stator and rectifier pack from the slip ring end housing ...

11.8D ... and remove the plastic sleeves

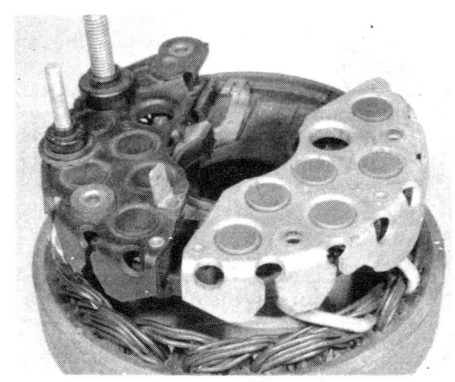

11.8E Stator and rectifier pack

Chapter 12 Electrical system

12 Starter motor – general description

All engines are fitted with pre-engaged starter motors of either Lucas or Bosch manufacture. These motors are all of series wound, four pole, four brush design and are fully serviceable.

The operation of the starter motor is as follows: With the ignition switched on, current flows to the solenoid which is mounted on the top of the starter motor body. Contained within the solenoid is a plunger which moves inward so causing a centrally pivoted engagement lever to move in such a manner that the forked end pushes the drive pinion into mesh with the starter ring gear. When the solenoid plunger reaches the end of its travel, it closes an internal contact and full starting current flows to the starter field coils. The armature is then able to rotate the crankshaft, so starting the engine.

A one way clutch is fitted to the starter drive pinion so that when the engine fires and starts to operate on its own, it does not drive the starter motor.

13 Starter motor – testing on engine

1 If the starter motor fails to operate, check the condition of the battery by turning on the headlamps. If they glow brightly for several seconds then gradually dim, the battery is in an uncharged condition.
2 If the headlights continue to glow brightly and it is obvious that the battery is in good condition, check the tightness of the battery wiring connections (and in particular the earth lead from the battery terminal to its connection on the body frame). If a terminal on the battery becomes hot when an attempt is made to work the starter this is a sure sign of a poor connection. To rectify, remove the terminal, clean the mating faces thoroughly and reconnect. Check the connections on the rear of the starter solenoid. Check the wiring with a meter, or test lamp, for breaks or shorts.
3 Test the continuity of the solenoid windings by connecting a test lamp circuit, consisting of a 12 volt battery and a low wattage bulb, between the 'STA' terminal and the solenoid body. If the two windings are in order the lamp will light. Next connect the test lamp between the solenoid main terminals. Energise the solenoid by applying a 12 volt supply between the unmarked Lucar terminal and the solenoid body. The solenoid should be heard to operate and the test bulb light. This indicates full closure of the solenoid contacts.
4 If the battery is fully charged, the wiring in order, the starter/ignition switch working and the starter motor still fails to operate, it will have to be removed from the car for examination. Before this is done ensure that the starter motor pinion has not jammed in mesh with the flywheel by engaging a gear (not automatic) and rocking the car to and fro. This should free the pinion if it is stuck in mesh with the flywheel teeth.

14 Starter motor – removal and refitting

1 Apply the handbrake then jack up the front of the car and support on axle stands.
2 Disconnect the battery.
3 Disconnect the wiring from the starter solenoid (photo).
4 Unbolt the starter motor from the transmission and remove it from under the car (photos).
5 Refitting is a reversal of removal.

15 Starter motor – dismantling and reassembly

Note: *The procedure for Lucas starter motors is similar to the procedure described in this Section for Bosch starters but refer to Fig. 12.12.*

1 Clamp the starter motor in a vice with protective soft jaws. Undo and remove the nut and washer securing the field winding cable to the

14.3 The wiring on the starter solenoid

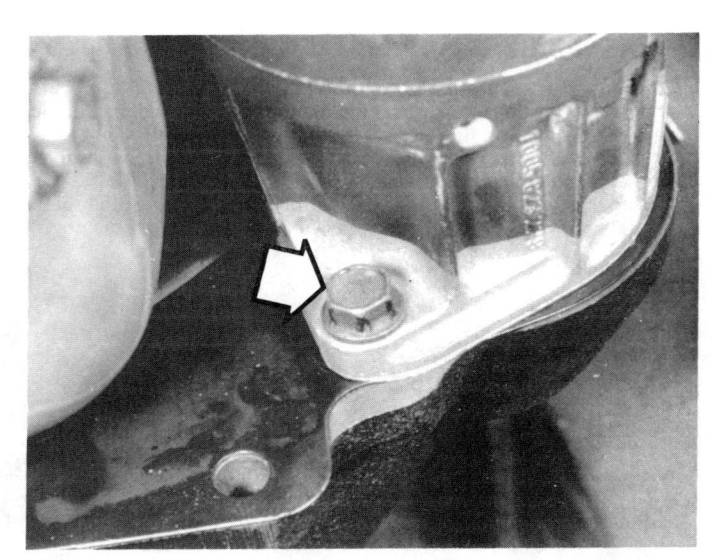

14.4A Unscrew the bolts (arrowed) ...

14.4B ... and remove the starter motor

Chapter 12 Electrical system

solenoid and detach the cable from the stud (photo).
2 Undo and remove the two screws securing the solenoid to the starter motor body (photo). Unhook the solenoid from the actuating arm and lift clear. Note the solenoid and armature are a complete unit and cannot be dismantled.
3 Take out the two screws that secure the commutator end housing cover and remove the cover and rubber seal (photos).
4 Wipe clean the exposed end of the armature shaft and prise off the circlip and shims (photos).
5 Unscrew and remove the two through-bolts and withdraw the end cover (photos).
6 The brushes and plate may now be removed from the motor. Carefully lift up the brush tensioning springs and pull the brushes clear of their guides. Now carefully slide the brush plate from its location on the armature (photos).
7 Separate the armature and drive end housing from the starter motor body by gently tapping the end housing (photo).
8 Remove the rubber insert and plate from the drive end housing (photo).
9 Unscrew the nut and remove the actuating arm pivot screw.
10 Withdraw the armature assembly and actuating arm from the drive end housing then remove the arm (photos).
11 The drive pinion assembly may be removed from the armature by tapping the thrust collar down the shaft with a suitable tube. This will expose the securing C-clip which may be withdrawn allowing the thrust collar and drive pinion to be slid off the end of the armature shaft (photo). **Note.** *Do not grip the one-way clutch during this operation as it is easily damaged.*
12 Carefully inspect the dismantled components of the starter motor for wear and damage. Check the brushes for sticking in their holders. If necessary clean the brushes and brush plate with a petrol-soaked cloth. If the brushes are worn down to the specified minimum or less they should be renewed as a set. To renew the brushes, cut the leads of the old brushes at a point midway between the brush and its base. Solder the leads of the new brushes at this point. Clean the commutator with a petrol-soaked rag and inspect for burning and scoring (photo). If cleaning the commutator with petrol fails to remove all the burnt areas and spots, then wrap a piece of glass paper round

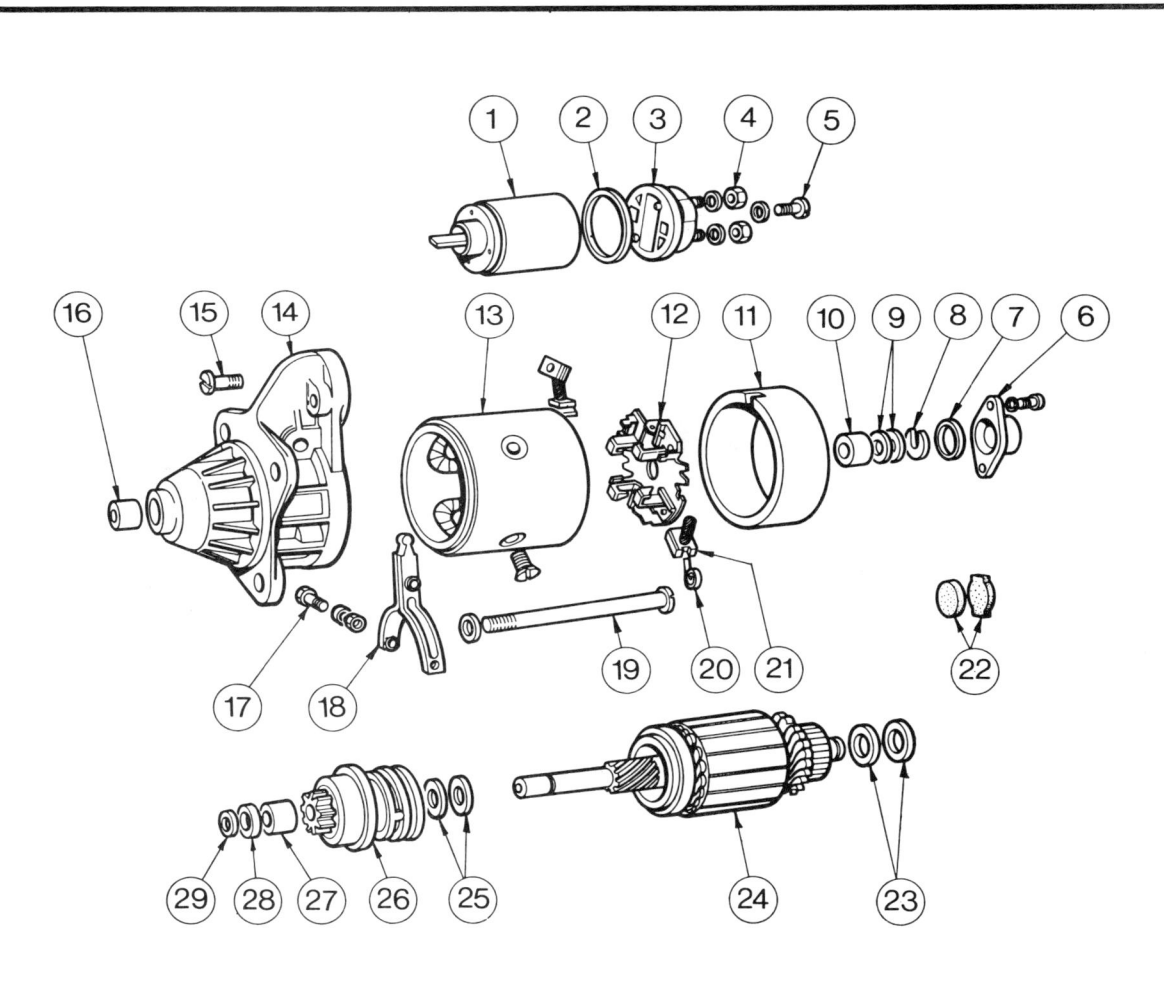

Fig. 12.11 Exploded diagram of the Bosch starter motor (Sec 15)

1 Solenoid assembly
2 Packing ring
3 Switch contacts and cover
4 Nut
5 Screw
6 Cover
7 Washer
8 Circlip
9 Thrust washer
10 Commutator end bearing
11 Commutator end housing
12 Brush plate
13 Yoke
14 Drive end housing
15 Screw
16 Bush
17 Pivot pin
18 Pivot lever
19 Through-bolt
20 Brush spring
21 Brush
22 Lubricating pads
23 Thrust washers
24 Armature assembly
25 Packing rings
26 Pinion
27 Bush
28 Stop ring
29 Stop ring

15.1 Disconnecting the field winding cable

15.2 Removing the solenoid screws

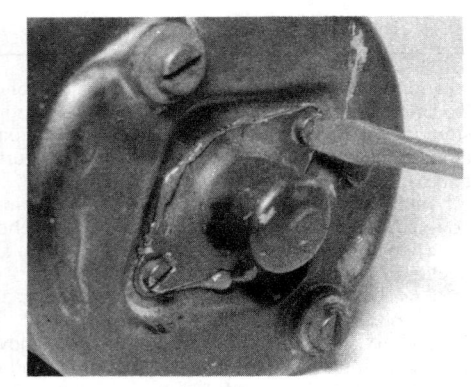

15.3A Remove the screws ...

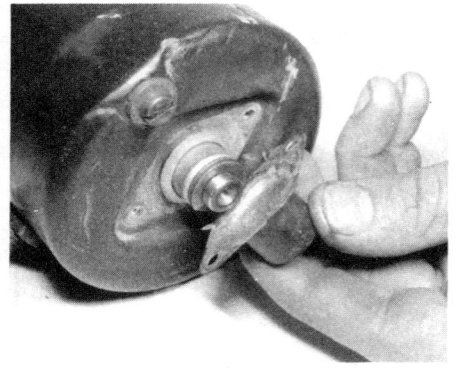

15.3B ... and withdraw the end cover

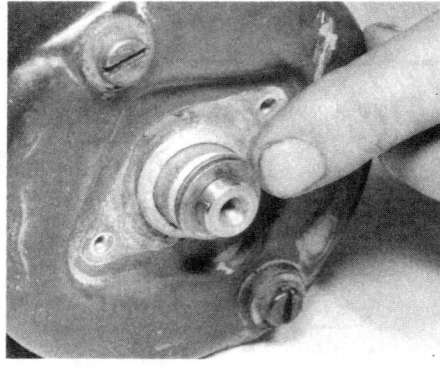

15.4A Circlip located in the groove

15.4B Removing the circlip ...

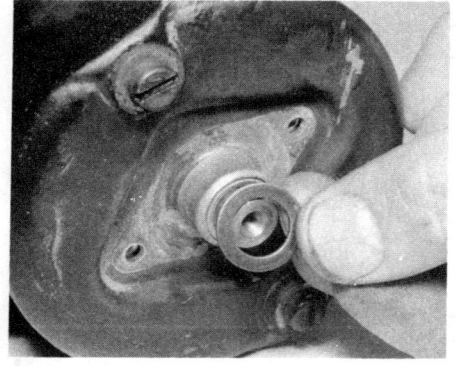

15.4C ... and shim

15.5A Remove the through-bolts ...

15.5B ... and withdraw the end cover

15.6A Starter motor brush (A) and tension spring (B)

15.6B Brush plate assembly

15.7 Separating the armature and drive end housing from the starter motor body

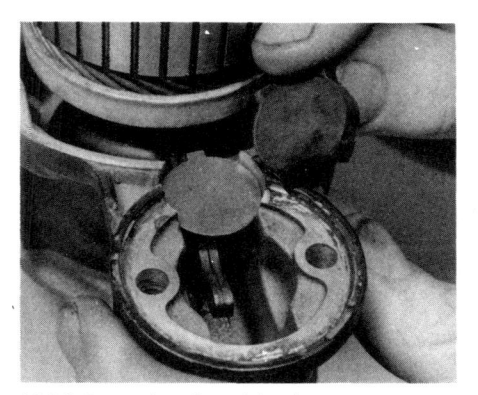

15.8A Removing the rubber insert ...

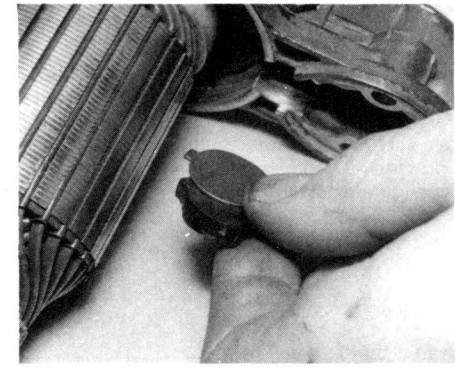

15.8B ... and plate

15.10A Removing the armature assembly and actuating arm

15.10B Armature assembly removed from the starter motor

15.11 Thrust collar moved to expose the C-clip

15.12 Starter commutator

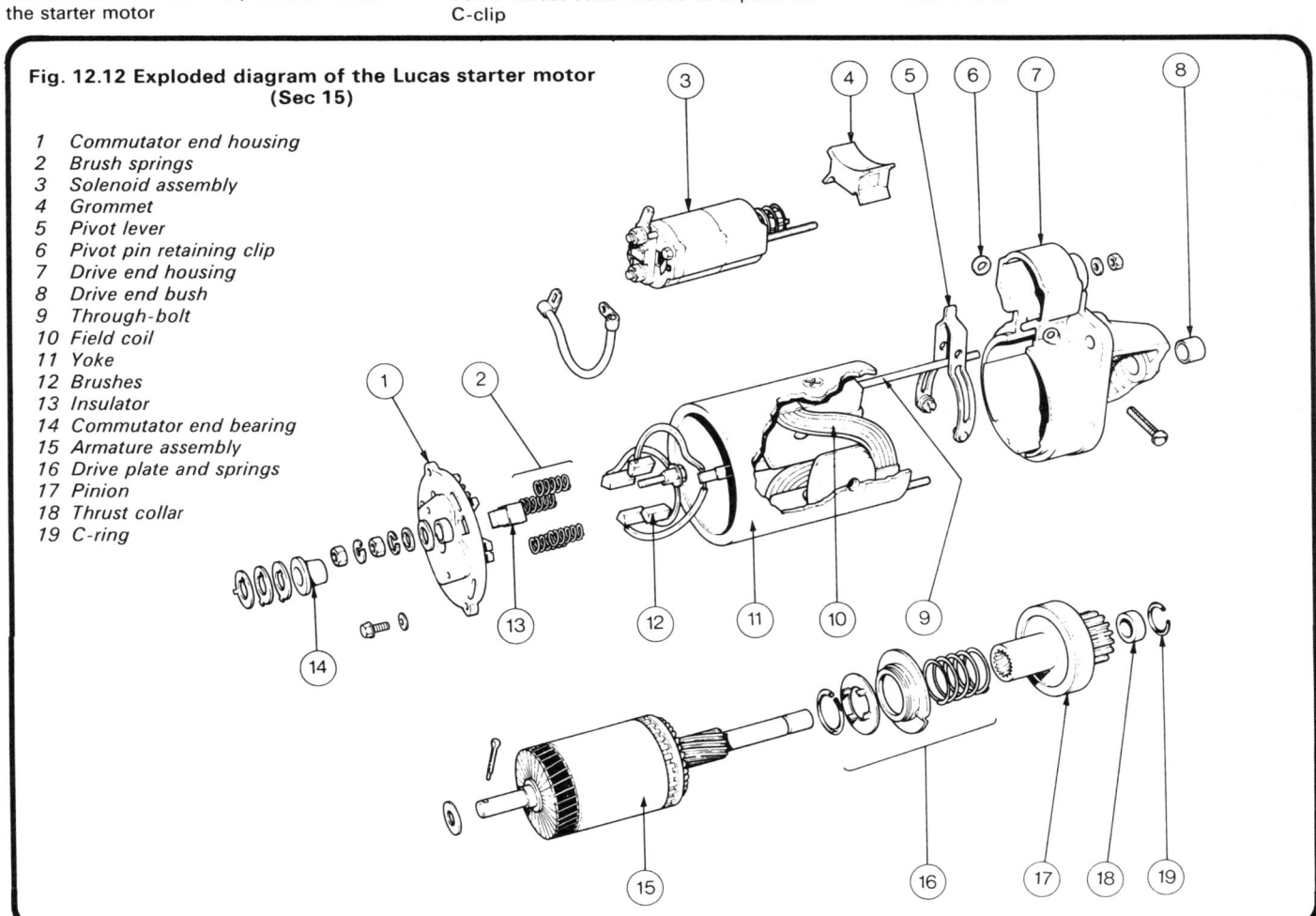

Fig. 12.12 Exploded diagram of the Lucas starter motor (Sec 15)

1 Commutator end housing
2 Brush springs
3 Solenoid assembly
4 Grommet
5 Pivot lever
6 Pivot pin retaining clip
7 Drive end housing
8 Drive end bush
9 Through-bolt
10 Field coil
11 Yoke
12 Brushes
13 Insulator
14 Commutator end bearing
15 Armature assembly
16 Drive plate and springs
17 Pinion
18 Thrust collar
19 C-ring

Chapter 12 Electrical system

the commutator and rotate the armature. If the commutator is very badly worn, remove the drivegear (if still in place on the armature), and have the commutator refinished on a lathe – a job for an auto-electrical company. With the lathe turning at high speed take a very fine cut-out of the commutator and finish the surface by polishing with glass paper. The minimum permissible diameter to which the commutator may be skimmed is given in the Specifications.

13 With the starter motor dismantled, test the four field coils for an open circuit. Connect a 12 volt battery with a 12 volt bulb in one of the leads between the field terminal post and the tapping point of the field coils to which the brushes are connected. An open circuit is proved by the bulb not lighting.

14 If the bulb lights, it does not necessarily mean that the field coils are in order, as there is a possibility that one of the coils will be earthed to the starter yoke or pole shoes. To check this, remove the lead from the brush connector and place it against a clean portion of the starter yoke. If the bulb lights, then the field coils are earthing. The field coils, poles and yoke are not serviced as separate items. If a fault is suspected on these units the yoke assembly will have to be renewed.

15 If the armature is damaged this will be evident after visual inspection. Look for signs of burning, discolouration, and for conductors that have lifted away from the commutator.

16 With the starter motor stripped down, check the condition of the bushes. They should be renewed when they are sufficiently worn to allow visual side movement of the armature shaft.

17 The old bushes are simply driven out with a suitable drift and the new bushes inserted by the same method. As the bushes are of the phosphor bronze type it is essential that they are allowed to stand in engine oil for 24 hours or if time is limited, oil at 100°C (212°F) for 2 hours.

18 Reassembly is now the reverse sequence to dismantling, but the following points should be noted:

(a) *Smear all pivot points and bearing locations with grease before reassembly*
(b) *Fit sufficient shims to the armature shaft to eliminate excessive endfloat*

16 Headlight assembly – removal and refitting

1 Open the bonnet and disconnect the battery.
2 Remove the cover plate (1 screw) for access to the rear of the headlamp(s) (photos).

Series II models

3 Disengage the spring clip, then pull off the cap and multi-plug assembly (photo).
4 Remove the retaining screw and withdraw the headlamp assembly (photo). If necessary, remove the adjusters and retaining clips from the lens and reflector assembly.

16.2A Remove the screw ...

16.2B ... and lift the cover plate for access to the headlamps

16.3 Headlamp cap and multi-plug (Series II models)

16.4 Removing a headlamp retaining screw (Series II models)

Chapter 12 Electrical system

16.5 Removing the headlight bezel (Series III models)

16.6A Loosen the screws ...

16.6B ... remove the surround trim ...

16.6C ... and withdraw the headlight unit from the mounting bracket (Series III models)

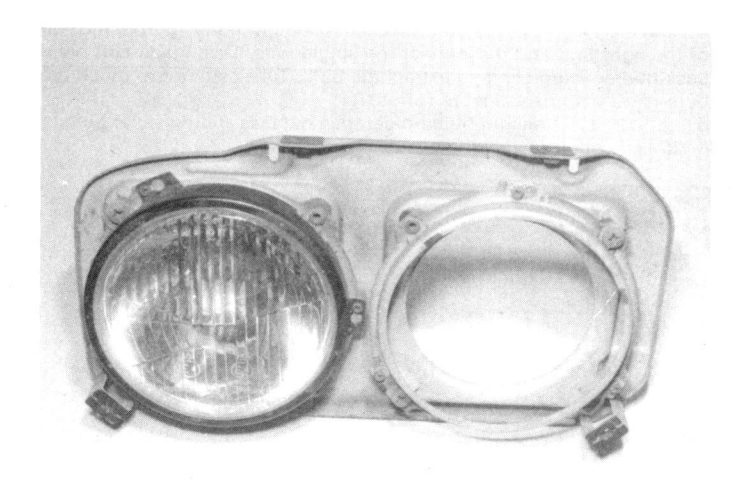

16.7 Headlight mounting bracket removed from the (Series III models)

Series III models

5 Detach the wire connector socket from the rear of the headlight unit then remove the four bezel retaining screws and withdraw the bezel (photo).

6 Loosen the headlight retaining screws, twist and remove the headlight unit and its surround trim (photos).

7 If required, the mounting bracket may be unbolted from the front panel (photo).

All models

8 Refitting is a reversal of removal but check and adjust the beam alignment as described in Section 17.

17 Headlight and auxiliary (driving) light alignment

Headlights

1 The headlights are adjustable individually for both horizontal and vertical alignment from within the engine compartment (after removing the cover plate).

2 Adjustments should not normally be necessary and, if their beam alignment is suspect, they should be checked by a Ford garage with optical alignment equipment.

3 A temporary adjustment can be made by turning the vertical and/or

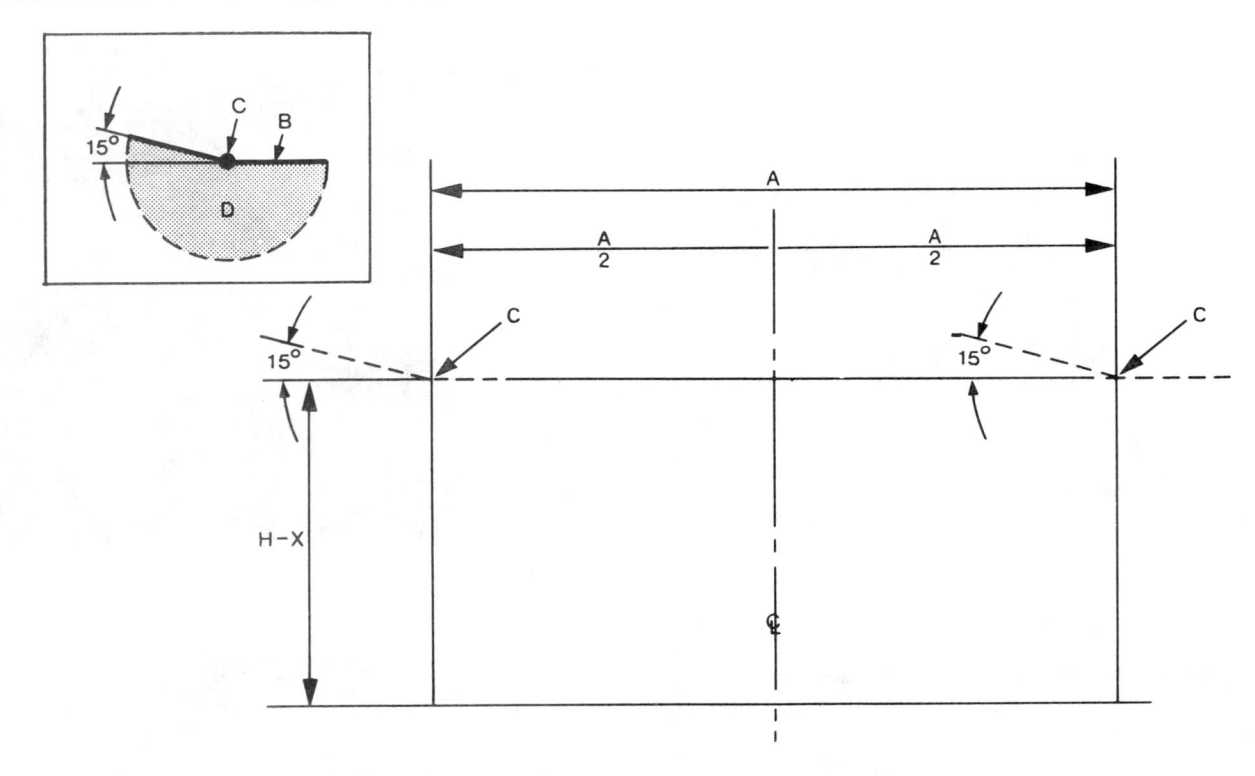

Fig. 12.13 Headlight beam alignment chart for RHD models (Sec 17)

A Distance between headlamp centres
B Light/dark boundary
C Dipped beam centre
D Dipped beam pattern
H Height from ground to centre of headlamps

Headlights X = 200 mm (8.0 in)
Driving lights X = 50 mm (2.0 in)

horizontal adjuster screws at the rear of each headlight unit. When making an alignment check, the car tyre pressure must be correct and the car standing unladen on level ground.

4 To assist in making an alignment check/adjustment the accompanying diagram shows the provisional headlamp beam alignment with the vehicle parked a distance of 10 m (33 feet) from a wall or aiming board (see Fig. 12.13). The headlamps are 'dipped'.

Driving lights

5 Temporary alignment of the driving lights is similar to that given for the headlights, except X = 50 mm (2 in).

18 Headlight and sidelight bulbs – renewal

1 Open the bonnet then remove the cover plate (1 screw) for access to the rear of the headlamp(s).

Series II models

2 Remove the retainer and extract the headlight bulb (photos).
3 Pull the sidelight holder from the bottom of the headlight and remove the bulb.

Series III models

4 Twist and remove the multi-socket connector housing from the rear of the light unit and disconnect the single wire. Disengage and pivot back the retaining clip and extract the bulb. Take care not to touch the bulb glass with bare fingers (photos).
5 Pull out the sidelight bulb holder then depress and twist the bulb to remove it (photos).

18.2A Remove the retainer ...

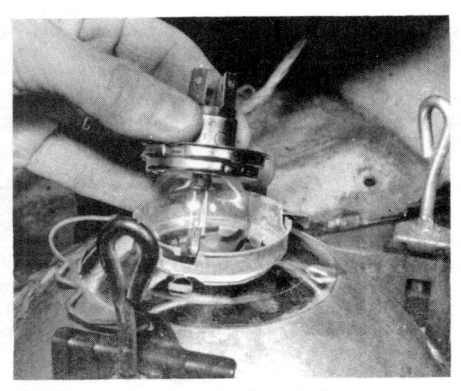

18.2B ... and extract the headlight bulb (Series II models)

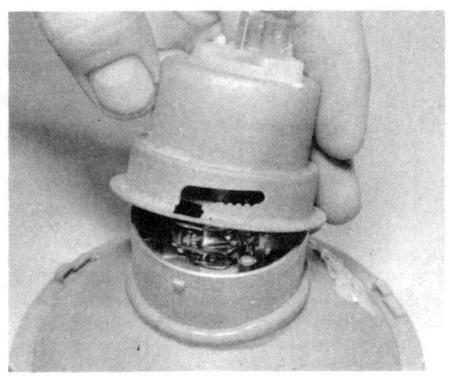

18.4A Twist and remove the connector ...

Chapter 12 Electrical system

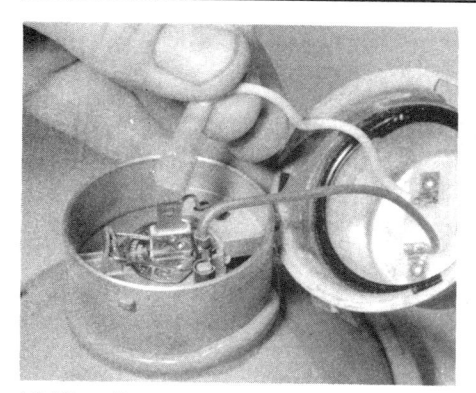

18.4B ... disconnect the single wire ...

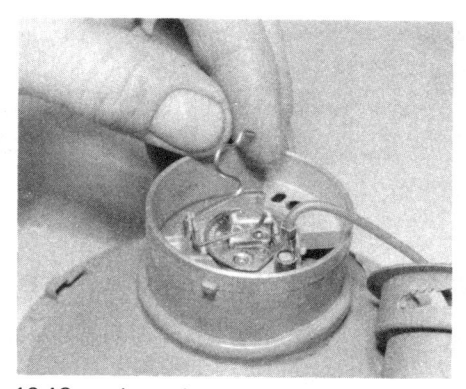

18.4C ... release the clip ...

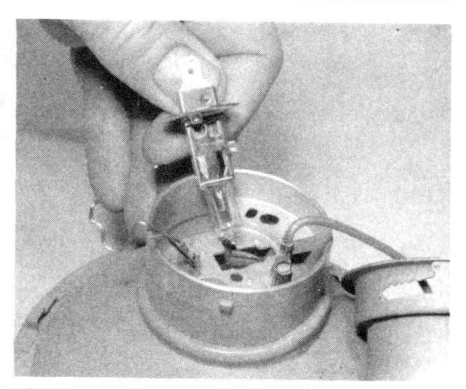

18.4D ... and extract the headlight bulb (Series III models)

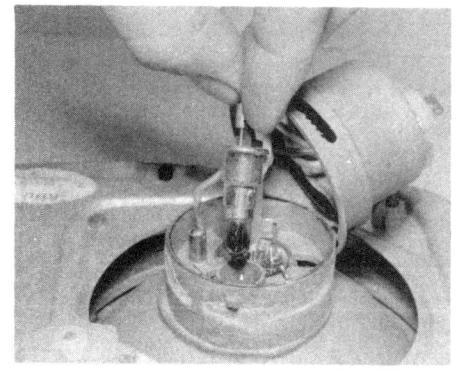

18.5A Pull out the bulbholder ...

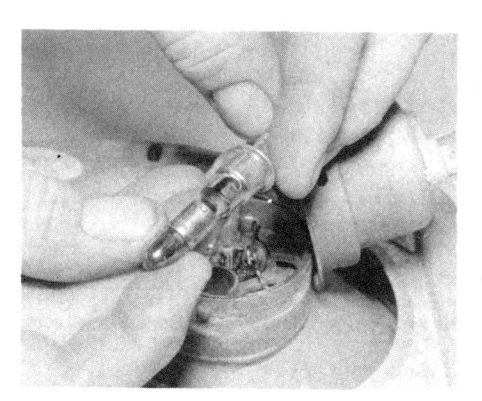

18.5B ... and remove the sidelight bulb (Series III models)

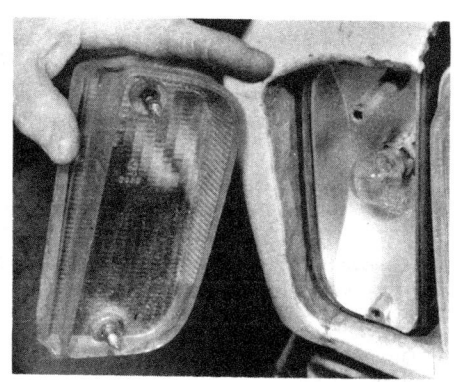

19.1 Removing the direction indicator lens (Series II models)

All models
6 Fit the new bulbs using a reversal of the removal procedure.

19 Front direction indicator lamp and bulb – removal and refitting

Series II models
1 Remove the indicator lens (2 screws) and remove the bulb (photo).
2 Remove the reflector. The lower retaining screw is accessible by inserting a screwdriver between the headlamp and reflector body.

Disconnect the wiring, to permit the assembly to be withdrawn.

Series III models
3 Reach up behind the lamp and turn the bulbholder anti-clockwise to release it (photos).
4 Depress and twist the bulb to remove it.
5 Where fitted remove the front foglamp.
6 Remove the two upper screws and withdraw the lamp from behind the bumper (photo).

All models
7 Refitting is a reversal of removal.

19.3A View of the front direction indicator lamp from behind the bumper (Series III models)

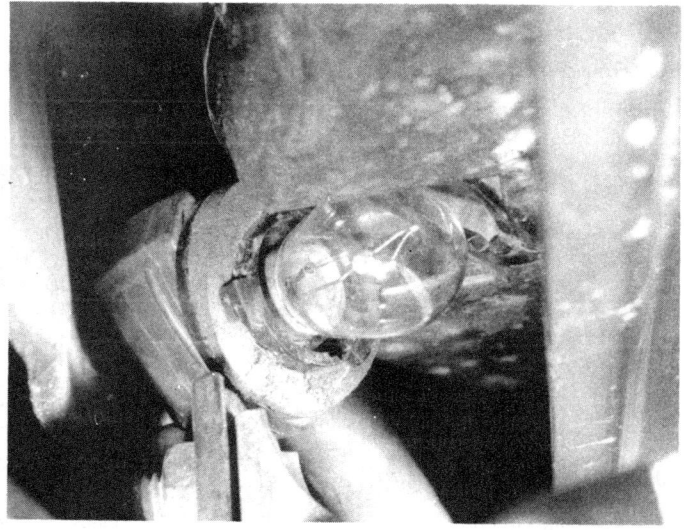

19.3B Removing the bulbholder from the front direction indicator lamp (Series III models)

Chapter 12 Electrical system

19.6 Removing the front direction indicator lamp screws (Series III models)

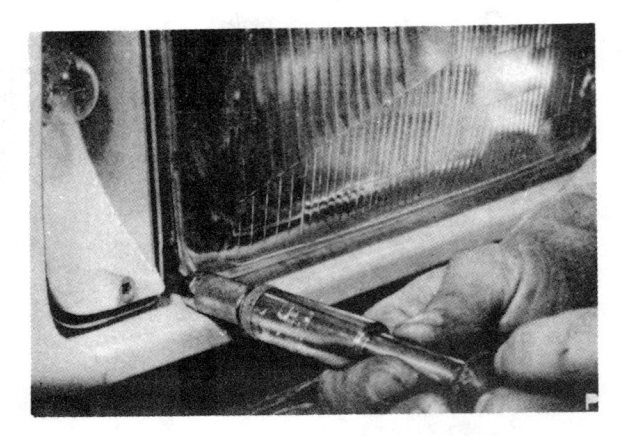

Fig. 12.14 Access to the front direction indicator lamp lower screw on Series II models (Sec 19)

20 Front foglamp and bulb – removal and refitting

1 Remove the screws from the rear of the lamp and withdraw the lens to the extent of the wiring (photo).
2 Release the clips so that the lens can be withdrawn from the bulb (photo).
3 Disconnect the wiring and remove the bulb, taking care not to touch the glass.
4 Unscrew the nuts from the mounting bracket and remove the foglamp from the front valence (photo).
5 Refitting is a reversal of removal.

21 Front driving lamp and bulb – removal and refitting

1 Remove the screw and clamp from the bottom of the lens (photo).
2 Release the lens from the upper retainer and disconnect the wiring (photos).
3 Release the clips and remove the bulb, taking care not to touch the glass.
4 Refitting is a reversal of removal.

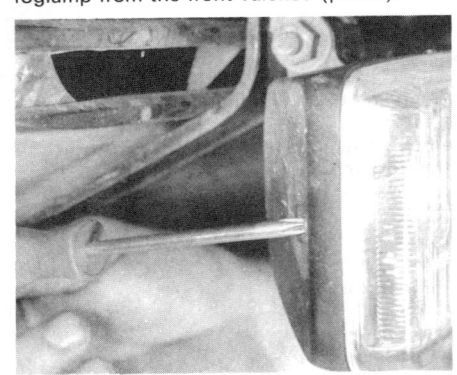

20.1 Front foglamp lens removal

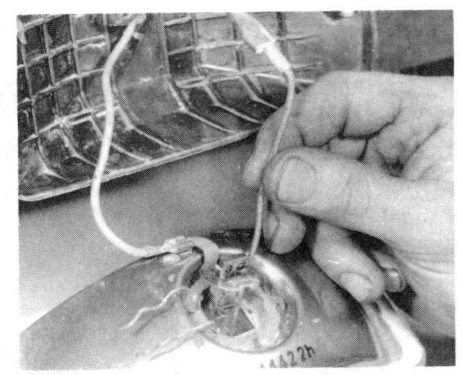

20.2 Removing the bulb from the lens

20.4 Removing the front foglamp and mounting bracket

21.1 Removing the driving lamp bottom clamp

21.2A Release the driving lamp lens from the retainer ...

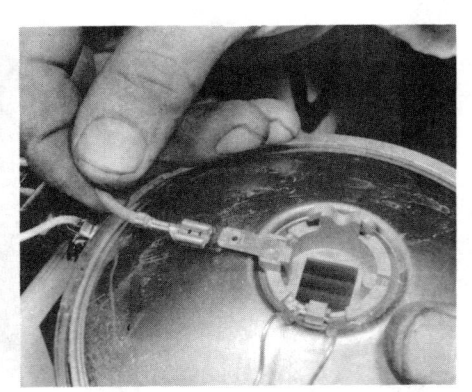

21.2B ... and disconnect the wiring

Chapter 12 Electrical system

22 Rear lamp and bulb – removal and refitting

1 Remove the four screws and withdraw the lens (photos).
2 Depress and twist the relevant bulb to remove it (photo).
3 To remove the rear lamp, open the tailgate and lift up the carpet and spare wheel cover.
4 Remove the trim panel where fitted.
5 Remove the two screws and withdraw the rear lamp from the rear panel (photo). On Series II models it will be necessary to lever out the lamp with a screwdriver to release it from the sealant, but take care not to damage the paintwork.
6 Disconnect the wiring (photos).
7 Refitting is a reversal of removal but apply sealant to the rear of the lamp on Series II models.

23 Rear number plate lamp and bulb – removal and refitting

1 Open the tailgate then remove the screws and withdraw the lens (photo).
2 Depress and twist the bulb to remove it (photo).
3 To remove the lamp, lift up the carpet and spare wheel cover, and remove the trim panel where fitted.
4 Disconnect the wiring and connect a piece of cord to it so that the wire can be pulled back through the body when refitting.
5 Refitting is a reversal of removal.

22.1A Remove the screws ...

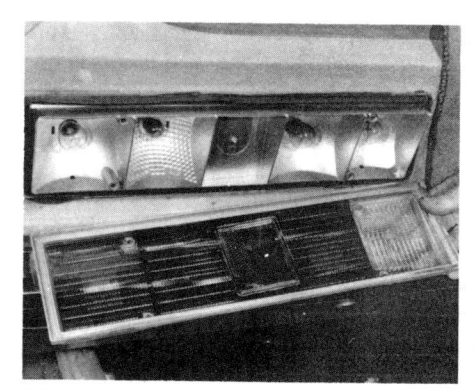

22.1B ... and withdraw the rear lamp lens

22.2 Rear lamp bulb removal

22.5 Removing a rear lamp retaining screw

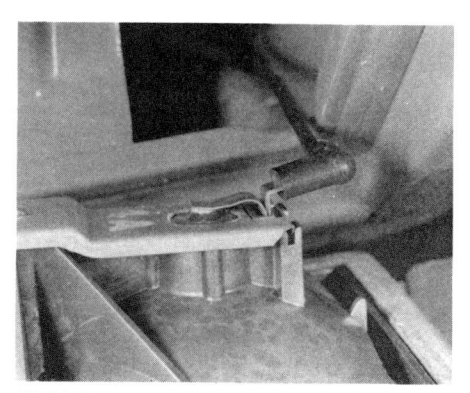

22.6A Rear lamp wiring for direction indicator

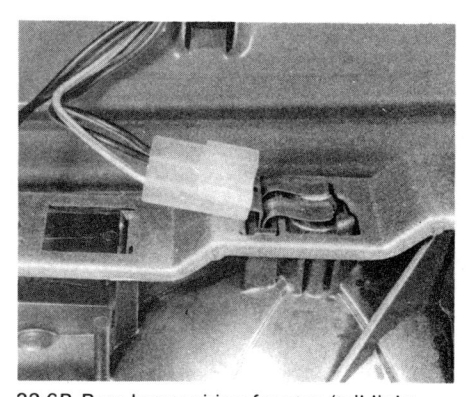

22.6B Rear lamp wiring for stop/tail light

23.1 Rear number plate lamp lens removal

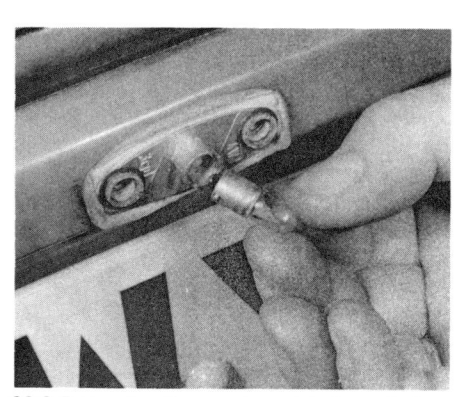

23.2 Removing the number plate lamp bulb

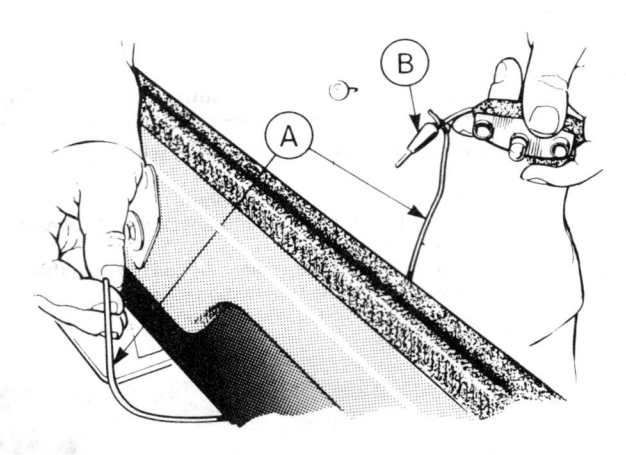

Fig. 12.15 Rear number plate lamp removal (Sec 23)

A Cord to facilitate refitting B Lamp wiring

24 Interior lamp and bulb – removal and refitting

1 Prise the lamp from the headlining or rear trim (photos).
2 Depress and twist the bulb to remove it.
3 To remove the lamp, first disconnect the battery earth lead then disconnect the lamp wiring and withdraw the lamp.
4 Refitting is a reversal of removal.

25 Tailgate wiper motor and linkage – removal and refitting

1 Disconnect the battery earth lead.
2 Remove the wiper arm and blade
3 Open the tailgate, and remove the tailgate trim panel (photo).
4 Disconnect the wiring at the wiper motor, noting the respective position of the leads (photo).
5 Remove the wiper spindle retaining nut and the three motor bracket retaining screws. Remove the motor and linkage assembly from the tailgate.
6 Remove the drive spindle nut and the three retaining bolts to detach the motor from the bracket.
7 Remove the circlip at the wiper spindle end and detach the linkage from the bracket.
8 Refitting is the reverse of the removal procedure; adjustment of the motor bracket being made before the bolts are finally tightened.

26 Tailgate washer pump – removal and refitting

1 Disconnect the battery earth lead.
2 Open the tailgate and remove the spare wheel cover.
3 Remove the washer pipes and leads, noting their installed positions to prevent mix-up when refitting (photo).
4 Remove the pump mounting screws and lift off the pump.
5 Refitting is the reverse of the removal procedure.

27 Tailgate washer nozzle – removal and refitting

1 Open the tailgate, remove the weather strip and pull down the headlining for access to the washer nozzle. Remove the nozzle by

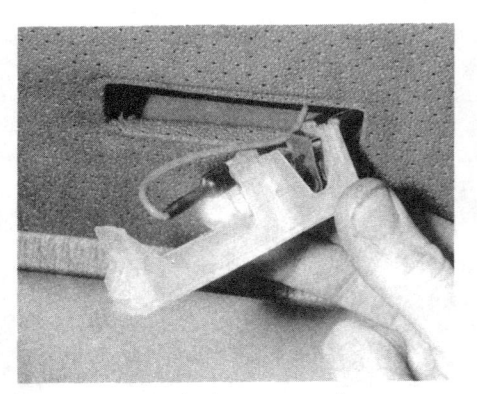

24.1A Front interior lamp removal

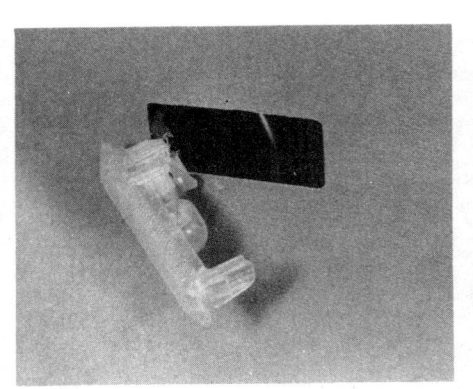

24.1B Rear interior lamp removal

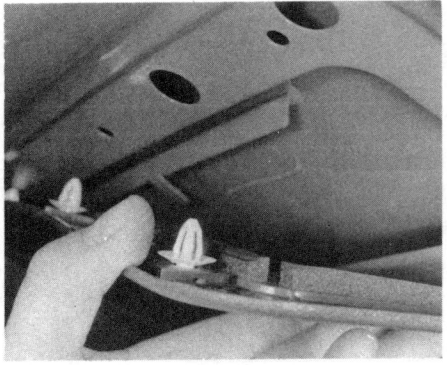

25.3 Removing the tailgate trim panel

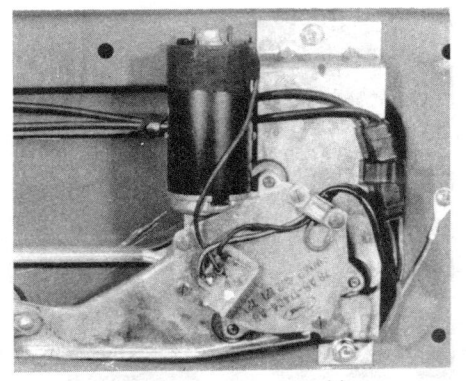

25.4 Tailgate wiper motor and wiring

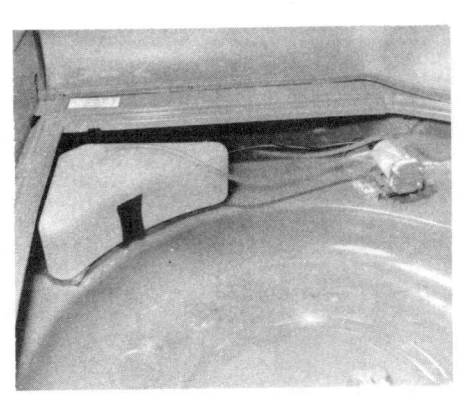

26.3 Tailgate washer pump and reservoir

Chapter 12 Electrical system

3 Remove the air cleaner and disconnect the choke cable at the carburettor.
4 Remove the steering column shroud.
5 Remove the retaining screws, and pull the lower dash insulating panel and cover panel assembly clear of the dash panel.
6 Disconnect the cigar lighter wiring and withdraw the panel assembly, complete with choke cable, away from the vehicle.
7 Remove the instrument cluster bezel and the instrument cluster.
8 Remove the glovebox catch striker and glovebox assembly. Disconnect the light wiring.
9 Disconnect the cable from the heater controls.
10 Disconnect the driver's side demister tube connector from the heater box and remove the connector and tube.
11 Disconnect and remove the driver's side face level vent tube.
12 Disconnect the wiring at the wiper motor and heater.
13 Remove the driver's side demister vent (1 screw).
14 Remove the wiper spindle retaining nuts and the motor bracket retaining screw. Remove the motor and linkage from the vehicle.
15 If necessary, separate the motor from the linkage.
16 Refitting is the reverse of the removal procedure, but ensure that the heater control cable and the choke operating cable are correctly adjusted.

Series III models
17 Disconnect the battery earth lead.
18 Remove the windscreen wiper arms and blades.
19 Remove the instrument cluster (Section 36) (photo).
20 Remove the mounting screws and withdraw the radio or cassette player unit. Disconnect the wiring and aerial.
21 Remove the heater control panel (4 screws).
22 Remove the glovebox and disconnect the illumination wiring.
23 Disconnect the air duct from the right of the heater and from the facia.
24 Disconnect the air duct from the right-hand face level vent.
25 Disconnect the wiring from the heater and wiper motor.
26 Remove the right-hand demister vent (1 screw).
27 Unscrew both wiper spindle retaining nuts.
28 Remove the wiper mounting bracket screws and withdraw the motor and linkage.

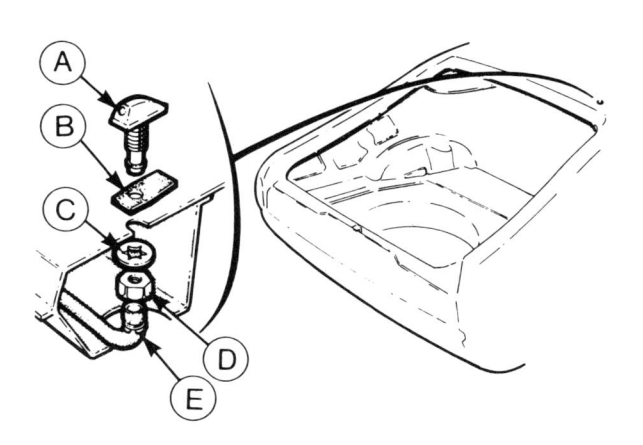

Fig. 12.16 Tailgate washer nozzle components (Sec 27)

A Jet
B Seal
C Washer
D Locknut
E Tube

unscrewing the locknut.
2 Refitting is the reverse of the removal procedure.

28 Windscreen wiper motor and linkage – removal and refitting

Series II models
1 Disconnect the battery earth lead.
2 Remove the windscreen wiper arms and blades.

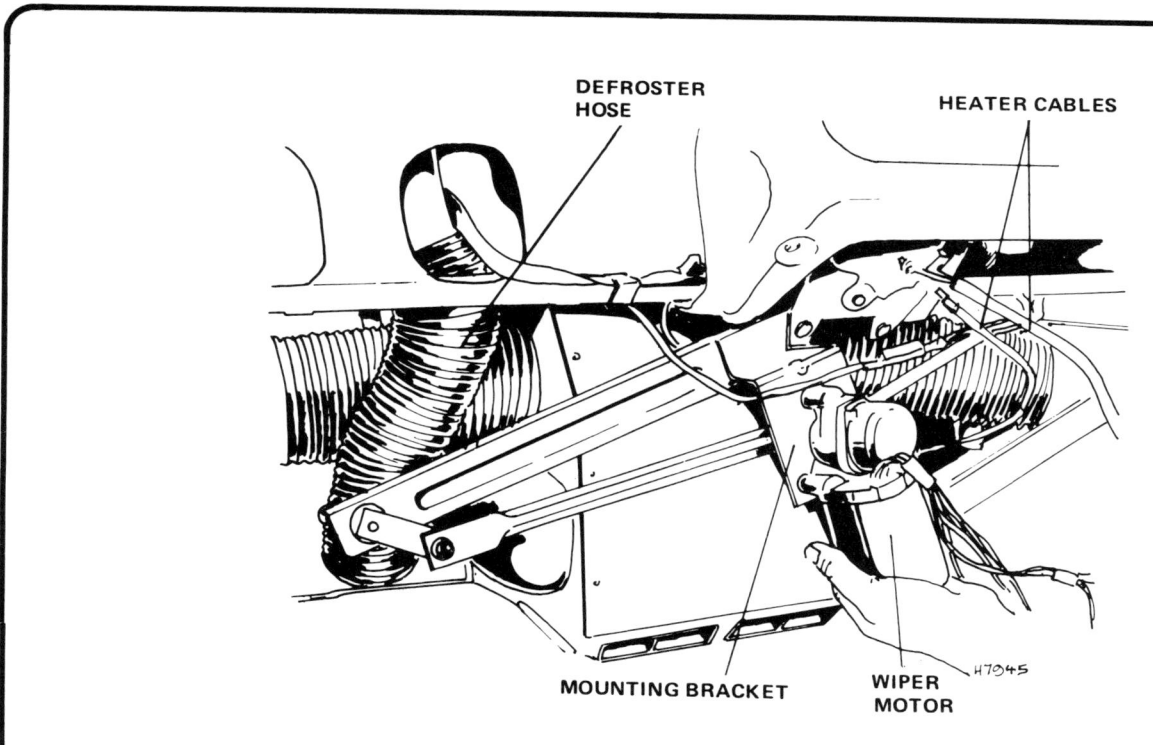

Fig. 12.17 Windscreen wiper motor removal on Series II models (Sec 28)

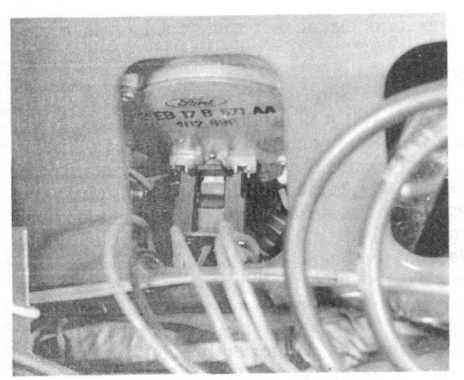

216   Chapter 12 Electrical system

28.19 The windscreen wiper motor can be seen through the bulkhead aperture with the instrument cluster removed (Series III models)

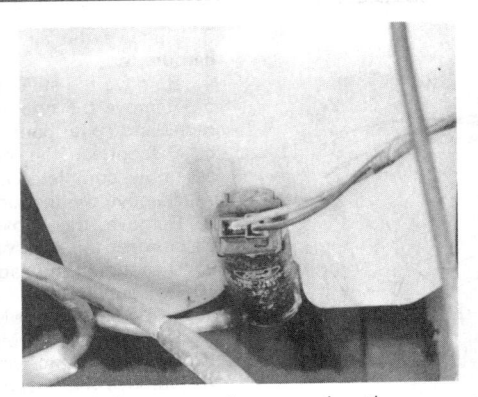

29 Separate the linkage from the motor by prising off the cranked arm and removing the three bolts.
30 Refitting is a reversal of removal.

29 Windscreen washer pump – removal and refitting

1 The windscreen washer pump is attached to the reservoir located in the engine compartment (photo).
2 To remove the pump and reservoir, pull off the electrical connections, lift up the reservoir and disconnect the flexible pipe. The pump can be removed from the reservoir if necessary.
3 Refitting is a reversal of removal, and fill the reservoir with the corect solution of washer fluid (photo).

30 Windscreen washer nozzle – removal and refitting

1 Remove the retaining screw, withdraw the nozzle and disconnect the pipe (photo).
2 Refitting is the reverse of the removal procedure.

31 Windscreen wipers – fault tracing

1 Should the windscreen wipers fail, or work very slowly, check the terminals on the motor for loose connections, and make sure the insulation of all the wiring is not cracked, or broken, thus causing a short circuit. If this is in order, check the current the motor is taking by connecting an ammeter in the circuit and turning on the wiper switch. Consumption should be between 2.3 and 3.1 amp.
2 If no current is passing through the motor, check that the switch is operating correctly.
3 If the wiper motor takes a very high current, check the wiper blades for freedom of movement. If this is satisfactory, check the gearbox cover and gear assembly for damage.
4 If the motor takes a very low current, ensure that the battery is fully charged. Check the brush gear and ensure the bruses are bearing on the commutator. If not, check the brushes for freedom of movement and if necessary, renew the tension springs. If the brushes are very worn they should be replaced with new ones. Check the armature by substitution if this unit is suspect.

32 Windscreen wiper motor – dismanting and reassembly

Series II models
1 Remove the two motor case/gear housing screws and withdraw the case and armature together.
2 Withdraw the brushes from the holders and remove the springs.
3 Remove the three brush mounting plate-to-wiper gear housing screws. Pull the wiring plug out of the side of the housing and remove the brush mounting plate.

29.1 Windscreen washer pump location

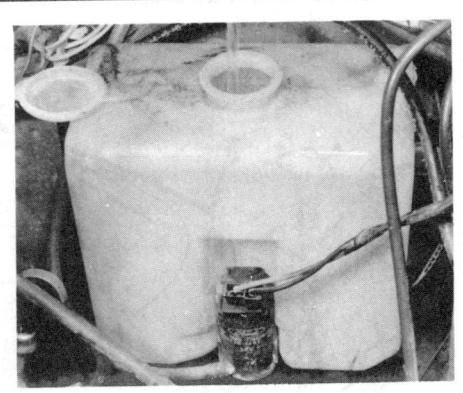

29.3 Topping-up the windscreen washer reservoir

4 Remove the screw and earth wire in the gear housing cover plate. Loosen the second screw and slide the cover plate away
5 Disconnect the white/green and black/green leads from the terminals on the switch cover assembly, then remove the wiring assembly from the motor.
6 Disconnect the motor multi-pin connector from the harness and remove the motor feed wires and brushes.

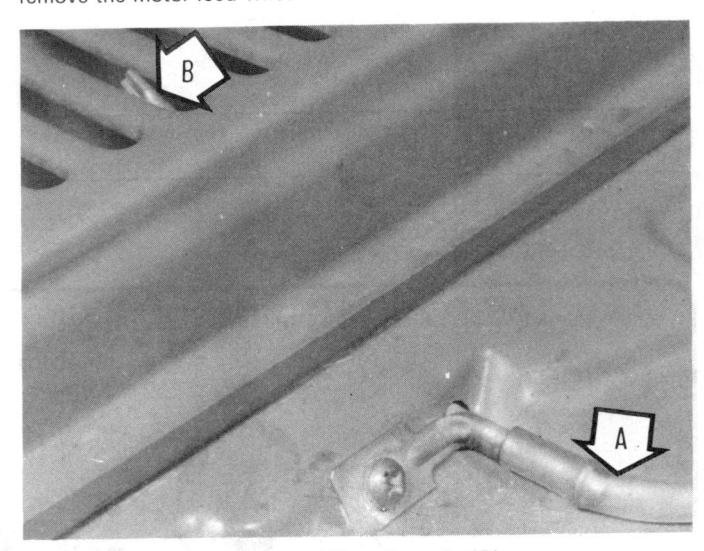

30.1 Windscreen washer pipe (A) and nozzle (B)

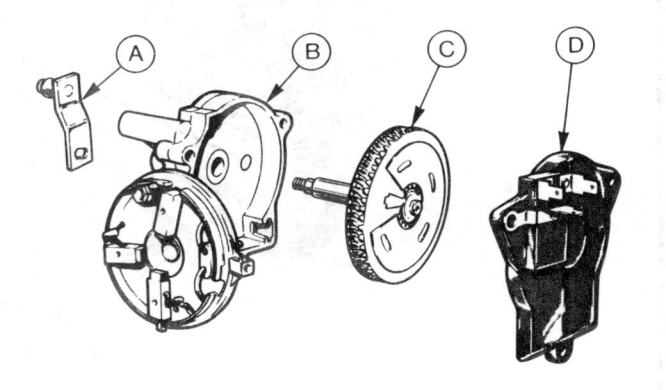

Fig. 12.18 Windscreen wiper motor components (Sec 32)

A Operating arm
B Gearbox
C Gear and shaft
D Cover

Chapter 12 Electrical system

7 Remove the remaining screw securing the gear housing cover plate and switch assembly. Remove the assembly.
8 Remove the spring steel armature stop from the gear housing.
9 Remove the spring clip and washer, which secure the pinion gear. Withdraw the gear and washers.
10 Remove the nut securing the motor output arm. Remove the arm, spring and flat washers.
11 Remove the output gear, the parking switch assembly and washer from the gear housing.
12 Reassembly is the reverse of the dismantling procedure. When refitting the brush gear, connect the black/green wire to the terminal marked 'black' and the white/green wire to the terminal marked 'green' on the switch cover assembly. Slide the gear housing cover plate into position, ensuring that the wires are correctly positioned in the cut-out on the cover plate.

Series III models

13 Disconnect the three clips and remove the case and armature.
14 Unscrew the nut and remove the operating arm.
15 Remove the gearbox cover, then scoop out the grease and slide out the gear and shaft.
16 Remove the brush mounting plate (1 screw), then disconnect the loom plug and remove the three bushes.
17 Clean the components and examine them for wear and damage.
18 Reassembly is a reversal of dismantling, but pack the gearbox with grease.

33 Windscreen wiper arms and blades – removal and refitting

1 To remove a wiper blade, raise the wiper arm away from the windscreen then depress the clip and either slide the blade out of the hooked end of the arm or remove it from the spring clip in the centre of the blade (photos). Refitting of the blade is a reversal of the removal procedure.
2 To remove a wiper arm, lift up the cap at the spindle, remove the nut and carefully prise off the arm (photo). When refitting, position the arm as necessary to obtain a satisfactory sweep on the windscreen.

34 Horn – fault tracing, removal and refitting

1 If the horn works badly or fails completely, first check the wiring leading to the horn for short circuits and loose connections. Also check that the horn is firmly secured.
2 Using a test lamp, check the wiring to the relevant fuse on the fusebox located in the engine compartment. Check that the fuse has not blown.
3 If the fault is an internal one it will be necessary to obtain a new horn.
4 To remove the horn, disconnect the battery and remove the radiator grille.
5 Detach the lead at the rear of the horn and then undo and remove the retaining bolt (photo).
6 Refitting the horn is the reverse sequence of removal.

35 Fuses

1 The main fusebox is located in the engine compartment on the right-hand side of the bulkhead.
2 Access to the fuses is gained by pulling off the outer cover and the inner clear plastic cover (photos).
3 There are also additional in-line fuses for the headlamp main and dipped beams located in or beneath the headlight relay (photo), and for the heated rear window, radio, and driving and foglamps located behind the facia. Fuse/relay units for driving lamps and foglamps are normally located in the engine compartment (photo).
4 If a fuse blows, always trace the cause before renewing it with one of the same rating.

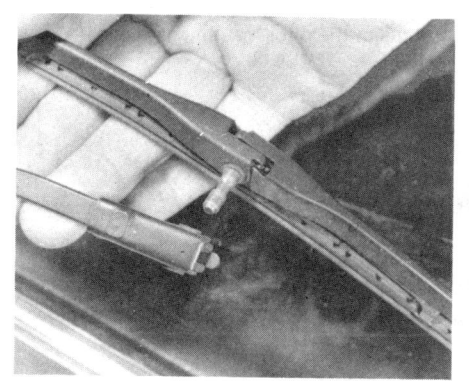

33.1A Early wiper blade removal

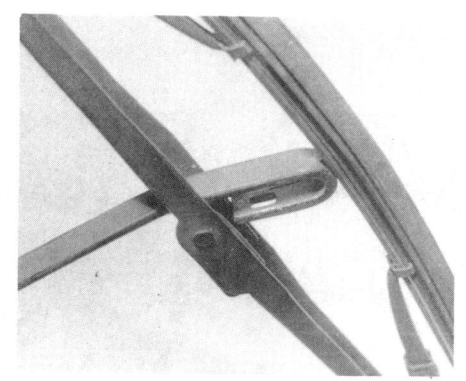

33.1B Later wiper blade removal

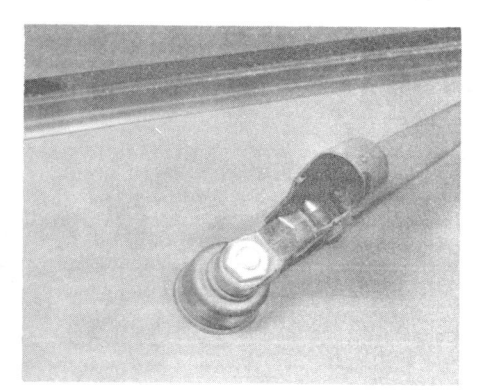

33.2 Lift the cap for access to the wiper arm retaining nut

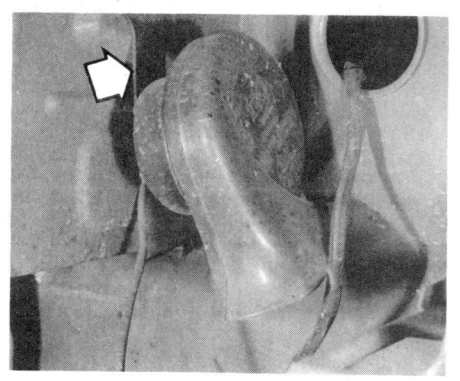

34.5 Horn and mounting bracket (arrowed)

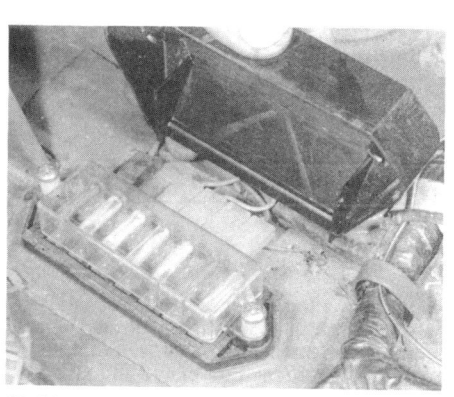

35.2A Fusebox outer cover removal

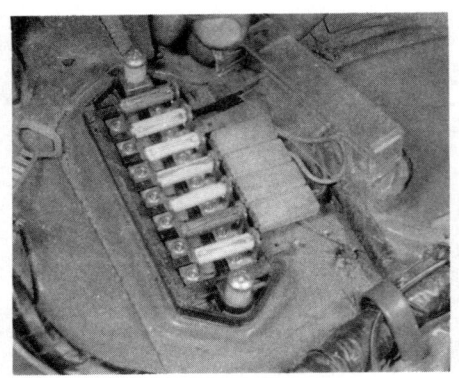

218 Chapter 12 Electrical system

35.2B Fusebox inner cover removal

35.3A Headlight relay located on the bulkhead

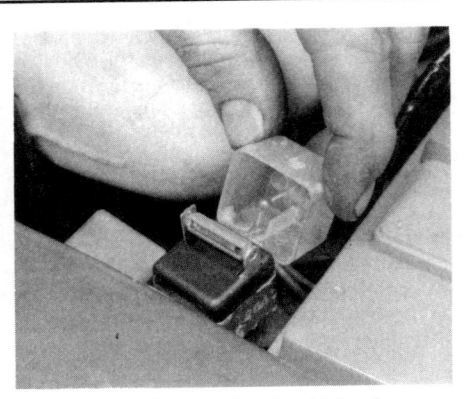

35.3B Fuse/relay unit for the driving lamps

36.4A Centre lower facia trim panel removal

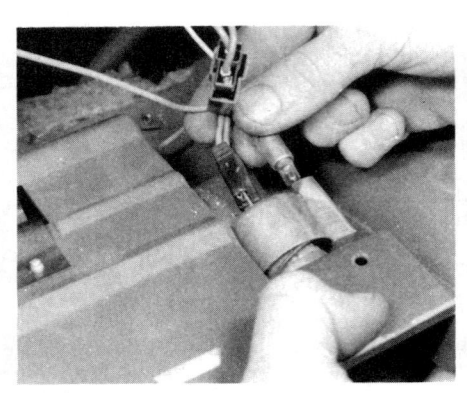

36.4B Removing the cigar lighter illumination bulbholder

36.4C Wiring to switches on the right-hand lower facia trim panel – no need to disconnect them for instrument cluster removal

36.5A Removing the control knobs from the radio/cassette player ...

36.5B ... and instrument illumination control

36 Instrument cluster – removal and refitting

1 Disconnect the battery.
2 Remove the steering column shrouds (2 screws).
3 Remove the lower facia panel.
4 Remove the screws and withdraw the lower facia trim panels. On Series III models first remove the ashtray and, after removal, disconnect the wiring and illumination bulb from the centre panel (photos).
5 Pull the control knobs from the radio/cassette player and also the instrument panel illumination control where applicable (photos).
6 Unscrew the nuts and remove the surround from the radio/cassette player (photos).
7 Remove the screws and withdraw the instrument cluster surround by unhooking it at the top (photos). Disconnect the switch wiring.
8 Remove the screws and ease out the instrument cluster to provide access to the rear connections (photos).
9 Unscrew the nut and disconnect the oil pressure gauge feed pipe (photo).
10 Disconnect the wiring multi-plug and the speedometer cable (photos).
11 With the instrument cluster removed, the warning and illumination bulbs can be removed by first pulling out the bulbholder then removing the push-fit bulb (photos).
12 Refitting is a reversal of removal.

36.6A Unscrew the nuts (arrowed) ...

36.6B ... and remove the radio/cassette player surround

36.7A Removing screws from instrument cluster surround

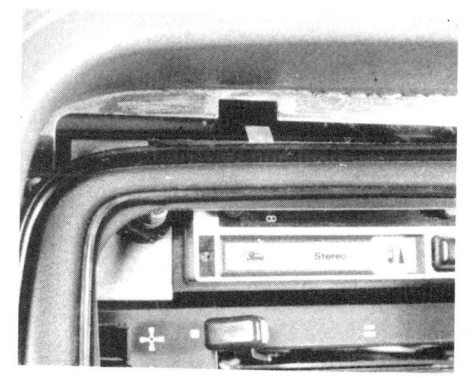

36.7B Releasing the top of the instrument cluster surround

36.8A Remove the screws ...

36.8B ... and withdraw the instrument cluster

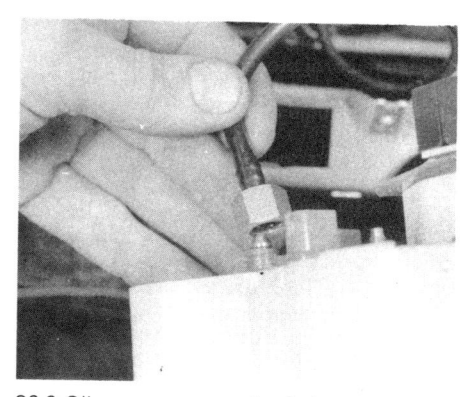

36.9 Oil pressure gauge feed pipe removal

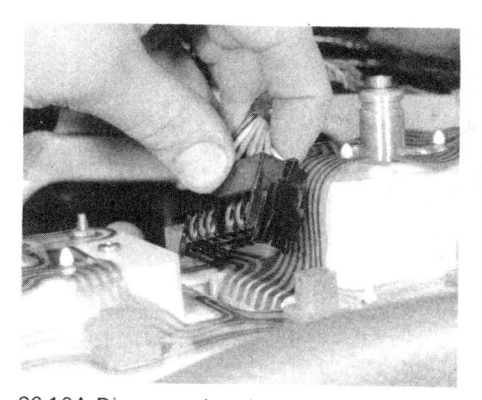

36.10A Disconnecting the wiring multi-plug

36.10B The upper end of the speedometer cable

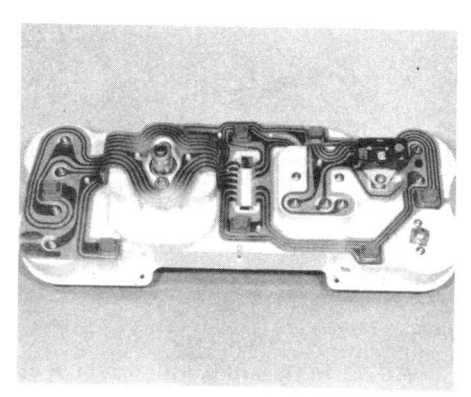

36.11A Rear of the instrument cluster

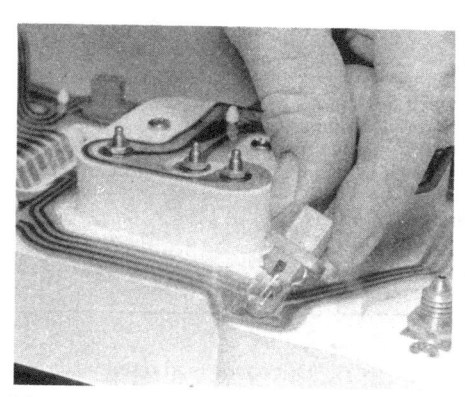

36.11B Instrument cluster bulb removal

220 Chapter 12 Electrical system

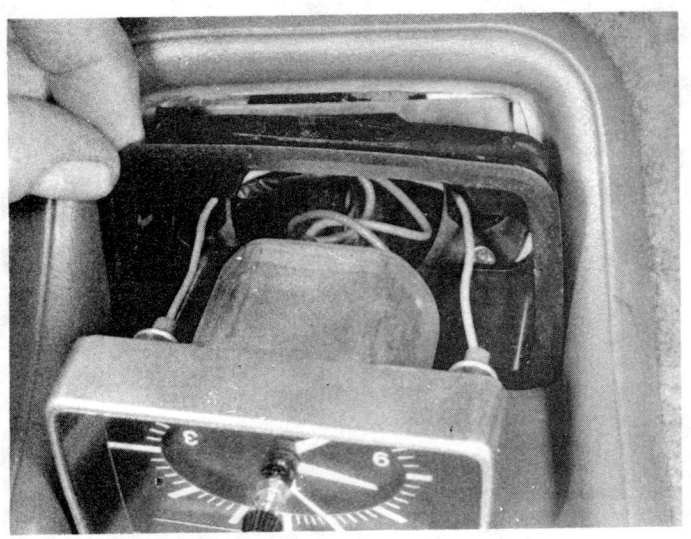

37.2 Removing the clock from the centre console

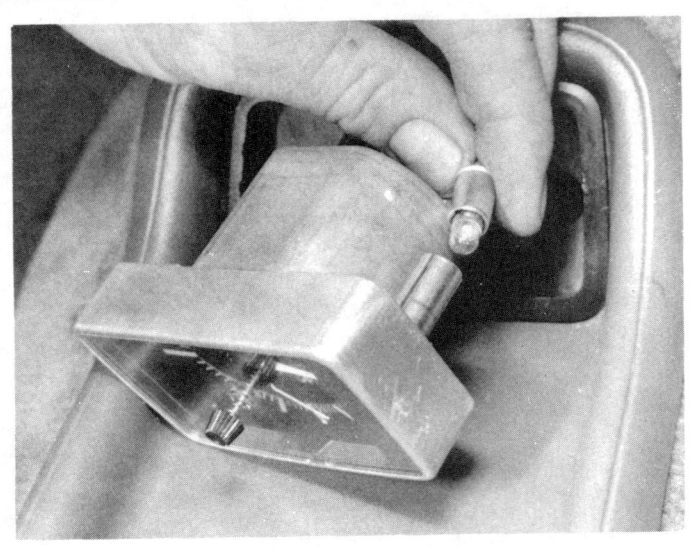

37.3 Clock illumination bulbholder removal

37.4 Disconnecting the clock wiring

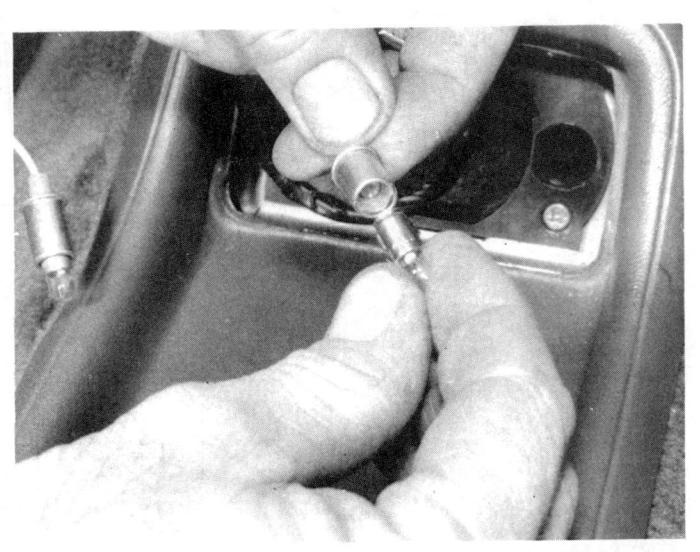

37.5 Removing a clock illumination bulb from its bulbholder

37 Clock (console mounted) – removal and refitting

1 Disconnect the battery.
2 Pull the clock out from the centre console (photo). To do this make up two brackets as described in Chapter 11, Section 32.
3 Pull out the illumination bulbholders (photo).
4 Disconnect the wiring and remove the clock (photo).
5 Depress and twist the bulbs to remove them from the bulbholders (photo).
6 Refitting is a reversal of removal.

38 Hazard warning switch – removal and refitting

1 Disconnect the battery earth lead.
2 Pull the switch from the lower facia trim panel and disconnect the wiring harness. The bulb may be removed after unscrewing the lens.
3 When refitting, connect the wiring harness and press the switch into the trim panel to retain it.

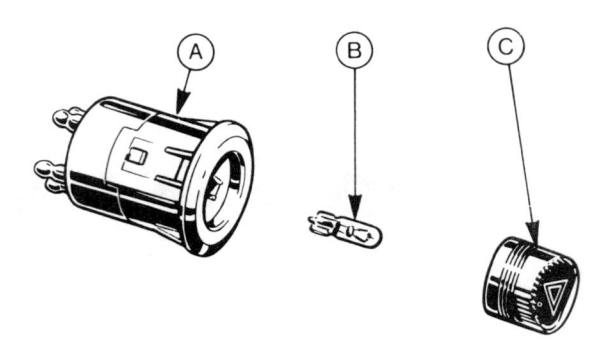

Fig. 12.19 Hazard warning switch components (Sec 38)

A Switch
B Bulb
C Lens

Chapter 12 Electrical system

39 Steering column multi-function switch – removal and refitting

1 Disconnect the battery earth lead.
2 Remove the steering column shrouds. The bottom half is retained by two screws and the top half can then be pulled out.
3 Remove the switch retaining screws (photo). Disconnect the multi-plug, then detach the switch from the steering column.
4 Refitting is the reverse of the removal procedure.

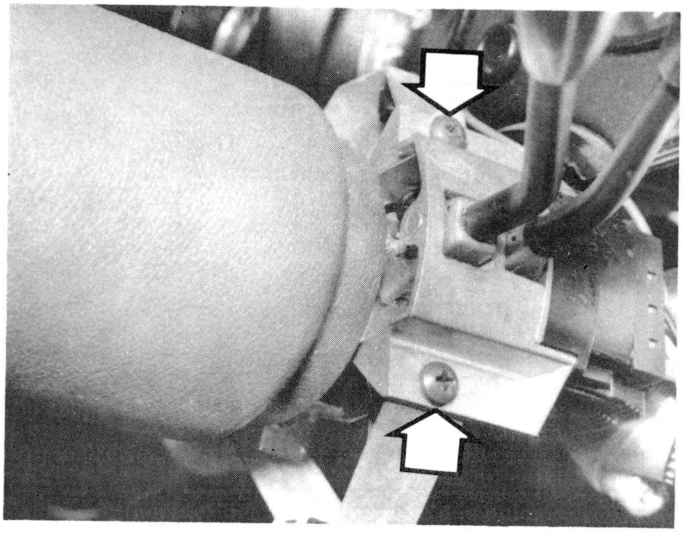

40 Ignition switch and lock – removal and refitting

1 To remove the ignition switch, first remove the steering column shrouds.
2 Disconnect the battery earth lead then disconnect the switch wiring.
3 Remove the screws and detach the ignition switch from the lock housing.
4 To remove the lock (photo) the complete steering column must be removed as described in Chapter 10.
5 Mount the column in a vice and drill out the shear bolts. The lock and clamp may then be removed from the column.
6 Remove the remains of the shear bolt.
7 Refitting is a reversal of removal, but check that the lock operates correctly before tightening the shear bolts until their heads break off.

41 Door pillar switch – removal and refitting

1 Prise the switch from the door pillar (Series II models) or remove the screw and pull out the switch (Series III models) (photo).
2 Disconnect the lead but make sure it does not fall back into the pillar.
3 Refitting is a reversal of removal but, if necessary, first clean the pillar to ensure good electrical contact is made.

42 Intermittent cluster illumination switch – removal and refitting

1 Remove the instrument cluster surround with reference to Section 36.
2 Unscrew the nut, remove the switch from the bulkhead and disconnect the wiring (photo).
3 Refitting is a reversal of removal.

39.3 Steering column multi-function switch retaining screws (arrowed)

Fig. 12.20 Ignition switch removal (Sec 40)

A Wiring multi-plug C Ignition switch
B Screws D Lock

40.4 Ignition switch and steering column lock (installed in car)

41.4 Door pillar switch removal (Series III models)

42.2 Instrument cluster illumination switch (arrowed)

43.2A Prise the switch from the surround  ...   43.2B ... and disconnect the wiring 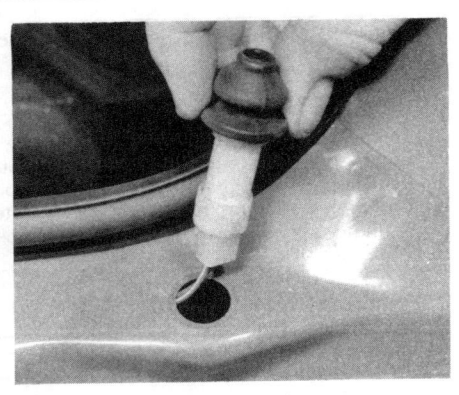   44.2 Tailgate switch removal

43 Central facia switches – removal and refitting

1 Disconnect the battery earth lead.
2 Prise the switch from the instrument cluster surround and disconnect the wiring (photos).
3 Where applicable the illumination bulb may now be removed.
4 Refitting is a reversal of removal.

44 Load space lamp switch – removal and refitting

1 Disconnect the battery earth lead.
2 Open the tailgate and pull the switch from its location (photo).
3 Disconnect the wiring and remove the swtich.
4 Refitting is a reversal of removal.

45 Handbrake warning light switch – removal and refitting

1 Remove the centre console as described in Chapter 11.
2 Remove the screws, lift the retainer, and pull the rubber gaiter up the handbrake lever.
3 Disconnect the wiring then remove the screws and withdraw the switch.
4 Refitting is a reversal of removal.

46 Light and windscreen wiper switches (Series II models) – removal and refitting

1 Disconnect the battery earth lead.
2 Slacken the three screws on the lower panel.
3 Fully depress one of the switches of the pair to be removed, then insert a suitably cranked tool such as a piece of bent welding rod into the exposed hole in the switch centre web (Fig. 12.22).
4 Hold down the lower panel and gently pull out the switches.
5 When refitting, first connect the plug, then refit the switch and tighten the panel screws.
6 Finally reconnect the battery earth lead.

47 Flasher unit and relays – general

1 The flasher unit is located behind the instrument cluster.
2 With the exception of the headlamp flasher relay and, on automatic transmission models, the inhibitor switch relay, all relays are located behind the lower facia panel.
3 In the event of failure always check the connecting wiring, bulbs and fuses before assuming that it is the relay or flasher unit that is at fault. Take the relay or flasher unit to your dealer for testing, or check the circuit by substituting a new component.

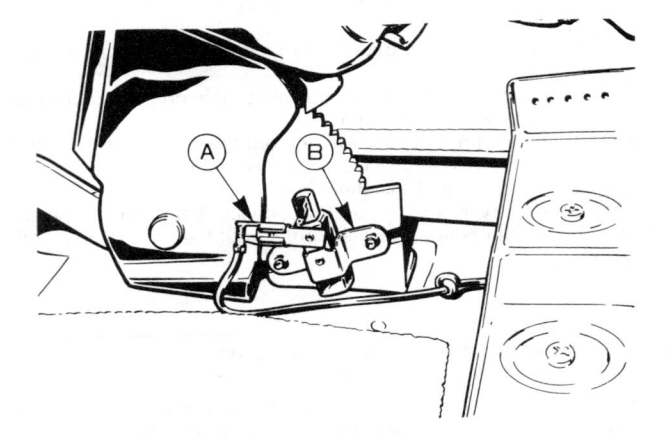

Fig. 12.21 Wire connection (A) and handbrake warning light switch (B) (Sec 45)

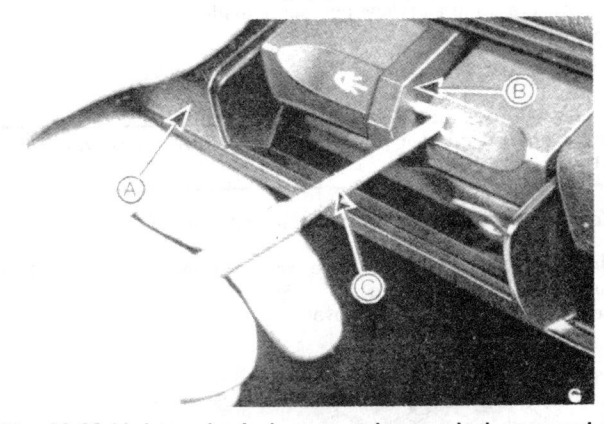

Fig. 12.22 Light and windscreen wiper switch removal on Series II models (Sec 46)

A Lower panel
B Centre web
C Cranked tool

48 Cigar lighter – removal and refitting

1 Disconnect the battery earth lead.
2 Remove the steering column shrouds.
3 Remove the lower facia and facia trim panels.
4 Disconnect the wiring then depress the rear collar and turn it to align the cut-outs and lugs.
5 Remove the collar and spring, then withdraw the cigar lighter from the panel.
6 Refitting is a reversal of removal.

Chapter 12 Electrical system

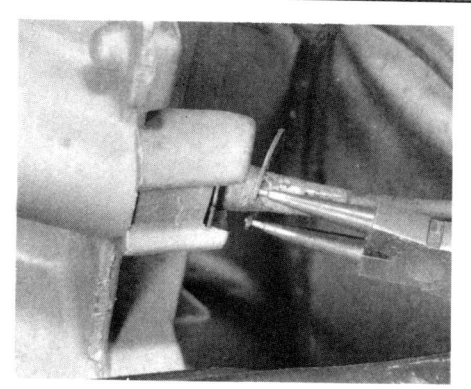

223

49.3A Extracting the circlip ...

49.3B ... when disconnecting the speedometer cable from the manual transmission

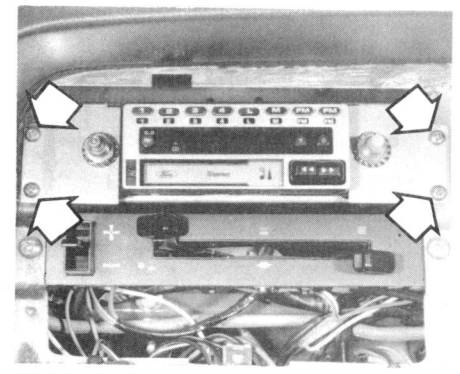

52.2 Radio/cassette player mounting screws (arrowed)

49 Speedometer cable – renewal

1 Remove the instrument cluster as described in Sction 36.
2 Apply the handbrake then jack up the front of the car and support on axle stands.
3 On manual transmission models, extract the circlip from the extension housing and disconnect the cable (photos).
4 On automatic transmission models, unscrew the clamp bolt, remove the clamp and disconnect the cable.
5 Release the cable from the body clips then withdraw it through the bulkhead.
6 Refitting is a reversal of removal but after inserting the bulkhead grommet pull the cable through until the colour band on the cable is positioned in the grommet.

50 Instrument voltage regulator – removal and refitting

1 Remove the instrument cluster as described in Section 36.
2 Unscrew and remove the single screw that retains the instrument voltage regulator to the rear of the instrument cluster and withdraw the regulator.
3 Refitting is a reversal of the removal procedure.

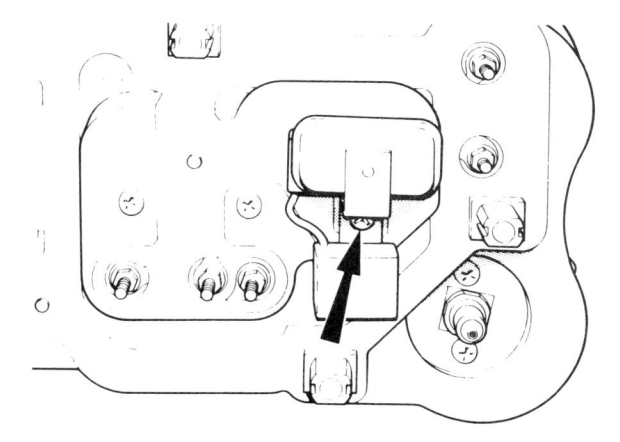

Fig. 12.23 Instrument voltage regulator and retaining screw – arrowed (Sec 50)

51 Dim-dip lighting system – general

As from late 1986 additional relays are fitted to operate the dim-dip lighting system. With the ignition on and the light switch in the first 'on' position, the headlights will be on dimmed-dipped beam. Turning the light switch to the second 'on' position will operate the normal dipped beam. The system is being introduced to prevent driving on sidelights only.

52 Radio/cassette player – removal and refitting

1 Remove the instrument cluster surround with reference to Section 36.
2 Remove the mounting panel screws and withdraw the radio/cassette player unit (photo).
3 Disconnect the wiring and aerial.
4 Refitting is a reversal of removal.

53 Mobile radio equipment – interference-free installation

Aerials – selection and fitting

The choice of aerials is now very wide. It should be realised that the quality has a profound effect on radio performance, and a poor, inefficient aerial can make suppression difficult.

A wing-mounted aerial is regarded as probably the most efficient for signal collection, but a roof aerial is usually better for suppression purposes because it is away from most interference fields. Stick-on wire aerials are available for attachment to the inside of the windscreen, but are not always free from the interference field of the engine and some accessories.

Motorised automatic aerials rise when the equipment is switched on and retract at switch-off. They require more fitting space and supply leads, and can be a source of trouble.

There is no merit in choosing a very long aerial as, for example, the type about three metres in length which hooks or clips on to the rear of the car, since part of this aerial will inevitably be located in an interference field. For VHF/FM radios the best length of aerial is about one metre. Active aerials have a transistor amplifier mounted at the base and this serves to boost the received signal. The aerial rod is sometimes rather shorter than normal passive types.

A large loss of signal can occur in the aerial feeder cable, especially over the Very High Frequency (VHF) bands. The design of feeder cable is invariably in the co-axial form, ie a centre conductor surrounded by a flexible copper braid forming the outer (earth) conductor. Between the inner and outer conductors is an insulator material which can be in solid or stranded form. Apart from insulation, its purpose is to maintain the correct spacing and concentricity. Loss of signal occurs in this insulator, the loss usually being greater in a poor quality cable. The quality of cable used is reflected in the price of the aerial with the attached feeder cable.

The capacitance of the feeder should be within the range 65 to 75 picofarads (pF) approximately (95 to 100 pF for Japanese and American equipment), otherwise the adjustment of the car radio aerial trimmer may not be possible. An extension cable is necessary for a long run between aerial and receiver. If this adds capacitance in excess of the above limits, a connector containing a series capacitor will be required, or an extension which is labelled as 'capacity-compensated'.

Fitting the aerial will normally involve making a $7/8$ in (22 mm) diameter hole in the bodywork, but read the instructions that come with the aerial kit. Once the hole position has been selected, use a centre punch to guide the drill. Use sticky masking tape around the area for this helps with marking out and drill location, and gives

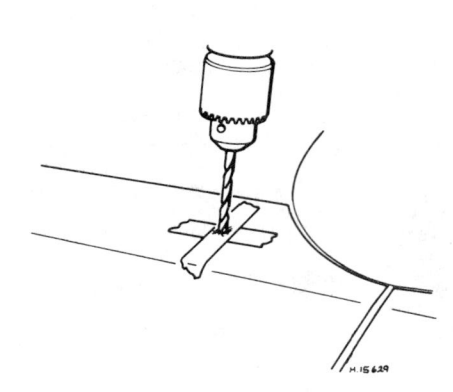

Fig. 12.24 Drilling the bodywork for aerial mounting (Sec 53)

protection to the paintwork should the drill slip. Three methods of making the hole are in use:

(a) Use a hole saw in the electric drill. This is, in effect, a circular hacksaw blade wrapped round a former with a centre pilot drill.
(b) Use a tank cutter which also has cutting teeth, but is made to shear the metal by tightening with an Allen key.
(c) The hard way of drilling out the circle is using a small drill, say 1/8 in (3 mm), so that the holes overlap. The centre metal drops out and the hole is finished with round and half-round files.

Whichever method is used, the burr is removed from the body metal and paint removed from the underside. The aerial is fitted tightly ensuring that the earth fixing, usually a serrated washer, ring or clamp, is making a solid connection. *This earth connection is important in reducing interference.* Cover any bare metal with primer paint and topcoat, and follow by underseal if desired.

Aerial feeder cable routing should avoid the engine compartment and areas where stress might occur, eg under the carpet where feet will be located. Roof aerials require that the headlining be pulled back and that a path is available down the door pillar. It is wise to check with the vehicle dealer whether roof aerial fitting is recommended.

Loudspeakers

Speakers should be matched to the output stage of the equipment, particularly as regards the recommended impedance. Power transistors used for driving speakers are sensitive to the loading placed on them.

Before choosing a mounting position for speakers, check whether the vehicle manufacturer has provided a location for them. Generally door-mounted speakers give good stereophonic reproduction, but not all doors are able to accept them. The next best position is the rear parcel shelf, and in this case speaker apertures can be cut into the shelf, or pod units may be mounted.

For door mounting, first remove the trim, which is often held on by 'poppers' or press studs, and then select a suitable gap in the inside door assembly. Check that the speaker would not obstruct glass or winder mechanism by winding the window up and down. A template is often provided for marking out the trim panel hole, and then the four fixing holes must be drilled through. Mark out with chalk and cut cleanly with a sharp knife or keyhole saw. Speaker leads are then threaded through the door and door pillar, if necessary drilling 10 mm diameter holes. Fit grommets in the holes and connect to the radio or tape unit correctly. Do not omit a waterproofing cover, usually supplied with door speakers. If the speaker has to be fixed into the metal of the door itself, use self-tapping screws, and if the fixing is to the door trim use self-tapping screws and flat spire nuts.

Rear shelf mounting is somewhat simpler but it is necessary to find gaps in the metalwork underneath the parcel shelf. However, remember that the speakers should be as far apart as possible to give a good stereo effect. Pod-mounted speakers can be screwed into position through the parcel shelf material, but it is worth testing for the best position. Sometimes good results are found by reflecting sound off the rear window.

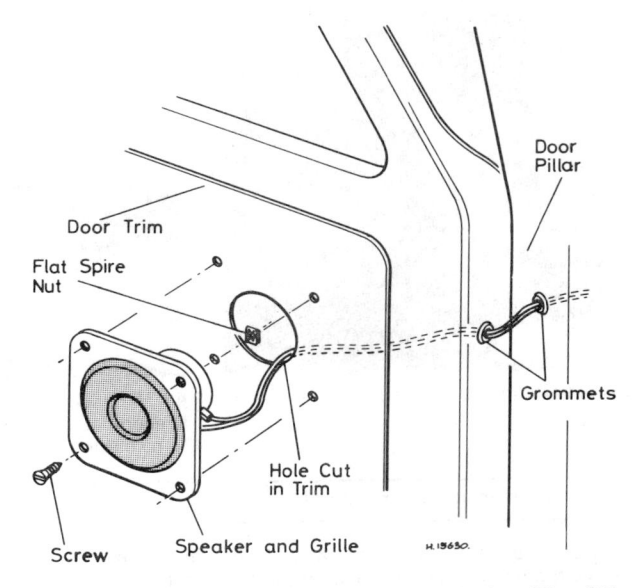

Fig. 12.25 Door-mounted speaker installation (Sec 53)

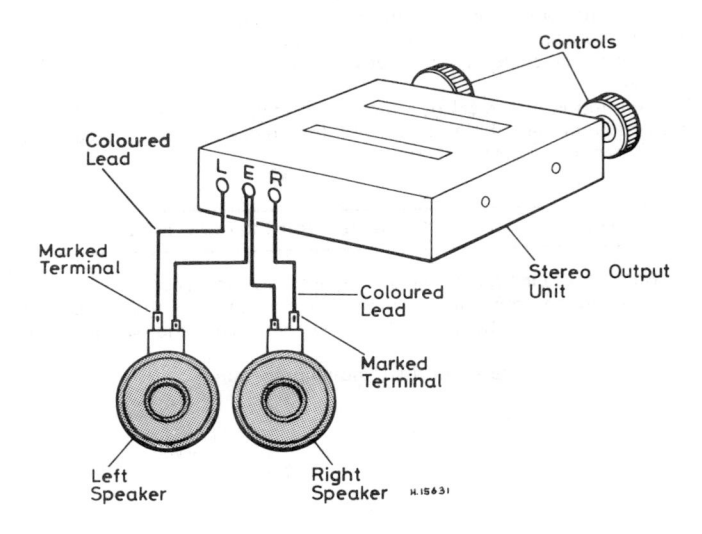

Fig. 12.26 Speaker connections must be correctly made as shown (Sec 53)

Unit installation

Many vehicles have a dash panel aperture to take a radio/audio unit, a recognised international standard being 189.5 mm x 60 mm. Alternatively a console may be a feature of the car interior design and this, mounted below the dashboard, gives more room. If neither facility is available a unit may be mounted on the underside of the parcel shelf; these are frequently non-metallic and an earth wire from the case to a good earth point is necessary. A three-sided cover in the form of a cradle is obtainable from car radio dealers and this gives a professional appearance to the installation; in this case choose a position where the controls can be reached by a driver with his seat belt on.

Installation of the radio/audio unit is basically the same in all cases, and consists of offering it into the aperture after removal of the knobs (*not* push buttons) and the trim plate. In some cases a special mounting plate is required to which the unit is attached. It is worthwhile supporting the rear end in cases where sag or strain may occur, and it is usually possible to use a length of perforated metal strip attached between the unit and a good support point nearby. In general it is recommended that tape equipment should be installed at or nearly horizontal.

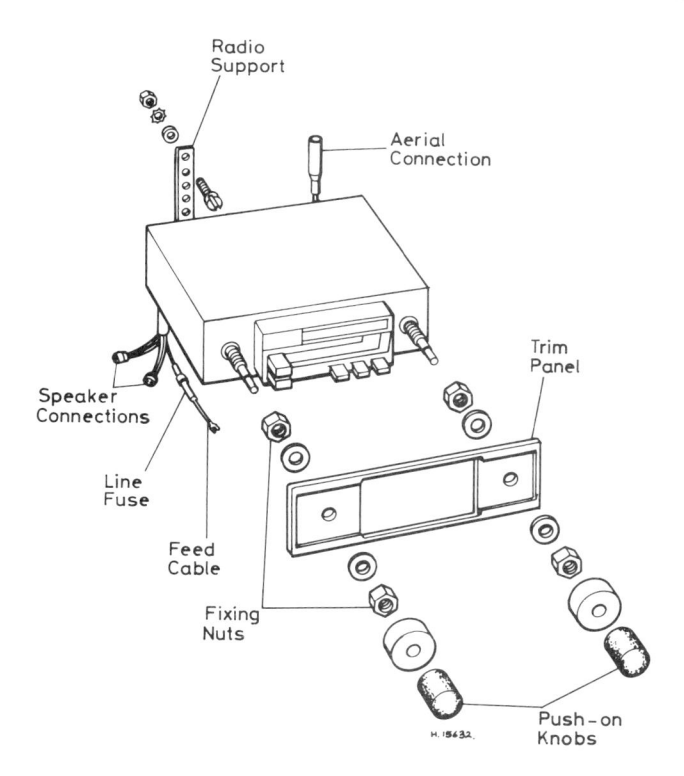

Fig. 12.27 Mounting component details for radio/cassette unit (Sec 53)

Connections to the aerial socket are simply by the standard plug terminating the aerial downlead or its extension cable. Speakers for a stereo system must be matched and correctly connected, as outlined previously.

Note: *While all work is carried out on the power side, it is wise to disconnect the battery earth lead.* Before connection is made to the vehicle electrical system, check that the polarity of the unit is correct. Most vehicles use a negative earth system, but radio/audio units often have a reversible plug to convert the set to either + or – earth. *Incorrect connection may cause serious damage.*

The power lead is often permanently connected inside the unit and terminates with one half of an in-line fuse carrier. The other half is fitted with a suitable fuse (3 or 5 amperes) and a wire which should go to a power point in the electrical system. This may be the accessory terminal on the ignition switch, giving the advantage of power feed with ignition or with the ignition key at the 'accessory' position. Power to the unit stops when the ignition key is removed. Alternatively, the lead may be taken to a live point at the fusebox with the consequence of having to remember to switch off at the unit before leaving the vehicle.

Before switching on for initial test, be sure that the speaker connections have been made, for running without load can damage the output transistors. Switch on next and tune through the bands to ensure that all sections are working, and check the tape unit if applicable. The aerial trimmer should be adjusted to give the strongest reception on a weak signal in the medium wave band, at say 200 metres.

Interference

In general, when electric current changes abruptly, unwanted electrical noise is produced. The motor vehicle is filled with electrical devices which change electric current rapidly, the most obvious being the contact breaker.

When the spark plugs operate, the sudden pulse of spark current causes the associated wiring to radiate. Since early radio transmitters used sparks as a basis of operation, it is not surprising that the car radio will pick up ignition spark noise unless steps are taken to reduce it to acceptable levels.

Interference reaches the car radio in two ways:

(a) by conduction through the wiring.
(b) by radiation to the receiving aerial.

Initial checks presuppose that the bonnet is down and fastened, the radio unit has a good earth connection (*not* through the aerial downlead outer), no fluorescent tubes are working near the car, the aerial trimmer has been adjusted, and the vehicle is in a position to receive radio signals, ie not in a metal-clad building.

Switch on the radio and tune it to the middle of the medium wave (MW) band off-station with the volume (gain) control set fairly high. Switch on the ignition (but do not start the engine) and wait to see if irregular clicks or hash noise occurs. Tapping the facia panel may also produce the effects. If so, this will be due to the voltage stabiliser, which is an on-off thermal switch to control instrument voltage. It is located usually on the back of the instrument panel, often attached to the speedometer. Correction is by attachment of a capacitor and, if still troublesome, chokes in the supply wires.

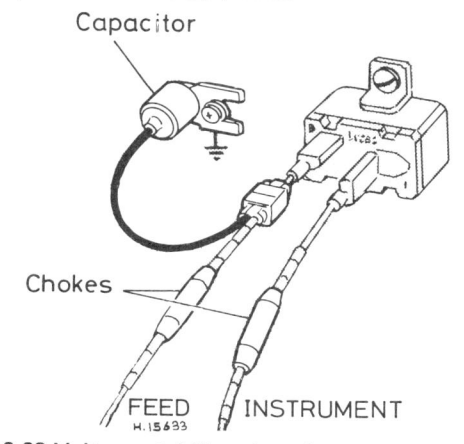

Fig. 12.28 Voltage stabiliser interference suppression (Sec 53)

Switch on the engine and listen for interference on the MW band. Depending on the type of interference, the indications are as follows.

A harsh crackle that drops out abruptly at low engine speed or when the headlights are switched on is probably due to a voltage regulator.

A whine varying with engine speed is due to the dynamo or alternator. Try temporarily taking off the fan belt – if the noise goes this is confirmation.

Regular ticking or crackle that varies in rate with the engine speed is due to the ignition system. With this trouble in particular and others in general, check to see if the noise is entering the receiver from the wiring or by radiation. To do this, pull out the aerial plug, (preferably shorting out the input socket or connecting a 62 pF capacitor across it). If the noise disappears it is coming in through the aerial and is *radiation noise*. If the noise persists it is reaching the receiver through the wiring and is said to be *line-borne*.

Interference from wipers, washers, heater blowers, turn-indicators, stop lamps, etc is usually taken to the receiver by wiring, and simple treatment using capacitors and possibly chokes will solve the problem. Switch on each one in turn (wet the screen first for running wipers!) and listen for possible interference with the aerial plug in place and again when removed.

Electric petrol pumps are now finding application again and give rise to an irregular clicking, often giving a burst of clicks when the ignition is on but the engine has not yet been started. It is also possible to receive whining or crackling from the pump.

Note that if most of the vehicle accessories are found to be creating interference all together, the probability is that poor aerial earthing is to blame.

Component terminal markings

Throughout the following sub-sections reference will be found to various terminal markings. These will vary depending on the manufacturer of the relevant component. If terminal markings differ from those mentioned, reference should be made to the following table, where the most commonly encountered variations are listed.

Chapter 12 Electrical system

Alternator	Alternator terminal (thick lead)	Exciting winding terminal
DIN/Bosch	B+	DF
Delco Remy	+	EXC
Ducellier	+	EXC
Ford (US)	+	DF
Lucas	+	F
Marelli	+B	F

Ignition coil	Ignition switch terminal	Contact breaker terminal
DIN/Bosch	15	1
Delco Remy	+	–
Ducellier	BAT	RUP
Ford (US)	B/+	CB/–
Lucas	SW/+	–
Marelli	BAT/+B	D

Voltage regulator	Voltage input terminal	Exciting winding terminal
DIN/Bosch	B+/D+	DF
Delco Remy	BAT/+	EXC
Ducellier	BOB/BAT	EXC
Ford (US)	BAT	DF
Lucas	+/A	F
Marelli		F

Suppression methods – ignition

Suppressed HT cables are supplied as original equipment by manufacturers and will meet regulations as far as interference to neighbouring equipment is concerned. It is illegal to remove such suppression unless an alternative is provided, and this may take the form of resistive spark plug caps in conjunction with plain copper HT cable. For VHF purposes, these and 'in-line' resistors may not be effective, and resistive HT cable is preferred. Check that suppressed cables are actually fitted by observing cable identity lettering, or measuring with an ohmmeter – the value of each plug lead should be 5000 to 10 000 ohms.

A 1 microfarad capacitor connected from the LT supply side of the ignition coil to a good nearby earth point will complete basic ignition interference treatment. *NEVER fit a capacitor to the coil terminal to the contact breaker – the result would be burnt out points in a short time.*

If ignition noise persists despite the treatment above, the following sequence should be followed:

(a) Check the earthing of the ignition coil; remove paint from fixing clamp.
(b) If this does not work, lift the bonnet. Should there be no change in interference level, this may indicate that the bonnet is not electrically connected to the car body. Use a proprietary braided strap across a bonnet hinge ensuring a first class electrical connection. If, however, lifting the bonnet increases the interference, then fit resistive HT cables of a higher ohms-per-metre value.

(c) If all these measures fail, it is probable that re-radiation from metallic components is taking place. Using a braided strap between metallic points, go round the vehicle systematically – try the following: engine to body, exhaust system to body, front suspension to engine and to body, steering column to body (especially French and Italian cars), gear lever to engine and to body (again especially French and Italian cars), Bowden cable to body, metal parcel shelf to body. When an offending component is located it should be bonded with the strap permanently.
(d) As a next step, the fitting of distributor suppressors to each lead at the distributor end may help.
(e) Beyond this point is involved the possible screening of the distributor and fitting resistive spark plugs, but such advanced treatment is not usually required for vehicles with entertainment equipment.

Electronic ignition systems have built-in suppression components, but this does not relieve the need for using suppressed HT leads. In some cases it is permitted to connect a capacitor on the low tension supply side of the ignition coil, but not in every case. Makers' instructions should be followed carefully, otherwise damage to the ignition semiconductors may result.

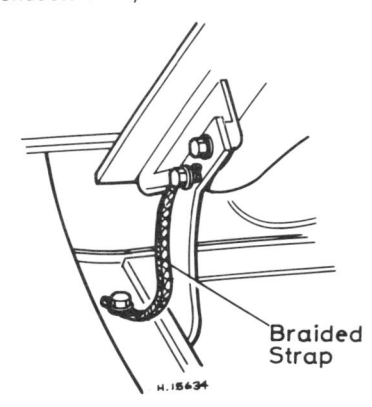

Fig. 12.29 Braided earth strap between bonnet and body (Sec 53)

Suppression methods – generators

For older vehicles with dynamos a 1 microfarad capacitor from the D (larger) terminal to earth will usually cure dynamo whine. Alternators should be fitted with a 3 microfarad capacitor from the B+ main output terminal (thick cable) to earth. Additional suppression may be obtained by the use of a filter in the supply line to the radio receiver.

It is most important that:

(a) Capacitors are never connected to the field terminals of either a dynamo or alternator.
(b) Alternators must not be run without connection to the battery.

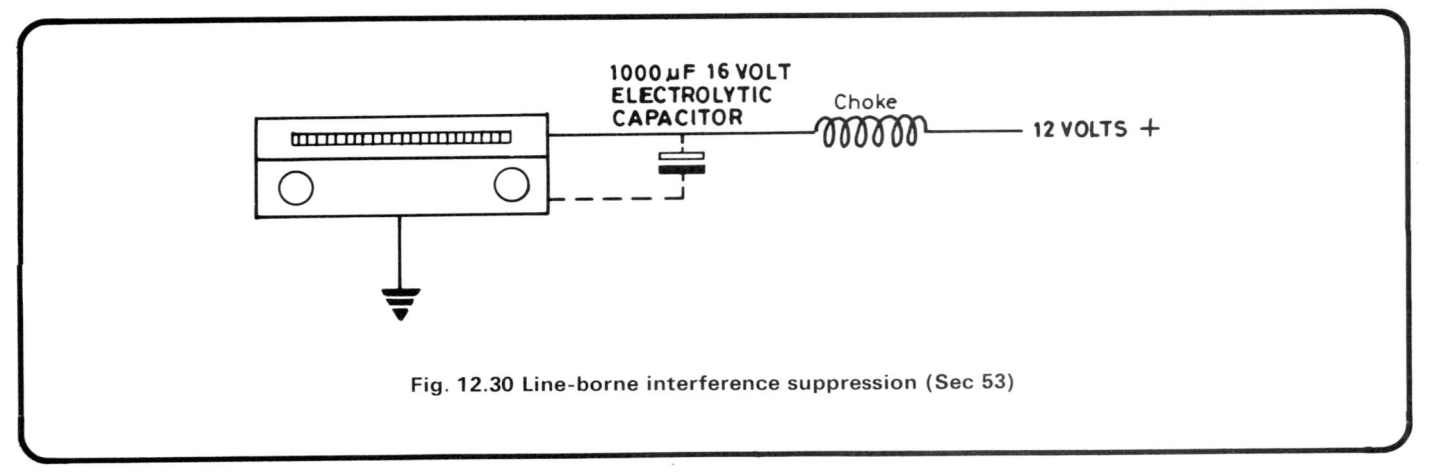

Fig. 12.30 Line-borne interference suppression (Sec 53)

Chapter 12 Electrical system

Suppression methods – voltage regulators

Voltage regulators used with DC dynamos should be suppressed by connecting a 1 microfarad capacitor from the control box D terminal to earth.

Alternator regulators come in three types:

(a) Vibrating contact regulators separate from the alternator. Used extensively on continental vehicles.
(b) Electronic regulators separate from the alternator.
(c) Electronic regulators built-in to the alternator.

In case (a) interference may be generated on the AM and FM (VHF) bands. For some cars a replacement suppressed regulator is available. Filter boxes may be used with non-suppressed regulators. But if not available, then for AM equipment a 2 microfarad or 3 microfarad capacitor may be mounted at the voltage terminal marked D+ or B+ of the regulator. FM bands may be treated by a feed-through capacitor of 2 or 3 microfarad.

Electronic voltage regulators are not always troublesome, but where necessary, a 1 microfarad capacitor from the regulator + terminal will help.

Integral electronic voltage regulators do not normally generate much interference, but when encountered this is in combination with alternator noise. A 1 microfarad or 2 microfarad capacitor from the warning lamp (IND) terminal to earth for Lucas ACR alternators and Femsa, Delco and Bosch equivalents should cure the problem.

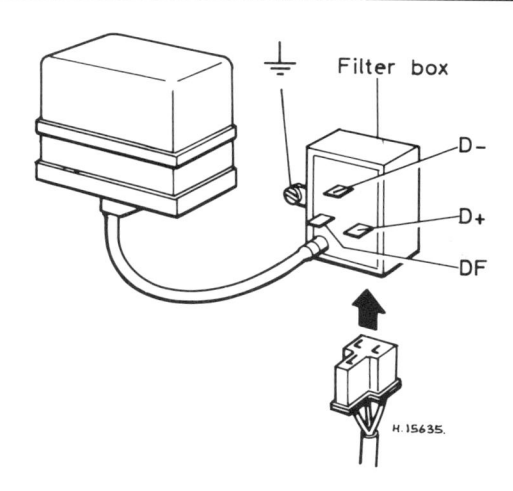

Fig. 12.31 Typical filter box for vibrating contact voltage regulator (alternator equipment) (Sec 53)

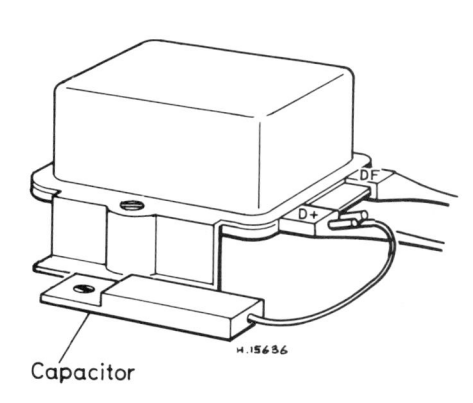

Fig. 12.32 Suppression of AM interference by vibrating contact voltage regulator (alternator equipment) (Sec 53)

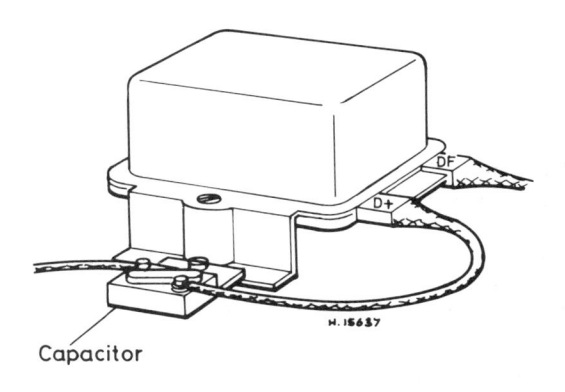

Fig. 12.33 Suppression of FM interference by vibrating contact voltage regulator (alternator equipment) (Sec 53)

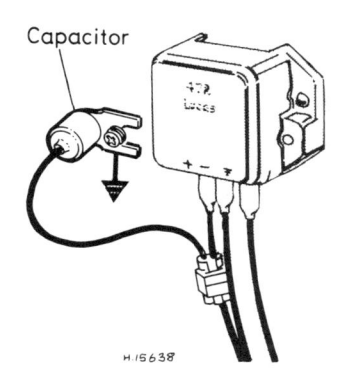

Fig. 12.34 Electronic voltage regulator suppression (Sec 53)

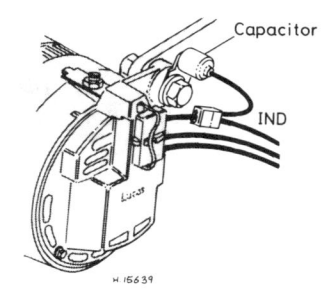

Fig. 12.35 Suppression of interference from electonic voltage regulator when integral with alternator (Sec 53)

Suppression methods – other equipment

Wiper motors – Connect the wiper body to earth with a bonding strap. For all motors use a 7 ampere choke assembly inserted in the leads to the motor.

Heater motors – Fit 7 ampere line chokes in both leads, assisted if necessary by a 1 microfarad capacitor to earth from both leads.

Electronic tachometer – The tachometer is a possible source of ignition noise – check by disconnecting at the ignition coil CB terminal. It usually feeds from ignition coil LT pulses at the contact

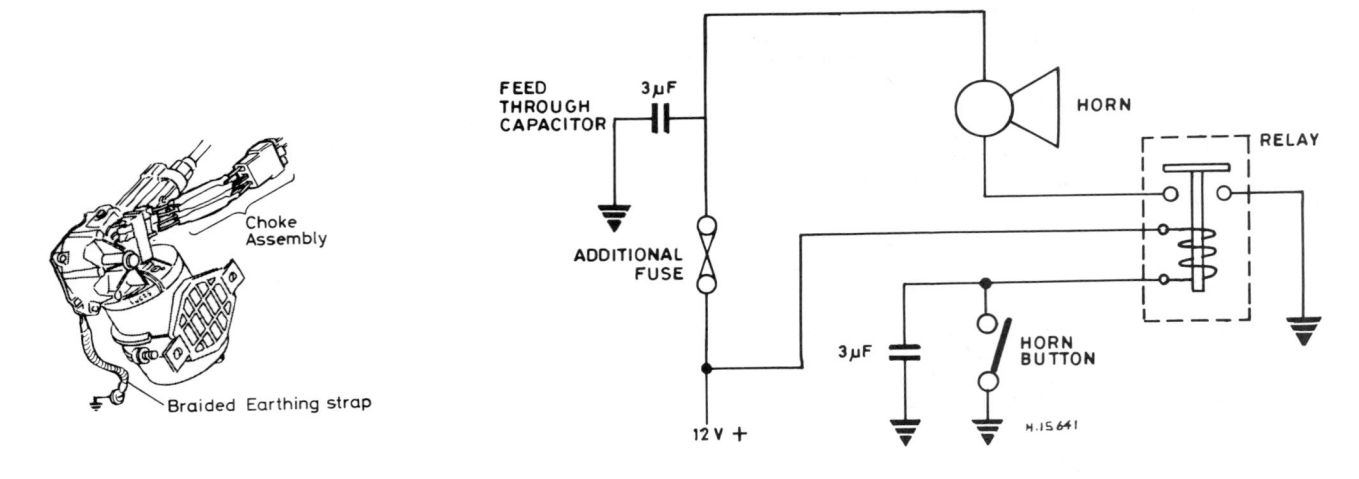

Fig. 12.36 Wiper motor suppression (Sec 53)

Fig. 12.37 Use of relay to reduce horn interference (Sec 53)

breaker terminal. A 3 ampere line choke should be fitted in the tachometer lead at the coil CB terminal.

Horn – A capacitor and choke combination is effective if the horn is directly connected to the 12 volt supply. The use of a relay is an alternative remedy, as this will reduce the length of the interference-carrying leads.

Electrostatic noise – Characteristics are erratic crackling at the receiver, with disappearance of symptoms in wet weather. Often shocks may be given when touching bodywork. Part of the problem is the build-up of static electricity in non-driven wheels and the acquisition of charge on the body shell. It is possible to fit spring-loaded contacts at the wheels to give good conduction between the rotary wheel parts and the vehicle frame. Changing a tyre sometimes helps – because of tyres' varying resistances. In difficult cases a trailing flex which touches the ground will cure the problem. If this is not acceptable it is worth trying conductive paint on the tyre walls.

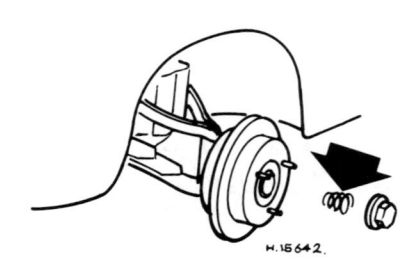

Fig. 12.38 Use of spring contacts at wheels (Sec 53)

Fuel pump – Suppression requires a 1 microfarad capacitor between the supply wire to the pump and a nearby earth point. If this is insufficient a 7 ampere line choke connected in the supply wire near the pump is required.

Fluorescent tubes – Vehicles used for camping/caravanning frequently have fluorescent tube lighting. These tubes require a relatively high voltage for operation and this is provided by an inverter (a form of oscillator) which steps up the vehicle supply voltage. This can give rise to serious interference to radio reception, and the tubes themselves can contribute to this interference by the pulsating nature of the lamp discharge. In such situations it is important to mount the aerial as far away from a fluorescent tube as possible. The interference problem may be alleviated by screening the tube with fine wire turns spaced an inch (25 mm) apart and earthed to the chassis. Suitable chokes should be fitted in both supply wires close to the inverter.

Radio/cassette case breakthrough

Magnetic radiation from dashboard wiring may be sufficiently intense to break through the metal case of the radio/cassette player. Often this is due to a particular cable routed too close and shows up as ignition interference on AM and cassette play and/or alternator whine on cassette play.

The first point to check is that the clips and/or screws are fixing all parts of the radio/cassette case together properly. Assuming good earthing of the case, see if it is possible to re-route the offending cable – the chances of this are not good, however, in most cars.

Next release the radio/cassette player and locate it in different positions with temporary leads. If a point of low interference is found, then if possible fix the equipment in that area. This also confirms that local radiation is causing the trouble. If re-location is not feasible, fit the radio/cassette player back in the original position.

Alternator interference on cassette play is now caused by radiation from the main charging cable which goes from the battery to the output terminal of the alternator, usually via the + terminal of the starter motor relay. In some vehicles this cable is routed under the dashboard, so the solution is to provide a direct cable route. Detach the original cable from the alternator output terminal and make up a new cable of at least 6 mm² cross-sectional area to go from alternator to battery with the shortest possible route. *Remember* – do not run the engine with the alternator disconnected from the battery.

Ignition breakthrough on AM and/or cassette play can be a difficult problem. It is worth wrapping earthed foil round the offending cable run near the equipment, or making up a deflector plate well screwed down to a good earth. Another possibility is the use of a suitable relay to switch on the ignition coil. The relay should be mounted close to the ignition coil; with this arrangement the ignition coil primary current is not taken into the dashboard area and does not flow through the ignition switch. A suitable diode should be used since it is possible that at ignition switch-off the output from the warning lamp alternator terminal could hold the relay on.

Connectors for suppression components

Capacitors are usually supplied with tags on the end of the lead, while the capacitor body has a flange with a slot or hole to fit under a nut or screw with washer.

Connections to feed wires are best achieved by self-stripping connectors. These connectors employ a blade which, when squeezed down by pliers, cuts through cable insulation and makes connection to the copper conductors beneath.

Chokes sometimes come with bullet snap-in connectors fitted to the wires, and also with just bare copper wire. With connectors, suitable female cable connectors may be purchased from an auto-accessory shop together with any extra connectors required for the cable ends after being cut for the choke insertion. For chokes with bare wires, similar connectors may be employed together with insulation sleeving as required.

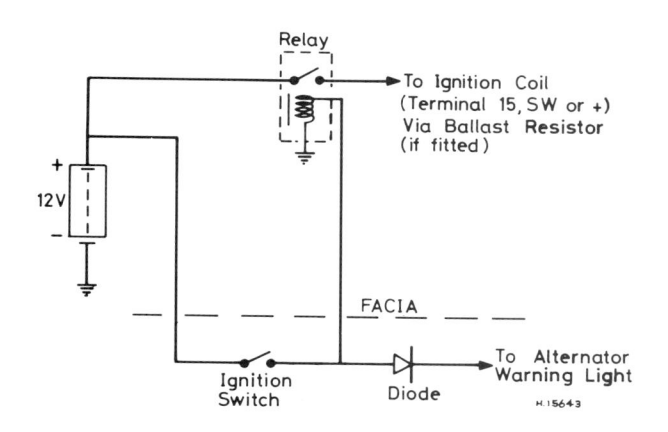

Fig. 12.39 Use of ignition coil relay to suppress case breakthrough (Sec 53)

VHF/FM broadcasts

Reception of VHF/FM in an automobile is more prone to problems than the medium and long wavebands. Medium/long wave transmitters are capable of covering considerable distances, but VHF transmitters are restricted to line of sight, meaning ranges of 10 to 50 miles, depending upon the terrain, the effects of buildings and the transmitter power.

Because of the limited range it is necessary to retune on a long journey, and it may be better for those habitually travelling long distances or living in areas of poor provision of transmitters to use an AM radio working on medium/long wavebands.

When conditions are poor, interference can arise, and some of the suppression devices described previously fall off in performance at very high frequencies unless specifically designed for the VHF band. Available suppression devices include reactive HT cable, resistive distributor caps, screened plug caps, screened leads and resistive spark plugs.

For VHF/FM receiver installation the following points should be particularly noted:

(a) Earthing of the receiver chassis and the aerial mounting is important. Use a separate earthing wire at the radio, and scrape paint away at the aerial mounting.
(b) If possible, use a good quality roof aerial to obtain maximum height and distance from interference generating devices on the vehicle.
(c) Use of a high quality aerial downlead is important, since losses in cheap cable can be significant.
(d) The polarisation of FM transmissions may be horizontal, vertical, circular or slanted. Because of this the optimum mounting angle is at 45° to the vehicle roof.

Citizens' Band radio (CB)

In the UK, CB transmitter/receivers work within the 27 MHz and 934 MHz bands, using the FM mode. At present interest is concentrated on 27 MHz where the design and manufacture of equipment is less difficult. Maximum transmitted power is 4 watts, and 40 channels spaced 10 kHz apart within the range 27.60125 to 27.99125 MHz are available.

Aerials are the key to effective transmission and reception. Regulations limit the aerial length to 1.65 metres including the loading coil and any associated circuitry, so tuning the aerial is necessary to obtain optimum results. The choice of a CB aerial is dependent on whether it is to be permanently installed or removable, and the performance will hinge on correct tuning and the location point on the vehicle. Common practice is to clip the aerial to the roof gutter or to employ wing mounting where the aerial can be rapidly unscrewed. An alternative is to use the boot rim to render the aerial theftproof, but a popular solution is to use the 'magmount' – a type of mounting having a strong magnetic base clamping to the vehicle at any point, usually the roof.

Aerial location determines the signal distribution for both transmission and reception, but it is wise to choose a point away from the engine compartment to minimise interference from vehicle electrical equipment.

The aerial is subject to considerable wind and acceleration forces. Cheaper units will whip backwards and forwards and in so doing will alter the relationship with the metal surface of the vehicle with which it forms a ground plane aerial system. The radiation pattern will change correspondingly, giving rise to break-up of both incoming and outgoing signals.

Interference problems on the vehicle carrying CB equipment fall into two categories:

(a) Interference to nearby TV and radio receivers when transmitting.
(b) Interference to CB set reception due to electrical equipment on the vehicle.

Problems of break-through to TV and radio are not frequent, but can be difficult to solve. Mostly trouble is not detected or reported because the vehicle is moving and the symptoms rapidly disappear at the TV/radio receiver, but when the CB set is used as a base station any trouble with nearby receivers will soon result in a complaint.

It must not be assumed by the CB operator that his equipment is faultless, for much depends upon the design. Harmonics (that is, multiples) of 27 MHz may be transmitted unknowingly and these can fall into other user's bands. Where trouble of this nature occurs, low pass filters in the aerial or supply leads can help, and should be fitted in base station aerials as a matter of course. In stubborn cases it may be necessary to call for assistance from the licensing authority, or, if possible, to have the equipment checked by the manufacturers.

Interference received on the CB set from the vehicle equipment is, fortunately, not usually a severe problem. The precautions outlined previously for radio/cassette units apply, but there are some extra points worth noting.

It is common practice to use a slide-mount on CB equipment enabling the set to be easily removed for use as a base station, for example. Care must be taken that the slide mount fittings are properly earthed and that first class connection occurs between the set and slide-mount.

Vehicle manufacturers in the UK are required to provide suppression of electrical equipment to cover 40 to 250 MHz to protect TV and VHF radio bands. Such suppression appears to be adequately effective at 27 MHz, but suppression of individual items such as alternators/dynamos, clocks, stabilisers, flashers, wiper motors, etc, may still be necessary. The suppression capacitors and chokes available from auto-electrical suppliers for entertainment receivers will usually give the required results with CB equipment.

Other vehicle radio transmitters

Besides CB radio already mentioned, a considerable increase in the use of transceivers (ie combined transmitter and receiver units) has taken place in the last decade. Previously this type of equipment was fitted mainly to military, fire, ambulance and police vehicles, but a large business radio and radio telephone usage has developed.

Generally the suppression techniques described previously will suffice, with only a few difficult cases arising. Suppression is carried out to satisfy the 'receive mode', but care must be taken to use heavy duty chokes in the equipment supply cables since the loading on 'transmit' is relatively high.

54 Fault diagnosis – electrical system

Symptom	Reason(s)
Starter motor fails to turn engine	Battery discharged Battery defective internally Battery terminal leads loose or earth lead not securely attached to body Loose or broken connection in starter motor circuit Starter motor switch or solenoid faulty Starter brushes badly worn, sticking, or brush wires loose Commutatior dirty, worn or burnt Starter motor armature faulty Field coils earthed
Starter motor turns engine very slowly	Battery in discharged condition Starter brushes badly worn, sticking, or brush wires loose Loose wires in starter motor circuit
Starter motor noisy or excessively rough engagement	Pinion or ring gear broken or worn Starter motor retaining bolts loose
Battery will not hold charge for more than a few days	Battery defective internally Electrolyte level too low or electrolyte too weak due to leakage Plate separators no longer fully effective Battery plates severely sulphated Fan/alternator drivebelt slipping Battery terminal connections loose or corroded Alternator not charging properly Short in lighting circuit causing continual battery drain
Ignition light fails to go out, battery runs flat in a few days	Fan drivebelt loose and slipping or broken Alternator faulty
Fuel gauge gives no reading	Fuel tank empty! Electric cable between tank unit and gauge broken Fuel gauge case not earthed Fuel gauge supply cable interrupted Fuel gauge unit broken
Fuel gauge registers full all the time	Electric cable between tank and gauge earthed or shorting
Horn operates at the time	Horn push earthed or stuck down Horn cable to horn push earthed
Horn fails to operate	Blown fuse Cable or cable connection loose, broken or disconnected Horn has an internal fault
Horn emits intermittent or unsatisfactory noise	Cable connections loose
Lights do not come on	Blown fuse If engine not running, battery discharged Light bulb filament burnt out or bulbs broken Wire connections loose, disconnected or broken Light switch shorting or otherwise faulty
Lights come on but fade out	If engine not running, battery discharged
Lights give very poor illumination	Poor earth connection Reflector tarnished or dirty Lamp badly out of adjustment Incorrect bulb with too low wattage fitted Existing bulbs old and badly discoloured Electrical wiring too thin not allowing full current to pass
Lights work erratically – flashing on and off, especially over bumps	Battery terminals or earth connection loose Lights not earthing properly Contacts in light switch faulty
Wiper motor fails to work	Blown fuse Wire connections loose, disconnected or broken Brushes badly worn Armature worn or faulty

Chapter 12 Electrical system

Symptom	Reason(s)
Wiper motor works very slowly and takes excessive current	Commutator dirty, greasy or burnt Wiper linkage bent or unlubricated Wiper spindle binding or damaged Armature bearings dry or unaligned Armature badly worn or faulty
Wiper motor works slowly and takes little current	Brushes badly worn Commutator dirty, greasy or burnt Armature badly worn or faulty
Wiper motor works but wiper blades remain static	Wiper linkage broken Wiper motor gearbox parts badly worn

Wiring diagrams – general
At the time of writing, no wiring diagrams for 2.8 litre models were available

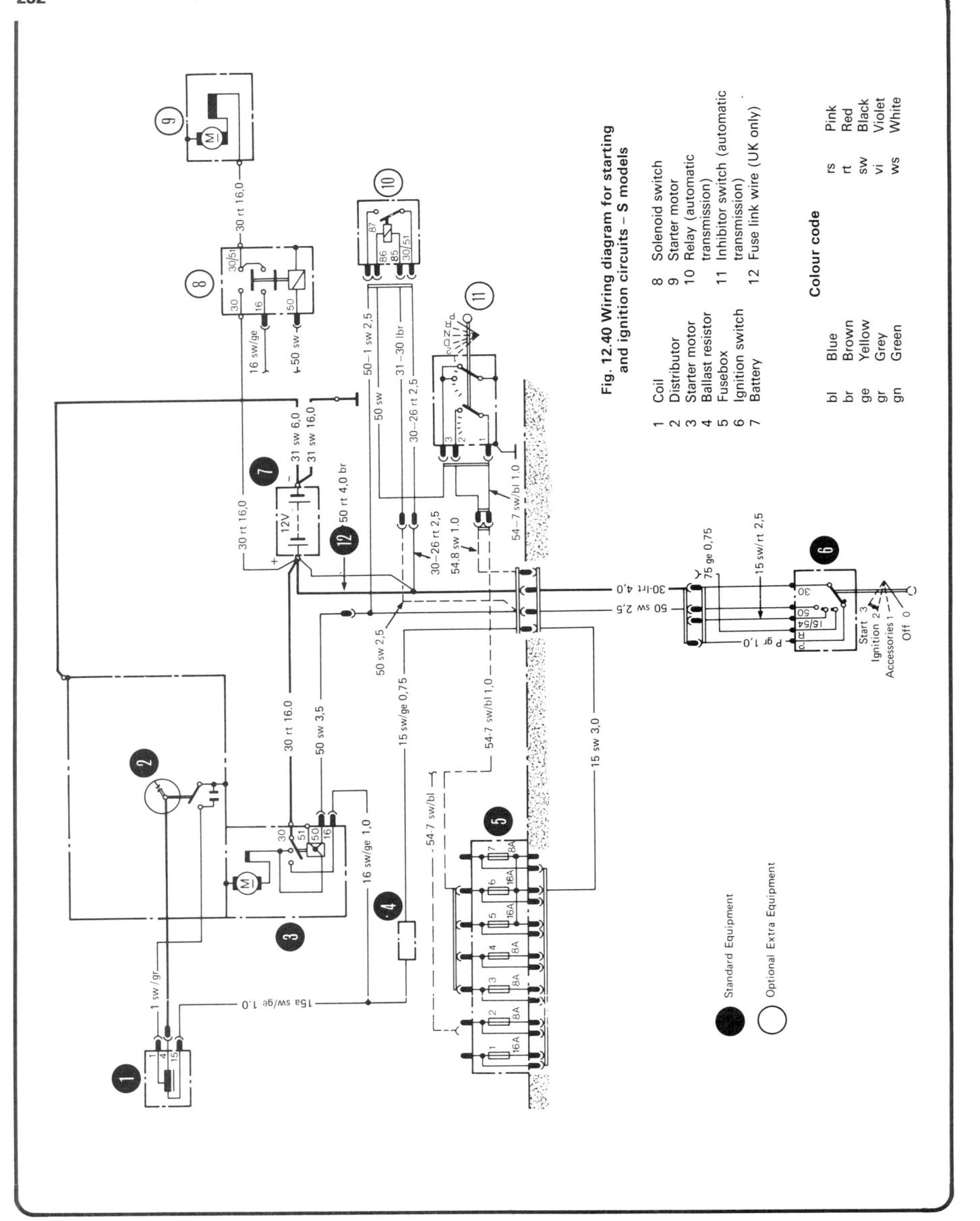

Fig. 12.40 Wiring diagram for starting and ignition circuits – S models

1 Coil
2 Distributor
3 Starter motor
4 Ballast resistor
5 Fusebox
6 Ignition switch
7 Battery
8 Solenoid switch
9 Starter motor
10 Relay (automatic transmission)
11 Inhibitor switch (automatic transmission)
12 Fuse link wire (UK only)

Colour code

bl	Blue	rs	Pink
br	Brown	rt	Red
ge	Yellow	sw	Black
gr	Grey	vi	Violet
gn	Green	ws	White

● Standard Equipment
○ Optional Extra Equipment

Fig. 12.41 Wiring diagram for charging circuit – S models

15 Alternator (Bosch and Femsa)
16 Regulator (Bosch and Femsa)
17 Battery
18 Ignition switch
19 Instrument cluster
20 Alternator (Lucas)

● Standard Equipment
○ Optional Extra Equipment

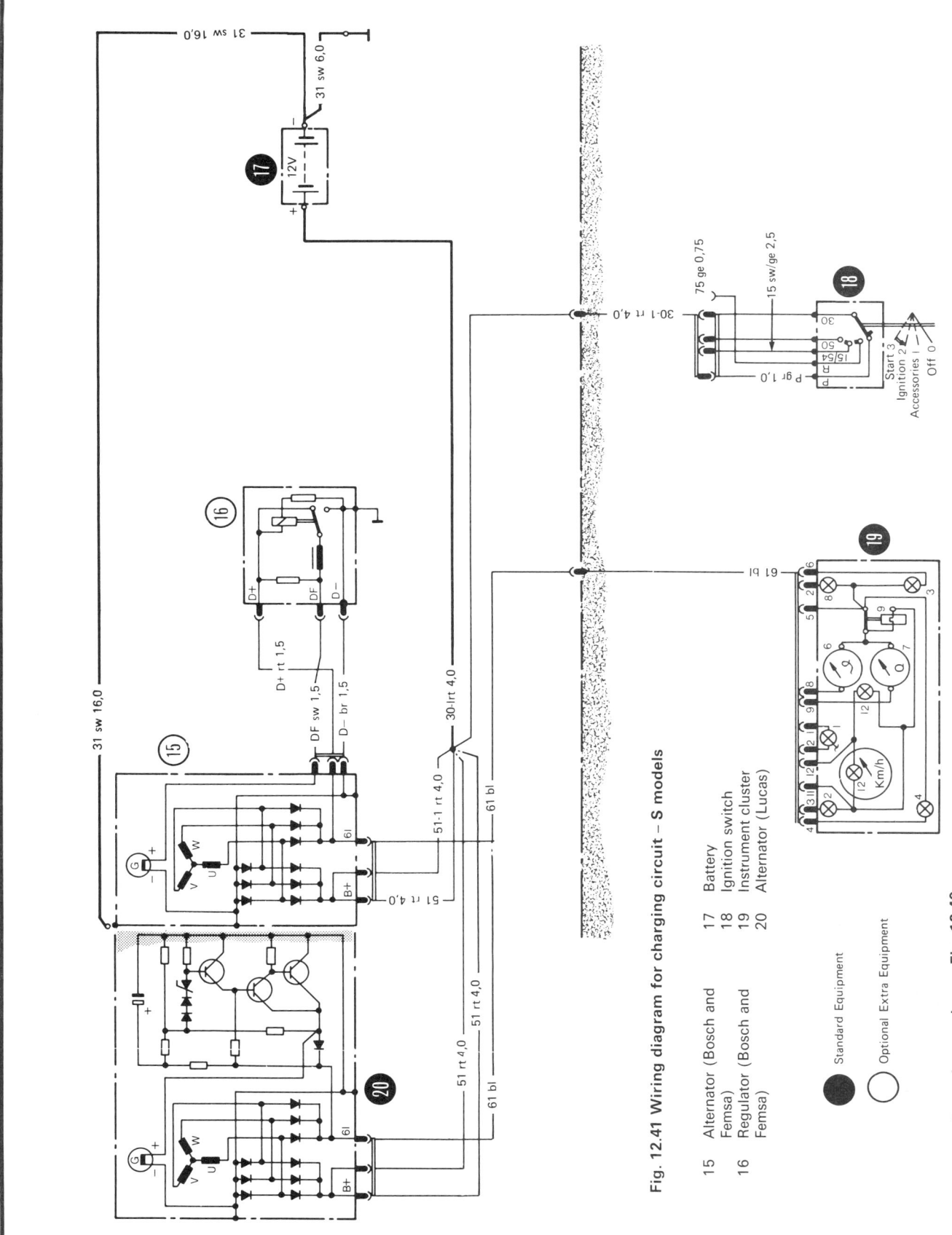

For colour code see Fig. 12.40

Fig. 12.42 Wiring diagram for exterior lights circuits – S models

24 Headlamp
25 Starter motor
26 Ballast resistor
27 Battery
28 Fusebox
29 Headlamp 'dip' relay
30 Rear lamp assembly
31 Light switch
32 Flasher switch
33 Number plate lamp
34 Ignition switch
35 Instrument cluster
36 Foglamp
37 Relay (foglamp)
38 Relay (foglamp)
39 'Dip' relay (foglamp)
40 'Dip' relay (headlamp)
41 Foglamp switch

For colour code see Fig. 12.40

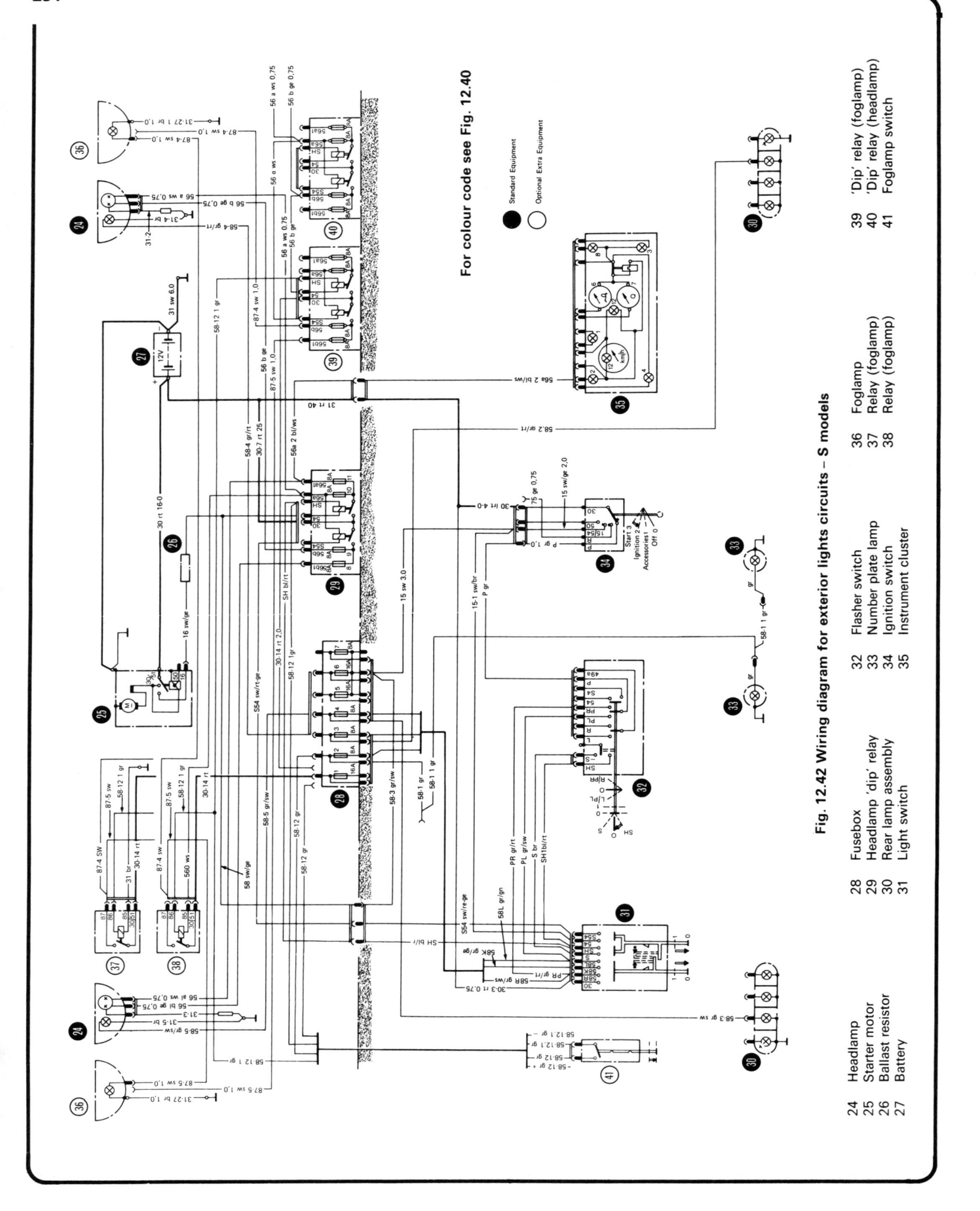

● Standard Equipment
○ Optional Extra Equipment

235

Fig. 12.43 Wiring diagram for interior lights circuits – S models

45 Starter motor
46 Ballast resistor
47 Battery
48 Fusebox
49 Interior light
50 Courtesy light switch
51 Ignition switch
52 Cigarette lighter
53 Glovebox lamp
54 Glovebox lamp switch
55 Interior light

Standard Equipment

For colour code see Fig. 12.40

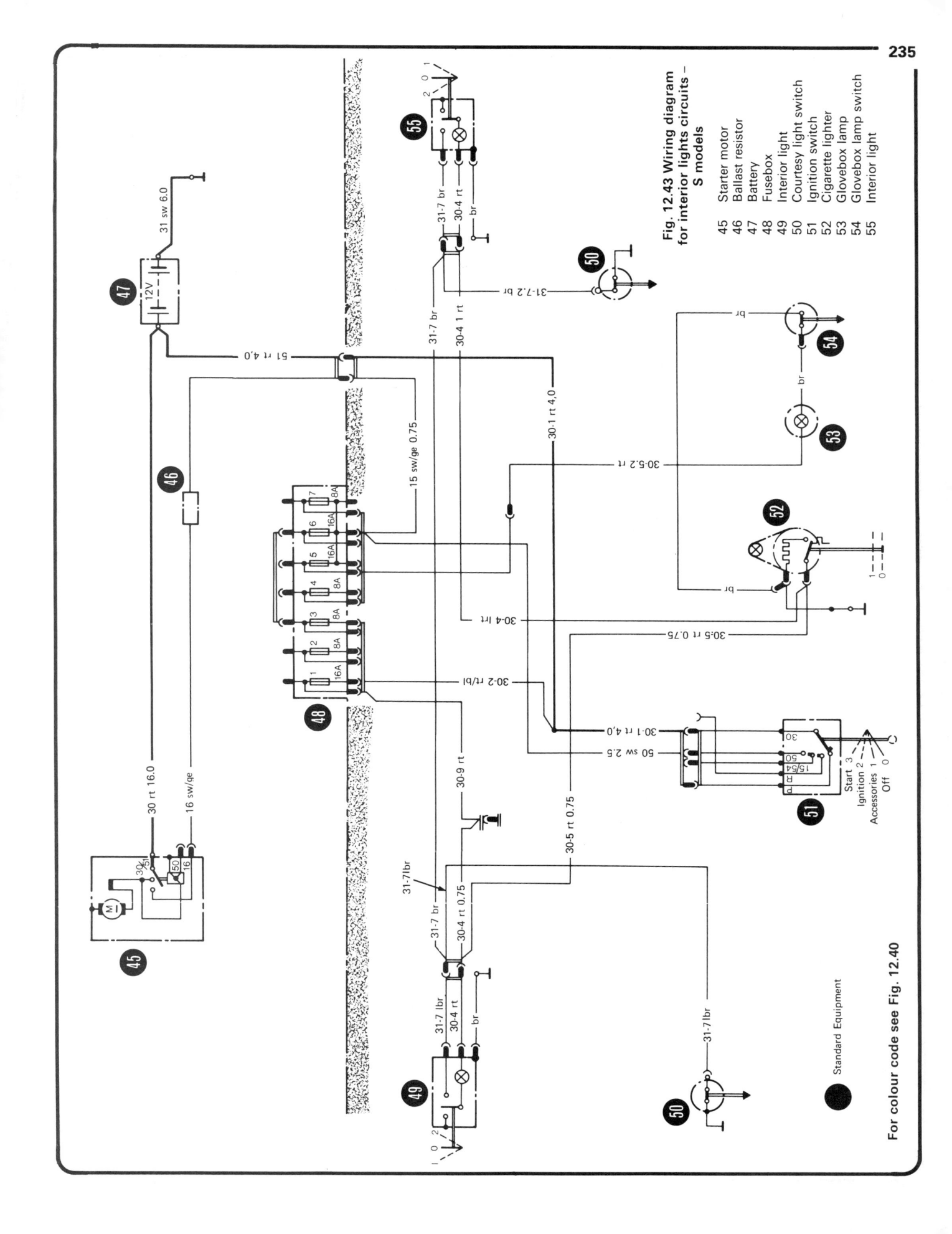

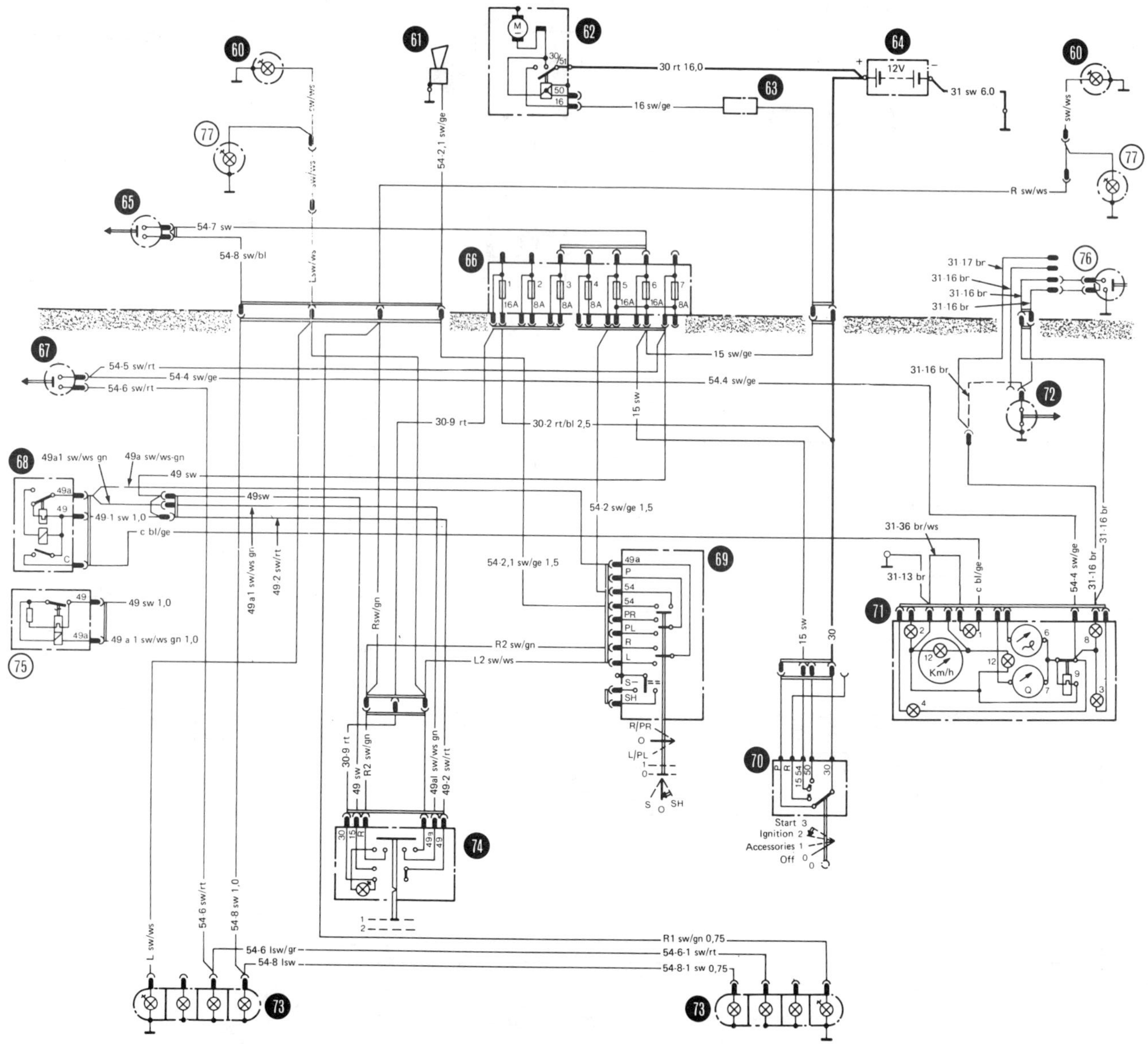

236

Fig. 12.44 Wiring diagram for horn, direction indicator and hazard warning lights circuits – S models

60 Front flasher lamp
61 Horn
62 Starter motor
63 Ballast resistor
64 Battery
65 Reversing lamp switch
66 Fusebox
67 Stop-light switch
68 Flasher unit
69 Indicator switch
70 Ignition switch
71 Instrument cluster
72 Handbrake warning light switch
73 Rear lamp assembly
74 Flasher switch warning system
75 Flasher unit (UK only)
76 Dual circuit brake warning system switch
77 Side repeater flasher lamp

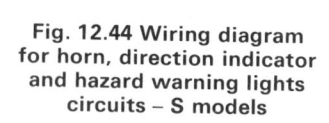

● Standard Equipment
○ Optional Extra Equipment

For colour code see Fig. 12.40

237

Fig. 12.45 Wiring diagram for heater, wiper and ancillary circuits – S models

80 Water temperature sender
81 Oil pressure control switch
81 Starter motor
83 Battery
84 Ballast resistor
85 Electric washer pump motor
86 Fusebox
87 Heated rear window
88 Wiper/washer system rear window
89 Heater blower motor
90 Wiper motor
91 Series resistor (heater motor)
92 Light switch
93 Ignition switch
94 Instrument cluster
95 Fuel gauge sender unit
96 Dimmer potentiometer
97 Wiper motor switch
98 Heater blower switch
99 Heater blower switch
100 Cigarette lighter
101 Electric wash pump switch

For colour code see Fig. 12.40

● Standard Equipment

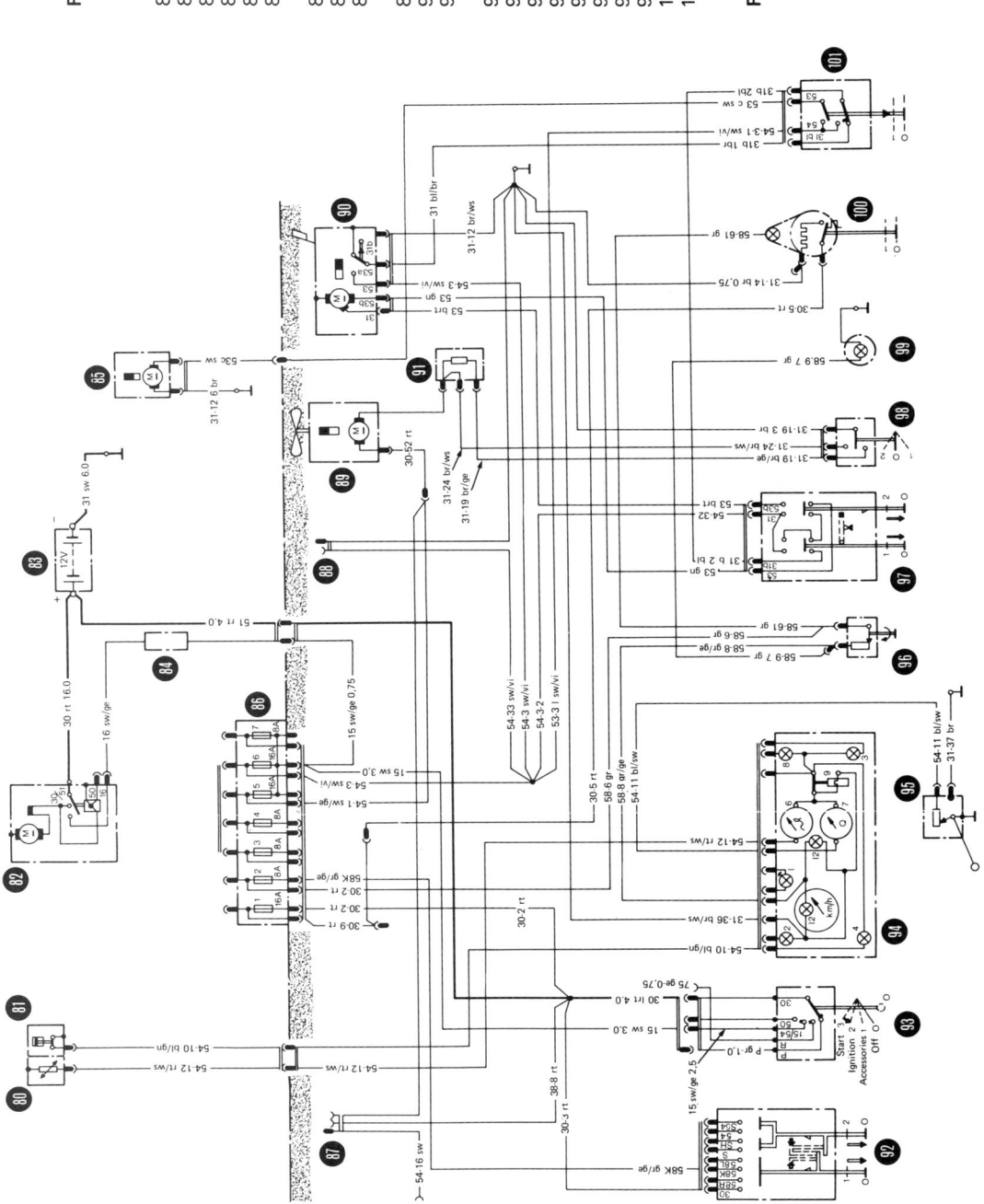

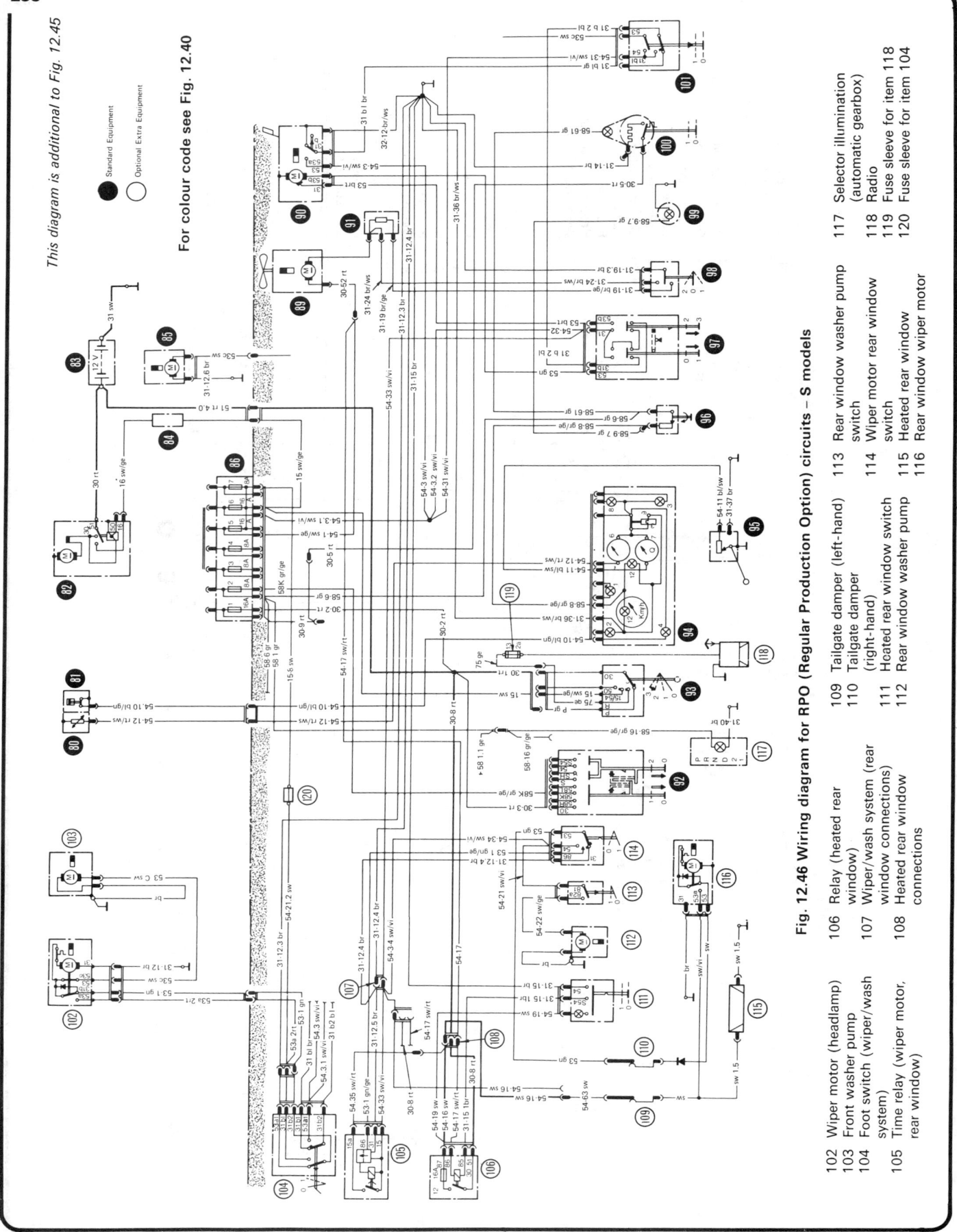

Fig. 12.46 Wiring diagram for RPO (Regular Production Option) circuits – S models

102 Wiper motor (headlamp)
103 Front washer pump
104 Foot switch (wiper/wash system)
105 Time relay (wiper motor, rear window)
106 Relay (heated rear window)
107 Wiper/wash system (rear window connections)
108 Heated rear window connections
109 Tailgate damper (left-hand)
110 Tailgate damper (right-hand)
111 Heated rear window switch
112 Rear window washer pump
113 Rear window washer switch
114 Wiper motor rear window switch
115 Heated rear window
116 Rear window wiper motor
117 Selector illumination (automatic gearbox)
118 Radio
119 Fuse sleeve for item 118
120 Fuse sleeve for item 104

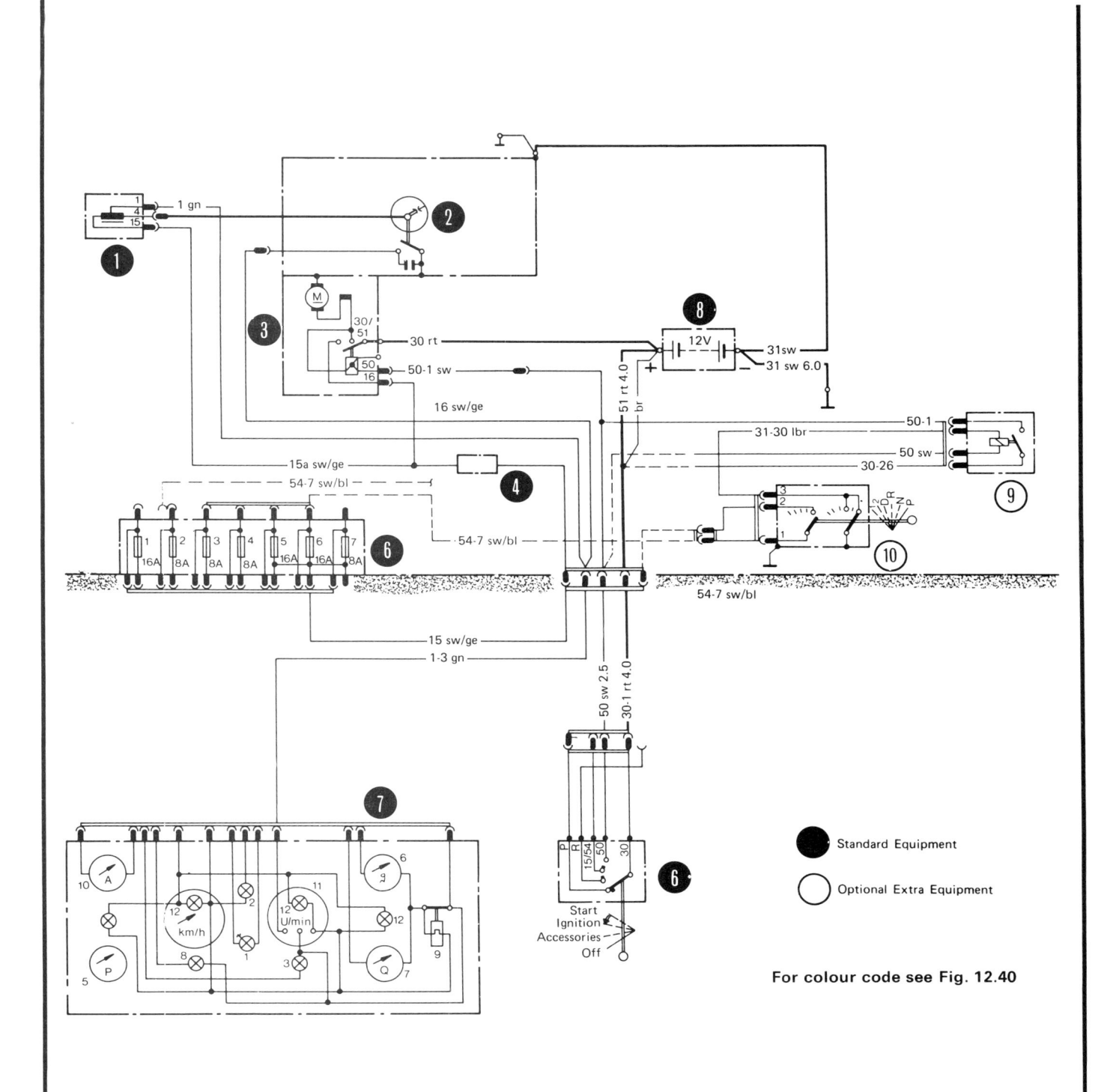

Fig. 12.47 Wiring diagram for starting and ignition circuits – GT and Ghia models

1 Coil
2 Distributor
3 Starter motor
4 Ballast resistor
5 Instrument cluster
6 Fusebox
7 Ignition switch
8 Battery
9 Relay (automatic transmission)
10 Inhibitor switch (automatic transmission)
11 Fuse link wire (UK only)

Standard Equipment

Optional Extra Equipment

For colour code see Fig. 12.40

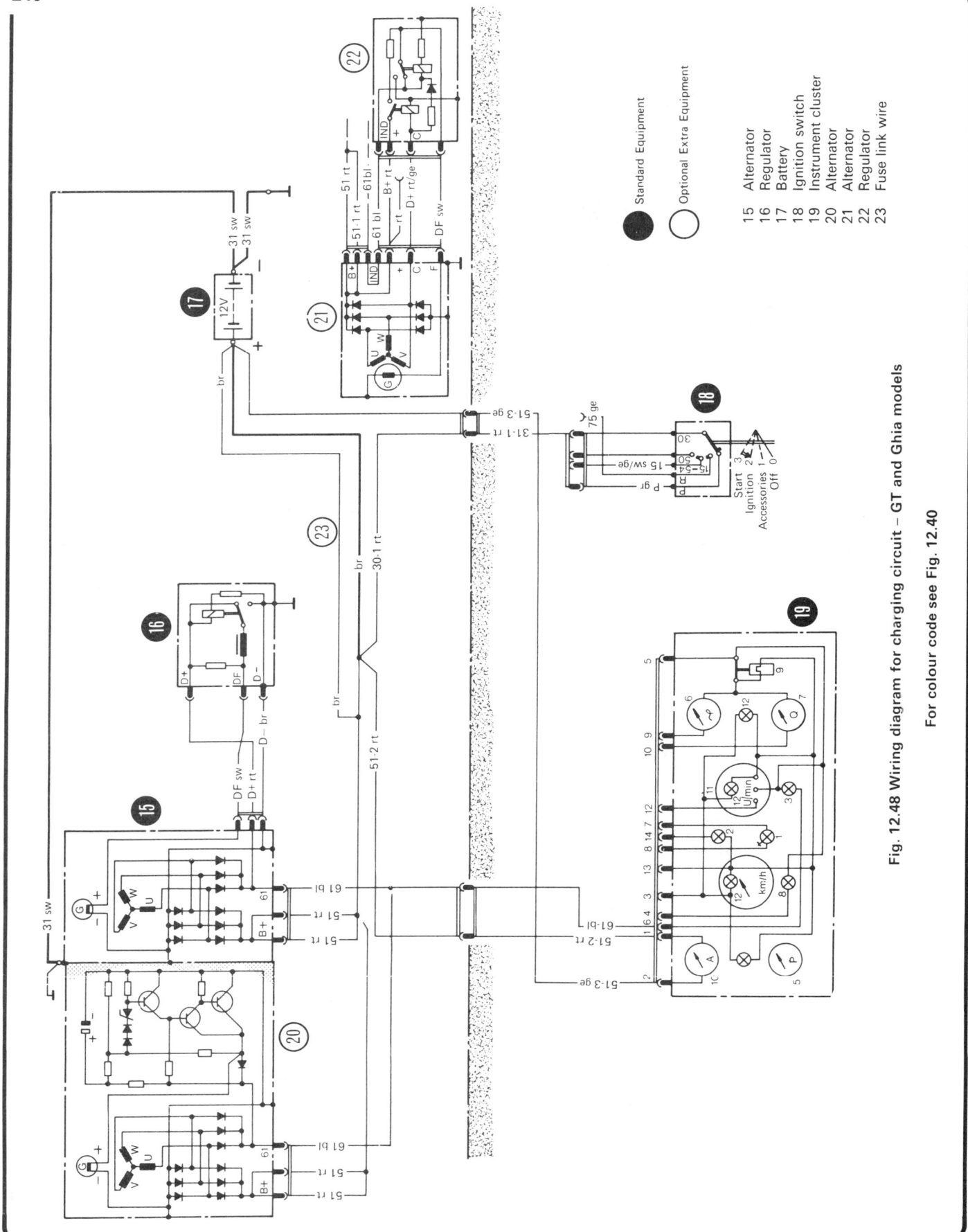

Fig. 12.48 Wiring diagram for charging circuit – GT and Ghia models

For colour code see Fig. 12.40

- Standard Equipment
- Optional Extra Equipment

15 Alternator
16 Regulator
17 Battery
18 Ignition switch
19 Instrument cluster
20 Alternator
21 Alternator
22 Regulator
23 Fuse link wire

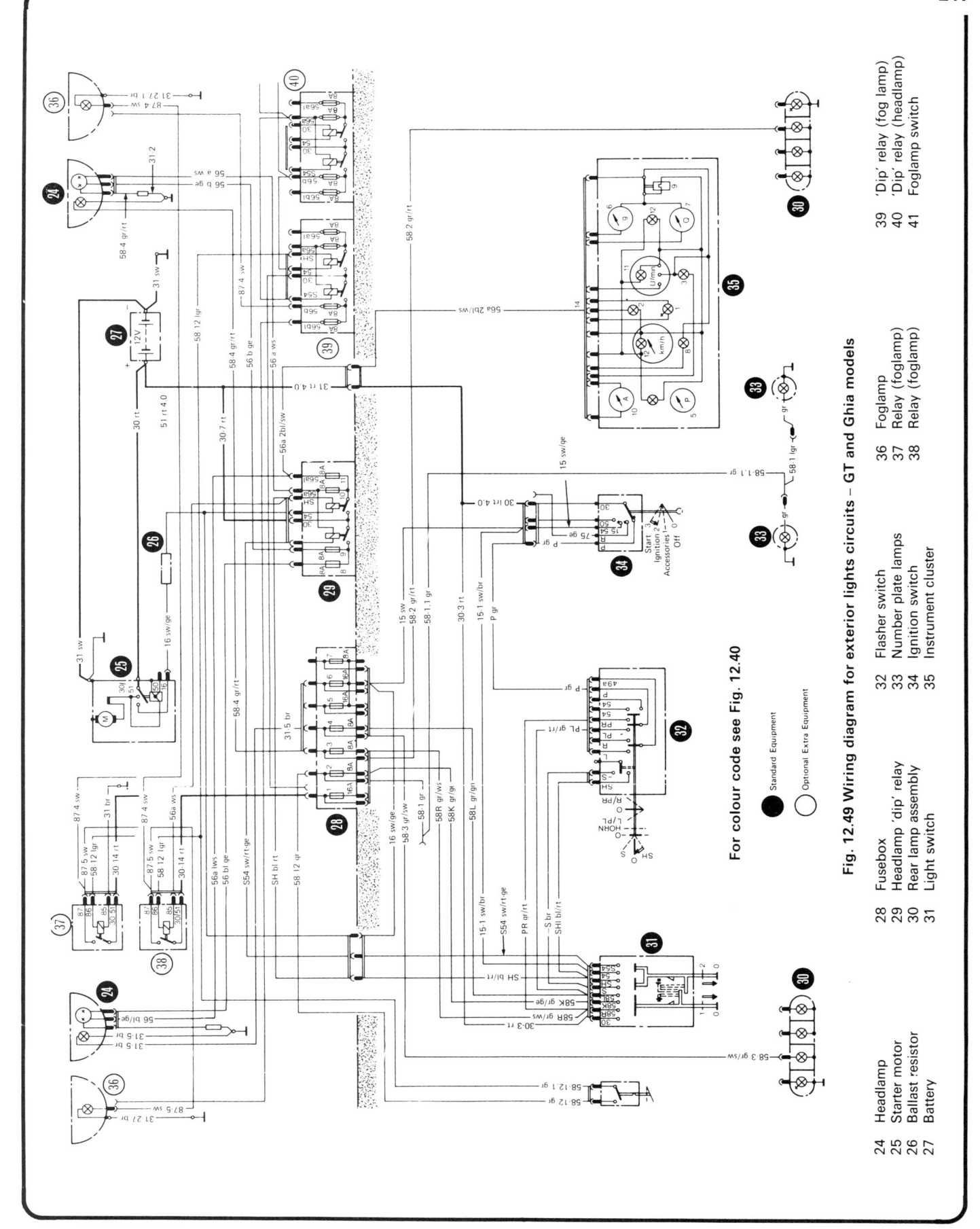

Fig. 12.49 Wiring diagram for exterior lights circuits – GT and Ghia models

24 Headlamp
25 Starter motor
26 Ballast resistor
27 Battery
28 Fusebox
29 Headlamp 'dip' relay
30 Rear lamp assembly
31 Light switch
32 Flasher switch
33 Number plate lamps
34 Ignition switch
35 Instrument cluster
36 Foglamp
37 Relay (foglamp)
38 Relay (foglamp)
39 'Dip' relay (fog lamp)
40 'Dip' relay (headlamp)
41 Foglamp switch

For colour code see Fig. 12.40

● Standard Equipment
○ Optional Extra Equipment

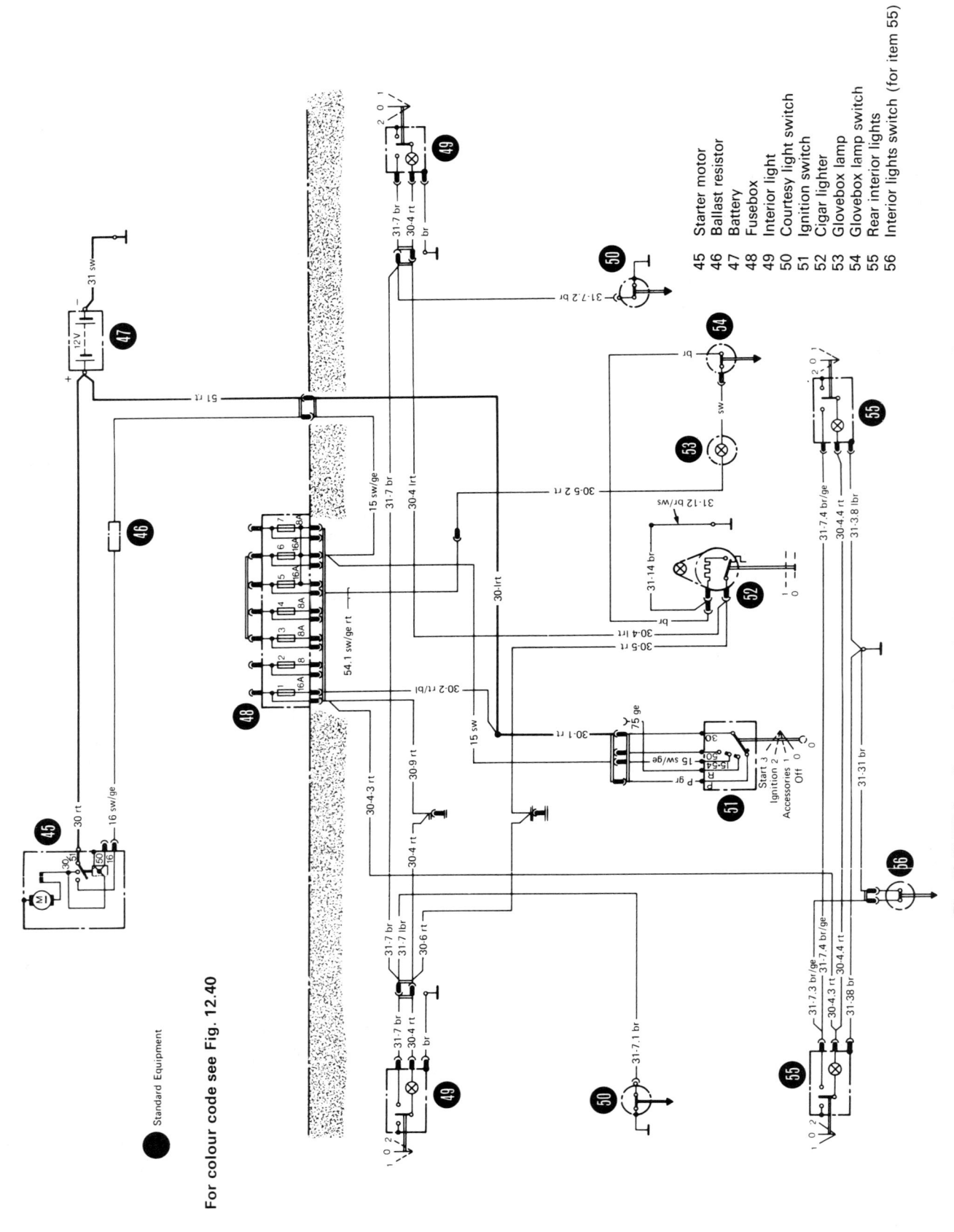

Fig. 12.50 Wiring diagram for interior lights circuits – GT and Ghia models

45 Starter motor
46 Ballast resistor
47 Battery
48 Fusebox
49 Interior light
50 Courtesy light switch
51 Ignition switch
52 Cigar lighter
53 Glovebox lamp
54 Glovebox lamp switch
55 Rear interior lights
56 Interior lights switch (for item 55)

Standard Equipment

For colour code see Fig. 12.40

243

Fig. 12.51 Wiring diagram for horn, direction indicator and hazard warning lights circuits – GT and Ghia models

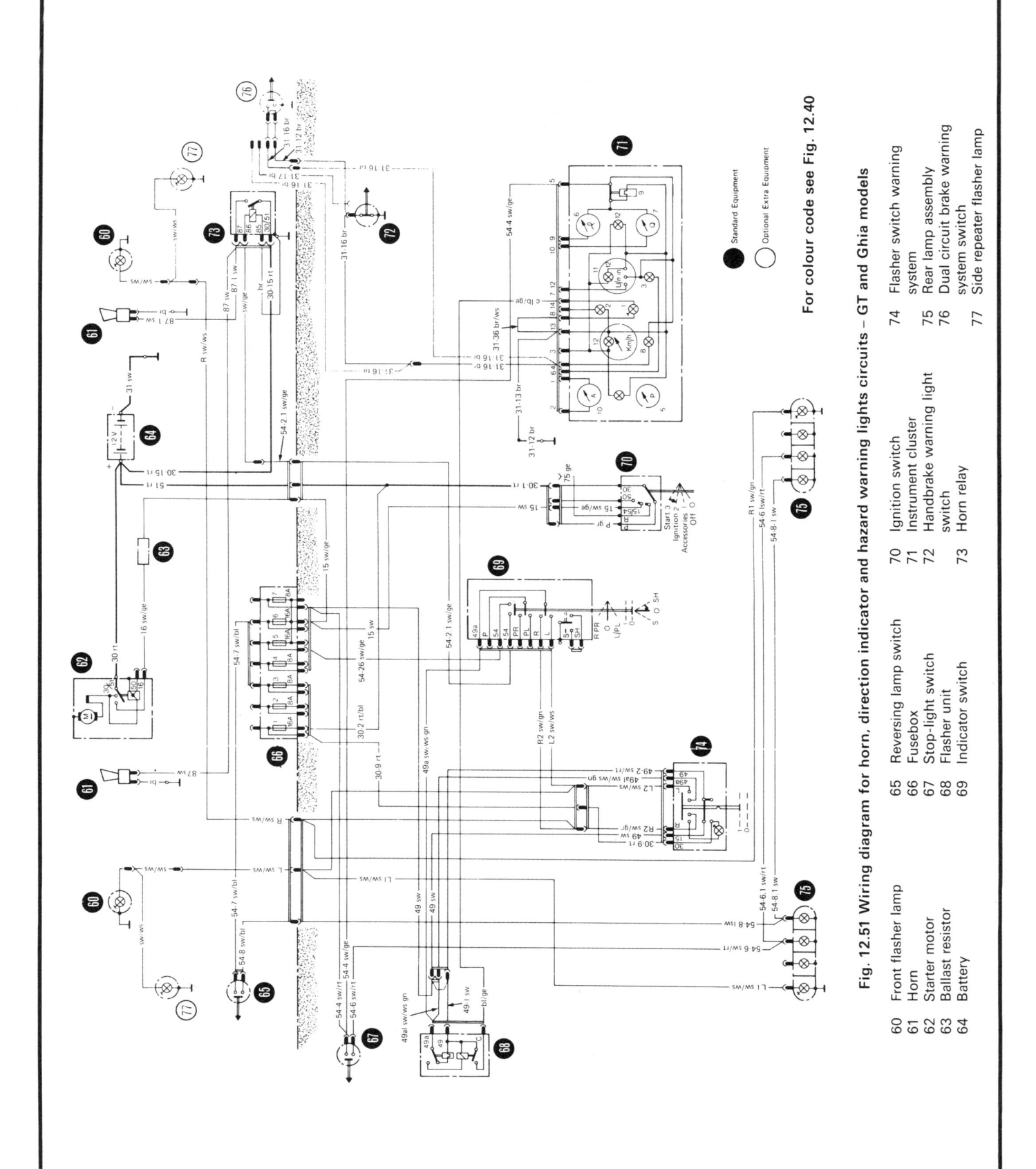

60 Front flasher lamp
61 Horn
62 Starter motor
63 Ballast resistor
64 Battery
65 Reversing lamp switch
66 Fusebox
67 Stop-light switch
68 Flasher unit
69 Indicator switch
70 Ignition switch
71 Instrument cluster
72 Handbrake warning light switch
73 Horn relay
74 Flasher switch warning system
75 Rear lamp assembly
76 Dual circuit brake warning system switch
77 Side repeater flasher lamp

For colour code see Fig. 12.40

Fig. 12.52 Wiring diagram for heater, wiper and ancillary circuits – GT and Ghia models

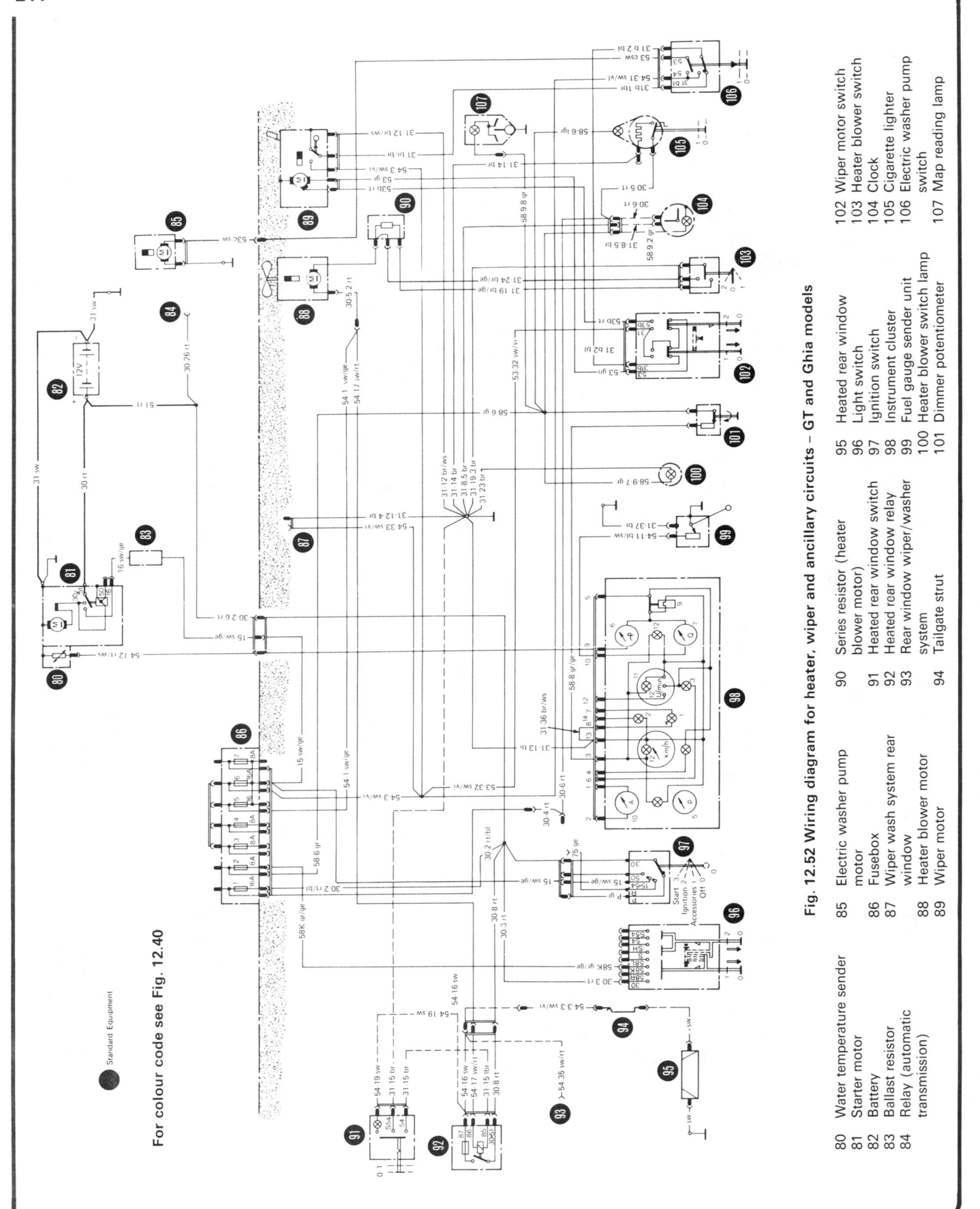

80 Water temperature sender
81 Starter motor
82 Battery
83 Ballast resistor
84 Relay (automatic transmission)
85 Electric washer pump motor
86 Fusebox
87 Wiper wash system rear window
88 Heater blower motor
89 Wiper motor
90 Series resistor (heater blower motor)
91 Heated rear window switch
92 Heated rear window relay
93 Rear window wiper/washer system
94 Tailgate strut
95 Heated rear window
96 Light switch
97 Ignition switch
98 Instrument cluster
99 Fuel gauge sender unit
100 Heater blower switch lamp
101 Dimmer potentiometer
102 Wiper motor switch
103 Heater blower switch
104 Clock
105 Cigarette lighter
106 Electric washer pump switch
107 Map reading lamp

For colour code see Fig. 12.40

● Standard Equipment

Fig. 12.53 Wiring diagram for RPO (Regular Production Option) circuits – GT and Ghia models

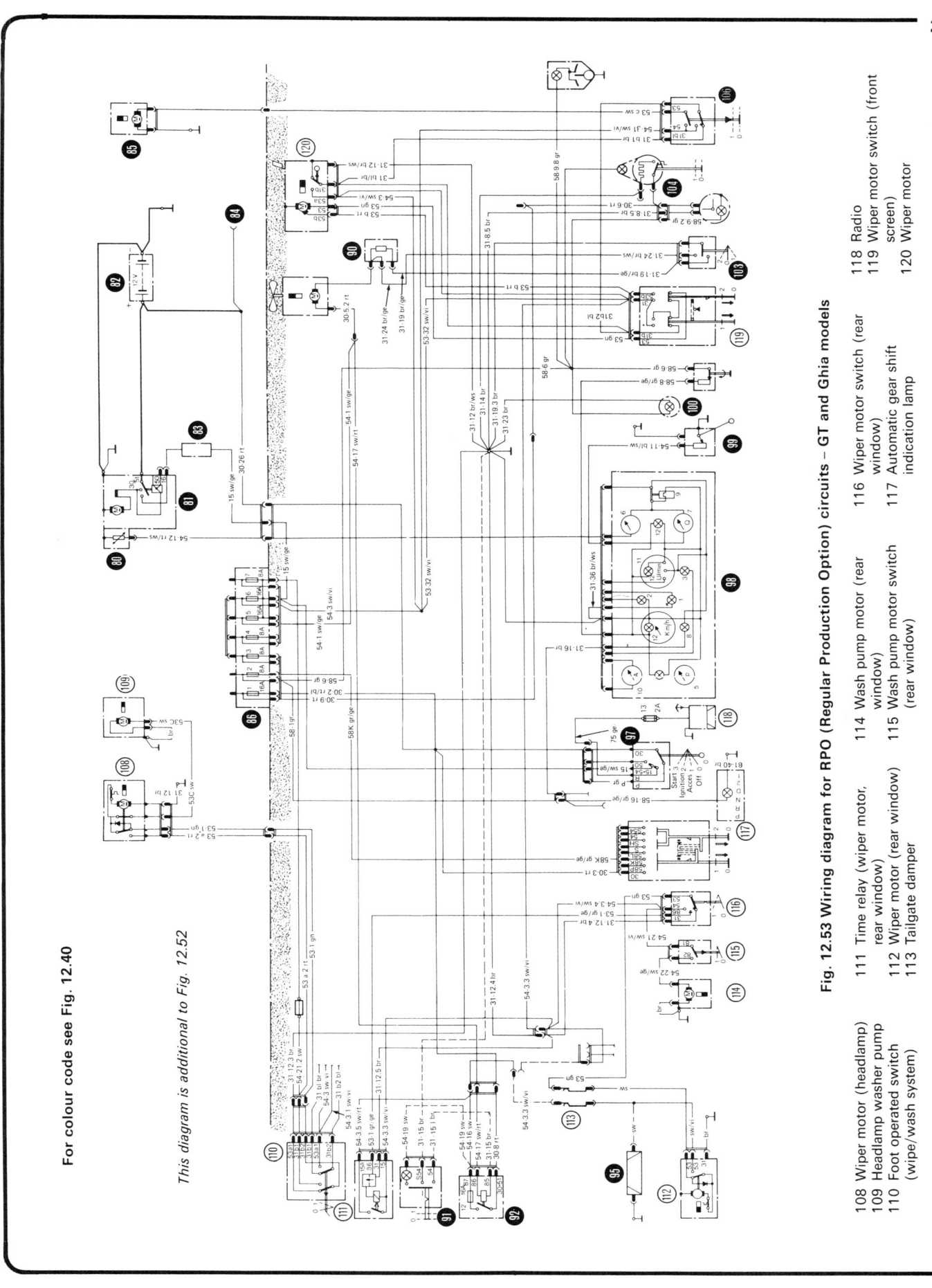

108 Wiper motor (headlamp)
109 Headlamp washer pump
110 Foot operated switch (wipe/wash system)
111 Time relay (wiper motor, rear window)
112 Wiper motor (rear window)
113 Tailgate damper
114 Wash pump motor (rear window)
115 Wash pump motor switch (rear window)
116 Wiper motor switch (rear window)
117 Automatic gear shift indication lamp
118 Radio
119 Wiper motor switch (front screen)
120 Wiper motor

For colour code see Fig. 12.40

This diagram is additional to Fig. 12.52

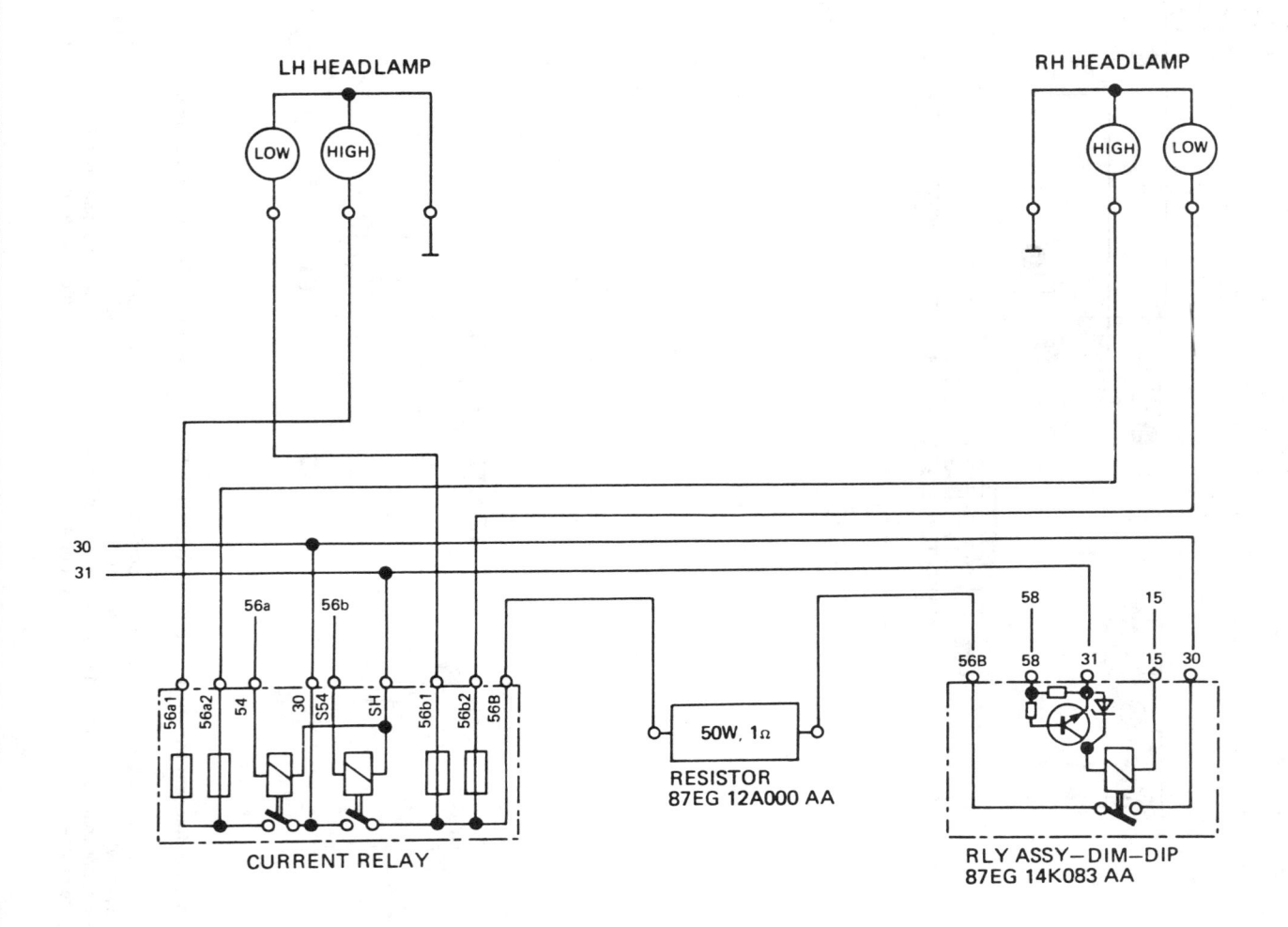

Fig. 12.54 Wiring diagram for dim-dip circuit – all models

Index

A

Accelerator
    cable removal and refitting – 74
    shaft and pedal removal and refitting – 75
Air cleaner
    removal and refitting
        carburettor models – 68
        fuel injection models – 80
Alternator
    brushes inspection, removal and refitting
        Bosch – 199
        Femsa – 200
        Lucas – 199
    dismantling and reassembly – 200
    removal and refitting – 198
Antifreeze mixture – 60
Automatic transmission
    description – 121
    downshift cable
        removal, refitting and adjustment – 126
    extension housing oil seal renewal – 126
    fault diagnosis – 126
    maintenance, routine – 122
    removal and refitting – 122
    selector mechanism
        adjustment – 125
        removal, overhaul and refitting – 124
    specifications – 102
    starter inhibitor/reverse lamp switch
        removal and refitting – 126
    torque wrench settings – 102
Auxiliary (driving) light
    alignment – 209

B

Battery
    charging – 198
    description and testing – 197
    removal and refitting – 198
Big-end bearings – 37, 52
Bleeding the hydraulic system
    brakes – 151
    power steering – 167
Bodywork and fittings – 172 *et seq*
Bodywork and fittings
    bonnet – 183
    bumpers – 175
    centre console – 186
    demister nozzles removal and refitting – 194
    description – 173
    doors – 175, 177, 179 to 182
    face level vents removal and refitting – 194
    facia panel – 185
    fuel filler flap – 185
    heater – 189 to 191, 193
    load space trim panel removal and refitting – 183
    maintenance
        bodywork and underframe – 173
        routine – 173
        upholstery and carpets – 173
    mirrors – 189
    open rear quarter glass assembly
        removal and refitting – 183
    rear quarter trim panel removal and refitting – 182
    repair
        major damage – 175
        minor damage – 173
    seat belts – 188
    seats – 189
    sunroof – 186 to 188
    tailgate – 177, 184, 185
    torque wrench settings – 172
    windscreen – 177
Bodywork repair sequence – *see colour section between pages 32 and 33*
Bonnet
    release cable removal and refitting – 183
    removal, refitting and adjustment – 183
Braking system – 141 *et seq*
Braking system
    description – 141
    fault diagnosis – 155
    footbrake pedal removal and refitting – 151
    front disc brakes – 143, 144
    handbrake – 152 to 154
    hydraulic system
        bleeding – 151
        pipes and hoses – 150
    maintenance, routine – 142
    master cylinder – 147, 148
    pressure differential switch removal and refitting – 149
    pressure reducing valve removal and refitting – 150
    rear drum brakes – 145 to 147
    specifications – 141
    stop lamp switch
        removal, refitting and adjustment – 151
    torque wrench settings – 141
    vacuum servo unit – 155
Bulbs, lamp
    renewal – 210 to 214
    specifications – 196
Bumpers
    removal and refitting – 175

C

Camshaft (3.0 litre engine)
    refitting – 54
    removal – 50
Camshaft and camshaft bearings
    examination and renovation
        2.8 litre engine – 35
        3.0 litre engine – 51

Index

Camshaft and front intermediate plate (2.8 litre engine)
  refitting – 41
Capacities, general – 6
Carburettor (Weber)
  automatic choke
    adjustment – 71
    removal, overhaul and refitting – 71
  dismantling and reassembly – 69
  general – 68
  idling adjustment – 73
  removal and refitting – 69
  specifications – 66
Carpets
  maintenance – 173
Centre console
  removal and refitting – 186
Choke, automatic – 71
Cigar lighter
  removal and refitting – 222
Clock (console mounted)
  removal and refitting – 220
Clutch – 96 *et seq*
Clutch
  assembly
    inspection – 98
    refitting – 98
    removal – 98
  cable renewal – 99
  description – 96
  fault diagnosis – 100
  maintenance, routine – 97
  pedal removal and refitting – 100
  release bearing removal and refitting – 99
  specifications – 96
Condenser (3.0 litre models)
  testing, removal and refitting – 90
Connecting rods (3.0 litre engine) – 51, 52, 54
Connecting rods and gudgeon pins
  examination and renovation – 37, 52
Connecting rods and pistons
  refitting – 42, 54
Contact breaker points (3.0 litre models)
  adjustment – 89
  removal and refitting – 89
Conversion factors – 18
Cooling system – 57 *et seq*
Cooling system
  antifreeze mixture – 60
  description – 58
  draining – 59
  drivebelts – 63
  expansion tank – 64
  fan hub bearing (3.0 litre models) – 64
  fault diagnosis – 65
  filling – 60
  flushing – 69
  maintenance, routine – 59
  radiator – 60
  specifications – 57
  temperature gauge sender unit – 64
  thermostat – 61
  torque wrench settings – 57
  viscous drive fan – 63
  water pump – 61
Crankcase ventilation system
  2.8 litre engine – 39
  3.0 litre engine – 53
Crankshaft
  2.8 litre engine
    examination and renovation – 37
    refitting – 40
  3.0 litre engine
    examination and renovation – 52
Crankshaft and main bearings (3.0 litre engine)
  removal – 51

Crankshaft pulley
  refitting (2.8 litre engine) – 44
  removal (3.0 litre engine) – 50
Crankshaft rear oil seal (3.0 litre engine)
  refitting – 55
  removal – 51
Cylinder bores
  examination and renovation – 36, 51
Cylinder heads
  2.8 litre engine
    dismantling, renovation and reassembly – 35
    refitting – 45
  3.0 litre engine
    reassembly – 55
    refitting – 55
    removal with engine – 49
    removal with engine in car – 49

D

Decarbonising – 52
Dimensions, vehicle – 6
Dim-dip lighting system – 223
Disc brakes *see* **Front disc brakes**
Distributor
  overhaul
    2.8 litre engine – 92
    3.0 litre engine – 91
  removal and refitting – 90
Doors
  exterior handle removal and refitting – 180
  lock assembly removal and refitting – 181
  pillar switch – 221
  private lock removal and refitting – 182
  rattles tracing and rectification – 175
  remote control handle removal and refitting – 177
  removal and refitting – 182
  trim panel removal and refitting – 179
  window frame
    moulding and door weatherstrips
      removal and refitting – 182
    removal and refitting – 182
  window glass removal and refitting – 181
  window regulator assembly
    removal and refitting – 180
Drivebelts
  removal and refitting – 63
Drum brakes *see* **Rear drum brakes**

E

Electrical system – 195 *et seq*
Electrical system
  alternator – 198 to 200
  auxiliary driving light – 209
  battery – 197, 198
  bulbs – 196, 210 to 214
  cigar lighter – 222
  clock – 220
  description – 197
  dim-dip lighting system – 223
  fault diagnosis – 19, 230
  flasher units and relays – 222
  fuses – 196, 217
  headlights – 208 to 210
  horn – 217
  instrument cluster – 218
  instrument voltage regulator – 223
  lamps – 211 to 214
  maintenance, routine – 197

Index

mobile radio equipment – 223
precautions – 197
radio/cassette player – 223
specifications – 195
speedometer cable – 223
starter motor – 204
switches – 220 to 222
tailgate washer – 214
tailgate wiper – 214
torque wrench settings – 196
windscreen washer – 216
windscreen wiper – 215 to 217
wiring diagrams – 232 to 246

Engine – 23 *et seq*
Engine (2.8 litre)
ancillaries – 29, 36
camshaft – 35, 41
components examination for wear – 34
connecting rods – 37, 42
crankcase ventilation system – 39
crankshaft – 37, 40
crankshaft pulley – 44
cylinder bores – 36
cylinder heads – 35, 45
description – 26
dismantling – 29
fault diagnosis – 20, 56
flywheel – 38, 42
front intermediate plate – 41
gudgeon pins – 37
lubrication system – 39
main and big-end bearings – 37
maintenance, routine – 26
oil pump – 38, 42
operations possible with engine in car – 22
operations requiring engine removal – 22
piston rings – 36
pistons – 36, 42
reassembly – 40
refitting – 46
removal – 22
removal methods – 22
rocker assembly – 34, 45
specifications – 23
start-up after overhaul – 46
sump – 34, 44
tappets and pushrods – 35
timing cover – 43
timing gears – 38, 43
torque wrench settings – 24
valves – 45

Engine (3.0 litre)
ancillaries – 49, 55
camshaft – 50, 51, 54
components examination for wear – 34, 51
connecting rods – 37, 42, 51, 52, 54
crankcase ventilation system – 53
crankshaft – 52, 54
crankshaft pulley – 50
crankshaft rear oil seal – 51, 55
cylinder bores – 36, 51
cylinder heads – 49, 55
decarbonising – 52
description – 47
dismantling – 29, 49
fault diagnosis – 20, 56
flywheel/driveplate – 50, 52, 55
front cover – 50
front plate – 54
gudgeon pins – 37, 52
lubrication system – 53
main and big-end bearings – 37, 52
maintenance routine – 48
oil pump – 50, 52, 54
operations possible with engine in car – 48
operations requiring engine removal – 48

piston rings – 36, 51
pistons – 36, 42, 51, 54
pushrods – 35, 51
reassembly – 40, 54
refitting – 56
removal – 48
removal methods – 48
rocker arms – 49, 51, 55
specifications – 25
sump – 50, 55
tappets – 35, 50, 51
timing gears – 38, 50, 52, 54
torque wrench settings – 25
valves – 46, 49, 52, 55

Exhaust system
removal and refitting – 76
Expansion tank (cooling system)
removal and refitting – 64

F

Facia panel
removal and refitting – 185
Fan hub bearing (3.0 litre models)
renewal – 64
Fan, viscous drive
removal and refitting – 63
Fault diagnosis – 19 *et seq*
Fault diagnosis
automatic transmission – 126
braking system – 155
clutch – 100
cooling system – 65
electrical system – 19, 230
engine – 20, 56
fuel system
  carburettor – 85
  fuel injection – 85
ignition system – 95
manual gearbox – 121
propeller shaft – 130
rear axle – 140
suspension and steering – 171
Firing order – 87
Flasher unit – 222
Flywheel/driveplate
2.8 litre engine
  examination and renovation – 38
  refitting – 42
3.0 litre engine
  examination and renovation – 52
  refitting – 55
  removal – 50
Front cover (3.0 litre engine)
removal – 50
Front direction indicator lamp and bulb
removal and refitting – 211
Front disc brakes
caliper
  overhaul – 144
  removal and refitting – 143
discs inspection, removal and refitting – 144
pads removal – 143
Front driving lamp and bulb
removal and refitting – 212
Front foglamp and bulb
removal and refitting – 212
Front suspension
coil spring removal and refitting – 161
description – 157
fault diagnosis – 171
hubs – 160, 161
maintenance, routine – 160
stabiliser bar removal and refitting – 162
strut removal and refitting – 161

Index

torque wrench settings – 157
track control arm removal and refitting – 163
Fuel and exhaust systems – 66 *et seq*
Fuel filler cap
    removal and refitting – 185
Fuel injection system
    accelerator cable – 74, 83
    air cleaner – 80
    description – 77
    exhaust manifold – 85
    fault diagnosis – 85
    fuel tank, sender unit and filler pipe – 83
    idle mixture adjustment – 82
    inlet manifold – 83
    main system components removal and refitting – 80
    maintenance, routine – 79
    specifications – 67
    torque wrench settings – 67
Fuel pump
    carburettor system
        cleaning – 68
        removal and refitting – 68
        testing – 68
    fuel injection system
        removal and refitting – 80
Fuel system (carburettor)
    accelerator cable – 74
    accelerator shaft and pedal – 75
    air cleaner – 68
    carburettor – 66, 68 to 73
    description – 67
    exhaust manifold – 76
    fault diagnosis – 85
    fuel pump – 68
    fuel tank – 73, 74
    fuel tank filler pipe – 74
    fuel tank sender unit – 74
    inlet manifold – 75
    maintenance, routine – 67
    specifications – 66
    torque wrench settings – 67
Fuel tank
    cleaning and repair – 74
    filler pipe removal and refitting – 74
    removal and refitting – 73
    sender unit removal and refitting – 74
Fuel tank sender unit (fuel injection models) – 83
Fuses
    general – 217
    specificatons – 196

G

Gearbox *see* **Manual gearbox**
Gudgeon pins – 37, 52

H

Handbrake
    adjustment – 152
    cable(s) and rod removal and refitting – 154
    lever removal and refitting – 153
    warning light switch – 222
Headlights
    alignment – 209
    assembly: removal and refitting – 208
    bulbs renewal – 210
Heater
    assembly: dismantling and reassembly
        Behr – 191
        Smiths standard and heavy duty – 193

assembly: removal and refitting – 190
controls
    adjustment – 189
    removal and refitting – 190
water valve (heavy duty heater)
    removal and refitting – 190
History of the model – 5
Horn
    fault tracing, removal and refitting – 217
Hubs, front
    bearings renewal – 161
    removal, refitting and bearing adjustment – 160
Hydraulic system (brakes)
    bleeding – 151
    pipes and hoses removal and refitting – 150

I

Ignition switch and lock – 221
Ignition system – 87 *et seq*
Ignition system
    amplifier module (2.8 litre models)
        removal and refitting – 93
    condenser (3.0 litre models) – 90
    contact breaker points (3.0 litre models) – 89
    description – 88
    distributor – 90, 91, 92
    fault diagnosis – 95
    firing order – 87
    maintenance, routine – 88
    spare plugs – 91, 94
    specifications – 87
    timing – 93
    torque wrench settings – 87
Instrument cluster
    illumination switch – 221
    removal and refitting – 218
Instrument voltage regulator
    removal and refitting – 223
Interior lamp and bulb
    removal and refitting – 214

J

Jacking – 7

L

Lubricants and fluids recommended – 17
Lubrication chart – 17
Lubrication system
    2.8 litre engine – 39
    3.0 litre engine – 53

M

Main and big-end bearings
    examination and renovation – 37, 52
Main bearings (3.0 litre engine)
    removal – 51
Maintenance, routine
    bodywork and fittings
        door check straps security check – 16
        locks, door check straps, fuel filler cap flap
            lubrication – 16

Index

locks operation check – 16
seat belts wear/damage/security check – 16
underbody protective coating check – 16, 173
upholstery and carpets – 173
windscreen and windows cleaning – 16
braking system
   handbrake linkage lubrication – 16, 143
   hydraulic fluid level check/top up – 16, 142
   hydraulic fluid renewal – 16, 143
   lines and hoses check – 16
   pads and shoes wear check – 16, 142
   servo check – 16, 143
   vacuum hose check – 16
clutch adjustment check – 16, 97
cooling system
   coolant level check/top up – 16, 59
   coolant renewal – 16, 59
   drivebelt condition/tension check – 16, 59
   leaks check – 59
electrical system
   battery electrolyte level check/top up – 16, 197
   equipment operation check – 16, 197
   headlamps cleaning – 16
   lights operation check – 16
   washer fluid levels check/top up – 16, 214, 216
engine
   crankcase emission valve renewal – 16, 27
   fluid leaks check – 16
   oil and filter change – 16, 26
   oil filler cap cleaning – 16, 26
   oil level check/top up – 16
   valve clearances check/adjust – 16, 46
exhaust system condition check – 16
fuel system
   air cleaner temperature control check – 16, 67
   air filter element renewal – 16, 67, 79
   fuel filter renewal (fuel injection models) – 16, 80
   fuel hoses condition check – 67
   idling speed/mixture check/adjust – 16, 73, 79
   inlet manifold nuts/bolts tighten – 16, 27, 83
   vacuum hoses check – 16
ignition system
   contact breaker points renewal (3.0 litre engine) – 16, 89
   distributor cap, rotor, HT leads and coil clean/check – 16, 88
   distributor lubrication – 16
   dwell angle check/adjust (3.0 litre engine) – 16, 89
   spark plugs clean/re-gap – 16, 94
   spark plugs renew – 16, 94
   timing check – 16, 93
rear axle oil level check/top up – 16, 132
safety precautions – 12
service schedules – 13
suspension and steering
   power steering drivebelt condition/tension check – 16, 160
   power steering fluid level and hoses check – 160
   steering components condition check – 16, 160
   suspension condition check – 16, 160
transmisson
   automatic transmission brake band adjustment – 16, 122
   automatic transmission linkage lubrication – 16, 122
   oil/fluid level check/top up – 16, 103, 122
tyres
   pressures check/adjust – 16, 157
   wear/damage check – 16, 170
wheels condition check – 170

Manifolds (carburettor models)
removal and refitting
   exhaust – 76
   inlet – 75

Manifolds (fuel injection models)
removal and refitting
   exhaust – 85
   inlet – 83

Manual gearbox
description – 103
fault diagnosis – 121
five-speed
   disengagement of 5th gear – 120
   dismantling – 110
   examination and renovation – 118
   input shaft dismantling and reassembly – 108, 118
   mainshaft dismantling and reassembly – 114
   reassembly – 118
four-speed
   dismantling – 105
   examination and renovation – 108
   input shaft dismantling and reassembly – 108
   laygear dismantling and reassembly – 108
   reassembly – 109
gearchange lever (four-speed) modifications – 110
maintenance, routine – 103
removal and refitting – 103
specifications – 101
torque wrench settings – 102

Manual gearbox and automatic transmission – 101 et seq

Master cylinder (braking system)
overhaul – 148
removal and refitting – 147

Mirror
removal and refitting
   adjustable exterior – 189
   interior – 189

O

Oil pump
2.8 litre engine
   dismantling, examination and reassembly – 38
   refitting – 42
3.0 litre engine
   dismantling, examination and reassembly – 52
   refitting – 54
   removal – 50

P

Pedal
accelerator – 75
brake – 151
clutch – 100

Piston and connecting rods
refitting – 42, 54

Pistons and piston rings
examination and renovation – 36, 51

Pistons, connecting rods and bearings (3.0 litre engine)
removal – 51

Propeller shaft – 127 et seq

Propeller shaft
centre bearing renewal – 128
description – 127
fault diagnosis – 130
removal and refitting – 127
torque wrench settings – 127
universal joints inspection – 129

Pushrods – 35, 51

R

Radiator
removal, inspection, cleaning and refitting – 60

Radiator grille
removal and refitting – 176

Radio/cassette player
removal and refitting – 223

Index

Radio equipment, mobile
    interference-free installation – 223
Rear axle – 131 *et seq*
Rear axle
    axle rear cover removal and refitting – 134
    axleshaft
        bearing renewal – 134
        removal and refitting – 133
    description – 131
    fault diagnosis – 140
    limited slip differential – 139
    maintenance, routine – 132
    overhaul – 135
    pinion oil seal renewal – 134
    removal and refitting – 132
    specifications – 131
    torque wrench settings – 131
    wheel stud renewal – 139
Rear drum brakes
    rear backplate removal and refitting – 147
    shoes renewal – 145
    wheel cylinder removal, overhaul and refitting – 146
Rear lamp and bulb
    removal and refitting – 213
Rear number plate lamp and bulb
    removal and refitting – 213
Rear suspension
    description – 157
    fault diagnosis – 171
    leaf spring removal, bush renewal and refitting – 164
    maintenance, routine – 160
    shock absorber removal and refitting – 164
    stabiliser bar removal and refitting – 164
    torque wrench settings – 157
Relays – 222
Repair procedures, general – 9
Rocker arms (3.0 litre engine)
    examination and renovation – 51
    removal – 49
Rocker arms and covers (3.0 litre engine)
    refitting – 55
Rocker assembly (2.8 litre engine)
    dismantling, examination and reassembly – 34
Routine maintenance *see* **Maintenance, routine**

S

Safety precautions – 12
Seat belts – 188
Seats
    removal and refitting – 189
Sidelight
    bulb renewal – 210
Spare parts
    buying – 8
    to carry in car – 19
Spark plug conditions – *see colour section between pages 32 and 33*
    removal, servicing and refitting – 94
Speedometer cable
    renewal – 223
Stabiliser bar
    front – 162
    rear – 164
Starter motor
    description – 204
    dismantling and reassembly – 204
    removal and refitting – 204
    testing on engine – 204
Steering
    column removal, overhaul and refitting – 169
    description – 157
    fault diagnosis – 171
    front wheel alignment and steering angles – 170
    gear (manual)
        overhaul and adjustment – 165
        removal and refitting – 165
    maintenance, routine – 160
    power steering
        bleeding – 167
        gear removal and refitting – 167
        pump removal and refitting – 167
    specifications – 156
    torque wrench settings – 157
    track rod end balljoint renewal – 166
    wheel removal and refitting – 167
Sump
    2.8 litre engine
        refitting – 44
        removal and refitting with engine in car – 34
    3.0 litre engine
        refitting – 55
        removal – 50
Sunroof
    bracket and drive assembly
        removal and refitting – 188
    panel adjustments – 186
    removal and refitting – 187
Suspension and steering – 156 *et seq*
Suspension *see* **Front suspension** *and* **Rear suspension**
Switches
    removal and refitting
        central facia – 222
        door pillar – 221
        handbrake warning light – 222
        hazard warning – 220
        ignition – 221
        instrument cluster illumination – 221
        light and windscreen wiper (Series II) – 222
        load space lamp – 222
        steering column multi-function – 221
        stop lamp – 151

T

Tailgate
    private lock removal and refitting – 184
    removal and refitting – 184
    striker plate removal and refitting – 185
    washer
        nozzle removal and refitting – 214
        pump removal and refitting – 214
    window glass removal and refitting – 177
    wiper motor and linkage removal and refitting – 214
Tappets (3.0 litre engine)
    removal – 50
Tappets and pushrods
    examination and renovation – 35, 51
Temperature gauge sender unit
    removal and refitting – 64
Thermostat
    removal, testing and refitting – 61
Timing gears
    examination and renovation – 38, 52
Timing gears (3.0 litre engine)
    removal – 50
Timing gears and front plate (3.0 litre engine)
    refitting – 54
Timing gears and timing cover (2.8 litre engine)
    refitting – 43
Timing, ignition
    adjustment – 93
Tools
    general – 10
    to carry in car – 19
Towing – 7

Index

Tyres
 care and maintenance – 170
 pressures – 157
 size – 157

U

Underframe
 maintenance – 16, 173
Upholstery
 maintenance – 173

V

Vacuum servo unit (braking system)
 description – 155
 removal and refitting – 155
Valves
 clearances checking and adjustment – 46, 55
Valves (3.0 litre engine)
 removal – 49

Valves and valve seats (3.0 litre engine)
 examination and renovation – 52
Vehicle identification numbers – 8

W

Water pump
 refitting – 44, 61
 removal – 61
Weights, vehicle – 6
Wheels
 care and maintenance – 170
 nuts/bolts torque wrench setting – 157
 stud renewal (rear) – 139
Windscreen
 removal and refitting – 177
Windscreen washer
 nozzle removal and refitting – 216
 pump removal and refitting – 216
Windscreen wipers
 arms and blades removal and refitting – 217
 fault tracing – 216
 motor and linkage removal and refitting – 215
 motor dismantling and reassembly – 216
 switch (Series II) – 222
Wiring diagrams – 232 to 246
Working facilities – 11